Business Web Strategy:
Design, Alignment, and Application

Latif Al-Hakim
University of Southern Queensland, Australia

Massimo Memmola
Catholic University, Italy

Information Science
REFERENCE

INFORMATION SCIENCE REFERENCE
Hershey • New York

Director of Editorial Content: Kristin Klinger
Managing Development Edtior: Kristin Roth
Editorial Assistant: Rebecca Beistline
Director of Production: Jennifer Neidig
Managing Editor: Jamie Snavely
Assistant Managing Editor: Carole Coulson
Typesetter: Lindsay Bergman
Cover Design: Lisa Tosheff
Printed at: Yurchak Printing Inc.

Published in the United States of America by
Information Science Reference (an imprint of IGI Global)
701 E. Chocolate Avenue, Suite 200
Hershey PA 17033
Tel: 717-533-8845
Fax: 717-533-8661
E-mail: cust@igi-global.com
Web site: http://www.igi-global.com

and in the United Kingdom by
Information Science Reference (an imprint of IGI Global)
3 Henrietta Street
Covent Garden
London WC2E 8LU
Tel: 44 20 7240 0856
Fax: 44 20 7379 0609
Web site: http://www.eurospanbookstore.com

Library of Congress Cataloging-in-Publication Data

Business web strategy : design, alignment and application / Latif Al-Hakim and Massimo Memmola, editors.

p. cm.

Includes bibliographical references and index.

Summary: "This book addresses the gap in business Web strategy through a collection of concentrated managerial issues, gathering the latest theoretical frameworks, case studies, and research pertaining to maximizing the power of the Web"--Provided by publisher.

ISBN 978-1-60566-024-0 (hbk.) -- ISBN 978-1-60566-025-7 (ebook)

1. Information technology--Management. 2. Electronic commerce. 3. Internet. I. Al-Hakim, Latif, 1946- II. Memmola, Massimo.

HD30.2.B88 2009

658.8'72--dc22

2008024387

British Cataloguing in Publication Data
A Cataloguing in Publication record for this book is available from the British Library.

Editorial Advisory Board

Table of Contents

Foreword xvi

Preface xviii

Section I
Design Web Strategy

Chapter I
Using Patterns for Engineering High-Quality E-Commerce Applications 1
Pankaj Kamthan, Concordia University, Canada
Hsueh-Ieng Pai, Concordia University, Canada

Chapter II
Informing Industry via Academic Research in ICT Skill and Capability Development 26
Krassie Petrova, Auckland University of Technology, New Zealand
Dawn Medlin, Appalachian State University, USA

Chapter III
The Impact of New Trends in the Delivery and Utilization of Enterprise ICT on Supplier and User Organizations 46
Jiri Vorisek, University of Economics Prague, Czech Republic
George Feuerlicht, University of Economics Prague, Czech Republic

Chapter IV
Enterprise 2.0: Collaboration and Knowledge Emergence as a Business Web Strategy Enabler 61
Javier Soriano, Universidad Politécnica de Madrid, Spain
David Lizcano, Universidad Politécnica de Madrid, Spain
Marcos Reyes, Telefónica I+D, Spain
Fernando Alonso, Universidad Politécnica de Madrid, Spain
Genoveva López, Universidad Politécnica de Madrid, Spain

Chapter V
Customer Relationship Management (CRM): An In-Depth Analysis 94
Mahesh Raisinghani, TWU School of Management, USA
Abdu Albur, Ministry of Education in the Kingdom of Saudi Arabia, Dhahran, Eastern Province
Sue Leferink, Montana Department of Commerce, USA
Thomas Lyle, PNC, USA
Stephen Proctor, CSC, USA

Chapter VI
Different Web Strategies for Different E-Marketplaces 118
L. Geppert, Catholic University of Milan, Italy

Section II
Aligning Web Strategy to Corporate Strategy

Chapter VII
Trends of Web Services Adoption: A Synthesis... 134
Vincent C. Yen, Wright State University, USA

Chapter VIII
Web & RFId Technology: New Frontiers in Costing and Process Management for Rehabilitation Medicine 145
Massimo Memmola, Catholic University, Italy
Giovanna Palumbo, Ospedale Valduce, Italy
Mauro Rossini, Ospedale Valduce, Italy

Chapter IX
The Web Strategy Development in the Automotive Sector 170
Massimo Memmola, Catholic University, Italy
Alessandra Tzannis, Catholic University, Italy

Chapter X
Adaptive Mobile Web Browsing Using Web Mining Technologies.. 198
Wen-Chen Hu, University of North Dakota, USA
Yanjun Zuo, University of North Dakota, USA
Lei Chen, Sam Houston State University, USA
Chyuan-Huei Thomas Yang, Hsuan Chuang University, Taiwan

Chapter XI
Integration of Public University Web Sites and Learning Management Systems 208
Bernard Ostheimer, University of Giessen, Germany
Axel C. Schwickert, University of Giessen, Germany

Chapter XII
Innovating through the Web: The Banking Industry Case 219
Chiara Frigerio, Universitá Cattolica del Sacro Cuore, Italy

Section III
Applications of Web Strategy

Chapter XIII
An Action Research Case Study of the Facilitators and Inhibitors of E-Commerce Adoption 236
Orla Kirwan, National University of Ireland, Ireland
Kieran Conboy, National University of Ireland, Ireland

Chapter XIV
Acceptance of the Mobile Internet as a Distribution Channel for Paid Content in Germany 248
Svenja Hagenhoff, University of Goettingen, Germany
Christian Kaspar, University of Goettingen, Germany
Lutz Seidenfaden, University of Goettingen, Germany
Björn Ortelbach, University of Goettingen, Germany

Chapter XV
Information Quality Satisfaction of Communication Portals: A Study of Central Cyber Government Office (CCGO) of the Hong Kong Government 264
Kevin K.W. Ho, The Hong Kong University of Science and Technology, Hong Kong

Chapter XVI
The Evaluation of IT Investments through Real Options 277
Maria Alice Frontini, University of São Paulo, Brazil
Fernando José Barbin Laurindo, University of São Paulo, Brazil

Chapter XVII
Strategic Use of the Internet and Organizational Structure for E-Business: "Celta" Case at GM Brazil 298
Silvia Novaes Zilber, UNINOVE, Brazil

Chapter XVIII
On the Use of Soft Computing Techniques for Web Personalization 318
G. Castellano, University of Bari, Italy
A. M. Fanelli, University of Bari, Italy
M. A. Torsello, University of Bari, Italy

Compilation of References 340

About the Contributors 371

Index 380

Detailed Table of Contents

Foreword xvi

Preface xviii

Section I
Design Web Strategy

The objective of the book's first section, which is subdivided into six chapters, is to look into the different Web strategy planning routes in different corporate contexts, in order to obtain optimal use of the Internet's technology potentials.

Chapter I
Using Patterns for Engineering High-Quality E-Commerce Applications 1
Pankaj Kamthan, Concordia University, Canada
Hsueh-Ieng Pai, Concordia University, Canada

In this chapter, the authors view the development and maintenance of high-quality electronic commerce (e-commerce) applications from a Web engineering perspective. A methodology for deploying patterns as means for improving the quality of e-commerce applications is presented. To that regard, relevant quality attributes and corresponding stakeholder types for the e-commerce applications are identified. The role of development process, the challenges in making optimal use of patterns, and feasibility issues involved in doing so, are analyzed. The activities of a systematic selection and application of patterns are explored. Examples illustrating the use of patterns during macro- and micro-architecture design of business-to-consumer (B2C) e-commerce applications are given. The implications of the use of patterns in a Semantic Web context are briefly highlighted.

Chapter II
Informing Industry via Academic Research in ICT Skill and Capability Development 26
Krassie Petrova, Auckland University of Technology, New Zealand
Dawn Medlin, Appalachian State University, USA

In recent years significant changes have occurred in the skill sets underpinning the undergraduate information systems and information technology curricula. It is imperative that educators comprehend the needs and demands of the industry where their graduates are going to need to apply their acquired knowledge and skills. It may be argued that employers and job recruiters also need to be aware of what skill sets and capabilities new graduates may be expected to come equipped with, in order to develop successful strategies for retaining and growing staff in an environment where the demand for professionals in information and communications technologies (ICT) exceeds the supply. In this chapter, a research framework representing the dynamics of the ICT profession supply and demand of graduates with relevant skills and capabilities is used to facilitate the initiation of a dialogue between industry and academia with the objective to identify issues raised from the lack of alignment between the two and to suggest a way of using academic research results to address these issues. The discussion is supported by the findings of two relevant case studies.

Chapter III
The Impact of New Trends in the Delivery and Utilization of Enterprise ICT on Supplier and User Organizations 46
Jiri Vorisek, University of Economics Prague, Czech Republic
George Feuerlicht, University of Economics Prague, Czech Republic

Enterprise information systems have rapidly evolved over the last decade. We expect these changes to accelerate during this decade as a result of new trends in enterprise computing. We argue in this chapter that ICT remains strategically important to organizations in the 21st century despite the prevailing trend to outsource ICT and related business processes. We have identified a number of important trends that include the move towards the software as a service (SaaS) model for enterprise applications, increased commitment to process orientation, and emphasis on managing the relationship between business and ICT using services. These trends lead to more effective management of ICT and closer integration of ICT with entrepreneurial activities and business processes in organizations, resulting in improvements in return on investment. These trends will have dramatic impact on both the suppliers and users of ICT, and will necessitate the reevaluation of the approach to ICT education as both the composition and qualifications of ICT workforce will undergo a fundamental change.

Chapter IV
Enterprise 2.0: Collaboration and Knowledge Emergence as a Business Web Strategy Enabler 61
Javier Soriano, Universidad Politécnica de Madrid, Spain
David Lizcano, Universidad Politécnica de Madrid, Spain
Marcos Reyes, Telefónica I+D, Spain
Fernando Alonso, Universidad Politécnica de Madrid, Spain
Genoveva López, Universidad Politécnica de Madrid, Spain

This chapter explores the Internet's present and future potential in relation to information sharing, knowledge management, innovation management, and the automation of cross-organizational business transactions. It points out how a business Web strategy that takes into account this potential will help not only to improve the existing information sharing and knowledge management processes, but also to protect investments in technology that would otherwise have resulted in expensive failures and severe losses. The suggested approach is based on the emerging Web 2.0 vision and will help to minimize the risk of key information and knowledge being lost or simply not being available on time for the stakeholder, projects started and never finished, worse time to market, results not meeting expectations, failure of global, cross-organizational IT integration processes or even incoherencies between technology and company strategy or structure, and so on. All managers, and particularly IT leaders, must be aware of this new potential and its implications in order to come up with innovative and effective answers to both known and new problems related to information sharing and knowledge management within their organizations. The chapter's contents are designed to guide entrepreneurs, managers, and IT leaders through the adoption of the latest Internet technologies, such as Web 2.0, Enterprise 2.0, and the global service-oriented architecture, and their application to their everyday work with a view to setting up a business Web strategy. Musser and O'Reilly (2006) claim that by defining and following a set of architecture building blocks, architectural design decisions, and normative guidance, they can build flexible, extensible, and reusable solutions for exploiting the best features of the emerging Web 2.0 technology suite to achieve the best ROI by leveraging the upcoming Web of user-centered services.

Chapter V
Customer Relationship Management (CRM): An In-Depth Analysis 94
Mahesh Raisinghani, TWU School of Management, USA
Abdu Albur, Ministry of Education in the Kingdom of Saudi Arabia, Dhahran, Eastern Province
Sue Leferink, Montana Department of Commerce, USA
Thomas Lyle, PNC, USA
Stephen Proctor, CSC, USA

This chapter discusses customer relationship management (CRM) as a customer focused business strategy enhanced by technology that automates and enhances business processes to proactively manage profitable and long-term customer relationships. CRM solutions span a continuum of implementations from a narrow tactical implementation of a specific technical solution to a broad strategic implementation of a customer centric solution. Furthermore, the authors hope that understanding the underlying assumptions and theoretical constructs through the use of CRM will not only inform researchers of a better CRM design for studying e-commerce and Internet marketing, but also assist in the understanding of intricate relationships between different factors.

Chapter VI
Different Web Strategies for Different E-Marketplaces 118
L. Geppert, Catholic University of Milan, Italy

This chapter presents two possible models of electronic marketplaces put in place at the beginning of this century, which, after their introduction, the first incoming wave of connected economy-based paradigms was ended. Both the two models show a particular use of Web-based information technology in order to exploit their mission and represent meaningful cases of application of well-defined Web strategies. Even though, at the moment, the initially built up Web sites supporting those related business have been closed and merged with other Web sites, they may introduce examples of a Web strategy approach having a relevant historical meaning that may be still redefined in practical implementations once revised and adequately updated. The mentioned cases described in this chapter are usteel.com and up2gold.com, two examples of Web-based business in two well-defined supply chains: the "steel" chain and the "gold and silver" chain.

Section II
Aligning Web Strategy to Corporate Strategy

The second section is dedicated to the subject of the Web strategy alignment with the corporate strategy. The six articles hosted in this section discuss this theme focusing their attention not only on the aspects related to the development of the public presence on the Web (Memmola and Tzannis; Frigerio), but also through an analytical perspective which can either be a technique (Yen; Wen-Chen Hu, Yanjun Zuo, Lei Chen and Chyuan-Huei Thomas Yang), or integrate other technologies (Axel C. Schwickert, Bernard Ostheimer) or integrate cost measurement and performance management processes (Massimo Memmola, Giovanna Palumbo and Mauro Rossini).

Chapter VII
Trends of Web Services Adoption: A Synthesis 134
Vincent C. Yen, Wright State University, USA

The technology of Web services has been a hot area in the software industry for many years. Many organizations in the past 5 years have conducted surveys designed to get a profile of the state of Web services adoption in various subject areas. Some of those survey results are available free from the Internet. Since conducting a large scale Web services survey takes time and significant financial commitment, the research conducted in this chapter is a synthesis from published free survey results. All sources of surveys indicate Web services are being adopted more or less in all mid-size to large organizations because of realized benefits, and are anticipated to become a viable component of information systems infrastructure. Some of the current issues in Web services adoption and implementation are standards, training, and security.

Chapter VIII
Web & RFId Technology: New Frontiers in Costing and Process Management
for Rehabilitation Medicine 145
Massimo Memmola, Catholic University, Italy
Giovanna Palumbo, Ospedale Valduce, Italy
Mauro Rossini, Ospedale Valduce, Italy

Radio frequency identification (RFId) has recently begun to receive increased interest from practitioners and academics. This type of technology has been widely used in healthcare organizations for different purposes, like to localize patients, devices, and medical instruments. This chapter presents the results of a study in which we used RFId technology and modern systems of cost management methodologies (e.g., activity-based costing, activity-based management, and process management) in a "proof of application" aimed at defining some specific data on care needs of a person with a disability, costs of the main activities performed during the person's rehabilitation process, and level of performance which could be reached in order to improve the "disability management" process, from a clinical as well as a managerial perspective.

Chapter IX
The Web Strategy Development in the Automotive Sector 170
Massimo Memmola, Catholic University, Italy
Alessandra Tzannis, Catholic University, Italy

Especially in recent years, indeed, a transformation is ongoing: the Web, besides being a means of information sharing (internal-external), becomes a powerful tool for saving costs, reducing the distribution structure, initiating distance transactions, and ever more becomes a mechanism of integration with the external environment and a catalyst of experiences for all stakeholder. Starting from the identification of the key elements, potentialities, and of the impact of the Internet on firms' performance, competitiveness, effectiveness, and efficiency, this chapter is focused on the changes in the automotive sector due to the integration between business strategy and Web strategy. Therefore, starting from the consideration of a clear identification and subsequent sharing need of strategic goals, a research work will be presented exploring, on the basis of an interpretative model, the Internet potential in the automotive sector, in order to achieve the identification of an optimal path definition and development of Web strategy. This objective will be developed through a desk analysis focused on the strategic positioning of the current businesses in the automotive sector (i.e., complexity evaluation of the presence on Internet, strategic architecture, quality, and effectiveness of the presence).

Chapter X
Adaptive Mobile Web Browsing Using Web Mining Technologies 198
Wen-Chen Hu, University of North Dakota, USA
Yanjun Zuo, University of North Dakota, USA
Lei Chen, Sam Houston State University, USA
Chyuan-Huei Thomas Yang, Hsuan Chuang University, Taiwan

Using mobile handheld devices such as smart cellular phones and personal digital assistants (PDAs) to browse the mobile Internet is a trend of Web browsing. However, the small screens of handheld devices and slow mobile data transmission make the mobile Web browsing awkward. This research applies Web usage mining technologies to adaptive Web viewing for handheld devices. Web usage mining is the application of data mining techniques to the usage logs of large Web data repositories in order to produce results that can be applied to many practical subjects, such as improving Web sites/pages. A Web usage mining system must be able to perform five major functions: (i) usage data gathering, (ii) data preparation, (iii) navigation pattern discovery, (iv) pattern analysis and visualization, and (v) pattern applications. This approach improves the readability and download speed of mobile Web pages.

Chapter XI
Integration of Public University Web Sites and Learning Management Systems 208
Bernard Ostheimer, University of Giessen, Germany
Axel C. Schwickert, University of Giessen, Germany

Internet technology has found its way into all areas of business and research. The World Wide Web is also used at universities to achieve different goals. On the one hand, it acts as a means of outer appearance (target groups: potential, current, and former students, researchers, lecturers, press, the interested publicity, etc.), on the other hand, as an instrument of knowledge transfer and knowledge examination (target groups: potential, former, and current students and lecturers). There exist other purposes in addition to those named above. Often different systems are used to achieve the different goals: usually, Web content management systems (WCMS) are used for the outer appearance and learning management systems (LMS) for transfer and examination of knowledge. Although these systems use the same medium (i.e., the WWW), it can be stated that often there is a heterogeneous landscape of systems. Resultant is the object of investigation of the present chapter. It analyzes the challenges concerning the integration of public Web sites and learning management systems (LMS) a typical European university has to face. The research framework used for investigation thus can be divided into two categories regarding the system types used: WCMS and LMS. In praxi, there is more than one system per system category implemented at a university because of the organizational conditions explained in this chapter.

Chapter XII
Innovating through the Web: The Banking Industry Case 219
Chiara Frigerio, Universitá Cattolica del Sacro Cuore, Italy

In recent years, the financial services industry has been witness to considerable consolidation and organizational progress in order to sustain two main objectives: efficiency and commercial effectiveness. In order to sustain customer-oriented and efficiency strategies, banks have started to explore new ways of conducting their business, introducing areas of innovation in their services, practices, and structures to offer the most complete array of services possible. On the other hand, new services and products drive retail banks to explore new ways of producing or delivering these novelties. This is true especially for Internet banking services that offer services to customers 24/7, and it becomes clear that adding new services, that is, trading online or bill payments, is easily and quickly geared towards improving commercial effectiveness. The following chapter aims at describing to what extent the Internet has developed new services and businesses, and what are the main figures of the phenomenon in Europe. Moreover, the Tnternet has introduced new coordination processes within each financial institution. Let us think about Intranet portal, content management tools, and business process management suites, which are now quite spread in banks due mainly to their technological easy of use. Thus, the Internet is representing an innovation wave extremely relevant for the financial industry as a whole, and the effects on banks' performance is emerging. What do we expect in the near future? In all probability, the usage of Web-based application will be bigger and bigger also in other contexts of the bank processes, even if some risks could occur when clear strategies and change management practices do not direct the innovation.

Section III
Applications of Web Strategy

The six chapters of the third section present the results of research work and empirical evidence on the application of Web strategy's principles and methodologies in various operational contexts.

Chapter XIII
An Action Research Case Study of the Facilitators and Inhibitors of E-Commerce Adoption 236
Orla Kirwan, National University of Ireland, Ireland
Kieran Conboy, National University of Ireland, Ireland

This research has studied an established Irish retail business as it takes its first tentative steps into the e-commerce arena. Although the adoption of e-commerce is widely studied in the academic world, only a small percentage of these studies focus on the small to medium size enterprise (SME) retail sector. SMEs account for 97% of Irish companies and employ up to 800,000 people (Chamber of Commerce Ireland, 2006). While examining the SME's adoption of e-commerce, the factors that affected the adoption process were specifically identified and understood. This was achieved by conducting an action research case study. Action research merges research and practice, thus producing exceedingly relevant research findings. This chapter demonstrates how the research was undertaken, and also discusses the justification, benefits, and limitations of using action research. The research concluded that the adoption of e-commerce within the SME sector tends to be slow and fragmented, the presence of a "Web champion" is paramount to the success of the project, and Internet adoption is faster with the recognition of a business need. It also supported the evidence that an SME is more likely to adopt e-commerce when the SME owner has a positive attitude towards IT.

Chapter XIV
Acceptance of the Mobile Internet as a Distribution Channel for Paid Content in Germany 248
Svenja Hagenhoff, University of Goettingen, Germany
Christian Kaspar, University of Goettingen, Germany
Lutz Seidenfaden, University of Goettingen, Germany
Björn Ortelbach, University of Goettingen, Germany

This chapter is about a survey based on 7,178 valid responses which analyses the mobile content usage in Germany. Key findings are that paid mobile contents will not be a mass market in the medium term. Nevertheless, we found that respondents that are familiar with mobile radio and handset technology and read specialized printed media on a regular basis showed the highest acceptance of mobile paid contents. The mobile Internet is perceived as a chance for the media industry to generate additional revenues from paid contents. Successful business models for the mobile Internet will only be possible if mobile content formats generate added consumer value. In this context, media companies planning to establish mobile services for content distribution are facing the problem that acceptance of mobile services has not yet been researched thoroughly.

Chapter XV
Information Quality Satisfaction of Communication Portals: A Study of Central Cyber Government Office (CCGO) of the Hong Kong Government.. 264
Kevin K.W. Ho, The Hong Kong University of Science and Technology, Hong Kong

Information quality is critical for a communication portal because there are a myriad of information types, including textual, audio, video, and other complex information types, which an organization has to manage. In this study, we examine whether information generated from an in-house developed communication portal of the Hong Kong government would have higher quality than those sister portals developed by individual government departments using commercial packages. We conducted a survey-based study to understand how users evaluate the information quality of these communication portals. This portal case is interesting because (1) Hong Kong Government has invested millions of US dollars in its implementation and (2) the number of potential users is huge (over 140,000).

Chapter XVI
The Evaluation of IT Investments through Real Options 277
Maria Alice Frontini, University of São Paulo, Brazil
Fernando José Barbin Laurindo, University of São Paulo, Brazil

The decisions about IT investments are increasingly more complex, due to technical uncertainties and to the dynamics of organizational and strategic issues. One promising alternative for solving this problem would be the use of real options. Thus, this chapter intends to apply a relatively new methodology, called real options, used in corporate strategy for evaluating and deciding about new investments in IT. In order to do so, an analytic review of the literature is presented, The real options methodology is particularly recommended in two situations: in the case of existence of a significant level of uncertainty about the benefits to be achieved by IT investments, or when IT benefits do not impact directly the current business but create a platform for future investments, capable of producing future new business impacts.

Chapter XVII
Strategic Use of the Internet and Organizational Structure for E-Business:
"Celta" Case at GM Brazil 298
Silvia Novaes Zilber, UNINOVE, Brazil

The Internet provides a global network infrastructure that is shifting business models, strategies, and processes. Many authors reflect on the importance of incorporating e-business into the firm's global strategy. This chapter deals with these issues in discussing the introduction of e-business activities by General Motors Brazil, specifically in connection with the launch of the Celta car, an entry-level car designed to be sold on the Internet. A historical examination of e-business strategy shows that many organizations have formulated excellent conceptual strategies for e-business but failed to deliver sound execution. A key to successful Internet strategies is the leadership shown by senior management. Technological demands may also conflict with the successful implementation of e-business initiatives, requiring greater interaction between the CEO and CIO. The organizational structure implemented for the launching and sales of the Celta car warranted the integration between employees on the business side and in IT in the context of GM Brazil's strategic objective of growing the market share for lower-priced cars.

Chapter XVIII
On the Use of Soft Computing Techniques for Web Personalization 318
G. Castellano, University of Bari, Italy
A. M. Fanelli, University of Bari, Italy
M. A. Torsello, University of Bari, Italy

Due to the growing variety and quantity of information available on the Web, there is urgent need for developing Web-based applications capable of adapting their services to the needs of the users. This is the main rationale behind the flourishing area of Web personalization that finds in soft computing (SC) techniques a valid tool to handle uncertainty in Web usage data and develop Web-based applications tailored to user preferences. The main reason for this success seems to be the synergy resulting from SC paradigms, such as fuzzy logic, neural networks, and genetic algorithms. Each of these computing paradigms provides complementary reasoning and searching methods that allow the use of domain knowledge and empirical data to solve complex problems. In this chapter, we emphasize the suitability of hybrid schemes combining different SC techniques for the development of effective Web personalization systems. In particular, we present a neuro-fuzzy approach for Web personalization that combines techniques from the fuzzy and the neural paradigms to derive knowledge from Web usage data and represent the knowledge in the comprehensible form of fuzzy rules. The derived knowledge is ultimately used to dynamically suggest interesting links to the user of a Web site.

Compilation of References 340

About the Contributors 371

Index 380

Foreword

Strategy is about predicting the future and ensuring we know where we want to go, so that we can develop the appropriate plans to move in that direction.

In this fast changing world, it is not easy to predict the future. To fully grasp the way the Internet has changed our lives, consider how you worked and lived ten years ago and compare that to your current habits. Now try to extrapolate into the future and imagine living in this world in the year 2018.

Amazingly, in 1946 a science fiction author using the nom de plume of 'Murray Leinster' introduced the concept of the Internet in his short story 'A Logic Named Joe'. As highlighted by Franson (2001), Leinster's story included desktop computers in virtually all homes and businesses with country-wide networked databases (including video), keyboards and vision screens, natural language interface, interactive customizable software, and online financial and weather reports. Unfortunately we do not all have Leinster's crystal ball and so we need to take a formal approach to strategic planning.

The provision of IT services to business is more complex then ever before. Many organisations engage in an intricate network of contracts involving providers of software, services, infrastructure, and data sources. The global economy has opened endless options include outsourcing and off-shoring IT development and service management. There are many stakeholders involved in IT today – no longer does IT focus on supporting back-office processes. Increasingly, thanks to the Internet, customers and suppliers have direct access to corporate data, systems and services.

It has become more important than ever before for IT management to participate in business strategy formulation. An organisation's ability to implement corporate strategies and achieve corporate goals depends on its capacity to effectively use IT. Investment in IT, and especially in web technologies can provide six strategic business objectives: operational excellence; new products, services and business models; customer and supplier intimacy; better decision making; competitive advantage; and ultimately, survival (Laudon & Laudon 2006).

Organisations find it challenging to ensure adequate governance and compliance requirements are met whilst retaining the flexibility and agility to respond to changes in the market. It is difficult to balance the temptation to adopt 'bleeding' edge business models and technology while remaining cognizant of concerns about the business value and risks of such investment.

This book provides a wealth of information across the lifecycle of web-enabled business processes from strategy through design to implementation. Managers need to be aware of new technologies such as RFID and mobile devices which have the potential to extend the reach and richness of information to enhance the relationship between the organisations and its customers.

Another important contribution of this book is in its presentation of rigorous recent academic research to the practitioner community. Many chapters offer insights by describing real-life applications which span public and private sectors. These case studies describe best practices and also provide valuable lessons learned to industry leaders. Others suggest innovative business models and frameworks to evaluate the business benefits from web-based business systems.

I hope that this book provides valuable and timely assistance to researchers, IT practitioners and business managers to look to the future and see the possibilities from the increasing trends of Internet adoption and web-based applications.

Aileen Cater-Steel
University of Southern Queensland, Australia

REFERENCES

Franson, R. W. 2001 http://www.troynovant.com/Franson/Leinster/Logic-Named-Joe.html accessed 25 June 2008.

Laudon, KC & Laudon, JP 2006, Management Information Systems: Managing the Digital Firm, 10 edn, Prentice Hall, Upper Saddle River, New Jersey.

Preface

ABSTRACT

This preface introduces the book Business Web Strategy: Design, Alignment, and Application. The book covers theoretical and conceptual frameworks, concepts, trends, challenges and opportunities, performance measurement, and applications of Web strategies. It comprises 18 chapters organized into three sections: Web strategy design, Web strategy alignment, and Web strategy applications. The book provides insights and support for researchers, managers, and practitioners concerned with the subject of strategic approach to the Internet.

INTRODUCTION

The Internet is fully integrated into western societies, and is rapidly growing in developing economies. Via the Web we can search for information, make communicate, establish social relations, buy goods and services, play games, carry out financial transactions, study educational courses, and much more. The Web potential and its exploitation will certainly grow further thanks to the spread of mobile communication devices, a process which has already started.

The Web has changed our way of living, our habits, the way we work, the way we interact with the people we know, and, all in all, the way we deal with the thousands of tasks we are confronted with in our daily lives. We do not need to go personally to a bank any longer, we order bank transfers or check our balances by connecting to our bank's Web site. We just click on the site of a tour operator and carry out the required operations to book our holidays. If we are looking for software that meets our requirements but we do not want to bear the expenses of the related licenses, we just need to find it in one of the many open-source communities that are present on the Web. If we want to tell other Web users about ourselves, about what we do, our hobbies, and maybe even our dreams, what we need to do is to open a blog. Likewise, if we want to establish a social network or expand it, we can simply access one of the many free or pay-services present on the Web, such as Facebook, Myspace, and so forth. We could present an endless list of examples here.

We may even go further and say that people live increasingly more on the Web. To be more precise, actions, relations, and in general certain moments in our lives up to some time ago could have taken place only in the real world, but are now progressively moving onto the Web virtuality. Naturally, this transition from reality to virtuality inevitably modifies the way we interact with each others and with time and costs.

The following statistics provides an up-to-date picture of percentage and growth of Internet users population:

1. The number of Internet users in the world reached 1.4 billion persons in 2007; that is one fifth of the world population (Internetworldstats, 2008).
2. Web access is common nowadays in the most economically developed areas of the world: about two out of three U.S. citizens (71.1%) regularly access the Internet. In Australia and in Europe the percentage of Internet users is 57.1% and 43.4%, respectively. In Europe, however, there are some geographical areas where Internet penetration is considerably higher than the North American average; this is true for Norway (88%), the Netherlands (87.8%), and Iceland (85.4%) (Internetworldstats, 2008).
3. The Internet is growing exponentially. The number of Internet users increased globally by 265.6% during the period 2000-2007 (Internetworldstats, 2008).
4. Interestingly, the growth rate in the number of Internet users in the developing countries during 2000-2007 is higher than that in developed countries (Internetworldstats, 2008). The increase in the number of Internet users in North America, Australia, and Europe in the period 2000-2007 was equal to 120.2%, 151.6%, and 231.2%, respectively. The growth rates of Internet users in Middle East (920.2%), Africa (882.7%), Latin America (598.5%), and Asia (346.8%) are astoundingly high growth rates.
5. The global number of Web sites was approximately 550 million in January 2008, witnessing an increase by 25% as against the same period in 2007 and by roughly 648% in comparison with January 2000 (Internet System Consortium, 2008).
6. A survey carried out by Burst Media (2008) finds 67.7% of the respondents (13,000 Web users) emphasize that their daily routine would be disrupted if Internet access was not available for one week. Furthermore, 42.9% of respondents stress that the daily disruption would be significant.
7. The same survey finds that the majority of Internet users 45 years and older believe online content is focused on younger age segments (Burst Media, 2008). The survey finds that younger respondents are interested more on entertainment information (44.7%) in comparison to only 34.1% for shopping and product information. While older respondents indicate that they regularly seek local/national news (55.9%) followed by shopping/product information (44.0%).
8. A survey carried out by Nielsen (2008a) in the U.S. shows that 146.6 million users accessed Internet during March 2008, 51.35% of them women. The amount of time spent online nearly reaches 71 hours/month for men and 65 hours/month for women.
9. Nielsen (2008b) reports an estimated 8 billion searches were conducted in the U.S. during April 2008. An estimated of 62% of the search queries were conducted at Google Search.
10. The Internet in the UK is mainly used to send e-mails (91%), access general information (82%), purchase (72%), manage bank accounts (57%), download music, films, and videos (46%), play games (36%), chat or talk (40%), and trade online (16%) (Ofcom, 2005).

Virtually, all organizations in developed countries create Web pages to advertise their product/services. Literature indicates that firms rely on Web access for variety of reasons, including increasing efficiency, enhancing effectiveness, cost reduction, sharing of managerial know-how, time-to-market reduction, customer satisfaction, improved corporate image and branding policies, and increasing market share (Barua, Whinston, & Yin, 2000; Bovet & Martha, 2000; Cox & Koelzer, 2004; Evans & Wurster, 2000; Porter, 2001; Ries & Ries, 2002; Wilson, 2002).

Empirical studies indicate high satisfaction of organizations from the investment on Web technologies. A worldwide survey indicates that more than half surveyed organizations (2847 respondents) declare to be satisfied with the outcome of their investments on Web technologies over the past five years; almost three-quarters of the interviewed companies declare to be wanting to continue or increase the investments in Web technologies in the coming years (McKinsey, 2007). A recent survey (Awareness, 2008) shows that 96% of the surveyed organizations are happy with their investment on Web technologies. The survey concludes that the Web can be used to improve communication and collaboration (91% of surveyed organizations), selection and allocation of professional resources (81%), sharing information (78%), and increase brand awareness and customer retention (64%).

Notwithstanding the benefits of investment on Web technologies for organizations' well-being, studies highlight a concern among firms that the Web may create additional managerial problems and may not actually be the panacea for their existing problems and when an organization has managerial problems it will probably translate the same problems to the virtual space. Outcomes of some empirical studies reflect detrimental and adverse outcomes:

1. An empirical study conducted by Clearswift (2007) reveals that 23% of 700 directors of human resources involved in the study are not familiar with the Web technologies.
2. According to the research carried out by Awareness (2008), 53% of the respondent organizations state that they are not aware of how the social media tools of Web technologies could be useful to their company.
3. The empirical study carried by Clearswift (2007) shows that 72% of organization do not allow accessing social networking sites during business hours for a variety of reasons, including concerns about lowered productivity (85%), security and leak of confidential information (58%), sites content (39%), and the ownership of intellectual property placed on those sites (31%).

AIM OF THE BOOK AND TARGET AUDIENCE

There has been evidence which highlights how investments in technology have often resulted in expensive failures and severe losses because: projects started and never finished, results did not meet expectations, there was incoherence between technology and company strategy or structure; and of the inability to evaluate results, and, therefore, it was difficult to know if the initiative had been successful or not (Fattah, 2002; Minard, 2001, 2002; Iacovou & Dexter, 2005; Kappelman, McKeeman, & Zhang, 2006). These studies point to the lack of a sound Web strategy and a gap between the firm Web strategy and other strategies, including corporate strategy and ICT strategy (Figure 1).

Figure 1. The interaction of Web strategy with other strategies and actors

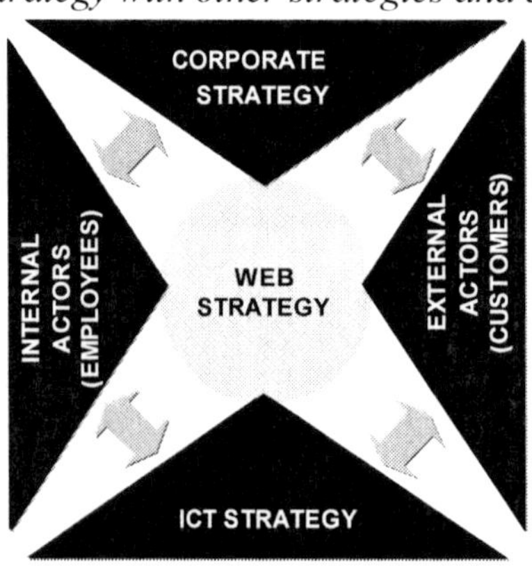

Like any other project entailing technology and changes in the way people work within an organization, the skills of employees and their competence are critical elements in the planning of a Web wtrategy (Internal actors—Figure 1). In addition, the positive interaction or satisfaction of customers (external actors—Figure 1) with the organization's Web forms one of the major challenges for the organization willing to develop a successful Web strategy.

This book addresses the gap in business Web strategy, considering positive interaction of external actors as well as skills and competence of internal actors. It aims to present cutting-edge research on the theory, applications, and challenges facing the design, development,t and implementation of the Web strategy.

This book is mainly oriented to people who, for different reasons, recognize that the Web is a powerful tool for supporting business strategy. Managers such as CEOs, general directors, plan and control responsible, and IT responsible will be considered as privileged actors.

In general, the natural users of this book will be not only the ICT vendors, but also those managers, consultants, and trainers who must offer innovative and effective answers in this new business field.

The Web strategy topic, both in industry and in academies, has grown rapidly over the past decade, and continues to grow. The book provides insights and support for the following groups of people:

- Professionals and researchers in the field of information system in general and in Web strategy in particular.
- Managers, practitioners, and consultants called to work out the most efficient and effective solutions for Web strategy, design, and implementation.

BOOK STRUCTURE

The book comprises 18 chapters organized into three sections: design, alignment, and applications. The following is a brief description of each section and the chapters included in them.

Section I. Design Web Strategy

The first section of this book features six chapters that deal with customer value and various trends, patterns, and methodologies affecting the design of Web strategy. The first chapter, "*Using Patterns for Engineering High-Quality E-Commerce Applications*" is authored by Pankaj Kamthan and Hsueh-Ieng Pai, who present a methodology for deploying patterns for improving the quality of e-commerce applications and identifying relevant quality attributes. The authors analyse the role of development processes and the challenges facing optimal use of patterns. They explore the use of patterns during macro and microarchitecture design of business-to-consumer (B2C) e-commerce applications.

Web technology has a strong impact on corporate processes and ultimately, requires new ICT skills and new training paths. "*Informing Industry via Academic Research in ICT Skill and Capability Development*" by Krassie Petrova and B. Dawn deal with these two aspects. This chapter designs a framework representing the dynamics of the ICT supply and demand for graduates with relevant skills and capabilities. The framework facilitates the dialogue between industry and academia, aiming to identify issues raised from the lack of alignment between the two and to suggest a pathway for using academic research results to address industry requirements. The framework is supported by the findings of two relevant case studies.

Internet does not only change processes, but also corporates logic and attitude toward ICT. Within the same scope of Chapter II, Jiri Vorisek and George Feuerlicht, in Chapter III, "*The Impact of New Trends in the Delivery and Utilization of Enterprise ICT on Supplier and User Organizations*" have identified a number of important trends within the framework of "software as a service (SaaS)" model. These trends lead to more effective ICT management of ICT. The authors emphasize that these trends will have dramatic impact on both the suppliers and users of ICT, and will necessitate the reevaluation of approaches to ICT education.

The logic for Web usage has been changing quickly over the past years. Javier Soriano, David Lizcano, Marcos Reyes, Fernando Alonso, and Genoveva López, in their chapter, entitled, "*Enterprise 2.0: Collaboration and Knowledge Emergence as a Business Web Strategy Enabler*," explore the Internet's present and future potential in relation to information sharing, knowledge management, innovation management, and the automation of cross-organizational business transactions. They point out how a business Web strategy protects investments in technology that would otherwise have resulted in expensive failures and severe losses. The chapter guides entrepreneurs, managers, and IT leaders through the adoption of the latest Internet technologies, such as Web 2.0, Enterprise 2.0, and global service oriented architecture with a view of setting up an effective business Web strategy.

The fifth chapter is this section, entitled, "*Customer Relationship Management (CRM): An In-Depth Analysis*" by Michael Raisinghani, Abdu Albur, Sue Leferink, Thomas Lyle, and Stephen Proctor, discusses customer relationship management (CRM) as a customer-focused business strategy that enhances business processes to proactively manage profitable and long-term customer relationships. The authors argue that understanding the underlying assumptions and theoretical constructs of CRM is an important factor for designing effective Web strategy.

Luigi Geppert in Chapter VI, "*Different Web Strategies for Different E-Marketplaces*," presents two possible models for electronic marketplaces. Both models show a particular use of Web-based information technology representing meaningful cases for application of well-defined Web strategies.

Section II. Aligning Web Strategy to Corporate Strategy

The second section of this book contains six chapters focusing on the alignment of Web strategy with other business strategies. The chapters discusses the Web strategy alignment from various perspectives, including public presence on the Web (Memmola, Tzannis, and Frigerio), analytical perspective (Yen, Hu, Zuo, Chen and Yang), integrating technologies (Schwickert and Ostheimer), or cost measurement and performance management (Memmola, Palumbo, and Rossini).

Many organizations in the past 5 years have conducted surveys designed to get a profile of the implementation of Web services in various subject areas. In the first chapter of this section, "*Trends of Web Services Adoption: A Synthesis*", Vincent C. Yen develops analyses of the results of these surveys. The chapter shows that all sources of surveys indicate that Web services are being adopted almost in all medium and large organizations because of realized benefits, and is to become a viable component of the information systems infrastructure. Some of the current issues in Web services adoption and implementation are standards, training and security.

Radio frequency identification (RFID) has recently begun to receive increased interest from practitioners and academics. This type of technology has been widely used in healthcare organizations for different purposes. In Chapter VIII, entitled, "*Web and RFId Technology: New Frontiers in Costing and Process Management for Rehabilitation Medicine,*" Massimo Memmola, Giovanna Palumbo, and Mauro Rossini present the results of a study in which they integrate RFID with Web technology for a test application aimed at defining some specific data on care needs of a disabled person, costs of the main activities performed during this person's rehabilitation process, and level of performance that could be reached in order to improve the "disability management" process, from a clinical as well as a managerial perspective.

Mobile handheld devices such as smart cellular phones and personal digital assistants (PDAs) are used to browse the mobile Internet. However, the small screens of handheld devices and the slow mobile data transmission make the mobile Web browsing awkward. In Chapter X, entitled "*Adaptive Mobile Web Browsing Using Web Mining Technologies,*" Wen-Chen Hu, Yanjun Zuo, Lei Chen, and Chyuan-Huei Thomas Yang present a research that applies Web usage mining technologies to handheld devices. Web usage mining is the application of data mining techniques to the usage logs of large Web data repositories in order to produce results that can be applied to many practical subjects, such as improving Web sites/pages. A Web usage mining system must be able to perform five major functions: (a) usage data gathering; (b) data preparation; (c) navigation pattern discovery; (d) pattern analysis and visualization; and (e) pattern applications. This approach improves the readability and download speed of mobile Web pages.

Axel Schwickert and Bernard Ostheimer, in Chapter XI, "*Integration of Public University Web Sites and Learning Management Systems,*" address the Web's potentials supporting learning pathways at universities. The World Wide Web is used as an instrument of knowledge transfer and knowledge examination. The chapter addresses the utilization of Web content management systems (WCMS) for learning management systems (LMS) to transfer and the examination of knowledge.

Chiara Frigerio, in her chapter named "*Innovating through the Web: The Banking Industry Case,*" deals with the Web's strategic role in the banking industry. The chapter provides statistics indicating significant increases in the utilization of Web technologies during 2001-2007. The chapter describes in details the role of Internet in facilitating services and developing new products for the banking industry.

Section III. Applications of Web Strategy

The third section of the book presents six chapters focusing in the Web strategy applications. It contains results from several case studies and empirical research.

In Chapter XIII, entitled, "*An Action Research Case Study of the Facilitators and Inhibitors of E-Commerce Adoption,*" Orla Kirwan and Kieran Conboy study the adoption of e-commerce by small and medium size enterprises (SMEs). The authors employ action research methodology and use an established Irish retail business as a case study. The authors demonstrate how the research was undertaken, and discuss the justification, benefits, and limitations of using action research. The research concludes that the adoption of e-commerce within the SME sector tends to be slow and fragmented. It also supports the evidence that an SME is more likely to adopt e-commerce when the SME owner has a positive attitude toward IT.

Chapter XIV, "*Acceptance of the Mobile Internet as Distribution Channel for Paid Content in Germany,*" by Svenja Hagenhoff, Christian Kaspar, Lutz Seidenfaden, and Björn Ortelbach presents a survey based on 7,178 valid responses which analyses the mobile content usage in Germany. The authors found that respondents who are familiar with mobile radio and handset technology and read specialized printed media on a regular basis show the highest acceptance of mobile paid contents. The mobile Internet is perceived as a chance for the media industry to generate additional revenues from paid contents. The authors conclude that media companies planning to establish mobile services for content distribution are facing the problem that acceptance of mobile services has not yet been researched thoroughly.

Information quality is critical for communication portals because there are myriads of information types, including textual, audio, video, and other complex information types. Kevin K.W. Ho, in his chapter entitled, "*Information Quality Satisfaction of Communication Portals: A Study of Central Cyber Government Office (CCGO) of the Hong Kong Government,*" presents a study that examines whether information generated from an in-house developed communication portal of the Hong Kong government would have higher quality than those portals developed by individual government departments using commercial packages. He conducted a survey-based study to understand how users evaluate the information quality of these communication portals. The case-study presented in the chapter is interesting because (1) the Hong Kong government has invested millions of US dollars in its implementation and (2) the number of potential users is huge (over 140,000 users).

Performance assessments of Web projects and IT investments in general is one of the most debated subjects in literature. Maria Alice Frontini and Fernando Jose Barbin, Laurindo present an interesting methodological approach to the subject, based on the "real options" method. The chapter, entitled "*The Evaluation of IT Investments through Real Options,*" underlines how chief information officers (CIOs) face prejudice from top management, and who are concerned with the real benefits that IT can bring to business. This chapter concludes that the technological uncertainties make it very difficult to estimate the value that can be extracted from the usage of new IT solutions.

Silvia Novaes Zilber, in her chapter, "*Strategic Use of the Internet and Organizational Structure for E-Business: "Celta" Case at GM Brazil,*" presents an interesting study about the introduction of e-business activities by General Motors Brazil, specifically in connection with the launch of the Celta car, which is an entry-level car designed to be sold on the Internet. The organizational structure implemented for the launching and sales of the Celta car warranted the integration between employees on both business and information technology (IT) sides, in the context of GM Brazil's strategic objective of growing the market share for lower-price cars.

The last chapter of the book, Chapter XVIII, is authored by G. Castellano, A. M. Fanelli, and M. A. Torsello and is entitled "*On the Use of Soft Computing Techniques for Web Personalization*". The authors discuss Web-based applications and emphasize that these applications become increasingly necessary, because of the growing variety and quantity of information available on the Web. In their chapter, the authors emphasise the suitability of hybrid schemes combining different soft computing (SC) techniques for the development of effective Web personalization systems. In particular, the authors present a neuro-fuzzy approach for Web personalization that combines the fuzzy and the neural paradigms to derive knowledge from Web usage data and represent it in the comprehensible form of fuzzy rules.

REFERENCES

Awareness. (2007). *Trends in adopting Web 2.0 for the enterprise in 2007*. Retrieved March 6, 2008, from http://www.awarenessnetworks.com/resources/

Barua, A., Whinston, A. B., & Yin, F. (2000). Value and productivity in the Internet economy. *Computer,* 33(5), 102-105.

Bovet, D., & Martha, J. (2000). *Value nets: Breaking the supply chain to unlock hidden profit*. New York: John Wiley & Sons.

Burst Media. (2008). *Online ageism: Content and advertising miss an important target*. Retrieved March 5, 2008, from http://www.burstmedia.com/research/archive.asp

Clearswift. (2007). *Internet and Web 2.0 creates unfamiliar battleground for HR professionals.* Retrieved March 7, 2008, from http://resources.clearswift.com/main/pages/Clearswift/RSRCCTR/ShowCollateral.aspx?oid=15019

Cox, B., & Koelzer, W. (2004). *Internet marketing.* New Jersey: Prentice Hall.

Evans, P., & Wurster, T. S. (2000). *Blown to bits: How the new economy of information transform strategy.* Boston: Harvard Business School Press.

Fattah, H. (2000). Failing health. *Media Week, 10*(17), 100-103.

Iacovou, C. L., & Dexter, A. S. (2005, April). Surviving IT project cancellations. *Communications of the ACM, 48*(4), 83-86.

Internet System Consortium. (2008). *ISC Internet domain survey.* Retrieved March 3, 2008, from http://www.isc.org/index.pl?/ops/ds/

Internetworldstats. (2008). *Internet usage statistics.* Retrieved March 3, 2008, from http://www.internetworldstats.com/stats.htm.

Kappelman, L. A., McKeeman R., & Zhang, L. (2006, Fall). Early warning signs of IT project failure: The dominant dozen. *Information Systems Management,* 31-36.

McKinsey. (2008). How businesses are using Web 2.0: A McKinsey global Survey. *The McKinsey Quarterly.* Retrieved March 5, 2008, from http://www.mckinseyquarterly.com/How_businesses_are_using_Web_20_A_McKinsey_Global_Survey_1913_abstract

Minard, B. (2001). CIO longevity: IT project selection and initiation in the Health Care Industry. *IT Health Care Strategist, 3*(12), 1-12.

Minard, B. (2002). CIO longevity and IT project leadership. *IT Health Care Strategist, 4*(1), 3-7.

Nielsen. (2008a). *Nielsen online reports topline U.S. data for March 2008.* Retrieved March 4, 2008, from www.nielsen-online.com

Nielsen. (2008b). *Nielsen online announces April US search share rankings.* Retrieved May 31, 2008, from http://www.netratings.com/pr/pr_080519.pdf

Ofcom. (2005). *Internet statistics compendium.* Retrieved March 7, 2008, from http://www.ofcom.org.uk/about/accoun/reports_plans/annrep0607/coreareas/research/

Porter, M. E. (2001, March). Strategy and the Internet. *Harvard Business Review, 79*(3), 63-78.

Ries, A., & Ries, L. (2002). *The 22 immutable laws of Internet branding.* UK: Collins.

Wilson, R. (2002). *Planning your Internet marketing.* New York: John Wiley & Sons.

Section I
Design Web Strategy

The objective of the book's first section, which is subdivided into six chapters, is to look into the different Web strategy planning routes in different corporate contexts, in order to obtain optimal use of the Internet's technology potentials.

Chapter I
Using Patterns for Engineering High-Quality E-Commerce Applications

Pankaj Kamthan
Concordia University, Canada

Hsueh-Ieng Pai
Concordia University, Canada

ABSTRACT

In this chapter, we view the development and maintenance of high-quality electronic commerce (e-commerce) applications from a Web engineering perspective. A methodology for deploying patterns as means for improving the quality of e-commerce applications is presented. To that regard, relevant quality attributes and corresponding stakeholder types for the e-commerce applications are identified. The role of development process, the challenges in making optimal use of patterns, and feasibility issues involved in doing so, are analyzed. The activities of a systematic selection and application of patterns are explored. Examples illustrating the use of patterns during macro- and micro-architecture design of business-to-consumer (B2C) e-commerce applications are given. The implications of the use of patterns in a Semantic Web context are briefly highlighted.

INTRODUCTION

Over the past decade, electronic commerce (e-commerce) (Kalakota & Whinston, 1996) has opened new vistas for many sectors of society including businesses and has revolutionized the way business is conducted. In particular, e-commerce applications have revolutionized the way business is conducted today, and reduced the gap between small-and-medium size enterprises (SMEs) and large corporations.

Although e-commerce applications have offered rich prospects, they have also brought various concerns to the providers (Kamthan, 1999).

Indeed, the successes have come with their share of failures (Nguyen, Johnson, & Hackett, 2003), many of which have been attributed to issues of quality (Pertet & Narasimhan, 2005).

A commitment to "high-quality" is both an imperative and a challenge to the providers of e-commerce applications. Indeed, empirical studies have shown (Sharkey, Scott, & Acton, 2006) that the quality of an e-commerce application is directly related to its "dimensions" of success. At the same time, it is crucial that efforts towards assuring the quality of these applications remain predictable and feasible in the long-term.

In this chapter, our interest is in a *systematic* approach of engineering large-scale and quality-centric e-commerce applications based on the knowledge garnered from past experience and expertise (Kamthan, 2008). For that, we rely on the notion of the problem-solving approach of patterns (Appleton, 1997; Buschmann, Henney, & Schmidt, 2007). As discussed later, the use of patterns has several advantages over other approaches in terms of quality improvement, including that the approach is preventative rather than curative (Dromey, 2003), is supported by developmental processes, and provides practical solutions along with their reasoning for problems that have been tackled in the past.

The rest of the chapter is organized as follows. We first outline the background necessary for the discussion that follows, and state our position in that regard. This is followed by the presentation of a *pattern-oriented electronic commerce applications methodology (POECAM)* for systematically addressing the quality of e-commerce applications. POECAM includes a model consisting of quality attributes at different tiers and the role of patterns as means for addressing them. Next, challenges and directions for future research are outlined. Finally, concluding remarks are given.

BACKGROUND

In this section, we present the motivation for a systematic approach to address the quality in e-commerce applications and a synopsis of patterns.

Challenges to Development of E-Commerce Applications

There are certain defining characteristics that make e-commerce applications unique compared to other software applications, and lead to special considerations towards their quality management.

The market may often dictate the direction of the development and the evolution of an e-commerce application. Typically, e-commerce applications have relatively short time-to-market. This additional constraint on scheduling can adversely impact the activities related to quality assurance and evaluation that the provider of an e-commerce application needs to contend with. For example, a provider may have to compromise the time allocated to verification (say, via inspections or testing). Furthermore, the information in e-commerce applications may also have to be frequently modified to reflect the state-of-the-art and/or to maintain competitive advantage.

The consumers of an e-commerce application need not be colocated; indeed, they may be in different jurisdictions in the same country or in different countries. The laws that govern the provider and the consumers of an e-commerce application may be different. For example, let A and B be two different jurisdictions and let P be a product available for purchase in jurisdiction A. Then, although browsing information on P may be legal in B, purchasing it may not be.

The consumers of an e-commerce application may reside in different locales with different time zones. The expectations of the consumers with respect to the uninterrupted servability of the application can place unique constraints on the providers, particularly towards maintenance.

There can also be stark variations in the personal preferences and abilities of consumers and their computing environments. For example, it cannot be assumed that every consumer may be familiar with the intricacies of the Web or that the consumer may be using a high-end computer with a specific user agent. The providers of e-commerce applications need to deal with such diversity.

The "faceless" nature of e-commerce can contribute to unfavorable perceptions related to trustworthiness, particularly in the time of crisis, such as (frequent) failed transactions. The providers of e-commerce applications once again need to take steps to establish credibility (Kamthan, 2007a) with the consumers.

An Engineering View of E-Commerce Applications

The need for managing increasing size and complexity of Web applications and the necessity of a planned development has led to the discipline of Web engineering (Ginige & Murugesan, 2001), which has been treated comprehensively in recent years (Kappel, Pröll, Reich, & Retschitzegger, 2006; Mendes & Mosley, 2006; Rossi, Pastor, Schwabe, & Olsina, 2008). We define Web engineering as a discipline concerned with the establishment and the use of sound scientific, engineering, and management principles and disciplined and systematic approaches to the successful development, deployment, and maintenance of Web applications.

For the sake of this chapter, we will consider an e-commerce application to belong to a special class of Web applications that is product of a Web engineering process. In particular, an e-commerce application can be viewed as a heterogeneous combination of views, that is, an interactive information system in a distributed (client-server) environment with a commercial intent.

Quality of E-Commerce Applications

There are different views of quality (Wong, 2006). From the ISO/IEC 9126-1: 2001 Standard, we define quality to be the totality of characteristics of an entity that bear on its ability to satisfy stated and implied needs. In our case, this entity is an e-commerce application.

The need for e-commerce applications to exhibit "high-quality" is critical to all stakeholders involved. If unaddressed, there is a potential for a resource in an e-commerce application to be inaccessible by a visually impaired customer, abruptly crash upon customer interaction, or be prohibitive to maintenance by an engineer. This could seriously impact the viability and evolvability of the e-commerce application in question, and undermine the bottom line of the business that owns the application.

There have been a few initiatives for addressing the quality of e-commerce applications (Albuquerque & Belchior, 2002; Hasan & Abuelrub, 2006; Offutt, 2002; Oliveira & Joia, 2005; Stefani & Xenos, 2001). However, these efforts are limited by one or more of the following issues: although the quality attributes relevant to e-commerce applications are given, the relationships between them including trade-offs are not always indicated; the results are restricted to empirical studies based on one class of e-commerce applications; means for addressing the quality attributes are either suggested informally or not at all; or the focus is less on assurance (prevention) and more on evaluation (cure).

The Interplay of Patterns, Development Process, and Quality of E-Commerce Applications

The reliance on past experience and expertise is critical to any development. A pattern is an entity of knowledge that provides a proven solution to a recurring problem in a given context (Appleton, 1997).

Formally, a pattern is typically described (Meszaros & Doble, 1998) using an ordered list of elements (highlighted in italics in the rest of the chapter) labeled as (pattern) *name*, *author*, *context*, *problem*, *forces*, *solution*, *example*, and *related patterns*. At times, the labels may vary across community, and other (optional) elements, such as those related to metadata, may be included to enrich the description. The *name* element of a pattern is often a metaphor reflecting what the *solution* may be about, the *author* element gives the identity of the pattern author(s), the *context* element provides the situation or preconditions within which the *problem* occurs, the *forces* element provides the constraints that are resolved to arrive at a *solution*, the *solution* element provides an abstract solution to the problem and is shown to work in practice via an *example* element, and the *related patterns* element outlines any other pattern(s) to which a pattern is related to in some way.

In the past decade, patterns have been discovered in a variety of domains of interest, including those that are applicable to the development of Web applications in general and e-commerce in particular. There are collections of patterns available in print form (Adams, Koushik, Vasudeva, & Galambos, 2001; Marks & Hong, 2006; Montero, Lozano, & González, 2002; Rossi & Koch, 2002; Rossi, Lyardet, & Schwabe, 2000; Van Duyne, Landay, & Hong, 2003; Weiss, 2003). There are collections of patterns also available in electronic form (over the Web) such as the Portland Pattern Repository, the Amsterdam Collection of Interaction Design Patterns, and the Yahoo! Design Pattern Library.

Indeed, patterns have been used for the development of certain e-commerce applications (Garzotto, Paolini, Bolchini, & Valenti, 1999; Markiewicz, Lucena, Alencar, & Cowan, 2002; Montero et al., 2002; Montero, López-Jaquero, & Molina, 2003). However, in these cases, the relation of patterns to any development process and/or to the improvement of quality is not discussed explicitly.

In general, patterns are especially suitable as means for addressing quality concerns. They go through a comprehensive review by other experts in the domain that enables them to be appropriately described and documented. They provide *solutions* that are specific to *problems* in a given *context*, balance competing *forces*, are based on established principles, are rationalized, and are proven to work in practical situations. The evaluation of usability of e-commerce applications by determining the absence/presence of design patterns has been carried out (Georgiakakis, Psaromiligkos, & Retalis, 2006; Sartzetaki, Psaromiligkos, Retalis, & Avgeriou, 2003).

There are some patterns available specifically for addressing maintainability concerns of e-commerce applications (Weiss, 2003). However, in some cases the *solutions* are highly technology-specific, not all the mandatory elements of a pattern are appropriately documented, and the integration of patterns into any development process is not mentioned.

There are also some patterns available for addressing usability concerns of e-commerce applications (Graham, 2003; Perzel & Kane, 1999). However, usability is viewed as an atomic (non-decomposable) concept, the patterns are strongly oriented towards user-interface design, and their integration into any user-centered development process is not shown explicitly.

A SYSTEMATIC APPROACH FOR INTEGRATING PATTERNS IN E-COMMERCE APPLICATIONS

In this section, we introduce POECAM, which is inspired by previous work on integrating patterns and Web engineering (Kamthan, 2008). POECAM consists of the following interrelated and nonlinear sequence of steps:

1. Setting Goals
2. Identifying Stakeholders
3. Selecting the Development Process Model
4. Identifying and Organizing Quality Concerns
5. Acquiring and Selecting Suitable Patterns
6. Applying Patterns

From a practical standpoint of integrating patterns in e-commerce applications, each of the steps must be feasible. If it is not, we revert back to it and move forward once the necessary modifications are made. The feasibility study could be a part of the overall e-commerce application project management planning activity. Figure 1 gives an overview of the steps in POECAM.

We now discuss the work involved in each of these six steps in detail.

Step One: Setting Goals

It is known that a project without clear goals will not achieve its goals clearly (Gilb, 1988). A POECAM approach must be based upon high-level organizational (business), social, and/or technical goals that need to be accomplished.

For instance, a technical goal for integrating patterns in the overall development strategy could be *to rely on past experience and expertise in the domain to produce high-quality e-commerce ap-*

Figure 1. A high-level view of a pattern-oriented approach for the development of large-scale e-commerce applications

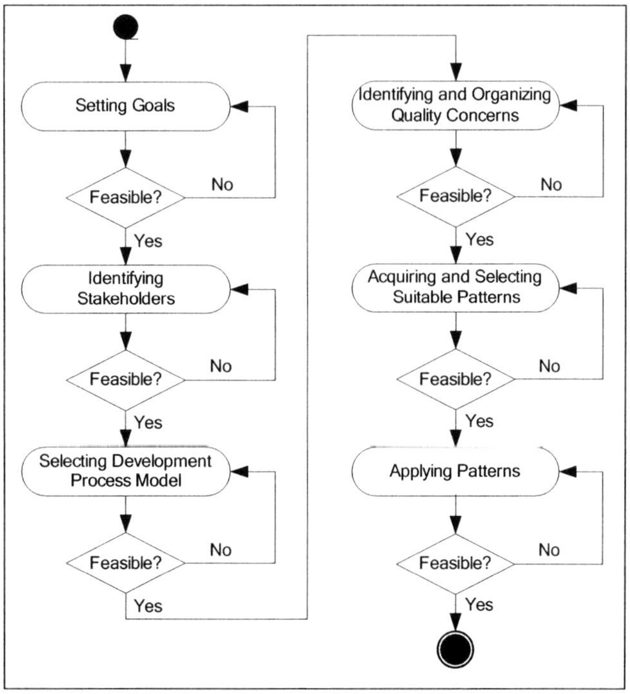

plications from stakeholders' perspective. For a goal to be realized in practice, however, it must also be feasible.

Step Two: Identifying Stakeholders

There are systematic approaches for identification and refinement of stakeholder classes (Sharp, Galal, & Finkelstein, 1999). We identify two broad classes of stakeholders with respect to their *roles* in relationship to an e-commerce application: a *producer* (provider, manager, engineer, or maintainer) is the one who develops, deploys, or maintains the e-commerce application, and a *consumer* (novice or expert user) is the one who uses the e-commerce application for some purpose.

We emphasize that the above classification is based on the *role* of a stakeholder. For example, both the engineer and the user could be the same person, but their roles with respect to the interaction with an e-commerce application are different.

The viewpoint-oriented requirements definition (VORD) (Kotonya & Sommerville, 1998) also gives us a way of classifying aforementioned stakeholders of interactive systems in a client-server environment such as e-commerce applications. From a VORD perspective, the provider and the manager are indirect viewpoints, while the engineer and the maintainer are direct viewpoints in their relationship to the e-commerce application.

Step Three: Selecting the Development Process Model

The inclusion of patterns in the development of e-commerce applications cannot be ad-hoc or an afterthought. We recommend POECAM within the framework of an *existing* process environment that already has broad community and tool support in order to keep the cost to a minimum and the learning curve low.

In light of the unique characteristics of e-commerce applications, is crucial that the development process be flexible and user-centered. A flexible user-centric process for developing e-commerce applications will typically be nonlinear (iterative *and* incremental) and, to that regard, we recommend the adoption of one of the following process models. Extreme programming (XP) (Beck & Andres, 2005) is a mature and broadly-used agile methodology for software development. XP places "lightweight" requirements on resources and is suitable for SMEs. On the other hand, the unified process (UP) (Jacobson, Booch, & Rumbaugh, 1999) is an archetype of model-based and uses a case-driven process *framework*, of which the rational unified process (RUP) (Kruchten, 2004) is an instance. RUP requires heavy business and technical modeling and documentation, and is especially suited for large businesses.

Both XP (Kappel et al., 2006; Wallace, Raggett, & Aufgang, 2002) and RUP (Kappel et al., 2006) have been "customized" for Web applications and by reference to e-commerce applications. There is provision of the use of patterns during the design phase in both XP and RUP.

Step Four: Identifying and Organizing Quality Concerns

For the purpose of this chapter, we focus on the *semiotic* quality of e-commerce applications. From a semiotics (Shanks, 1999; Stamper, 1992) viewpoint, we can view an e-commerce application as an information system on six interrelated levels: physical, empirical, syntactic, semantic, pragmatic, and social. In this chapter, we shall restrict ourselves to the discussion of the pragmatic level, which is responsible for the relation of signs to their interpreters.

Next, we contend that pragmatic quality is a multidimensional concept, and decompose it into granular levels that consist of known attributes that can be addressed directly or indirectly. For the definitions of these quality attributes, we resort to the IEEE Standard 1061-1998, the ISO 9241-11:1998 Standard, and the ISO/IEC 9126-1: 2001 Standard.

Finally, we assign patterns as means for improving the quality attributes. Table 1 summarizes this construction.

We contend that the quality attributes in Table 1 are necessary, but make no claim of their sufficiency. We also note that the pragmatic quality attributes (discussed later) in Table 1 cannot (at least mathematically), with respect to stakeholders, be *completely* satisfied. For example, an *a priori* guarantee that an e-commerce application will be usable to *all* users at *all* times in *all* computing environments that the users deploy, is not realistic.

The quality attributes in Table 1 are not mutually exclusive. Indeed, the quality attributes in Tier 3 depend on that in Tier 2, which in turn depend on Tier 1. For example, if a user cannot read, the user cannot comprehend the information in an e-commerce application, and thereby cannot use it to its full potential. Similarly, for an e-commerce application to be reliable, it must be available.

Furthermore, the quality attributes within the same tier in Table 1 are not necessarily mutually exclusive. For example, the steps taken towards improving reliability (say, fault tolerance) may lead to redundant source code or data (that can be unfavorable to maintainability) but enable ease-of-use (that can be favorable to usability).

The Quality-Stakeholder Contract

For the sake of this chapter, we view pragmatic quality as a *contract* between an e-commerce application and a stakeholder. For simplicity, we will limit ourselves to the discussion of (not necessarily mutually exclusive) stakeholders of the type *end-user* and *engineer.*

The relevance of quality attributes in Table 1 varies with respect to stakeholder types:

- **Pragmatic-Tier 1.** The quality attributes of direct concern to an end-user are aesthetics, availability, familiarity, and readability. The quality attribute of direct concern to an engineer is space and time efficiency.

Table 1. A model for the pragmatic quality of e-commerce applications

Semiotic Level	Quality Attributes	Means for Quality Assurance
Social Quality Concerns		Patterns
Pragmatic	[Tier 3] Maintainability, Usability	
	[Tier 2] Comprehensibility, Performance, Reliability	
	[Tier 1] Aesthetics, Availability, Efficiency, Familiarity, Readability	
Physical, Empirical, Syntactic, and Semantic Quality Concerns		

- **Pragmatic-Tier 2.** The quality attributes of direct concern to an end-user are comprehensibility, performance, and reliability. The quality attribute of direct concern to an engineer is comprehensibility.
- **Pragmatic-Tier 3.** The quality attribute of direct concern to an end-user is usability. We will view accessibility as a special case of usability (Mendes & Mosley, 2006). The quality attribute of direct concern to an engineer is maintainability. We will consider modifiability, portability, and reusability as special cases of maintainability (Buschmann, Meunier, Rohnert, Sommerlad, & Stal, 1996).

Finally, we note that the significance of quality attributes will vary across business models (Bambury, 1998) as well as different types of e-commerce applications. For example, the quality needs of a music shopping portal will vary from that of a fine art auction application.

Step Five: Acquiring and Selecting Suitable Patterns

The challenges in the patterns' acquisition and selection process stem from a variety of factors:

- **Availability of Patterns.** For an adoption of a pattern-based approach to the development of e-commerce applications, it is important to have design and implementation patterns that can sufficiently "map" the solution space. However, there is no *a priori* guarantee that for every quality-related problem, there will be suitable pattern(s) available for it.
- **Findability of Patterns.** There are currently several challenges that inhibit our ability of detect and find desirable patterns (Manolescu, Kozaczynski, Miller, & Hogg, 2007). We consider three of these challenges:
    - **Classification of Patterns.** There is no currently unique classification of patterns. There are some patterns that may be classified into familiar categories (like "structural" or "behavioral") while others may be presented as a loosely related informal collection. Apart from some isolated efforts (Van Welie & Van Der Veer, 2000; Schumacher et al., 2006), patterns are currently not organized by quality attributes and in only very few cases organized as directly relevant to e-commerce. This could adversely impact the task of locating desirable patterns (Segerståhl & Jokela, 2006).
    - **Relationships among Patterns.** Often, patterns do not exist in isolation and are part of an overall vocabulary (e.g., a "pattern system" or a "pattern language") that attempts to solve a larger problem than possible with an individual pattern. The relationships among patterns can be implicit or explicit. Indeed, due to the context-driven relationships among them, the use of one pattern can lead to the commitment of using other pattern(s). For example, selection and application of a pattern for a shopping system would place us in a context that would require the selection of a pattern for payment, which in turn would require the selection of a pattern specific to authentication.
    - **Similarity among Patterns.** The *name* of a pattern plays a crucial role in acquiring and selecting desirable candidates. There are some patterns that have similar or same names but have semantically different functionality. There are also other patterns in different pattern collections that have similar intent or functionality but have

different names. For example, the NEWS pattern (Rossi et al., 2000) is similar to the WHAT'S NEW PAGE pattern (Van Duyne et al., 2003). These patterns may have been (re)discovered independently.

- **Cost of Deploying Patterns.** There is evident cost associated with the deployment of patterns that cannot be disregarded. It is known that patterns are means of *conceptual* reuse, and there is a cost in terms of time, effort, and resources of learning and adaptation involved in any reuse. The issue of cost is particularly acute in educational contexts where affordability of commercial resources (such as books on patterns) is a constant concern.

Step Six: Applying Patterns

There are three main nonmutually exclusive concerns in the application of patterns: the understanding of the pattern description, the order in which patterns are applied, and the result upon the composition of patterns.

1. **Understanding the Pattern.** The understanding of the underlying *problem*, the *context* in which it occurs, and the trade-offs and consequences of the proposed *solution* are imperative. For example, patterns that suggest *solutions* involving the use of different colors will not be applicable in contexts where the underlying monitor does not support it (such as when the screen is monochrome) or if the user is color blind.
2. **Order of Application of Patterns.** In the design phase, the patterns for a high-level (macro-architecture) design are applied first, followed by the patterns for a low-level (micro-architecture) design. In each of these levels, the order of the application of patterns does matter and, if ignored, it may not be at all possible to apply other patterns and/or lead to a nonoptimal result.
3. **Composition of Patterns.** The relationships among patterns can be favorable, unfavorable, or neutral. Even when a certain order in the application of patterns is followed, it is the result upon the composition of patterns that should appear suitable. Although the application of *individual* patterns may appear acceptable, the result with a *multiplicity* of patterns may not.

For the remainder of this section, we will assume that, among the given possibilities (Adams et al., 2001), the e-commerce application under development is of the type business-to-consumer (B2C) and being accessed via stationary devices (e.g., desktop computers) or pseudo-stationary devices (e.g., notebook computers). Furthermore, even though patterns are applicable to all the phases of a development process, including requirements, design, and implementation for considerations of space, we will limit ourselves in this chapter to addressing the role of patterns in the design phase of e-commerce applications.

As evident from the discussion that follows, the patterns presented here form a skeleton sequence that traverses through several existing collections of patterns. Our selection of patterns is based on their generality, neutrality with respect to any specific application domain, broad availability, parity to the quality attribute at hand, suitability of the *context* and the *forces* (where available), and the reputation of the authors.

In order to distinguish the patterns from the main text, their *names* are listed in uppercase. The Appendix summarizes the quality attributes and patterns mentioned in this section.

Macro-Architecture Design of E-Commerce Applications

The macro-architecture design is the place where high-level design decisions, independent of any implementation paradigm or technology, are made.

Patterns Applied from a Business Viewpoint

An e-commerce application will implicitly or explicitly target some domain. The choice of the domain name (such as `.net` or `.com`) does not always or automatically reveal the *nature* of the domain. There are patterns available for certain common genres, like ENABLING INTRANETS (for businesses in general), PERSONAL E-COMMERCE (for small businesses), VALUABLE COMPANY SITES (for large organizations), and STIMULATING ARTS AND ENTERTAINMENT (for news organizations) (Van Duyne et al., 2003). The use of such genre-specific patterns can increase user familiarity with the e-commerce application.

Furthermore, the organization owning an e-commerce application may wish to serve (potential) consumers in diverse cultural and/or geopolitical situations (i.e., in different countries and using different natural languages). This could be done using the LOCALE HANDLING pattern (Busse, 2002). In retrospect, it increases maintenance responsibilities.

Patterns Applied from a Technical Viewpoint

The macro-architecture patterns that we suggest are based on the fact that e-commerce applications are a class of distributed request-response-type interactive systems. Specifically, the applicable patterns are the CLIENT-SERVER pattern (Schmidt, Stal, Rohnert, & Buschmann, 2000), followed by the APPLICATION SERVER pattern (Manolescu & Kunzle, 2001), which in turn is followed by the MODEL-VIEW-CONTROLLER (MVC) pattern (Buschmann et al., 1996).

The CLIENT-SERVER pattern supports maintainability. For example, a server or resources on

Figure 2. A view of the macro-architecture design patterns in the development of e-commerce applications

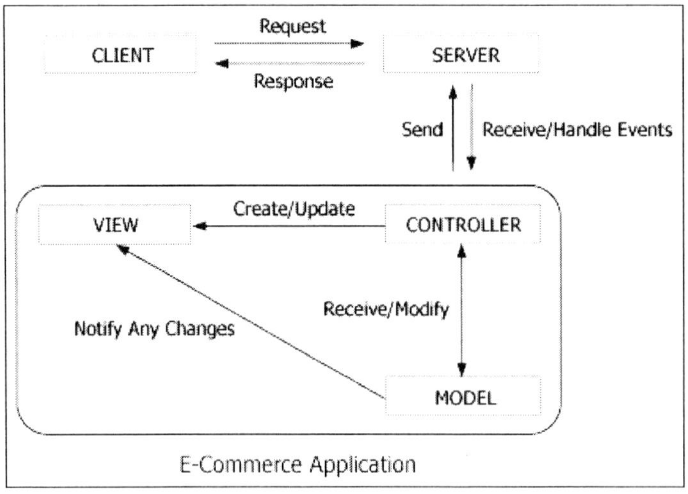

the server-side could be modified without impacting the client. Also, a single server can support multiple clients simultaneously, or a client could make simultaneous requests for resources residing on multiple servers. The APPLICATION SERVER pattern also supports maintainability: it isolates the e-commerce application from other aspects on the server-side such that the communication between the application itself and the Web server takes place via the SINGLE POINT OF ACCESS (Yoder & Barcalow, 1997) pattern. This separation allows the e-commerce application to evolve independently.

In applying the MVC pattern, the principle of separation of structure from presentation of content in a markup document leads to a separation of semantically-different aspects into three components, namely model, view, and controller. This minimizes the coupling between these components. Therefore, the same model in a MVC could be used with multiple views and multiple controllers. For example, the same information could be transformed and delivered to different browser environments or user needs. This improves the maintainability of an e-commerce application.

Figure 2 presents an abstract view of the aforementioned macro-architecture design patterns.

There are several implementations of MVC available in a variety of programming languages such as Java and hypertext preprocessor (PHP), and application frameworks like asynchronous JavaScript and XML (AJAX) and Rails.

The CUSTOMIZED PRESENTATION TO HOST pattern (Adams et al., 2001) mimics some aspects of MVC. It aims to improve usability by providing a user-friendly interface. However, there is a tight coupling between the customized presentation view and the host application. Hence, any changes to the host application may require changes to the presentation view. This is unfavorable to maintainability.

Reliability Design

For addressing reliability (specifically, availability) concerns, the macro-architecture design of server-side components of an e-commerce application could use a number of patterns (Ahluwalia & Jain, 2006; Manolescu & Kunzle, 2001). For example, the extra measures to support the availability of an e-commerce application (unrelated to the functionality of the application) could be included by using the INTRODUCE REDUNDANCY pattern.

One way to introduce redundancy is to have a cluster of multiple servers such as suggested by the FAIL-OVER THROUGH CLUSTERING pattern, where if one (primary) server fails, the other (secondary) server takes over the responsibility. In retrospect, redundancy also increases maintenance responsibilities.

If and when the need arises, a failure message could be relayed directly using the FAILURE NOTIFICATION pattern or indirectly using the HEARTBEAT pattern, where an engineer is informed via periodic broadcasts that a specific Web server is available; the absence of such a message would then imply its unavailability.

Micro-Architecture Design of E-Commerce Applications

The micro-architecture design is the place where low-level design decisions are cast.

For the rest of the section, we will focus only on the design aspects that impact pragmatic quality. As such, in the following, our attention is geared more towards client-side rather than server-side concerns.

Interaction design (Preece, Rogers, & Sharp, 2002) is perhaps the most crucial client-side concern among e-commerce applications. The goal of interaction design is to make both the content and the user interface useful, easy-to-use, and enjoyable. Many of the patterns available for interaction design for interactive systems in general are also applicable to e-commerce applications.

We now consider four of the most critical interaction design aspects of e-commerce applications, namely information design, navigation design, search design, and presentation design independent of any specific domain. We note that these aspects of design are not mutually exclusive.

Information Design

The delivery of information could either be static or dynamic. The dynamic delivery can take place via conversion of a *single source* of information in, for example, a database, to a format suitable for the Web, which enables adaptive maintenance. This can be accomplished by using the GENERIC CONTENT FORMAT pattern (Zdun & Vogel, 2002) for representing the information inside the "model" in MVC.

It is evident that not all users are the same; the properties of an e-commerce application like structure, content, and links could be "personalized" for each user. This is possible via patterns for personalization (Rossi, Schwabe, Danculovic, & Miaton, 2001), namely STRUCTURE PERSONALIZATION, CONTENT PERSONALIZATION, and LINK PERSONALIZATION. This will improve the user's experience with (and thus the usability of) the application. However, an optimal personalization would usually require the user to release some degree of personal information (say, via registration on a Web Portal), which in turn would lead to loss of privacy.

The information presented on a single "Web page" is often aggregated from several sources. For example, the entry point (from user-perspective) to an e-commerce application can be realized using the HOME PAGE pattern (Graham, 2003). Then, the "home page" of a news organization served from the main source may include a latest news ticker from one server, weather information from another server, the stock market information from a financial Web service, and periodically changing advertisements from yet another source.

The delivered information needs to be organized. This can be systematically realized by the use of the WHOLE-PART pattern (Buschmann et al., 1996), which enables a hierarchical organization of objects. Since each of these objects can be modified or replaced independently, the WHOLE-PART pattern supports maintainability. Also, since a "part" can correspond to more than one "whole," the WHOLE-PART pattern also supports reusability. However, multiple indirections stemming from client requests and responses for fulfilling them can lead to a loss of performance, particularly when each "part" itself is structured as WHOLE-PART.

Next, we look at the classification of information, which is a conventional approach by humans for understanding information. The information organization patterns (Van Duyne et al., 2003), when used appropriately, aid readability, comprehensibility, and usability. For example, the GRID LAYOUT pattern suggests the organization of information in a single document into a grid of rows and columns, where each atomic information element is made to fit within this grid. For special cases, such as a shopping system, the information of products could be further organized using the CATALOG pattern (Fernandez, Liu, & Pan, 2001). The WHAT'S NEW PAGE pattern provides newly added information and could include the CHRONOLOGICAL ORGANIZATION pattern.

The users of an e-commerce application can vary in their capabilities and preferences, and may find one view of information to be more usable than another. The MIRRORWORLD pattern (Germán & Cowan, 2000) provides two or more views of the same information. Specifically, information in these views could be presented (Tidwell, 2006) in TWO-PANEL SELECTOR pattern when we have two different views that are to be presented simultaneously, or CLOSABLE PANELS or CARD STACK patterns when we have several different views to be presented in such as way that only one view is visible at a time in each panel or stack, respectively.

Now, documents in an e-commerce application may contain images for presenting some information such as the corporate logo or product pictures. The FAST-DOWNLOADING IMAGES pattern (Van Duyne et al., 2003) suggests the creation of images optimized for color and size in an appropriate format, and thus aids accessibility and performance. The REUSABLE IMAGES pattern (Van Duyne et al., 2003) suggests caching images that appear at multiple places in an e-commerce application, and thereby aids performance.

To improve usability, there should be a provision in the information design to support an internal locus of control (thereby provide options to a user) and for users to recover (say, from inadvertent errors). The MULTI-LEVEL UNDO pattern (Tidwell, 2006) provides a way to easily reverse a series of actions performed by the user in an e-commerce application that can track user session and maintain state.

Navigation Design

Navigation is traversal in information space (Lynch & Horton, 2002) for some purpose, such as casual or targeted browsing for information or complementing a reading sequence (like in electronic books). Both the intra- and the inter-document navigation within the context of an e-commerce application are realized by the use of hypermedia (Germán & Cowan, 2000).

There are various patterns for navigating through an e-commerce application that have been proposed over the years (Lyardet & Rossi, 1998; Marks & Hong, 2006; Tidwell, 2006; Van Duyne et a., 2003). The navigation patterns, when use appropriately, aid usability. For example, the BREADCRUMBS pattern (Marks & Hong, 2006; Van Duyne et al., 2003) could be used to inform the user of the user's location. The Yahoo! Directory was one of the earliest users of the BREADCRUMBS pattern. The CLEAR ENTRY POINTS pattern (Tidwell, 2006) presents only a few entry points into the interface, which can restrict the navigation to a specific category and make it task-oriented. For special cases, such as a shopping system, the products could be navigated through well-defined steps using the SHOPPING PROCESS pattern (Fernandez et al., 2001).

The FLY-OUT MENU pattern (Marks & Hong, 2006) could be used to present content organized in a "compound" menu where each menu item itself has a submenu that expands only upon interaction and when the user desires. This enables a large amount of navigation information to be "hidden" from the user and presented only "on-demand," thereby improving both (spatial) efficiency and readability. The FLY-OUT MENU pattern could itself be arranged horizontally or vertically as suggested by the HORIZONTAL NAVIGATION or VERTICAL NAVIGATION patterns (Marks & Hong, 2006), respectively.

Any navigation design must take exceptional behavior into consideration to support usability. The SESSION pattern (Weiss, 2003) can help maintain the state of the e-commerce application in the event of an interruption of navigation flow. The MISSING LINK pattern (German & Cowan, 2000) informs the user that certain hyperlink does not exist and suggests alternatives.

There are navigation design patterns that enable efficient use of space and aid comprehensibility (Tidwell, 2006). For example, the WIZARD pattern leads the user through the interface step by step for carrying out tasks in a prescribed order. It can also be used to implement a context-sensitive help on a given functionality. The RESPONSIVE DISCLOSURE pattern starts with a very minimal interface, and guides a user through a series of steps by showing more of the interface as the user completes each step. These two patterns could, for example, be used for carrying out a registration process. Now, during such a process, the user may have to be presented with several options (e.g., multiple mailing addresses or credit cards).

However, the designer is faced with the problem that the information on all of them would not fit on a single panel and a user does not have to see the details of all the options simultaneously. In such a case, the CARD STACK pattern could be used where one option is visible by default and the others are hidden.

Search Design

The goal of searching is finding information. Even though searching is not native to e-commerce applications, it has become ever more challenging as the amount of information to be searched through increases.

Various patterns for searching have been proposed over the years (Lyardet, Rossi, & Schwabe, 1999; Marks & Hong, 2006; Van Duyne et al., 2003;). When used appropriately, searching patterns aid comprehensibility and performance.

The use of a STRAIGHTFORWARD SEARCH FORMS pattern (Van Duyne et al., 2003) with a SIMPLE SEARCH INTERFACE pattern (Lyardet et al., 1999), that requires minimal technical background on part of the user, will contribute towards comprehensibility. The use of a SELECTABLE SEARCH SPACE pattern (Lyardet et al., 1999) that can restrict the search to a specific category, the SELECTABLE KEYWORDS pattern (Lyardet et al., 1999) that can suggest keywords for improving subsequent search results based on the past experience, and the ORGANIZED SEARCH RESULTS pattern (Van Duyne et al., 2003) that presents a summary of the most relevant search results, can all improve the effectiveness of the searching activity. The Google search engine implements some of these search patterns.

When search results are too numerous such that they could cognitively overload a user, they could be split into multiple sections that could be navigated sequentially using the PAGING pattern (Marks & Hong, 2006). Finally, we note that the ACCOUNT SETUP pattern (Marks & Hong, 2006) allows reuse of a user's personal information and can assist in realizing the efficacy of the SELECTABLE KEYWORDS pattern, especially on Web portals.

Any search design must also take exceptional behavior into consideration to support usability. For example, long keywords may be difficult for a user to remember and are prone to typographical

Figure 3. An assembly of some interaction design patterns for an abstract e-commerce application

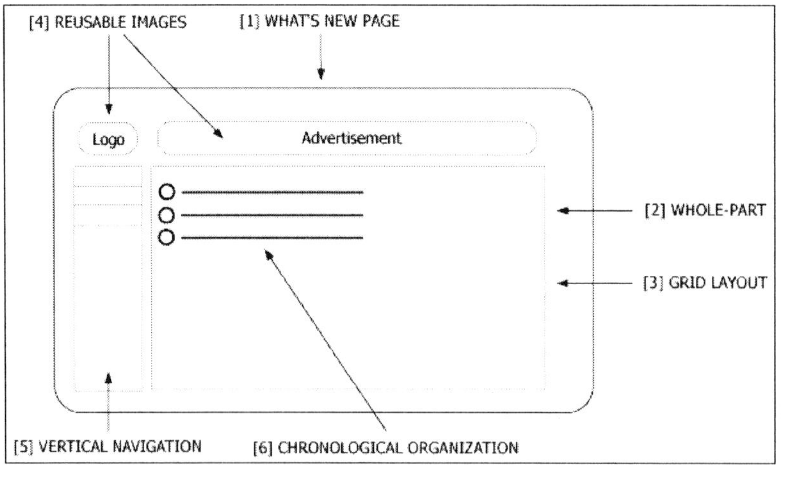

errors. The use of the AUTOCOMPLETE pattern (Yahoo! Design Pattern Library) can circumvent this problem.

Presentation Design

It has been shown in surveys (Fogg, Soohoo, Danielson, Marable, Stanford, & Tauber, 2002) that users highly value the aesthetics or the "look and feel" of an e-commerce application. An e-commerce application that is appropriately presented also gives the perception of "professionalism" that is essential for establishing credibility. The elements of presentation apply to all aspects of design discussed previously.

For the purpose of aesthetics, a variety of patterns could be used. For example, using the DEEP BACKGROUND pattern (Tidwell, 2006), an image or gradient could be placed in the background of a document in such a way that it visually recedes behind the foreground elements.

In traditional "brick and mortar" shops, certain products that are seasonal, for special occasions, on sale or otherwise recommended, are often displayed in a special manner. This selected list of products could be highlighted via the FEATURED PRODUCTS pattern (Van Duyne et al., 2003).

It is known that colors can have a positive impact both cognitively and aesthetically if used appropriately, in particular taking into the considerations of the color blind (Rigden, 1999). Often, organizations have their favorite color(s) that are prominently reflected in their logo, and placed on their business cards and office supplies. Using patterns like FEW HUES, MANY VALUES, or COLOR-CODED SECTIONS (Tidwell, 2006), an e-commerce application could be given a unique "identity." It is known that the user interface must reflect the state of the software. So, for example, the option in a navigation bar selected by a user could be reflected uniquely upon visitation by the user. There are of course other presentation issues that are crucial, such as choice of fonts and their properties, layout and positioning of user interface components, use of white space, and so on. These could also be addressed via an appropriate use of patterns.

Example

Figure 3 gives an example of a sequence of structural (specifically, visual) patterns for interaction design in an abstract commercial setting. This could serve as a typical manifestation of the "what's new page" or a "products page" of an organization.

This sequence of patterns in the example can be enriched by extension in two different directions: (1) if necessary, further patterns can be added to the sequence, and (2) the example itself could be a part of the overall information architecture of an e-commerce application, the design of which is also based on a pattern.

POECAM in Perspective

In this section, we briefly outline the scope and limitations of POECAM for e-commerce applications.

First of all, for POECAM to be applicable, the organization should be at a level of maturity so that it is able to adopt a systematic process for developing the e-commerce application. However, achieving such a level has an associated cost that is neither trivial nor automatic (Paulk, Weber, Curtis, & Chrissis, 1995). Furthermore, engineers must be trained to make appropriate use of patterns, which often comes with experience. For example, not being able to understand the *context* in which a *problem* occurs can lead to a misapplication of the *solution* of the pattern, and hence a nonoptimal or even undesirable result. Finally, although it is not an inherent limitation of POECAM, we note that the list of patterns presented in this chapter is by no means complete and is subject to evolution.

FUTURE TRENDS

The work presented in this chapter can be extended in a few different directions, which we discuss briefly.

Extensions of POECAM

One possible extension of the model presented in Table 1 is increasing the granularity of the quality attributes at the level Pragmatic-Tier 1, and thereby adding another level (say, Tier 0) underneath it. In that case, for example, fault tolerance and recoverability could be candidate quality attributes that belong to the level Pragmatic-Tier 0. Some patterns for fault tolerance and recoverability have been suggested (Ahluwalia & Jain, 2006). Other possible extensions of the model are the use of patterns for improving higher- or lower-level semiotic quality concerns.

Use of Patterns in E-Commerce Applications for the Social Web

In recent years, the term social Web, or more commonly known by the pseudonym Web 2.0 (O'Reilly, 2005), has been put forth to describe the apparent "humanization" and "socialization" of the Web as it moves towards becoming a means of participation and collaboration driven by "collective intelligence." An elaboration of the social level of Table 1 could help towards accommodating e-commerce applications targeted for Web 2.0.

For example, one extension of interest would be addressing the social quality concerns, namely credibility, legality, privacy, and security. (We note that not all issues pertaining to these attributes, such as those related to the underlying platform or operating system, are within the scope of the development of e-commerce applications.) These can be organized in multiple different tiers, with credibility placed at a higher tier than legality, privacy, and security.

There are patterns that partly address credibility and legal concerns, for example, those for writing policy statements (Perzel & Kane, 1999) such as information, reputation, and warranty policies (Kaluscha & Grabner-Kräuter, 2003), and for licensing (Perry & Kaminski, 2005), respectively. For instance, including a pattern for a policy statement on advertising would contribute to the overall credibility of the e-commerce application in which the example from the previous section resides. Also, patterns for privacy (Hafiz, 2006; Romanosky, Acquisti, Hong, Cranor, & Friedman, 2006; Van Duyne et al., 2003; Weiss, 2003) and security (Adams et al., 2001; Marks & Hong, 2006; Schumacher, Fernandez-Buglioni, Hybertson, Buschmann, & Sommerlad, 2006; Yoder & Barcalow, 1997; Van Duyne et al., 2003; Weiss, 2003) have been proposed in recent years, many of which apply to the development of e-commerce applications.

The aforementioned extensions, however, would require that the aspects of micro-architecture design to which the patterns are applied are essentially different than those dealt with in this chapter. For example, in case of applying patterns for privacy and security, the attention would be more on the server-side rather than on the client-side components. Furthermore, since steps taken towards the aforementioned social quality concerns are not always favorable to maintainability and/or usability or some of the quality attributes they depend upon, care would need to be taken in selecting and applying these patterns in the development of e-commerce applications.

Use of Patterns in E-Commerce Applications for the Semantic Web

The Semantic Web has recently emerged as an extension of the current Web that adds technological infrastructure for better knowledge representation, interpretation, and reasoning (Hendler, Lassila, & Berners-Lee, 2001). It thus makes information far more amenable to machine-consumption than that is possible within the current Web. An elaboration of the semantic level of Table 1 could help towards accommodating applications for the Semantic Web.

An ontology is an explicit formal specification of a conceptualization that consists of a set of concepts in a domain and relations among them (Gruber, 1993). Ontologies provide precise means of representing knowledge and, arguably, form one of the most important layers in the Semantic Web infrastructure. By enabling better opportunities for organization and inferencing from given information (via navigation or searching), ontologies can play a crucial role in e-commerce applications of the future (Kamthan & Pai, 2006a, 2006b).

The ontology for Web application patterns (OWAP) (Kamthan & Pai, 2006a) is an ontology of typical patterns, many of which occur in e-commerce applications. With the help of a reasoner, it is possible to devise complex query formulations to make *implicit* knowledge in OWAP *explicit*. Thus, we can obtain more "intelligent" answers than that possible in a setting of, say, a database-backed e-commerce application where the search results are typically based on a simple keyword-based fetch and return.

As ontologies increase in size and complexity their methodical development becomes a necessity, and for which several approaches have been suggested (Gómez-Pérez, Fernández-López, & Corcho, 2004). The efforts towards a systematic approach towards addressing the quality of ontologies for the Semantic Web are at present scarce (Burton-Jones, Storey, Sugumaran, & Ahluwalia, 2005) and patterns for developing ontologies are currently in their infancy (Gangemi, 2005).

Use of Patterns in M-Commerce Applications

As mobile devices proliferate, they are increasingly being used for B2C commerce. The need for a systematic approach towards the engineering of mobile commerce (m-commerce) applications (Kamthan, 2007b; Mahmoud & Maamar, 2006) then arises.

The constrained environment of mobile devices (as compared to desktop computers) poses unique challenges pertaining to quality of applications available on them, and is of concern to both the producers and the consumers. For example, the restricted user interface and limited processor speed require that the producers take special measures to ensure usability and performance of m-commerce applications. Thus, the interplay between the stakeholder quality concerns (Chan & Fang, 2001; Kamthan, 2007c) and patterns (Ihme & Abrahamsson, 2005; Mazhelis, Markkula, & Jakobsson, 2005) for m-commerce applications, similar to POECAM, would be of interest.

POECAM in E-Commerce Education

There are various challenges being faced in e-commerce education (Chan, 2001), including the orientation of courses. The curriculum at educational institutions related to e-commerce applications has traditionally been dominated by courses that focus on technological (Yan & Fang, 2005) rather than theoretical aspects of development. In recent years, there has been a movement towards embracing the "engineering" aspects of e-commerce applications (Treese & Stewart, 2003), although a commitment to quality in general and patterns in particular is currently lacking.

POECAM presents an opportunity for integration of patterns in e-commerce-related course projects in such a way that theory and practice meet, thrive, and harmonically coexist. However, such integration cannot be ad hoc; it needs to be aligned with the e-commerce body of knowledge adopted and the teaching strategies and learning theories being followed. In this regard, an examination of the positioning of POECAM among the options for e-commerce education systems (Toraskar & Lee, 2006) would be of interest.

CONCLUSION

A lasting commitment towards quality has both ethical (Tavani, 2004) and business implications. For longevity and for acceptance by stakeholders, a disciplined but sufficiently agile approach towards the development of large-scale e-commerce applications is necessary. Integral to this is a systematic and lasting view towards quality, and means for addressing it.

As reflected by POECAM, patterns provide one practical means for addressing the quality of e-commerce applications, if they are found, adopted, and applied with feasibility issues in consideration. Also, for patterns to continue being useful as the source of guidance and knowledge, they must be adequately described and documented, freely available and readily findable, and evolve with the needs of the e-commerce domain.

In conclusion, we believe that an initial investment in a quality-centric approach to e-commerce applications using patterns is in interest of all stakeholders where the benefits can outweigh the costs in the long-term. For that to be realized, a reflection on and a shift in the current organizational culture (Wiegers, 1996) may be necessary. Furthermore, for a strategic change in an organization to be effective, it needs to be predictable and in turn systematic. Therefore, the integration of patterns in general and the adoption of POECAM in particular cannot be ad-hoc/spontaneous, and immune/isolated from its organizational context (Pettigrew, 1985). It has to be a part of the overall organizational vision/willingness/capability to change, and needs to be planned in advance and executed on demand.

ACKNOWLEDGMENT

The authors would like to thank the reviewers for their feedback and suggestions for improvement.

REFERENCES

Adams, J., Koushik, S., Vasudeva, G., & Galambos, G. (2001). *Patterns for e-business: A strategy for reuse*. IBM Press.

Ahluwalia, K. S., & Jain, A. (2006, October 21-23). *High availability design patterns*. Paper presented at the Thirteenth Conference on Pattern Languages of Programs (PLoP 2006), Portland.

Albuquerque, A. B., & Belchior, A. D. (2002, May 7-11). *E-commerce Websites: A qualitative evaluation*. Paper presented at the Eleventh International World Wide Web Conference (WWW 2002), Honolulu.

Appleton, B. A. (1997). Patterns and software: Essential concepts and terminology. *Object Magazine Online, 3*(5), 20-25.

Bambury, P. (1998). A taxonomy of Internet commerce. *First Monday, 3*(10).

Beck, K., & Andres, C. (2005). *Extreme programming explained: Embrace change* (2nd ed.). Addison-Wesley.

Burton-Jones, A., Storey, V. C., Sugumaran, V., & Ahluwalia, P. (2005). A semiotic metrics suite for assessing the quality of ontologies. *Data and Knowledge Engineering, 55*(1), 84-102.

Buschmann, F., Henney, K., & Schmidt, D. C. (2007). *Pattern-oriented software architecture, volume 5: On patterns and pattern languages*. John Wiley and Sons.

Buschmann, F., Meunier, R., Rohnert, H., Sommerlad, P., & Stal, M. (1996). *Pattern oriented software architecture, volume 1: A system of patterns*. John Wiley and Sons.

Busse, D. (2002, April 21). *Usable Web design patterns for world-ready e-commerce sites*. Paper presented at the CHI 2002 Workshop on Patterns in Practice: A Workshop for UI Designers, Minneapolis.

Chan, S. (2001, August 5-8). *Challenges and opportunities in e-commerce education.* Paper presented at he The Seventh Americas Conference in Information Systems (AMCIS 2001), Boston.

Chan, S., & Fang, X. (2001, August 5-8). *Usability issues in mobile commerce.* Paper presented at the Seventh Americas Conference in Information Systems (AMCIS 2001), Boston.

Dromey, R. G. (2003). Software quality - prevention versus cure? *Software Quality Journal, 11*(3), 197-210.

Fernandez, E. B., Liu, Y., & Pan, R. (2001, September 11-15). *Patterns for Internet shops.* Paper presented at the Eighth Conference on Pattern Languages of Programs (PLoP 2001), Monticello.

Fogg, B. J., Soohoo, C., Danielson, D., Marable, L., Stanford, J., & Tauber, E. R. (2002, October 29). How do people evaluate a Web site's credibility? Results from a large study. *Consumer Reports WebWatch Research Report.*

Gangemi, A. (2005, November 6-10). *Ontology design patterns for Semantic Web content.* Paper presented at the Fourth International Semantic Web Conference (ISWC 2005), Galway, Ireland.

Garzotto, F., Paolini, P., Bolchini, D., & Valenti, S. (1999, November 15-18). *"Modeling-by-patterns" of Web applications.* Paper presented at the International Workshop on the World Wide Web and Conceptual Modeling (WWWCM 1999), Paris, France.

Georgiakakis, P., Psaromiligkos, Y., & Retalis, S. (2006, June 20-23). *The use of design patterns for evaluating personalisable Web-based systems.* Paper presented at the Fifth Workshop on User-Centred Design and Evaluation of Adaptive Systems, Dublin, Ireland.

Germán, D. M., & Cowan, D. D. (2000, January 4-7). *Towards a unified catalog of hypermedia design patterns.* Paper presented at the Thirty Third Hawaii International Conference on System Sciences (HICSS 2000), Maui.

Gilb, T. (1988). *Principles of software engineering management.* Addison-Wesley.

Ginige, A., & Murugesan, S. (2001). Web engineering: An introduction. *IEEE Multimedia, 8*(1), 14-18.

Gómez-Pérez, A., Fernández-López, M., & Corcho, O. (2004). *Ontological engineering.* Springer Verlag.

Graham, I. (2003). *A pattern language for Web usability.* Addison-Wesley.

Gruber, T. R. (1993). Toward principles for the design of ontologies used for knowledge sharing. *Formal ontology in conceptual analysis and knowledge representation.* Kluwer Academic Publishers.

Hafiz, M. (2006, October 21-23). *A collection of privacy design patterns.* Paper presented at the Thirteenth Conference on Pattern Languages of Programs (PLoP 2006), Portland.

Hasan, L. R., & Abuelrub, E. (2006, June 19-21). *Criteria for evaluating quality of Websites.* Paper presented at the Sixth IBIMA Conference on Managing Information in Digital Economy, Bonn, Germany.

Hendler, J., Lassila, O., & Berners-Lee, T. (2001). The Semantic Web. *Scientific American, 284*(5), 34-43.

Ihme, T., & Abrahamsson, P. (2005, July 27-28). *The use of architectural patterns in the agile software development of mobile applications.* Paper presented at the 2005 International Conference on Agility (ICAM 2005), Otaniemi, Finland.

Jacobson, I., Booch, G., & Rumbaugh, J. (1999). *The unified software development process.* Addison-Wesley.

Kalakota, R., & Whinston, A. B. (1996). *Frontiers of electronic commerce.* Addison-Wesley.

Kaluscha, E. A., & Grabner-Kräuter, S. (2003, June 25-29). *Towards a pattern language for consumer trust in electronic commerce.* Paper presented at the Eighth European Conference on Pattern Languages of Programs (EuroPloP 2003), Irsee, Germany.

Kamthan, P. (1999, March 29). E-commerce on the WWW: Prospects and concerns. *Internet related technologies.* Retrieved June 15, 2008, from http://www.IRT.org

Kamthan, P. (2007a). Towards a systematic approach for the credibility of human-centric Web applications. *Journal of Web Engineering, 6*(2), 99-120.

Kamthan, P. (2007b). Extreme programming for mobile applications. In D. Taniar (Ed.), *Encyclopedia of mobile computing and commerce* (Vol. 1). Hershey, PA: IGI Global, Inc.

Kamthan, P. (2007c). Accessibility of mobile applications. In D. Taniar (Ed.), *Encyclopedia of mobile computing and commerce* (Vol. 1). Hershey, PA: IGI Global, Inc.

Kamthan, P. (2008). Patterns for improving the pragmatic quality of Web information systems. In C. Calero, M. Á. Moraga, & M. Piattini (Eds.), *Handbook of research on Web information systems quality.* Hershey, PA: IGI Global, Inc.

Kamthan, P., & Pai, H.-I. (2006a). Representation of Web application patterns in OWL. In D. Taniar, & J. W. Rahayu (Eds.), *Web semantics and ontology* (pp. 41-67). Hershey, PA: IGI Global, Inc.

Kamthan, P., & Pai, H.-I. (2006b, May 21-24). *Semantic Web-enabled Web engineering: The case of patterns.* Paper presented at the Seventeenth Annual Information Resources Management Association International Conference (IRMA 2006), Washington, D.C.

Kappel, G., Pröll, B., Reich, S., & Retschitzegger, W. (2006). *Web engineering.* John Wiley and Sons.

Kotonya, G., & Sommerville, I. (1998). *Requirements engineering: Processes and techniques.* John Wiley and Sons.

Kruchten, P. (2004). *The rational unified process: An introduction* (3rd ed.). Addison-Wesley.

Lyardet, F., & Rossi, G. (1998, August 11-14). *Patterns for designing navigable information spaces.* Paper presented at the Fifth Conference on Pattern Languages of Programs (PLoP 1998), Monticello.

Lyardet, F., Rossi, G., & Schwabe, D. (1999, July 8-10). *Patterns for adding search capabilities to Web information systems.* Paper presented at the Fourth European Conference on Pattern Languages of Programming and Computing (EuroPLoP 1999), Irsee, Germany.

Lynch, P. J., & Horton, S. (2002). *Yale CAIM Web style guide* (2nd ed.). Yale University Press.

Mahmoud, Q. H., & Maamar, Z. (2006). Engineering wireless mobile applications. *International Journal of Information Technology and Web Engineering, 1*(1), 59-75.

Manolescu, D., Kozaczynski, W., Miller, A., & Hogg, J. (2007). The growing divide in the patterns world. *IEEE Software, 24*(4), 61-67.

Manolescu, D., & Kunzle, A. (2001, September 11-15). *Several patterns for eBusiness applications.* Paper presented at the Eighth Conference on Pattern Languages of Programs (PLoP 2001), Monticello.

Markiewicz, M. E., Lucena, C. J. P., Alencar, P. S. C., & Cowan, D. D. (2002). Views and patterns in e-commerce application design. *Annals of Software Engineering, 13*(1-4), 111-140.

Marks, M., & Hong, D. (2006). *Web design patterns collection technical design* (Center for Document Engineering Tech. Rep. CDE2006-TR09). University of California, Berkeley.

Mazhelis, O., Markkula, J., & Jakobsson, M. (2005, June 13-15). *Specifying patterns for mobile application domain using general architectural components.* Paper presented at the Sixth International Conference on Product Focused Software Process Improvement (PROFES 2005), Oulu, Finland.

Mendes, E., & Mosley, N. (2006). *Web engineering.* Springer-Verlag.

Meszaros, G., & Doble, J. (1998). A pattern language for pattern writing. In R. C. Martin, D. Riehle, & F. Buschmann (Eds.), *Pattern languages of program design 3* (pp. 529-574). Addison-Wesley.

Montero, F., López-Jaquero, V., & Molina, J. P. (2003, September 1-2). *Improving e-shops environments by using usability patterns.* Paper presented at the Second Workshop on Software and Usability Cross-Pollination, Zürich, Switzerland.

Montero, F., Lozano, M., & González, P. (2002, August 5-7). *Designing Web sites by using design patterns.* Paper presented at the Second Latin American Conference on Pattern Languages of Programming (SugarLoafPLoP 2002), Rio de Janeiro, Brazil.

Nguyen, H. Q., Johnson, R., & Hackett, M. (2003). *Testing applications on the Web: Test planning for mobile and Internet-based systems* (2nd ed.). John Wiley and Sons.

Offutt, J. (2002). Quality attributes of Web software applications. *IEEE Software, 19*(2), 25-32.

Oliveira, L. C. B., & Joia, L. A. (2005, May 26-28). *A model for evaluating B2C e-ocmmerce Websites: Application in the CD e-retailing industry in Brazil.* Paper presented at the Thirteenth European Conference on Information Systems (ECIS 2005), Regensburg, Germany.

Paulk, M. C., Weber, C. V., Curtis, B., & Chrissis, M. B. (1995). *The capability maturity model: Guidelines for improving the software process.* Addison-Wesley.

Perry, M., & Kaminski, H. (2005, July 6-10). *A pattern language of software licensing.* Paper presented at the Tenth European Conference on Pattern Languages of Programs (EuroPloP 2005), Irsee, Germany.

Pertet, S. M., & Narasimhan, P. (2005). *Causes of failure in Web applications* (PDL Tech. Rep. PDL-CMU-05-109). Carnegie Mellon University.

Perzel, K., & Kane, D. (1999, August 15-18). *Usability patterns for applications on the World Wide Web.* Paper presented at the Sixth Conference on Pattern Languages of Programs (PLoP 1999), Monticello.

Pettigrew, A. M. (1985). *The awakening giant: Continuity and change in ICI basil.* Blackwell.

Preece, J., Rogers, Y., & Sharp, H. (2002). *Interaction design: Beyond human-computer interaction.* John Wiley and Sons.

Rigden, C. (1999). The eye of the beholder: Designing for colour-blind users. *British Telecommunications Engineering Journal, 17.*

Romanosky, S., Acquisti, A., Hong, J., Cranor, L. F., & Friedman, B. (2006, October 21-23). *Privacy patterns for online interactions.* Paper presented at the Thirteenth Conference on Pattern Languages of Programs (PLoP 2006), Portland.

Rossi, G., & Koch, N. (2002, July 3-7). *Patterns for adaptive Web applications.* Paper presented at the Seventh European Conference on Pattern Languages of Programs (EuroPLoP 2002), Irsee, Germany.

Rossi, G., Lyardet, F., & Schwabe, D. (2000, July 5-9). *Patterns for e-commerce applications.* Paper presented at the Fifth European Conference on Pattern Languages of Programs (EuroPLoP 2000), Irsee, Germany.

Rossi, G., Pastor, O., Schwabe, D., & Olsina, L. (2008). *Web engineering: Modelling and implementing Web applications*. Springer-Verlag.

Rossi, G., Schwabe, D., Danculovic, J., & Miaton, L. (2001, July 4-8). *Patterns for personalized Web applications*. Paper presented at the Sixth European Conference on Pattern Languages of Programs (EuroPLoP 2001), Irsee, Germany.

Sartzetaki, M., Psaromiligkos, Y., Retalis, S., & Avgeriou, P. (2003, June 22-27). *Usability evaluation of e-commerce sites based on design patterns and heuristic criteria*. Paper presented at the Tenth International Conference on Human - Computer Interaction, Crete, Greece.

Schmidt, D. C., Stal, M., Rohnert, H., & Buschmann, F. (2000). *Pattern-oriented software architecture, volume 2: Patterns for concurrent and networked objects*. John Wiley and Sons.

Schumacher, M., Fernandez-Buglioni, E., Hybertson, D., Buschmann, F., & Sommerlad, P. (2006). *Security patterns: Integrating security and systems engineering*. John Wiley and Sons.

Segerståhl, K., & Jokela, T. (2006, April 22-27). *Usability of interaction patterns*. Paper presented at the CHI 2006 Conference on Human Factors in Computing Systems, Montréal, Canada.

Shanks, G. (1999, September 29). *Semiotic approach to understanding representation in information systems*. Paper presented at the Information Systems Foundations Workshop, Sydney, Australia.

Sharkey, U., Scott, M., & Acton, T. (2006, June 12-14). *The influence of quality on the success of e-commerce systems*. Paper presented at the Fourteenth European Conference on Information Systems (ECIS 2006), Gothenburg, Sweden.

Sharp, H., Galal, G. H., & Finkelstein, A. (1999, August 30-September 3). *Stakeholder identification in the requirements engineering process*. Paper presented at the Tenth International Conference and Workshop on Database and Expert Systems Applications (DEXA 1999), Florence, Italy.

Stamper, R. (1992, October 5-8). *Signs, organizations, norms and information systems*. Paper presented at the Third Australian Conference on Information Systems, Wollongong, Australia.

Stefani, A., & Xenos, M. (2001, December 7-9). *A model for assessing the quality of e-commerce systems*. Paper presented at the First Panhellenic Conference in Human-Computer Interaction (PC-HCI 2001), Patras, Greece.

Tavani, H. T. (2004). *Ethics and technology: Ethical issues in an age of information and communication* technology. John Wiley and Sons.

Tidwell, J. (2006). *Designing interfaces: Patterns for effective interaction design*. O'Reilly Media.

Toraskar, K. V., & Lee, P. C. B. (2006). A contextualist approach to the development of e-commerce education: An analysis of key issues and challenges. *Journal of Information Technology Management, XVII*(2), 1-12.

Treese, G. W., & Stewart, L. C. (2003). *Designing systems for Internet commerce* (2nd ed.). Addison-Wesley.

Van Duyne, D. K., Landay, J., & Hong, J. I. (2003). *The design of sites: Patterns, principles, and processes for crafting a customer-centered Web experience*. Addison-Wesley.

Van Welie, M., & Van Der Veer, C. (2000, April 2-3). *A structure for usability based patterns*. Paper presented at the CHI 2000 Workshop on Pattern Languages for Interaction Design: Building Momentum, The Hague, The Netherlands.

Wallace, D., Raggett, I., & Aufgang, J. (2002). *Extreme programming for Web projects*. Addison-Wesley.

Weiss, M. (2003, September 8-12). *Patterns for Web applications.* Paper presented at the Tenth Conference on Pattern Languages of Programs (PLoP 2003), Urbana.

Wentzlaff, I., & Specker, M. (2006, July 10). *Pattern based development of user friendly Web applications.* Paper presented at the Workshop on Model-Driven Web Engineering (MDWE 2006), Palo Alto.

Wiegers, K. (1996). *Creating a software engineering culture.* Dorset House.

Wong, B. (2006). Different views of software quality. In E. Duggan & J. Reichgelt (Eds.), *Measuring information systems delivery quality* (pp. 55-88). Hershey, PA: IGI Global, Inc.

Yan, Y., & Fang, X. (2005). Technology adoption in e-commerce education: A case study. *Information Systems Education Journal, 3*(40).

Yoder, J., & Barcalow, J. (1997, September 3-5). *Architectural patterns for enabling application security.* Paper presented at the Fourth Conference on Pattern Languages of Programs (PLoP 1997), Monticello.

Zdun, U., & Vogel, O. (2002, July 3-7). *Content conversion and generalisation on the Web: A pattern language.* Paper presented at the Seventh European Conference on Pattern Languages of Programs (EuroPLoP 2002), Irsee, Germany.

APPENDIX

Table 2 provides a mapping between quality attributes and corresponding patterns. A (+) symbol post-fixed to a pattern *name* reflects a positive impact on the corresponding quality attribute, whereas a (–) symbol reflects a negative impact. As mentioned earlier, the list of patterns is by no means complete and is subject to evolution. The rating scheme can also evolve to become more granular.

Table 2. Pragmatic quality attributes of an e-commerce application and corresponding patterns along with their impact ratings.

Pragmatic Quality Attribute	**Pattern(s)**
Aesthetics	COLOR-CODED SECTIONS (+) DEEP BACKGROUND (+) FEATURED PRODUCTS (+) FEW HUES, MANY VALUES (+)
Availability	FAIL-OVER THROUGH CLUSTERING (+) FAILURE NOTIFICATION (+) HEARTBEAT (+) INTRODUCE REDUNDANCY (+)
Comprehensibility	CATALOG (+) CHRONOLOGICAL ORGANIZATION (+) GRID LAYOUT (+) PAGING (+) RESPONSIVE DISCLOSURE (+) SIMPLE SEARCH INTERFACE (+) STRAIGHTFORWARD SEARCH FORMS (+) WIZARD (+)
Efficiency	CARD STACK (+) FLY-OUT MENU (+) ORGANIZED SEARCH RESULTS (+)
Familiarity	ENABLING INTRANETS (+) PERSONAL E-COMMERCE (+) VALUABLE COMPANY SITES (+) STIMULATING ARTS AND ENTERTAINMENT (+)
Maintainability	APPLICATION SERVER (+) CLIENT-SERVER (+) CUSTOMIZED PRESENTATION TO HOST (–) FAIL-OVER THROUGH CLUSTERING (–) INTRODUCE REDUNDANCY (–) LOCALE HANDLING (–) MODEL-VIEW-CONTROLLER (+) WHOLE-PART (+)
Performance	ACCOUNT SETUP (+) FAST-DOWNLOADING IMAGES (+) REUSABLE IMAGES (+) SELECTABLE KEYWORDS (+) SELECTABLE SEARCH SPACE (+) WHOLE-PART (–)

Continued on following page

Table 2. continued

Readability	FLY-OUT MENU (+) GRID LAYOUT (+) HORIZONTAL NAVIGATION (+) VERTICAL NAVIGATION (+)
Reliability	FAILURE NOTIFICATION (+) INTRODUCE REDUNDANCY (+)
Usability	AUTO COMPLETE (+) BREADCRUMBS (+) CARD STACK (+) CLEAR ENTRY POINTS (+) CLOSABLE PANELS (+) CONTENT PERSONALIZATION (+) CUSTOMIZED PRESENTATION TO HOST (+) FAST-DOWNLOADING IMAGES (+) HOME PAGE (+) LINK PERSONALIZATION (+) MISSING LINK (+) MIRRORWORLD (+) MULTI-LEVEL UNDO (+) NEWS (+) SESSION (+) SHOPPING PROCESS (+) STRUCTURE PERSONALIZATION (+) TWO-PANEL SELECTOR (+) LOCALE HANDLING (+) WHAT'S NEW PAGE (+)

Chapter II
Informing Industry via Academic Research in ICT Skill and Capability Development

Krassie Petrova
Auckland University of Technology, New Zealand

Dawn Medlin
Appalachian State University, USA

ABSTRACT

In recent years significant changes have occurred in the skill sets underpinning the undergraduate information systems and information technology curricula. It is imperative that educators comprehend the needs and demands of the industry where their graduates are going to need to apply their acquired knowledge and skills. It may be argued that employers and job recruiters also need to be aware of what skill sets and capabilities new graduates may be expected to come equipped with, in order to develop successful strategies for retaining and growing staff in an environment where the demand for professionals in information and communications technologies (ICT) exceeds the supply. In this chapter, a research framework representing the dynamics of the ICT profession supply and demand of graduates with relevant skills and capabilities is used to facilitate the initiation of a dialogue between industry and academia with the objective to identify issues raised from the lack of alignment between the two and to suggest a way of using academic research results to address these issues. The discussion is supported by the findings of two relevant case studies.

INTRODUCTION

The umbrella term information and communication technologies (ICT) is commonly used to cover a broad spectrum of disciplines starting from computer engineering, to computer science (CS), software engineering (SE), information systems (IS), information technology (IT), and e-business. A fast growing process of technological convergence has contributed to the renewal of the debate about the relevance of the academic curriculum to the needs of the ICT industry. In line with the industry trend, the IS/IT academic curriculum has become more inclusive and most undergraduate programs are positioned along the above continuum of discipline groupings (Petrova & Medlin, 2006).

Historically, two types of skill sets have underpinned the undergraduate IS and IT curricula. Engineering and science programs have primarily been aimed at the development of technical and technological capabilities ("hard" skills), such as programming and systems development. While IT and IS programs have placed sufficient emphasis on hard skills, they have also recognized the importance of and need to produce graduates who understand business processes, and possess the much-demanded "soft" skills, including interpersonal and communication skills (Litecky, Arnett, & Prabhakar, 2004).

Therefore, it is imperative that educators comprehend the needs and demands of the industry where their graduates are going to need to demonstrate their capabilities and apply both their knowledge and skills as acquired throughout the university experience. It may be argued that employers and job recruiters also need to be aware of what skill sets and capabilities new graduates may be expected to come equipped with, in order to develop successful strategies for retaining and growing staff in an environment where the ever-increasing demand for ICT professionals exceeds the supply. In fact, and according to the results of a study of the IT workforce trends for the period 2005-2008 (commissioned by the Society for Information Management), most companies will need to increase their number of full-time IT staff through 2008 (SIM Advocacy Research Team, 2006).

Demand: The Changing Nature of the Job Market

As previously mentioned, the IS/IT curriculum development is driven by the need to meet changing job market requirements and employer demands. However, academic response to changes in the work environment has not always been either quick or adequate. The traditional academic conservatism, and to some extent, a certain lack of intrinsic motivation (Chandra, March, Mukherjee, Pape, Ramesh, Rao, & Waddoups, 2000; Davis, Siau, & Dhenuvakonda, 2003), have contributed to the creation of a gap between academia and industry in relationship to the identified and necessary skill set a graduate must possess. A second factor supporting the existence of the gap may be that significant operational difficulties and academic resources are needed for effective curriculum adjustment or change.

However one more factor may be contributing to the gap between ICT curriculum and the demands of the ICT industry: The ongoing process of the redefinition of job requirements as posited in the job advertising space. Clearly the roles and job descriptions of ICT professionals reflect ICT progress and emerging trends that will continue to evolve and grow. Similarly, the outcomes of IS/IT education will need to be better aligned with the evolving definition of the ICT profession and the expectations of industries that employ ICT professionals in order to meet the demands of new job opportunities (Gallivan, Truex, & Kvasny, 2004; Lee, Trauth, & Farwell, 1995; Lee, 2002).

Supply: What Skills and Capabilities are Needed?

In previous research, the identification of critical ICT skills and capabilities, curriculum development problems and solutions, and issues arising from cultural and socioeconomic differences as related to bridging the industry-academia gap have been the major focal points of study. The literature on how academia should develop IS/IT graduates meeting both the prescribed academic profile of the programs of higher learning, and the needs of the ICT industry, has produced results highlighting existing contradictions.

An example is provided by one of the issues arising from previous findings, which can be formulated as "generalist vs. specialist." Moore and Streib (1989) argue that "academia needs to place a greater emphasis on the basis of education, leaving the job specific skills to the corporate training environment," whereas Lee et al. (1995) suggest that "current IS curricula ... are not well aligned with business needs," and conclude that "IS curriculum design must be driven by a clear vision of the career path of the graduates". The issue is further exacerbated by the ICT industry itself, which is also sending contradictory messages. On one side a concern is expressed about the lack of "solid [broad] foundational education in CS and IT principles" (Reisman, 2005), while the need for even more narrowly specialized graduates is also advocated. Some predict a lack of such highly professional graduates as early as 2010 (Williams, 2004).

Another issue is the one usually referred to as "soft skills vs. hard skills" dilemma. Prior research identified skills and capabilities most needed by the ICT industry as either soft or hard, business or technical (e.g., Couger, Davis, Dologite, Feinstein, Gorgone, Jenkins et al., 1995; Gallivan et al., 2004; Lee, Koh, Yen, & Tang, 2002). However, in a study of classified advertising for IT positions from 1988 to 2003, it was found that a recruiting gap existed whereby organizations said they wanted well-rounded individuals with business knowledge and soft skills, but in reality they advertised for individuals with hard technical skills (Gallivan et al., 2004). The paradox was explored in a study proposing a new model for IS recruiting called image theory (Litecky et al., 2004). This research posits a two stage hiring process with a selection/filtration stage and a choice/hiring stage. The selection/filtration stage is what is represented in job postings. The choice/hiring stage is where additional skills and experience on the "softer" side come into play in the hiring decision.

Even though within the larger global environment, IS/IT professional skill sets and capabilities are globally transferable (Lee, 2005) and ICT students and graduates have become internationally mobile (West & Bogumil, 2001), regional and perceptional differences might still exist. As an example, Lee (2002) studied data from published advertisements and found that contrary to assumptions in Singapore technical skills were more in demand compared to business/interpersonal skills, thereby suggesting that Singapore-based academic institutions might need to place a higher emphasis on technical core subject areas and less on business-oriented skills. However, this only may be true if the assumption can be made that Singapore educated professionals will be only be targeted by the Singapore job market.

Finally, future graduates, influenced by the environmental factors outside academia, may develop perceptions about the relative importance of certain skill sets, which may not match the expectations of the job market. As these perceptions may determine important decisions about course of study and program choices, academic educators need to be aware of the phenomenon (Medlin, 2004; Petrova & Claxton, 2005).

In this chapter, a research framework representing the dynamics of the ICT profession supply and demand skill set is used to facilitate the initiation of a dialogue between industry and academia with the objective to address the identified issues resulting from the lack of alignment between the two. It is proposed that academic research undertaken within this framework may be able to inform some of the stakeholders (e.g., employers and ICT recruiters) to better understand the dynamics of the relationship between industry and academia.

The rest of the chapter is organized as follows. First, prior research is reviewed in the background section. Next, a framework representing the relationships between academia and industry in the context of ICT is introduced. The framework is used further to identify some research perspectives and to present the findings of two relevant case studies. The last section discusses future trends and is followed by a conclusion.

BACKGROUND

Previous Research

Early research on IT job skills used newspaper position advertisements to identify which job skills were in demand. In one of the studies, position advertisements from 10 major city newspapers in 1989 and then in 1996 were collected to develop taxonomy of skills needed in IT (Athey & Plotnicki, 1998). Based on Athley and Plotnicki' results, Litecky et al. (2004) replicated the study to update the findings and to further refine and classify IT skills sought by businesses advertising IT job openings.

McLean and Schneberger (1997) did an exhaustive survey of IT job skill demand in the state of Georgia using localized IT job advertisements, and then compared them to what was being taught in the Georgia university system. The study grouped IT skills into technical (subdivided into "conceptual," such as networking and normalizing, and "work," such as C++ and SQL) and nontechnical (subdivided into "business," such as marketing or sales, and "personal," such as interpersonal communications and project management). The study showed that technical work skills were the most demanded by Georgia IT employers, based on the analysis of more than 10,000 Georgia IT job advertisements.

Other research has examined whether students are receiving the skills and education necessary to be successful in obtaining IT positions. A technical knowledge and skill requirements survey among 429 university students and faculty conducted by Tang, Lee, and Koh (2001) found educational gaps

Table 1. Summary of ICT skill sets according to stakeholder groups (Adapted from Petrova and Medlin, 2006)

	Education sector	Industry sector
Classification based on findings from primary data sources (published research results)	• Technical/Business systems development and design • Managerial • Analytical • Communication/Interpersonal/Generic.	• Technical/Technology management • Nontechnical/Soft (relationship management, conceptual/intellectual, interpersonal/management) • Nontechnical: Business/Business functional
Classification based on findings from secondary data sources (published research results)	[ICT curricula] • Applications • Business/organization and society • Analysis and design • Environment/Platform/Technology; • Programming/Computer Languages; • Interpersonal/Personal.	[Job advertisements] • Management • Business • Technical (software, hardware, development) • Social/Nontechnical (communication, interpersonal) • Systems/Problem solving.

in nine areas of IT knowledge and skills. Among the deficient areas identified three of them were in the technology category (systems development methodologies, implementation, and hardware).

Another educational and ICT skills concern addressed by Petrova and Claxton (2005) is that undergraduate academic programs are often too theoretical and the skill sets taught are out of date. As a result, some students may graduate with outdated skill sets that industry neither values nor requires (Lee, 2002). Other researchers have also found a technology adoption gap between academia and industry (Chandra et al., 2000; Davis et al., 2003).

The United States Department of Labor formed the Secretary's Commission on Achieving Necessary Skills (SCANS) to study the kinds of competencies and skills that workers must have to succeed in today's workplace. The results of the study were published in a document titled "What Work Requires of Schools: A SCANS Report for America 2000" (U. S. Department of Labor, 1999). The report identifies the following types of technology skills: 1) *selecting technology*, which requires choosing procedures, tools, or equipment including computers and related technologies; 2) *applying technology* to task, which requires an understanding of intent and proper procedures for setting up and operating equipment, and 3) *maintaining and troubleshooting equipment,* which involves preventing, identifying, or solving problems with equipment, including computers and other technologies. While the skill sets are in terms of process, not topic (as are most other studies including the previous Georgia study), the SCANS report nonetheless highlighted a number of topical skills.

Other research has highlighted the longitudinal changes in ICT skills demanded by employers. As information technologies have become more complex, so have employee skill levels (Alexander, 1999). The typical IT staff member has moved beyond data processing to embrace a variety of new programming languages and paradigms, networking protocols and devices, client-server distributed computing, and other business and technology-related skills (Goff, 2000). Due to shortages of qualified ICT professionals and the ever-changing business environment, ICT professionals are expected and required to remain on the cutting edge of technological advancements (Alexander, 1999).

But the question still remains of how ICT employees can stay current with new, in-demand information technologies if they do not know what ICT technical skills are being demanded. For example, and not found in information technology literature, are studies directly comparing what ICT skills students perceive to be necessary for getting an ICT-related job, and what advertised ICT-related jobs skills are being demanded. Similarly, while studies in ICT skill and capability development aim to inform academics and curriculum building processes, little or no attention has been paid to the information needs of ICT employers and job recruiters who are the also among the potential beneficiaries of academic research findings.

ICT Skill Set Types

As previously mentioned, current and past research results have revealed a significant change in the ICT skill demand patterns (Gallivan et al., 2004; Lee, 2002; Lee et al., 1995), and have placed a certain amount of pressure on academia to produce employable graduates with the necessary skill set. Meeting these evolving industry needs has also required most universities to complete a thorough and extensive curriculum redevelopment (Couger et al., 1995; Petersen & Wehmeyer, 2005; Shackelford, Cassel, Cross, Davies, Impagliazzo, Kamali et al., 2004). Academic research in the same area has used both data available in the public domain, and data collected from human participants, to identify and classify ICT skill sets. A specific example is provided by the study of Wade and Parent (2002).

Examples of publicly available data sources used in research studies include job advertisements in newspapers and journals (Gallivan, Truex, & Kvasny, 2002; Gallivan et al., 2004), online job advertisement sites (Davies et al., 2003), Web sites of IT corporations (Lee, 2005), and ICT curriculum documentation (Tang et al., 2001).

Stakeholder groups from industry and academia are most often used to gather data and to identify and compare perceptions and expectations (deploying techniques such as surveys, interviews, focus groups, and discussion groups). Frequently two stakeholder groups are included: employers and students (Petrova & Claxton, 2005), industry managers and academic school managers (Yen, Chen, Lee, & Koh, 2003), and academics and practitioners (Lee et al., 2002). ICT skill classifications have also been discussed from cross-cultural, gender, and regional perspectives (Burn, Ng Tye, Ma, & Poon, 1994; Lee, 2002; Pick, Mallen, Navarrette, Guthrie, & Valle, 1998; Von Hellens & Nielsen, 2001; Yen et al., 2003).

Table 1 shows a summary of prior work with respect to skill types, categorized according to data source and stakeholder group (Petrova & Medlin, 2006). As seen, some variations exist among the skill set classification, thus making it difficult to compare findings and results (as also noted by Gallivan et al. [2004]). Some of the reasons for these variations may be attributed to the type of data source used, or to structural organizational changes (Segars & Hendrickson, 2000).

A FRAMEWORK FOR THE STUDY ICT SKILL SUPPLY AND DEMAND

Despite the variations observed, there exists a significant commonality between the different skill set classifications, meaning that stakeholder groups still "speak the same language," and that a research-facilitated dialogue between stakeholders (e.g., academia and industry) is a possibility. The "supply and demand" framework proposed below aims to provide a structure for such a dialogue.

The framework (Figure 1) has two main components: the ICT skill supply side (academia) and the ICT skill demand side (job market). Here "academia" includes stakeholders involved in developing and supplying ICT skills (educators, and also students—the future IT practitioners), and the means of developing ICT skills, knowledge, and capabilities (as exemplified by the academic curricula of universities and other tertiary education institutions). "Job market" represents the stakeholders demanding ICT professionals with relevant skills and capabilities and hiring them as ICT staff (i.e., employers and senior IT professionals), and also the means to announce ICT job requirements (job advertisements). The framework distinguishes between "perceived" values with regard to ICT skills and capabilities (i.e., stakeholder' perceptions and/or expectations about ICT skills), and "stated" values (i.e., documented academic learning outcomes, documented job descriptions, and advertised job requirements).

The framework implies that the stakeholders on the supply side are the academic community while stakeholders on the demand side are ICT employ-

Figure 1. ICT skills: Supply and demand framework (Petrova & Medlin, 2006)

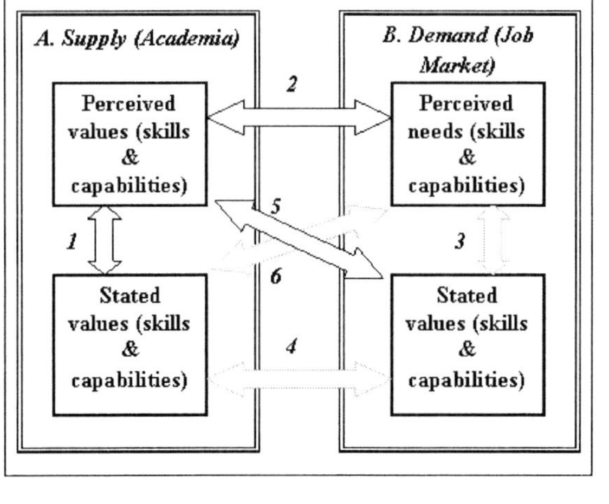

Figure 2. ICT industry information needs (Adapted from Gill and Bhattacherjee [2007]).

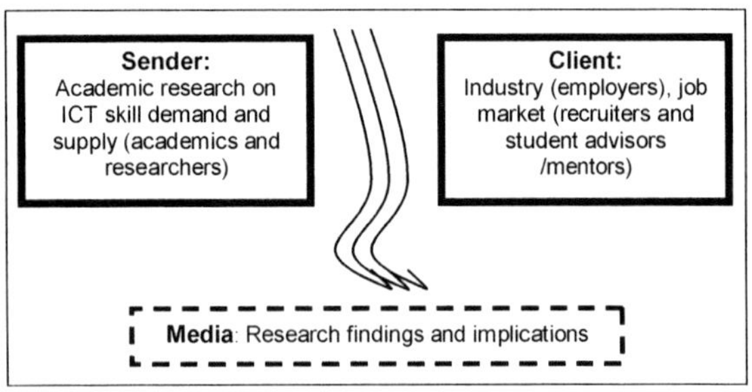

ers and ICT job recruiters. The third stakeholder group (students/graduates) is the subject of both the skill and capability building processes, and to the processes in the job market space.

The framework offers a model for academic research, informing the ICT industry (employers and job recruiters) about the results and implications following from the findings of academic research (Figure 2). The ICT industry stakeholder group plays the role of the "client" (consumer of research results). The relevant academic community has the role of the "sender" (producer of research results), while academic research itself constructs the media through which the sender "informs" the client (Gill & Bhattacherjee, 2007).

Each of the relationships identified can be viewed as a research perspective (Medlin, 2004; Petrova, & Claxton, 2005) with the potential to deliver outcomes valuable to the client. As pointed out by Clear (2006):

CS education represents an important research field, with the ability to contribute significantly to the quality of education of the million or more students globally who study IT each year ....[it] has significant implications for all stakeholders of global IT education and for the economies of the affected countries through the innovative potential of the resulting IT graduates.

Therefore, it will be important to investigate further the validity of the proposed framework by identifying the relative importance of the expected research outcomes with respect to stakeholders.

The Information Needs of the ICT Industry Stakeholder Group

The research perspectives derived from the framework were used to formulate a set of propositions (**P1-P6**) about the information needs of the ICT industry stakeholder group (Figure 1).

Proposition P1. A comparison between students' perceptions about the value of their ICT skills and capabilities acquired through academic study and the stated graduate outcomes of the course of study undertaken may be used as a measure of the success of a program in terms of producing employable graduates, especially when new programs are introduced under the pressure to align ICT education with ICT practice (Ng Tye, Poon, & Burn, 1995). Therefore, it may be important for academia to know whether graduates perceive the ICT skills and capabilities they have developed in the course of their study as matching the stated outcome in the official program documentation.

Proposition P2. A comparison between students' perceptions about the value of their acquired ICT skills and capabilities and the expectations of their future employers about the needed ICT skills and capabilities may be used to generate results informing ICT curricula (Petrova & Claxton, 2005). Therefore, it may be important for employers within the current recruiting conditions of insufficient supply to know whether graduates perceive the ICT skills and capabilities developed in the course of study as matching the expectations of their future employers.

Proposition P3. A review of the literature found that for employers, technical skills are important but are not by themselves sufficient as knowledge of analytical and managerial skills as well as communication skills was found necessary by both students and employers to function effectively in the area of information technology (Medlin, 2004). In addition, IT employer profiling as published by the Information Technology Association of America has indicated that "non-IT companies" are the largest employer of IT workers, meaning that there are different types of employers recruiting ICT students (Information Technology Association of America, 2001). While academic curriculum must meet the needs of both types of constituent industry stakeholders, it may also be important for job recruiters to understand better the differences between the requirements of IT and non-IT employers.

Proposition P4. Prior research has shown that an ever increasing number and variety of skills are required of newly hired ICT staff (Gallivan et al., 2002), and that job advertisements might over-emphasize certain skills (Gallivan et al., 2004; Lee, 2002). A comparison between the expectations of ICT employers about skills and capabilities needed in the workplace, and the skills and capabilities explicitly mentioned when job positions are advertised, may be useful in measuring the reliability of job advertisements as a predictor of the actual need of ICT skills and may contribute to the understanding of the "paradox of soft skills vs. hard skills" (Litecky et al., 2004). Therefore, it may be important for job recruiters and for employers to better understand the relationship between advertised job roles and descriptions and the actual skill sets required.

Proposition P5. A comparison between graduate profiles (as described in academic programs) and industry needs as reflected in job advertisements may show how to measure the alignment of ICT curricula with globally recognized ICT needs and skill and capability transferability (Lee, 2005). Therefore, it may be important for employers as well as for job recruiters to understand better the relationship between academic programs graduate outcomes and global ICT skill and capability needs.

Proposition P6. A comparison between students' perceptions of their ICT skills and the degree to which these match job market requirements, based on job advertisements (ICT graduates' employability as perceived by students), may influence the way academic programs are perceived by employers, including an international perspective (Gallivan et al., 2004). Therefore, it may be important for employers and job recruiters to know whether students perceive their acquired skills as matching advertised required skills and capabilities.

What Industry Stakeholders have Stated

To investigate the propositions further, a project was developed and administered. The aim of the study was to establish the relevance of academic research directions stemming from the framework in Figure 1 to ICT industry stakeholders. The study used the informing science theoretical paradigm (Figure 2), with New Zealand-based ICT employers and job recruiters identified as the "client" of academic research in the area of ICT skills demand and supply by university researchers (the designated "sender").

A series of semistructured interviews were conducted with representatives of the ICT industry. The interviewees were recruited at an annual event which met ICT employers and job marketers with students graduating in IS/IT field (organized by the New Zealand Computer Society, Auckland, New Zealand, September 2006). The group included two senior IT managers (M1 from a non-IT company and M2 from a technology solutions company), two senior job recruiters (J1 and J2), one well-known IT consultant (C1), and an experienced student advisor (A1). The sample was small (6 participants), but may be considered internally valid as it includes representatives of all identified ICT job market stakeholders, drawn from the largest city in New Zealand.

Table 2. Perceived research outcomes importance

Proposition and weighted average	Number of responses "very important"	Number of responses "very important	Number of responses "somewhat important"	Number of responses "not important"	Number of responses "not applicable"
P1 (4.75)	• 3	• 1	• 2	• 0	• 0
P2 (4.75)	• 1	• 5	• 0	• 0	• 0
P3 (4)	• 1	• 2	• 3	• 0	• 0
P4 (4.75)	• 2	• 3	• 1	• 0	• 0
P5 (4.5)	• 4	• 0	• 2	• 0	• 0
P6 (4.25)	• 3	• 1	• 1	• 0	• 1

Participants were asked first to consider a set of research directions as outlined in the propositions and to rank them on a scale from 1 to 5 (1=not important, 5=very important) according to the perceived importance of the research outcomes of these investigations, should the results be made available to them. In the subsequent interviews participants were asked to elaborate and comment on the reasons for their ranking. The summary of the responses with respect to ranking is shown in Table 2.

It was found that the most important research result would be the one providing information about how graduates could match themselves against industry requirements and academic outcomes (propositions **P1** and **P2**), and against requirements as advertised in job position descriptions (proposition **P4**).

Proposition P1. It emerged that despite all efforts, employers and job recruiters were not well acquainted with what universities actually offer. There were some knowledge gaps and misperceptions: "...I am not particularly well educated in the differing values of the course offered by the various educational bodies" (M1); "...employers... their ideas are currently *no clue*" and ..."some employers...might not necessarily understand well young employees" (J1); "...employers expectations are often unrealistic given the 3 year time frame for a degree and the complexity.."(C1).

Proposition P2. Participants felt that knowledge about what students thought about their own skills would be helpful as often employers found that new graduates had unrealistic perceptions of the level of their own skills and of the expected remuneration: "behave like a prima donna without having proved"; "money expectations might be difficult to satisfy" (M2). Recruiters also felt that understanding graduates better would be helpful: "[we are] ...far removed from academia, so we often get client view but not the student view" (J1).

Proposition P4. Participants pointed out that "many vacancies advertisement are too vague and do not give enough specific information" (M1), that students might not be able to work out what an advertisement says (J1), and that adverts "often fail to distinguish between skills required and those which are only 'desirable'" (C1). Communication skills are not the focus of advertisements (J1), and "for some reason demand and supply 'do not meet' at the job market place" (M2).

The remaining three research directions (propositions **P5**, **P6**, **P3**) were considered less likely to produce results applicable to the client's practice. As opposing views were expressed with respect to **P3** and to **P5**, the findings do not provide sufficient basis for conclusions about these two initial propositions. Similarly, research under proposition **P6** was considered important but the reasons for the rating did not give a clear direction.

Proposition P5. Four participants agreed that it would be very important to be informed how graduate profiles, as advertised by academia, match industry needs as advertised in job adverts: "to let employers know" (J2). The other two participants considered this only "somewhat important." An interesting observation was made by C2, that is, that "the local industry press [already has] the perception that universities are not 'aligned'."

Proposition P6. Three participants rated the proposed research direction as very important. The reasons for the rating included: "not important for me as employer" (M1), "important pure to academics" (J2), "important but because job descriptions are badly worded with marketing spin" (M2), and "students are [already] aware that their qualification do not have the mix" (C1).

Proposition P3. There were differing views about the requirements of ICT and non-ICT companies: some considered them "not so important" (J1) while others thought they were "substantial" (C1).

Discussion and Implications

The findings above and results from prior research allow the revealing of three different facets of the gap between industry and academia:

1. First, there is an already identified and recognized gap with respect to academic programs not being aligned with industry requirements.
2. Second, a gap exists with respect to industry understanding what academia supplies to the job market in terms of programs and graduates.
3. Third, a gap exists with respect to industry understanding what expectations graduates may formed based on their studies and influenced by the environment.

Therefore, academic research based on the research framework introduced earlier (propositions **P1**, **P2** and **P4**) may be able to offer useful results to inform the client ICT industry stakeholder group, and to provide recommendations with respect to resolving some of the identified misalignment issues as formulated in 1), 2), and 3) above. An empirical case study aligned with proposition **P4** is presented in the next subsection.

Framework propositions **P3**, **P5** and **P6** were not substantiated and their relevance as informing media was not established; however, the size of the sample is too small to conclude that studies based on them would not be relevant to the client.

Student Perceptions: An Empirical Study

A further empirical study to investigate hypotheses developed from **P4** illustrates the method. To measure student perceptions of needed IT skills, a survey was conducted in upper level IT classes within an accredited 4 year university program within the United States. While focusing on the technical skills needed for IT employment, the survey research instrument actually measured student perception of the technical skills necessary to be successful as an IT professional in today's global business environment. Demographic information (e.g., student age, gender, academic class, major, and frequency of college level information technology classes) was also collected.

The survey instrument used a five-point Likert-type scale to measure a respondent's degree of agreement or disagreement with the statements (1=strongly disagree, 5=strongly agree). The instrument was first pilot-tested using undergraduate level IT students who were not included in the actual survey, to eliminate survey question ambiguity and improve the instrument's reliability. The survey was modified after feedback from the pilot study. Survey variables were students' views of technical skills such as programming, networking, and knowledge of databases as well as software applications, computer languages, graphical interface design, computer networking, and computer hardware.

A useable sample size of 259 senior information systems majors was identified with a focus on students who had taken a wide range of ICT courses and who were actively involved in employment searches. The participants were therefore expected to be familiar with a broad spectrum of ICT skills and with what skills potential employers were seeking. Students did not receive credit or compensation for completing the survey.

The internal validity of the survey was conducted using Cronbach's Alpha. T-test analysis was used to identify those variables that suggested strong student perceptions for a particular ICT skill (SAS Institute, 1993). Student perceptions were then ranked by the six groups (i.e., software applications, programming languages, design of user-friendly graphical interfaces, databases, networking, and computer hardware) of technical ICT skills.

Employer Skill Demands

To answer the question of "What ICT skills and jobs are being demanded?," earlier research had shown that job placement advertisements can accurately show employer interest in those skills necessary for hiring IT workers; this research effort similarly used ICT position listings. Over 21,000 online position advertisements were analyzed using the keyword search approach. The numbers of instances of keywords in listings were used to rank the demand for each ICT skill.

The same six groups of technical skills used in the survey of student perceptions of skill demands were used in searching job listings. These were software applications, computer languages, graphical interface design, computer networking, and computer hardware. The keywords used for each group were taken from a widely used yearly survey linking ICT skills with current salaries (Foote Partners LLC Research Group, 1997). This survey is performed quarterly by Foote Partners, LLC, a general management consultancy and IT compensation and workforce management research firm that continuously monitors the compensation, attitudes, and workplace experiences of 50,000 ICT workers in the U.S., Canada, the United Kingdom, and Europe. The key words used in searching for ICT positions were:

- **Software Applications:** Access, Acrobat, software application, cc:Mail, Cold Fusion, Excel, Frontpage, Groupwise, Notes, Outlook, PaintShop, PowerPoint, and Word.
- **Programming Languages:** ActiveX, ASP, BASIC, C, CGI, COBOL, C+, C++, C#, Delphi, FORTRAN, HTML, JAVA, J2EE, J++, PERL, PHP, PowerBuilder, programming, RAD, RPG, Script, SE, Smalltalk, SQL, VB, XML, and .NET
- **Graphical Interface Design:** Graph, GUI, Swing, Webstart, and Rich-client.
- **Databases:** ABAP, Access, Apptivity, Baan, database, DBA, DB2, Edwards, Informix, MySQL, normalize, PeopleSoft, Oracle, relational, SAP, SQL, SQL Server, Sybase, Siebel, and 2000 Server.
- **Computer Networking:** Apache, APPC, ATM, Ethernet, Exchange, IIS, IPX, gateway, HTTP, LAN, networking, Netware, networks, NT, router, SAN, SIP, SMTP, SNA, TCP/IP, VoIP, WAP, and WAN.
- **Computer Hardware:** BIOS, bus, cache, circuit, clock, CPU, digital, disk, gateway, hardware, microprocessor, RAM, ROM, router, and server.

The searches were made in a comprehensive manner; advertisements that had any of the group keywords were counted for that group. Listings were limited to the previous 30 days, and again to be all-encompassing, all job categories were used to find every job seeking ICT skills, not just "IT jobs." All listings were searched and counted on the same day.

Determining which job sites to search was based on rankings from 11 authoritative sources on 10 online job sites, based on individual documents from Forbes, Business Week, PC Magazine, Workforce Management, About.com, Yahoo! Internet Life, ResumeHelp.org, Job Looks, Quint Careers, College Grad, and Epinions.

The top 10 sites were monster, careerjournal, dice, hotjobs, careerbuilder, americasjobbank, worktree, vaultsjobboard, directemployers, and headhunter. Nine of the 11 sources ranked monster.com as the best source for online job listings, followed by hotjobs.com and careerbuilder.com. Those top three sites were used to perform the keyword IT skill searches. Dice.com was added to the top three since it listed IT jobs exclusively and could offer a valuable focus on IT skills.

Analysis was performed by simply tallying the number of jobs within each group containing the ICT skill keywords used, and ranking them. Finally, the rankings from the survey instrument on student perceptions of what IT skills were needed to be successful in getting hired were compared directly with the rankings of IT skills from the online job listings.

Data Analysis of Student IT Skill Perceptions

Based on the demographic data from the surveys, the student sampling unit was comprised of 259 students that were composed of 74% males and 26% females; this was consistent with previous studies suggesting that upper level IT classes consist of a much higher number of male students (Medlin & McCarthy, 1999). A strong overall reliability of the survey instrument was shown by a Cronbach's Alpha value of 0.8272.

The t-test procedure identified the technical ICT skills groups where strong student perceptions suggested the group was necessary for IT success. The means of the variables "Strongly Agree" statistically exceeded 4.0, and for "Agree" these measures were between 3.5 and 4.0 on the Likert scale of 1 through 5 (1=strongly disagree, 5=strongly agree). Since none of the means were less than 2.5, the test for disagreement was not conducted. The results of the measured skill groups are shown in Table 3 and are presented in ascending order of the means.

Table 4 indicates the number of ICT jobs listed online for each site containing the ICT skill group keywords. The measures are listed in ascending order from the largest of the six groups to the smallest. Database skills were the most desired skill advertised followed by software applications. The lowest number of jobs listed was the ability to design user-friendly graphical interfaces.

Discussion and Implications

Table 5 shows the combined rankings in ascending order by perceptive rankings for comparison to the advertised ICT skills shown in Table 4.

Table 3. Respondents' strong agreement on needed technical skill sets

Associated Measures	Mean	p-value
Knowledge of standard software applications	4.51	0.0001*
Knowledge of programming languages	4.27	0.0083*
The ability to design user-friendly graphical interfaces	4.27	0.0102*
Knowledge of databases	4.15	0.0832*
Knowledge of networking	4.14	0.0001**
Knowledge of computer hardware	4.10	0.0001**

Notes: * Ho: $\mu \leq 4.0$ against Ha: $\mu > 4.0$
** Ho: $\mu \leq 3.5$ against Ha: $\mu > 3.5$

Table 4. Advertised demand for IT skill groups

Measure	Monster	Rank	Hotjobs	Rank	CareerBld	Rank	Dice	Rank	Total
Knowledge of databases	671	3	588	2	5938	1	905	1	8102
Knowledge of software	822	2	314	3	3330	2	637	4	5103
Knowledge of programming languages	899	1	788	1	1568	3	776	3	4031
Knowledge of networking	526	4	184	4	1065	4	842	2	2617
Knowledge of hardware	372	5	179	5	635	5	410	5	1596
Design graphical interfaces	16	6	6	6	29	6	17	6	68

Table 5. Combined perception and actual ICT skill demands

Measure	Perception mean rank	Advertised rank
Knowledge of standard software applications	1	2
Knowledge of programming languages	2 (tie)	3
Ability to design user-friendly graphical interfaces	2 (tie)	6
Knowledge of databases	4	1
Knowledge of networking	5	4
Knowledge of computer hardware	6	5

Interestingly, it appears that student perceptions of demanded ICT technical skills were within one ordinal ranking for four of the six skill sets: knowledge of standard software applications, programming languages, computer networking, and computer hardware. Two of these two skill sets, however, had significantly different ordinal rankings. The ability to design graphical interfaces was perceived by the students to be in second place (tied with programming languages) while in ICT advertisements it was ranked sixth, and in last place. Similarly, students perceived knowledge of databases to be fourth of the six technical skill sets, while ICT job advertisements ranked it number one. Student perceptions of the marketplace demands for ICT skill sets appeared to be fairly accurate except for the two previously mentioned skills.

Students may be gratified to know that the results suggest they are doing a good job of gauging employer skill demands, but they also suggest there may still be significant disconnects. Whatever means students are using to determine ICT skill demands may need to be adjusted or extended to more accurately reflect actual advertised demands. While the choices for senior IT students to rectify skill shortages are likely to be limited that late in the educational cycle, students can at least make employment marketing decisions (e.g., in choosing what to highlight in résumés) in the light of the skill sets in demand.

Business ICT employers who want to be proactive in ensuring a well trained supply of ICT workers to meet their ICT employment needs might seek ways to more accurately portray those needs to students, and early enough for students to make appropriate course selections. IT employers should likewise be aware of what students perceive to be the demanded skills sets so that employers can lobby educational institutions to provide the courses to meet their actual ICT skill demands.

Finally, educational institutions can continuously determine the ICT technical skill sets demanded in the IT employment marketplace and help to ensure students are kept abreast of those marketplace demands. Educational institutions may want to form alliances with key players in the IT marketplace to further that goal and improve the effectiveness of course curricula. Colleges and universities can actively join forces with business and industry to prepare the workers of tomorrow for jobs in the so-called "knowledge economy," emphasizing the need to keep a competitive edge in rapidly-changing national and global markets.

FUTURE TRENDS

According to the Occupational Outlook Handbook 2006-07, published by the United States Department of Labor, employment in the information supersector is expected to increase by 11.6%, adding 364,000 jobs by the year 2014 (U.S. Department of Labor, 2006). Industries included are software publishers, Internet publishing and broadcasting, Internet service providers, Web search portals, and data processing services.

There is a slight shift in IT related jobs, with demand in "pure" computer programmers predicted to decline but with an increased demand in the areas of database administration, systems analysis, networking, and software engineering (predicted to be among the top 30 jobs added category). In fact, network systems and data communications represent the fastest-growing job category among all occupations (53.4% growth). Database administrator, software engineer, and computer systems analyst jobs will grow rapidly (28.6%, 44.6%, and 29.0%, respectively) (Panko, 2007). Similarly, in New Zealand and according to the New Zealand Department of Labour, the demand for ICT professionals is continuing to grow. For example, there were 20% more IT vacancies in May 2007 than in the same month in 2006 (New Zealand Department of Labour, 2007).

The growing demand for network and communication jobs indicates that the trend towards increased use of network and Internet-based technologies will be strong, with all companies compelled to develop an e-business ("Internet") facet in order to survive, as famously predicted by Intel's former chairman Andy Grove (Symonds, 1999). Non-IT industries with IS/IT skill demands are also expected to be affected by information technology and see growth by 2014; for example, finance and insurance are expected to have an increase of 8.3% (496,000 jobs), public education and hospitals are expected to show a 10% growth from 21.6 million to 23.8 million jobs (U.S. Department of Labor, 2006). In New Zealand, the importance of ICT applications within businesses, including Internet commerce and other Internet-based applications, is also predicted to grow and to result in future growth in IS/IT skill demand (New Zealand Department of Labour, 2006).

In summary, there is evidence to show a strong trend of increased demand for ICT skills, with expected job growth clustered predominantly in the SE/IS/IT area. Software engineering emerges as a preferred role in software development, and the variety of roles in the IS/IT area will increase. Research has also supported the conclusion that a mix of skills both technical and managerial is necessary. Therefore, all stakeholder representatives involved (guidance counselors, job recruiters, employers) should advise college students about the growth trends in ICT job areas and the required skill sets.

However, in contrast to the projections above, the predominant trend observed on the supply side is a decrease in enrollment numbers in ICT related programs,most notably CS ones (VanLengen, 2007; Vegso, 2005), but also IS, IT, and e-business courses (Becker, Hassan, & Naumann, 2006; George, Valacich, & Valor, 2005; Granger, Dick, Jacobson, & Van Slyke, 2007). The decrease in enrollments has been attributed to a number of factors, including a delayed reaction to past downturns in the demand cycle (VanLengen, 2007), the current high market demand for ICT professionals (which "drains" the pool of eligible university candidates), and also the poor understanding of the skill set underlying IS/IT programs and ICT graduate employability (Granger et al., 2007). Strategies and approaches to reverse the trend have been suggested (Becker et al., 2006; Looney & Akbulut, 2007), and relevant skill sets aligned with industry needs have been identified (Ives, Valacich, Watson, & Zmud, 2002; Litecky et al., 2004; Litecky, Prabhakar, & Arnett, 2006).

Another dimension of the continuing gap between ICT supply and demand is the move towards "industry certification." The failure to produce an adequate number of graduates and the growing need for graduates with business-oriented skills to augment the technical capabilities acquired through academic study have contributed to the development of a "counter-trend" in the ICT industry. The adoption of a professional standard built around an ICT skills framework which can be also used to develop and endorse professional ICT certification programs offered by ICT training providers. An example of such a framework is skills framework for the information age (SFIA) endorsed by the British Computer Society. The framework places a special emphasis on business change, wireless and Web technology, service management, security, and IT governance and compliance ("How SFIA works," n.d.). It consists of six main categories: 1) strategy and planning, 2) development, 3) business change, 4) service provision, 5) procurement and management support, and 6) ancillary. The skill sets underlying the first five (main) categories compare well with the job predictions cited earlier as most of the skill groupings in the framework "belong" to IS/IT, with Category 2 also related to SE and Category 1 also related to e-business.

In the light of the trend towards developing independent professional training and certification programs, which may further exacerbate the downturn in ICT enrolments, universities as a stakeholder in the ICT sector will be forced to react to the influences of newer technological skill set needs, and the global marketplace. On one side, they may need to rethink the design of their CS programs (Boroni, Goosey, Grinder, & Ross, 1998), to refocus e-business programs (Duncan, 2006), and to redevelop SE programs jointly with the engineering discipline (Parnas, 1998). On the other hand, they may need to diversify and to innovate IS/IT program delivery methods. For example, they may need to add lifelong learning and workplace related components absent until now, such as internships and cooperative placements, the latter also offering students an opportunity to travel to foreign lands in order to demonstrate resourcefulness and awareness of the differing marketplace needs (Fincher, Clear, Petrova, Birch, Claxton, Hoskyn et al., 2004; Nixon, Smith, Stafford, & Camm, 2006; Petersen & Wehmeyer, 2005).

The contradicting trends of high demand and low supply of IS/IT graduates, and the trend to develop industry-led professional training and certification programs for fast-track building of ICT skills—as opposed to the slower traditional academic degree approach—demonstrate a significant lack of alignment among the values and perceptions of the stakeholders in the ICT sector. Suggested research directions may include comparative studies investigating the alignment among stakeholders' perceptions and stated values, such as:

1. Students' perceptions about the value of their ICT skills and capabilities acquired through academic study vs. stated learning outcomes of the course of study undertaken.

2. Students' perceptions about the value of their acquired ICT skills and capabilities vs. the expectations of their future employers about the needed ICT skills and capabilities.
3. Expectations of ICT employers about skills and capabilities needed in the workplace vs. skills and capabilities explicitly mentioned when job positions are advertised.
4. Graduate profiles (as described in academic programs) vs. industry needs (as reflected in job advertisements).
5. Students' perceptions of their ICT skills vs. the degree to which these match job market requirements as in job advertisements (ICT graduates' employability as perceived by ICT graduates).
6. Employers' perceptions of ICT skills and capabilities as advertised in academic documentation vs. the degree to which these match employers' expectations of (ICT graduates' employability as perceived by ICT employers).

Studies involving mult-party comparisons can be also conducted to validate conclusions reached and to test them in different country settings or in specific contexts. The outcomes of the work may give useful insights in what factors play a role in forming a balanced alignment among industry and academia stakeholder groups, and serve to inform decisions made by stakeholders.

CONCLUSION

This chapter captures the link between academia and the ICT job market, and offers a view of the dynamic relationships between the ICT skill supply and demand which could be used by academics, students, and future employers for long-term planning and decision making. The literature on ICT skill development and alignment of the curricula with industry needs focuses significantly on meeting the information needs of one stakeholder group, that is, academia (educators). The research presented here evaluates the significance of academic research with respect of the possible impact on other stakeholder groups (industry and students), using an informing sciences approach. The first study reported investigates the information needs of the client (ICT employer and job recruiters). It identifies information gaps which may be used to provide directions for further academic research. Illustrating the approach, the second study identifies the ICT technical skills employers seek through online position advertisements, and highlights the contrasts between the two categories that could be based on ICT skill set rankings.

Research has been conducted on the connectivity between job opportunities and the growth or lack of student enrollment in information systems and technology programs. In fact, it has been proposed by some that program enrollments can possibly be increased if the course of studies of the graduating major is focused more on why the major subject content is valuable to the organization rather than simply just emphasizing the technologies taught (Ives et al., 2002). Similarly, findings from academic research may play an important role in informing decision making about job marketing and advertising strategies and new employee hiring processes, to achieve a greater alignment among all stakeholder groups, and help address the issues arising from the gaps identified in the research studies presented above.

ACKNOWLEDGMENT

The authors would like to acknowledge the support of the Wyn Hoadley Chancellor's Fund (Auckland University of Technology) in the form of a publication grant.

REFERENCES

Alexander, S. (1999, June 21). Want to give your career its best skills boost? Invest in these honey pots, say top IT execs. *Computerworld*, 78-82.

Athey, S., & Plotnicki, J. (1998). A comparison of information system job opportunities for it professionals. *Journal of Computer Information Systems, 38*(3), 71-88.

Becker, J., Hassan, N., & Naumann, J. D. (2006). Combating the enrolment downturn in IS/IT programs. In *Proceedings of the Twelfth Americas Conference on Information Systems* (pp. 2289-2298). Atlanta: Association for Information systems.

Boroni, C. M., Goosey, F. W., Grinder, M. T., & Ross, R. J. (1998). A paradigm shift! The Internet, the Web, browsers, Java and the future of computer science education. *Association for Computing Machinery Special Interest Group Computer Education Bulletin, 30*(1), 145-152.

Burn, J. M., Ng Tye, E. M. W., Ma, L. C. K., & Poon, R. S. K. (1994). Job expectations of IS professionals in Hong Kong. In J. W. Ross (Ed.), *Proceedings of the 1994 Computer Personnel Research Conference on Reinventing IS: Managing Information Technology in Changing Organizations* (pp. 231-241). New York: Association for Computing Machinery.

Chandra, J., March, S. T., Mukherjee, S., Pape, W., Ramesh, R., Rao, H. R., & Waddoups, R. O. (2000). Information systems frontiers. *Communications of the Association for Computing Machinery, 43(1),* 71-79.

Clear, T. (2006). Valuing computer science education research? In A. Berglund & M. Wiggberg (Eds.), *Proceedings of the 6th Baltic Sea conference on Computing Education Research: Koli Calling 2006* (pp. 8-18). Uppsala, Sweden: University of Uppsala.

Couger, J. D., Davis, G. B., Dologite, D. G., Feinstein, D. L, Gorgone, J. T., Jenkins, A. M., et al. (1995). IS'95: Guideline for undergraduate IS curriculum. *Management Information Systems Quarterly, 19*(3), 341-359.

Davis, S., Siau, K., & Dhenuvakonda, K. (2003). A fit-gap analysis of e-business curricula vs. industry needs. *Communications of the Association for Computing Machinery, 46*(12), 167-177.

Duncan, D. G. (2006). E-Business curriculum: Literature review reveals opportunities and challenges. *Issues in Information Systems, 7*(1), 264-267.

Fincher, S., Clear, T., Petrova, K., Birch, R., Claxton, G., Hoskyn, C., et al. (2004). Cooperative education in information technology. In R. Coll & C. Eames (Eds.), *International handbook for cooperative education: An international perspective of the theory, research and practice of work-integrated learning* (pp. 111-121). Hamilton, New Zealand: WACE.

Foote Partners LLC Research Group. (2007). *Hot IT skills and certifications pay index*. Retrieved June 15, 2007, from http://www.footepartners.com/htscpi_latest.htm

Gallivan, M., Truex, D. P., & Kvasny, L. (2002). An analysis of the changing demand patterns for information technology professionals. In *Proceedings of the 2002 Association for Computing Machinery Specialist Interest Group Computer Personnel Research Conference on Computer personnel Research* (pp. 11-24). New York: Association for Computing Machinery.

Gallivan, M. J., Truex, D. P., & Kvasny, L. (2004). Changing patterns in IT skill sets 1988-2003: a content analysis of classified advertising. *The Association for Computing Machinery DATA BASE for Advances in Information Systems, 35*(3), 64-87.

George, J. F., Valacich, J. S., & Valor, J. (2005). Does information systems still matter? Lessons for a maturing discipline. *Communications of the Association for Information Systems, 16,* 219-232.

Gill, G., & Bhattacherjee, A. (2007). The informing sciences at a crossroads: The role of the client. *Informing Sciences: The International Journal of an Emerging Transdiscipline, 10,* 17-39.

Goff, L. (2000). The e-lusive staff. *Computerworld*, 44-46.

Granger, M. J., Dick, G., Jacobson, C. M., & Van Slyke, C. (2007). Information systems enrollments: Challenges and strategies. *Journal of Information Systems Education, 18* (3), 303-311.

How SFIA works. (n.d.). Retrieved December 15, 2007, from http://www.bcs.org/server.php?show=nav.9029

Information Technology Association of America. (2001). *When can you start? Building better information technology skills and careers.* Retrieved August 2, 2005, from http://www.it-careernet.com/betterskills.htm

Ives, B., Valacich, J., Watson, R., & Zmud, R. (2002). What every business student needs to know about information systems. *Communications of the Association for Information Systems, 9*, 467-477.

Lee, P. C. B. (2002). Changes in skill requirements of information systems professionals in Singapore. In *Proceedings of the 35th Hawaii International Conference on System Sciences* (pp. 3307-3315).

Lee, C. K. (2005). Transferability of skills over the ICT career path. In J. E. Moore & S. E. Yager (Eds.), *Proceedings of the 2005 Association for Computing Machinery Special Interest Group management Information Systems Computer Personnel Research Conference on Computer Personnel Research* (pp. 85-93). New York: Association for Computing Machinery.

Lee, S., Koh, S., Yen, D., & Tang, H.-L. (2002). Perception gaps between IS academics and IS practitioners: An exploratory study. *Information and Management, 40*(1), 51-61.

Lee, D. M. S., Trauth, E. M., & Farwell, D. (1995). Critical skills and requirements of IS professionals: A joint academic/industry investigation. *Management Information Systems Quarterly, 19*(3), 313-340.

Litecky, C. R., Arnett, K. P., & Prabhakar, B. (2004). The paradox of soft skills vs. technical skills in IS hiring. *Journal of Computer Information Systems, 45*(1), 69-76.

Litecky, C. R., Prabhakar, B., & Arnett, K. P. (2006). The IT/IS job market: A longitudinal perspective. In K. Kaiser & T. Ryan (Eds.), *Proceedings of the 2006 Association for Computing Machinery Special Interest Group Management Information Systems Conference on Computer Personnel Research* (pp. 50-52). New York: Association for Computing Machinery.

Looney, C. A., & Akbulut, A. Y. (2007). Combating the IS enrollment crisis: The role of effective teachers in introductory IS courses. *Communications of the Association for Information Systems, 19,* 781-805.

McLean, E., & Schneberger, S. (1997). *Georgia ICAPP information technology strategic industry needs assessment.* Unpublished manuscript, special report for the Board of Regents, University System of Georgia.

Medlin, B. D. (2004). Skills crucial to the information technology professionals in the global business environment: An empirical study in the United States. *International Journal of Human Resources Development and Management, 4*(2), 183-199.

Medlin, B. D., & McCarthy, P. (1999). A study of grade distribution differences between gender and use of presentation technology. *Palmetto Review, 2*(1), 35-38.

Moore, F. L., & Streib, J. T. (1989). Identifying the gaps between education and training. *Association for Computing Machinery Special Interest Group Computer Education Bulletin,* 52-55. New York: Association for Computing Machinery.

New Zealand Department of Labour. (2006). *Information technology professional: Occupational skill shortage assessment.* Retrieved July 22, 2007, from http://www.worklife.govt.nz/PDFs/jvm-prof-ict-2005.pdf

New Zealand Department of Labour. (2007). *Job vacancy monitoring programme reports.* Retrieved July 12, 2007, from http://dol.govt.nz/jvm/job-ad-monthly-report.asp?from=alert)

Ng Tye, E. M. W., Poon, R. S. K., & Burn, J. M. (1995). Information systems skills: Achieving alignment between the curriculum and the needs of the IS professionals. *Association for Computing Machinery Specialist Interest Group management Information Systems Database, 26*(4), 47-61.

Nixon, I., Smith, K., Stafford, R., & Camm, S. (2006). Work-based learning: Illuminating the higher education landscape. *The Higher Education Academy final report.* Retrieved December 5, 2007, from http://www.theksapartnership.co.uk/assets/downloads/KSA_HEA_work_based_learning_final_report.pdf

Panko, R. (2007, December 6). *Bureau of Labor Statistics 2006-2016 employment projection.* Message posted to ISWORLD Information Systems World Network electronic mailing list, archived at http://lyris.isworld.org/lyris.pl?sub=4931909&id=356489948

Parnas, D. L. (1998). Software engineering programmes are not computer science programmes. *Annals of Software Engineering, 6,* 19-37.

Petersen, A. W., & Wehmeyer, C. (2005). European qualification strategies in information and communications technologies (ICT)—towards a European (reference) ICT skills and qualification framework. *Studies in Informatics and Control, 14*(1), 37-46.

Petrova, K., & Claxton, G. (2005). Building student skills and capabilities in information technology and eBusiness: A moving target. *Journal of Information Systems Education, 16*(1), 27-43.

Petrova, K., & Medlin, D. (2006). Towards a general framework for the study of ICT skill supply and demand. In M. Khosrow-Pour (Ed.), *Emerging Trends and Challenges in Managing Information Technology: Proceedings of the 2006 Information Resources Management Association Conference* (pp. 255-263). Hershey, PA: International Resource Management Association. .

Pick, J. B., Mallen, G., Navarrette, C., Guthrie, R., & Valle, L. (1998). Cross cultural comparison of the concordance of IS education/training and IS careers in the United States and Latin America (panel). In *Proceedings of the 1998 Association for Computing machinery Special Interest Group Computer Personnel Research Conference on Computer Personnel Research* (pp. 108-109). New York Association for Computing Machinery.

Reisman, S. (2005, May-June). Quality IT education strikes out. *IT Professional,* 16-17.

SAS Institute. (1993). *SAS/STAT user guide, version 6* (4th ed.). Cary, NC: Autho.

Segars, A. H., & Hendrickson, A. R (2000). Value, knowledge, and the human equation: Evolutions of the information technology function in modern organizations. *Journal of Labor Research, 21*(3), 431-445.

Shackelford, R., Cassel, L, Cross, J., Davies, J., Impagliazzo, J., Kamali, R., et al. (2004). *A guide to undergraduate degree programs in computing (Strawman Draft).* Joint Task Force for Computing

Curricula 2004. Retrieved June 12, 2004, from http://www.acm.org/education/

SIM Advocacy Research Team. (2006). *Information technology workforce: Trends and implications 2005-2008*. Society for Information Technology. Retrieved November 5th, 2007, from http://www.simnet.org/Content/NavigationMenu/Programs/Advocacy_and_Communities/Advocacy_Downloads/SIMAdvocacyITWorkforceexecsummary3-10.pdf

Symonds, M. (1999, June 26-July 2). The Net imperative. *The Economist*. Retrieved December 5, 2007, from http://crab.rutgers.edu/~goertzel/economistnetbusiness.htm

Tang, H-L., Lee, S., & Koh, S. (2001). Educational gaps as perceived by IS educators: A survey of knowledge and skill requirements. *Journal of Computer Information Systems, 41*(2), 76-81.

U. S. Department of Labor. (1999). *Skills and tasks for jobs: A SCANS Report for America 2000. Retrieved June 12, 2005, from* http://wdr.doleta.gov/opr/fulltext/document.cfm?docn=6140

U. S. Department of Labor. (2006). *Occupational outlook handbook (OOH), 2006-07 edition*. Retrieved July 15, 2007, from http://www.bls.gov/oco/home.htm

VanLengen, C. (2007). Supply of computer graduates is not binary! In *Proceedings of Information Systems Educators' Conference 2007* (Vol. 24, p. 2714). Chicago: Association of Information Technology Professionals, Foundation for Information Technology Education.

Vegso, J. (2005). Interest in CS as a major drops among incoming freshmen. *Computing Research News, 17*(3). Retrieved December, 5 2007, from http://www.cra.org/CRN/articles/may05/vegso.html

Von Hellens, L., & Nielsen, S. (2001). Australian women in IT0. *Communications of the Association for Computing Machinery, 41*(7), 46-52.

Wade, M., & Parent, M. (2002). Relationships between job skills and performance: A study of Webmasters. *Journal of Management Information Systems, 18*(3), 71-96.

West, L. A., & Bogumil, W. A. (2001). Immigration and the global IT work force. *Communications of the Association for Computing Machinery, 44*(7), 34-38.

Williams, J. (2004, November-December). IT Education: More specialised in 2010. *IT Professional,* 18-20.

Yen, D. C., Chen, K., Lee, S., & Koh, S. (2003). Differences in perceptions of IS knowledge and skills between academia and industry: Findings from Taiwan. *International Journal of Information Management, 23*(16), 507-522.

Chapter III
The Impact of New Trends in the Delivery and Utilization of Enterprise ICT on Supplier and User Organizations

Jiri Vorisek
University of Economics Prague, Czech Republic

George Feuerlicht
University of Economics Prague, Czech Republic

ABSTRACT

Enterprise information systems have rapidly evolved over the last decade. We expect these changes to accelerate during this decade as a result of new trends in enterprise computing. We argue in this chapter that information and communication technology (ICT) remains strategically important to organizations in the 21st century despite the prevailing trend to outsource ICT and related business processes. We have identified a number of important trends that include the move towards the software as a service (SaaS) model for enterprise applications, increased commitment to process orientation, and emphasis on managing the relationship between business and ICT using services. These trends lead to more effective management of ICT and closer integration of ICT with entrepreneurial activities and business processes in organizations, resulting in improvements in return on investment. These trends will have dramatic impact on both the suppliers and users of ICT, and will necessitate the reevaluation of the approach to ICT education as both the composition and qualifications of ICT workforce will undergo a fundamental change.

INTRODUCTION

The high cost of information and communication technology (ICT) solutions combined with the fast evolution of hardware and software necessitating frequent and costly upgrades has led some observers to conclude that investment in IT does not provide a competitive advantage to organizations and in some cases can detract from the core business in which the organization is engaged. Nicolas Carr (2003), in his controversial article titled "IT Doesn't Matter," sparked a wide-ranging discussion about the benefits of IT in the context of enterprise computing, and in his follow up paper titled "The End of Corporate Computing" (Carr, 2005) he further argues that "IT is shifting from being an asset companies own to a service they purchase." Carr predicts a momentous shift from the present enterprise computing environment based on ownership of ICT infrastructure and licensed software to the world of utility computing where software services are delivered to organizations from a remote data center in much the same way as electricity.

There is little doubt that enterprise computing is undergoing a dramatic transition driven by organizations looking for ways to reduce the cost of ICT solutions and at the same time to increase the effectives of ICT in supporting their business goals. Among the alternatives that are becoming increasingly more popular are various forms of outsourcing, including the software as a service (SaaS) model with enterprise applications and the corresponding infrastructure hosted by an external application service provider (ASP). Enterprise computing is being transformed from the traditional model based on licensed and on-site installed software towards subscription-based models with software delivered as a service by external providers. We have previously analyzed these trends (Feuerlicht & Voříšek, 2003, 2004) and concluded that the SaaS model will become a dominant form of delivering enterprise applications in the near future and that the use of externally provided information services will shape the ICT market and determine management decisions about deployment of enterprise ICT. This conclusion was supported by leading ICT professionals in recent interviews conducted in the Czech Republic (M. Bednár, personal communication, May 2005; R. Hradílek, personal communication, April 2005; J. Kameníček, personal communication, May 2005; J. Polák, personal communication, June 2005). These experts confirm that the changes currently taking place will have dramatic impact on both user organizations and organizations supplying ICT products and services. Our analysis in this chapter is based on these interviews and other relevant literature sources. We first identify the main enterprise computing trends, and then discuss the impact of these trends on ICT user and supplier organizations, and finally comment on the resulting changes in university-level ICT education.

KEY ENTERPRISE COMPUTING TRENDS

The claims about waning importance of ICT in the context of enterprise computing and the commoditization of enterprise applications (Carr, 2003, 2005) leading to the wide-spread adoption of the utility computing model requires a closer analysis in the context of the dominant ICT trends.

Strategic Importance of ICT

Direct comparison of ICT services with the supply of electricity by utilities ignores the rather complex relationship between software applications and business processes that these applications support or implement, treating ICT in isolation from entrepreneurial activities and company culture. In general, ICT has an important influence over the effectiveness of business processes and the success of the business model. Competitive advantage arises from close integration of ICT with business processes and entrepreneurial activities within the organization. The unique character of the combination of a product or a service with the supporting ICT infrastructure and business applications can deliver significantly higher business value or a lower cost, giving the company a strategic advantage.

It is possible to argue that hardware, operating systems, and some types of applications (e.g., office applications) have the characteristics of a commodity, but enterprise applications such as enterprise resource planning (ERP),customer relationship management (CRM), and business intelligence (BI) require extensive customization and integration with other business application and this can create a strategic advantage for an organization (J. Kameníček, personal communication, May 2005). While there are some types of applications that no longer provide strategic advantage to organizations over their competitors (e.g., e-mail, office automation applications, etc.) there are many other types of applications that support specialized business processes that are the very source of competitive advantage. It is important to differentiate between applications that support business processes (e.g., back office, logistics, CRM, etc.) and applications that directly implement business processes, that is, where the process exists only in the form of software (e.g., electronic banking, mobile telephony, etc.). Applications that support business processes can result in competitive advantage when combined with specialized business processes and knowledge within an organization. This unique combination can result in improved effectiveness and agility of business activities. Applications that directly implement business processes can produce competitive advantage by timely deployment leading to increased market share (M. Bednár, personal communication, May 2005). For example, the success of Czech Courier founded in 1999 to focus on express delivery of consignments can be largely attributed to their unique online application system, eKurýr (www.ekuryr.cz). As a result the deployment of this system the company has become a pioneer in utilizing the Internet as a new commercial platform for courier services and gained a significant market share.

Another important factor that supports the argument for strategic importance of ICT is the increasing demand for timeliness and quality of information used for making decision by all levels of management. Largely due to globalization, reaction times to external events such as changed business conditions or new business opportunities are today measured in days; responding within such a short period of time requires an agile and highly effective ICT architecture.

It can be concluded that strategic advantage cannot be achieved today by simply deploying new technology. However, strategic advantage can be derived from the close integration of ICT with organizational business processes and entrepreneurial activities that produces an environment which facilitates rapid responses to important business events, and at the same time reduces the cost and increases the quality of products and services.

Process Orientation

An important prerequisite for maximizing the value of ICT investment is process orientation of the enterprise (Hamm, 2005; Hammer & Champy, 1993; Nevens, 2002; Scheer, 2000). It is becoming clear that managing an enterprise along functional boundaries results in difficulties that include problems with reconciling the interests of individual departments with organizational goals, ambiguous business processes, and poor predictability of response times to important events. Adopting process orientation addresses these issues and at the same time improves the strategic value of ICT solutions (Voříšek, Pavelka, & Vít, 2003).

Figure 1 depicts a process-managed enterprise identifying key layers of the business architecture:

Figure 1. Business architecture of a process-managed enterprise

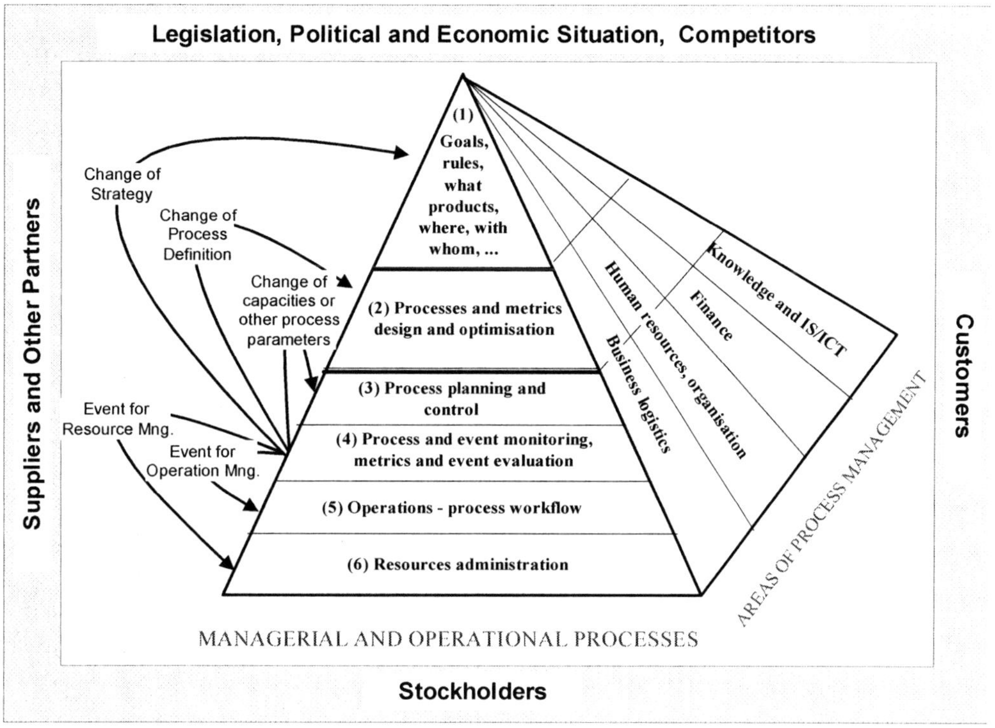

- Business strategy formulation and implementation (layer 1)
- Business processes and metrics definition and optimization (layer 2)
- Process planning and control, supply-chain management (from integrating with suppliers to connecting with customers) (layers 2, 4, and 5)
- Resources administration (acquisition, maintenance, utilization, etc.). (layer 6).

A process-managed organization responds to events such as the arrival of an order, or a production line failure in real-time using active sensors (layer 4). As soon as the event takes place, a corresponding process is activated.

Many enterprises are also adopting process-management of their ICT systems using methods such as ITIL and COBIT that are the de facto standards in this area. There is a relatively wide choice of tools for process modeling, optimization, monitoring, and management. The Organization for the Advancement of Structured Information Standards (OASIS) ("OASIS standards," 2005) has defined several standards in this area.

Implementing process management is a time-consuming activity that requires specific knowledge and skills. The success of process-management depends on a number of critical success factors. The most important of these are:

- **Appropriately skilled managers.** Process management requires different knowledge and skills than the classical functional management; some managers do not have these skills and are not able to accept new responsibilities in this area.

- **Suitable process granularity and alignment of the process definition with the knowledge of the employees supporting the process** (Vořísek, 2000). A detailed definition of a process enables less qualified but well-trained employees to support the business process. The disadvantage of this approach is that it prevents utilization of employees' creativity and reduces the flexibility of the process.
- **Suitably selected process maturity level.** Capability maturity model (CMM) defines six levels or process maturity (Software Engineering Institute, 2002). The lowest level is for a nonexistent process, the highest describes an optimizing process. However, it is not always required to plan for the highest level for each process, as this would result in excessive costs for processes that are not vital to the enterprise or are used infrequently.
- **Utilization of process methods and standards.** When implementing process management it is essential to use appropriate methods and standards (i.e., sector-based reference models of business processes, ITIL). Experience from process-oriented projects shows that it can be dangerous to apply standard methodologies mechanically, and that methods must be tailored to the specific conditions of the enterprise.

Management of the Relationship between Business and ICT using Services

Since the early days of enterprise computing, computer professionals and end users have been looking for optimal ways to link business processes and ICT to facilitate the communication among ICT and business professionals, and to allocate

Figure 2. SPSPR model

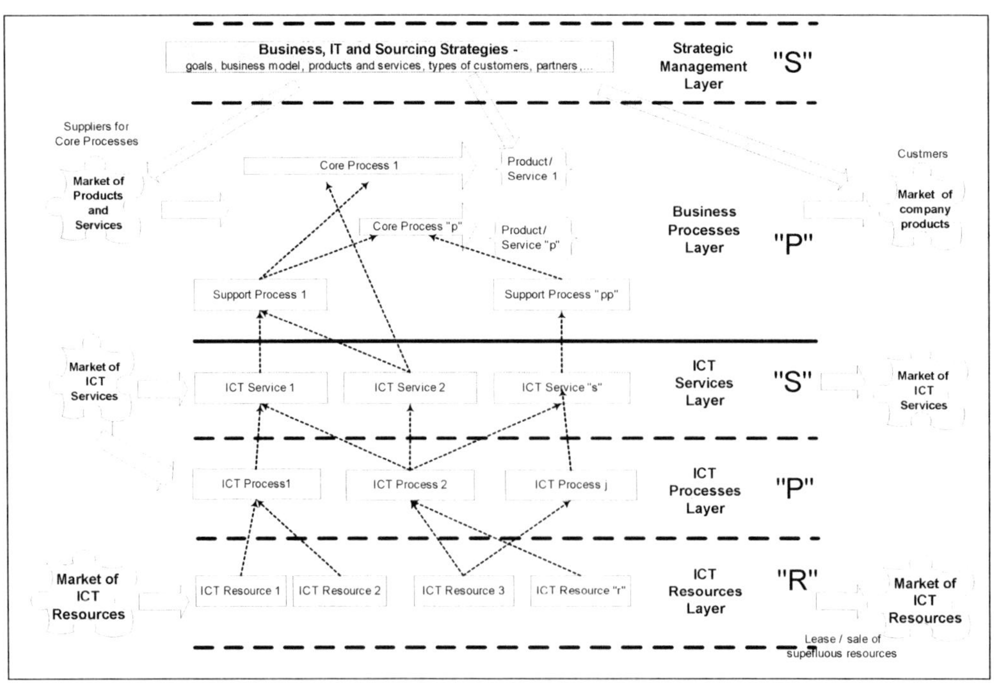

responsibilities in order to minimize the costs and maximize the benefits of ICT projects. New methods that utilize the concept of ICT services described using service level agreements (SLAs) show potential for addressing this problem. For example, the SPSPR method illustrated in Figure 2 uses services to define the boundary between business and ICT (Voříšek & Dunn, 2001).

An important benefit of the SPSPR method is that it defines the content of communications between end users and ICT professionals without undue use of technological concepts. Another benefit lies in the clear demarcation of responsibilities of different types of managers (i.e., top managers, business line managers, owners of a business processes, CIOs, owners of an ICT processes, and managers of ICT resources) and their responsibilities for costs and benefits of ICT.

Management of the relationship between business processes and ICT using ICT services has proved to be an effective solution to the long-standing problem of communication between business and ICT professionals (Hohpe & Wolf, 2004; OGC, 2007). However, the management of ICT services has a number of critical success factors. The most important of these are:

- Ability of the owners of business processes to define SLA that closely reflect ICT requirements.
- Focus on ICT services; ICT services should be derived from the requirements of business processes, and should not be based on the interests of enterprise departments. ICT services should focus on improving performance of the enterprise, on resolving business process bottlenecks, and on business continuity. When ICT service is not related to any specific business process, the requirements for the ICT service are often inappropriate and based on the interests of a particular department, not on the goals of the enterprise (e.g., the department requires a higher availability of the service than is absolutely necessary).
- The ability of ICT managers to specify and manage ICT infrastructure that facilitates provision of agreed-upon scalable ICT services.

Emphasis on Management of the Return on ICT Investment

The ICT crisis that occurred at the beginning of this decade resulted in increased emphasis on the management of the return on ICT investment. Well-managed enterprises no longer invest into ICT without a thorough analysis of the return on investment and refuse to finance risk-prone long-running ICT projects. Increasingly, management requires that investment into ICT correlates with enterprise turnover, and that no project is started unless an improvement in the performance of the enterprise can be assured. This requires a scalable ICT infrastructure and processes. When an enterprise operates its information systems using its own ICT infrastructure, scalability (up or down) is often limited. Typically, the ICT infrastructure is implemented to support the maximum anticipated load, and incremental increases and reductions in capacity (e.g., disposing of surplus hardware, software licenses, or reducing the number of ICT specialists) involve excessive costs. A potential solution to this problem is to purchase external services, as discussed in the next section.

Improvements in the enterprise performance cannot be assured by the ICT department alone. This requirement can be met only by an appropriate allocation of responsibilities of ICT and business managers, for example using the SPSPR model (see Figure 3). According to this model, the ICT services are the responsibility of business process managers. The owner of the business process has to add the cost of each ICT service to other process costs and then evaluate the effectiveness of the business process. If the cost of an ICT service is too high, the requirements should be reconsidered (e.g., reducing functionality, the number of users or availability). The CIO is responsible for ensuring that the cost of an ICT

Figure 3. SPSPR model – responsibilities of business and IT managers

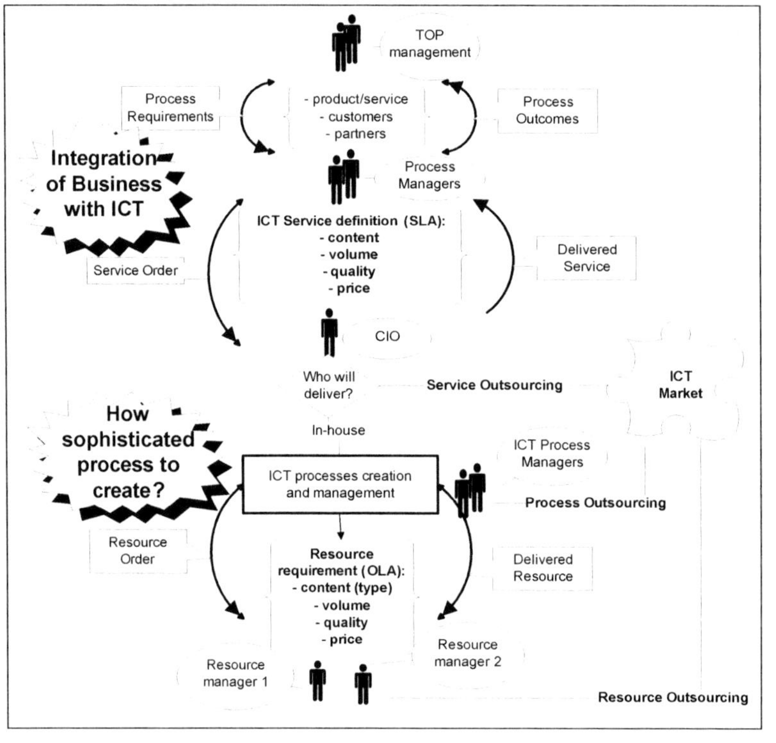

service is competitive with respect to similar ICT services on the market.

Global ICT trends, such as outsourcing the development of software, and manufacture of hardware to countries with lower labor costs, have resulted in reduced overall cost of ICT. However, the increased demand on ICT services will ensure that ICT investment will remain a significant component of expenditure in most organizations.

ICT Services vs. ICT Products

The need to focus enterprise activities on core business combined with the requirement for scalable ICT services (i.e., the cost of services should reflect the level of their usage, based on the "pay as you go" principle), lead enterprises in the direction of outsourcing ICT of supporting processes. In 2004, PMP Research carried out a survey into the extent of ICT outsourcing in end user organizations (Sweet, 2004). The results of this research show that more than a fifth of end user organizations spend more than half of their ICT budget on outsourcing and almost two thirds of organizations expect equal or greater expenditure on outsourcing services.

Deciding which software services to purchase from external services providers requires careful analysis. An enterprise should have a sourcing strategy to facilitate making such decisions. The development of a sourcing strategy as well as its use for decision making is a complex process since a large number of variants with different critical success factors need to be considered (Feuerlicht & Voříšek, 2003). The outsourcing variants include:

1. Business process outsourcing (BPO), where the entire business support process is removed (e.g., layer "P"- Support Process in Figure 2), accounting or transport including ICT services and their support.
2. Total outsourcing of IS/ICT, where a selected external partner takes over all the ICT services, processes and resources—see layers "S,""P," and "R" in Figure 2.
3. Partial outsourcing of ICT, where some ICT services, processes, and resources are provided externally. Partial outsourcing has a large number of variants, including the following:
   a. Traditional ASP, where the application is hosted and operated by an external provider, and used by many customers (companies). An example of such a service provider is SalesForce.com that currently supplies CRM applications for more than 38,000 customers.
   b. ASP that operates single-customer applications (e.g., SAP R/3) on its own ICT infrastructure.
   c. As b), but each customer has its own dedicated infrastructure.
   d. The customer operates an application on its own infrastructure but the administration of the infrastructure is outsourced.
   e. ICT infrastructure is outsourced by the customer who operates and administers its own applications.
   f. External administration of end-user workstations.
   g. External operation of a call centre.
4. Outsourcing the development of applications.

Recent analysis (Dignan, 2005; Voříšek & Feuerlicht, 2004) and predictions (Cohen, 2004) show that the prevailing forms of outsourcing will be those described in 1), 2), and 3a) above. According to Hradílek (2005), IBM expects that most of their customers will move to complete outsourcing of their IS/ICT by 2010.

Effective utilization of outsourcing depends on a number of critical success factors:

- Choosing an appropriate variant of outsourcing.
- Choosing an appropriate granularity of ICT services. At one extreme an ICT service can be all of the functionality of an ERP system, at the other extreme a service could be single transaction (e.g., ordering an airline ticket using a Web service).
- Monitoring of ICT services to be able to carry out a detailed analysis of the cost of the services, processes, and resources. Without effective monitoring it is not possible to compare the cost of internally-implemented services with the cost of externally-supplied services.
- Quality of information about the ICT market (services on offer) and the quality of the sourcing strategy.

According to M. Bednár (personal communication, May 2005), an outsourcing contract can significantly complicate enterprise restructuring and attempts at ICT innovation. When an enterprise disposes of all of its ICT expertise, it can find itself in a difficult situation regarding the future development of ICT.

IMPACT OF THE TRENDS ON END USER ORGANIZATIONS

Given the above ICT trends and critical success factors we can anticipate the following impact on end user organizations:

- Increasingly, the decision about utilizing ICT will be made by the owners of business processes and within the context of strategic management. This will require changes in their qualification. Managers, who understand how to use ICT to develop new products or services, or how to gain new customers, will become indispensable members of the top management team in most enterprises (Santosus, 2001).

- ICT professionals will need to be able to demonstrate the value of ICT for the business and offer new ways of utilizing ICT by the business. This requires a good understanding of the company's strategy, as well as marketing and commercial activities.
- Outsourcing will result in a decrease in the number of technical specialists (e.g., programmers, ICT administrators, etc.). However, the number of employees involved with the relationship between business and ICT services (e.g., requirements definition for ICT, SLA specification, monitoring of the supply of services, etc.) will increase. This trend is already evident from the results of the research conducted in the Czech Republic by the CSSI, SPIS, and CACIO associations (Doucek, Novotný, Pecáková, & Voříšek, 2007).
- The integrative and innovative role of ICT departments will grow (Levy, 2004; McCabe, 2001). This is particularly the case for the ICT-based industries. This is because ICT processes are not like standard support processes such as accounting or purchasing; ICT processes tend to have an immediate impact on the effectiveness of most core business processes.
- The volume of ICT services will be scalable, and ICT costs will correlate with the level enterprise activities and the turnover.

Even though many ICT services will be purchased from external suppliers, the number of employees concerned with the utilization of ICT will not decrease, but the structure of their qualifications will change. An enterprise must maintain the following key expertise:

- How to gain a competitive advantage using ICT, that is, how to use ICT to create new products and services, gain new customers, improve the response of the enterprise to external events, and reduce process costs. This involves supporting business processes with appropriate ICT services, that is, ICT services that provide appropriate functionality, quality, and volume at a competitive cost.
- How to design the overall architecture for ICT services.
- Which services, processes, and resources should be owned and which should be outsourced.
- Selection of the best supplier of an ICT service.
- Monitoring and control of the supply of ICT services.
- Monitoring of ICT services and measurement of the benefits of ICT for business processes.

IMPACT OF THE TRENDS ON SUPPLIER ORGANIZATIONS

If the above trends are realized, particularly outsourcing of ICT services and effective monitoring of the relationship between costs and benefits, we anticipate the following impact on supplier organizations:

- The sale of new software licenses to end user organizations will decrease.

  The information provided in Table 1, 2, and 3, and compiled using company annual reports confirm this trend. We can observe a decline of license revenues during the period of 2002-2003; however, the support revenues had been growing during the same period. In the subsequent years (2004-2006) license revenues started to grow again, however, the ratio of license/maintenance revenues

Table 1. Oracle – license and support growth rates

Oracle	1999	2000	2001	2002	2003	2004	2005	2006
License								
annual growth		20%	6%	- 25%	- 6%	8%	16%	20%
as a fraction of the total income	41%	43%	42%	36%	34%	34,9%	34,7%	34,1%
Support								
annual growth		27%	20%	8%	8%	15%	18%	25%
as a fraction of the total income	27%	29%	33%	40%	44%	44,6%	45,2%	46,1%

Table 2. SAP AG – license and support growth rates

SAP AG	1999	2000	2001	2002	2003	2004	2005	Q3 2006
Licence								
annual growth		27%	5%	-11%	-6%	10%	18%	15%
as a fraction of the total income	38%	39%	35%	31%	31%	31,4%	32,7%	28,4%
Maintenance								
annual growth		44%	27%	15%	6%	10%	12%	12%
as a fraction of the total income	23%	27%	29%	33%	37%	37,6%	37,3%	40,1%

Table 3. SAP AG: Software revenue – old vs. new customers

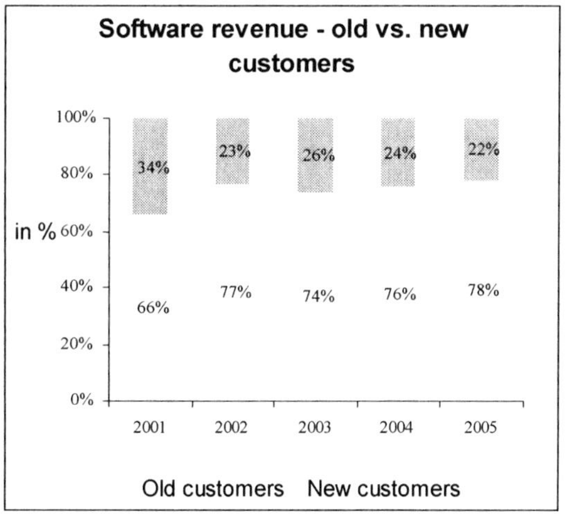

declined further. Haber (2004) came to similar results, concluding that 80% of software cost can be attributed to maintenance of applications and related activities. The increasing cost of maintenance is another factor that contributes to growing interest of end user organizations in outsourcing their enterprise applications.

- Because of the decrease in the sale of new licenses and the increase in outsourcing of enterprise applications, the total number of software companies is likely to fall. This trend is already well under way with many medium and small size software companies being bought out by large software vendors.

- Hardware and software products are "returning" to their producers who use them to offer services on a large scale (Dubie, 2004). According to F. Hoch, the vice-president of the Software & Information Industry Association, "The software industry, as we know it, is passing away and a new industry is being born" (Ulfelder, 2005). According to Hamm and Ante (2005), Accenture, thanks to BPO, added $2.2 billion to its revenue,

Table 4. IBM

IBM % revenue	**2000**	**2001**	**2002**	**2003**	**2004**	**2005**
servers	22,68%	22,32%	20,04%	18,72%	18,89%	14,2%
PCs	17,83%	14,51%	13,78%	12,97%	13,47%	12,48%
HW total	**40,51%**	**36,83%**	**33,82%**	**31,68%**	**32,35%**	**26,68%**
SW	14,81%	15,58%	16,10%	16,06%	15,68%	17,29%
IT services	**38,96%**	**42,08%**	**44,79%**	**47,83%**	**47,99%**	**51,96%**
Financial services	4,07%	4,12%	3,98%	3,17%	2,71%	2,64%
others	1,65%	1,39%	1,31%	1,26%	1,27%	1,43%

Table 5. Hewlett Packard

HP % revenue	**2000**	**2001**	**2002**	**2003**	**2004**	**2005**
servers	26,00%	25,26%	22,33%	21,03%	20,12%	19,26%
PCs	35,85%	33,04%	30,23%	29,03%	30,81%	30,84%
Printers, scanners	22,47%	24,01%	28,20%	30,89%	30,28%	29,02%
HW total	**84,32%**	**82,31%**	**80,75%**	**80,95%**	**81,22%**	**79,12%**
IT services	**14,05%**	**15,84%**	**17,10%**	**16,91%**	**17,24%**	**17,92%**
Financial services	2,00%	2,62%	2,89%	2,63%	2,37%	2,42%

Table 6. SUN Microsystems

SUN Microsystems % revenue	**2001**	**2002**	**2003**	**2004**	**2005**	**2006**
Servers	67,88%	59,19%	54,60%	52,34%	52,63%	45,89%
Storage	14,39%	13,58%	13,56%	13,42%	11,74%	18,17%
Products total	**82,27%**	**72,77%**	**68,16%**	**65,76%**	**64,37%**	**64,06%**
Support services	11,99%	20,31%	24,87%	26,81%	27,38%	28,15%
Professional and Knowledge services	5,74%	6,92%	6,97%	7,43%	8,25%	7,8%

50% more than in the previous year, while IBM achieved an income of $3 billion from outsourcing and related activities, representing an increase of 45%.

The gradual transition of large ICT suppliers from products to services is evident in the information provided in tables Table 4, 5, and 6.

Hamm and Ante (2005) provide further evidence about the transformation from product to service orientation of large ICT vendors:

The change is tangible. The number of employees focused more on the commerce than on pure technology has risen from 3,500 in 2002 to the today's more than 50,000 (out of the total of 330,000 worldwide), and this represents a growth of over 10,000 annually. As a part of a painful process, other employees are leaving in their thousands—for example from the administrative and computer maintenance divisions. With the sale of the loss-making PC manufacturing to Lenovo Group, Palmisano cut off a large part of the company's computer inheritance. At the same time he netted more than a dozen of companies in the area of support for entrepreneurial activities, including Daksh, an Indian customer relationship management company with six thousand employees.

- The approach to delivering ICT services to customers is changing. It is no longer a question of supplying as much as possible for the highest price without considering the real needs of the customer. The aim of the leading suppliers of ICT services is to understand what the customer requires, and then plan and manage the supply of ICT services accordingly.
- There will be changes in the structure and culture of ICT supplier companies that reflect the fact that supplying ICT services requires a different business model from the traditional business model used for the supply of products and software licenses.

INTERVIEW WITH REPRESENTATIVES OF IBM, SAP, HP, AND DELOITTE&TOUCHE

Recent interviews conducted in Prague, Czech Republic with leading providers of ICT products and services confirm that the above trends are already taking place. The following is an extract from the discussion between representatives of IBM, SAP, HP, and Deloitte&Touche:

According to J. Kameníček (personal communication, May 2005): "HP—formerly primarily a product company is increasing its focus on services. The traditional ASP is not yet part of the HP's services portfolio, but outsourcing IS/ICT is our fastest growing service segment. Outsourcing services are offered as a part of the 'utility computing' concept, that is, the cost of services is derived from unit costs (per user in a particular category, per server, etc.). The customer can thus easily change the volume of the service as required and pay only for the actual number of users or supported servers in a given month. In the Czech Republic, the transfer to ASP will not happen quickly. The reason is the conservatism of customers and unwillingness to relinquish control of data and key ICT infrastructure."

According to J. Polák (personal communication, June 2005), Deloitte&Touche has begun to offer ERP in the form of ASP. "From January 2005 we are offering SAP and Peoplesoft in this way. It will be possible to rent individual modules as an application service. However, it is still not a part of our core business."

According to R. Hradílek (personal communication, April 2005): "Recently, in the Czech Republic, there has been a decline in the demand for strategic and business consulting services, and an increase in the demand for ICT services and outsourcing. IBM has been strengthening its focus on supplying ICT services. We offer these under the label 'On-demand Services' and 'Managed Services.' In the near future large ICT companies, as well as offering ICT services, will be offering other types of services that require ICT (e.g., collecting parking fees and fines, accounting, etc.)."

M. Bednár (personal communication, May 2005) "Agrees that the number of technically oriented specialists in end-users organization will fall. However, outsourcing results in new problems for which the customers should be ready."

According to Polák (2005), "The increase in the required number of specialists who integrate business processes with ICT services will not affect just end-user organizations. It will be an opportunity for new consulting firms that specialize in this area. I do not think there will be less demand for developers. I can imagine that there will be tens of new software houses, each with tens of programmers. There are examples of outsourcing that involves a UK company that successfully bids for a development projects, but actual development will take place in the Czech Republic. Compared to China or India this form of outsourcing will have the advantage of the knowledge of cultural and legislative conditions in Europe."

J. Kameníček (personal communication, May 2005): "Agrees that there is a trend to decrease the number of ICT specialist in customer companies. However, this will not affect Czech economy in the next few years. Centralization and remote administration of ICT may make the operation of ICT more effective, but thanks to cheaper labor, there is a significant trend in relocating companies from Western Europe to Central and Eastern Europe. These transfers are creating many new work opportunities in ICT so that the number of ICT workers has been increasing, particularly in Prague and other large towns."

According to Bednár (2005), "The reduction in software implementation services is real. However, this is true more for financial and administrative business processes than for production and logistics."

R. Hradílek (personal communication, April 2005) agrees that the coming market changes will affect the organizational structure and company culture of large ICT suppliers. "Such changes have taken place in IBM during the merger with PWC. ICT companies will have to train many new specialists in the sale of services—we shall be looking at new graduates for this type of role."

According to M. Bednár (personal communication, May 2005), "The return of hardware and software products to their originators is more relevant for hardware than software companies. The question of who is interested in this sort of business depends on whether the service is focused on the operation of ICT infrastructure or whether it also includes a responsibility for outsourcing of business processes. In the latter case, there will be room for specialist companies. For example, outsourcing of payroll and human resources has consequences well beyond ICT. The main expertise of such providers will be in human resources and social sphere, not in ICT."

IMPACT OF THE TRENDS ON HIGHER EDUCATION

The changes in the orientation of suppliers and users of ICT will be reflected in the number of ICT graduates that will be required and the type of qualifications they will need. The following changes can be anticipated:

- Lower demand for purely technical professionals due to the decrease in the number of such specialists in end user organizations. However, technically oriented graduates will continue to be in demand by the ICT suppliers. The knowledge and skills required will include traditional skills such as the ability to develop new applications, but also ability to deploy and manage secure and highly available applications used by thousands of users. There is some evidence of lower demand for technically oriented specialists. For example, in 2004, the average level of unemployment in Australia was 5.7%, but in the ICT sector it was over 10% and for programmers 18% (Philipson, 2004). While there was a recovery in ICT employment recently, the demand is mainly for professionals with a combination of business and technical skills, rather than purely technical specialists.

The lower demand for technical professionals will not be so pronounced in the countries of Central and Eastern Europe. The reason for this is the transfer of a number of development and operations centers of large ICT supplier organizations to this area (including IBM, Sun Microsystems and Computer Associates International development and delivery centers).

- High demand for specialists who are able to integrate ICT with business processes has been noted above. Graduates with two specializations, for example, those who major in ICT and minor in production logistics are likely to be in very high demand.

CONCLUSION

We have argued in this chapter that ICT remains strategically important to organizations in the 21st century despite the prevailing trend to outsource ICT and related business processes. We have identified a number of important trends that include the transition to the SaaS model for enterprise applications, increased commitment to process orientation, and focus on managing the relationship between business and ICT using services. These trends lead to more effective management of ICT, closer integration of ICT with entrepreneurial activities and business processes in organizations, and a corresponding improvement in the return on investment. These trends will have a dramatic impact on both the suppliers and users of ICT, and will necessitate the reevaluation of the approach to ICT education as both the composition and qualifications of ICT workforce will need to change.

REFERENCES

Carr, N. G. (2003). IT doesn't matter. *Harvard Business Review, 81*(5).

Carr, N. G. (2005). The end of corporate computing. *MIT Sloan Management Review, Management of Information Systems, 46*(3), 67-73.

Cohen, P. (2004). *Twelve technical and business trends shaping the year ahead.* Retrieved May 13, 2004, from http://www.babsoninsight.com/contentmgr/showdetails.php/id/687

Dignan, L. (2005). *Outsourcing overseers needed.* Retrieved January 15, 2005, from http://www.baselinemag.com/article2/0,3959,1526051,00.asp

Doucek, P., Novotný, O., Pecáková, I., & Voříšek, J. (2007) *Lidské zdroje v ICT—Analýza nabídky a poptávky po IT odbornících v ČR* [Human resources in ICT—analysis of the demand and supply of ITCT professionals in the CR]. Praha: VŠE.

Dubie, D. (2004). Vendors make the utility computing grade. *Network World Fusion.* Retrieved March 13, 2004, from http://www.nwfusion.com/news/2004/0322dellsummit.html

Feuerlicht, G., & Voříšek, J. (2003). Key success factors for delivering application services. In *Proceedings of the Systems Integration 2003 Conference* (pp. 274-282). Praha: VŠE. ISBN 80-245-0522-3.

Feuerlicht, G., & Voříšek, J. (2004). Is it the right time for the enterprise to adopt software-as-a-service model. *Information Management, 17*(3/4), 18-21. ISSN 1080-286X.

Haber, L. (2004). *ASPs still alive and kicking.* Retrieved January 30, 2004, from http://www.aspnews.com /trends/article.php/3306221

Hamm, S., & Ante, S. E. (2005). Blue world. *BusinessWeek, 16*, 82

Hammer, M., & Champy, J. (1993). *Reengineering the corporation: A manifest for business revolution*. London: Nicholas Brealey Publishing.

Hohpe, G., & Woolf, B. (2004). *Enterprise integration patterns*. New York: Addison-Wesley.

Levy, A. (2004). *What do enterprises want from ASPs?* Retrieved February 13, 2004, from http://www.aspnews.com/analysis/analyst_cols/article.php/2217411

McCabe, L. (2001). *A winning combination: Software-as-services plus business consulting and process services* (Summit Strategies market strategy report). Retrieved October 13, 2001, from http://www.summitstrat.com/store/3ss07detail

Nevens, M. (2002). The real source of the productivity boom. *Harvard Business Review, 80*(3), 23-24.

OASIS standards. (2005). Retrieved April 20, 2005, from http://www.oasis-open.org

OGC. (2007). *ITIL: Service strategy*. London: TSO. ISBN 10: 0113310455.

Oracle financial reports. Retrieved October 13, 2006, from http://www.oracle.com/corporate/investor_relations/analysts/

Philipson, G. (2004) *Time to accept the grim reality of IT*. Retrieved March 3, 2004, from http://smh.com.au/articles/2004/03/01/1078117352254.html

Santosus, M. (2001). *Inferiority complex*. Retrieved April 14, 2005, from http://www.cio.com/archive/031504/ reality.html

SAP financial reports. Retrieved October 13, 2006, from http://www.sap.com/company/investor/reports/

Scheer, A. W. (2000). *ARIS—business process modeling*. Berlin: Springer-Verlag.

Software Engineering Institute. (2002). *Capability maturity model® integration (CMMISM)—version 1.1—staged representation* (Tech. Rep. CMU/SEI-2002-TR-012). Retrieved October 15, 2003, from http://www.sei.cmu.edu/cmmi/

Sweet, P. (2004). *Outsourcing is in PmP research*. Retrieved October 13, 2001, from http://www.conspectus.com/2004/september/ article1.asp

Ulfelder, S. (2005). *Apps on tap*. Retrieved May 13, 2005, from http://www.adtmag.com/print.asp?id=10535

Voříšek, J. (2000). Nová dimenze systémové integrace—integrace podnikových procesů a znalostí. In *Proceedings of the Systémová integrace'2000, sborník mezinárodní konference* (pp. 195-206). Praha,VŠE. ISBN 80-245-0041-8.

Voříšek, J., & Dunn D. (2001). Management of business informatics—opportunities, threats, solutions. In *Proceedings of Systems Integration 2001 Conference*. Praha: VŠE. ISBN 80-245-0169-4.

Voříšek, J., Pavelka, J., & Vít, M. (2003). *Aplikační služby IS/ICT formou ASP—Proč a jak pronajímat informatické služby*. Praha: Graga Publishing. ISBN 80-247-0620-2.

Chapter IV
Enterprise 2.0: Collaboration and Knowledge Emergence as a Business Web Strategy Enabler

Javier Soriano
Universidad Politécnica de Madrid, Spain

David Lizcano
Universidad Politécnica de Madrid, Spain

Marcos Reyes
Telefónica I+D, Spain

Fernando Alonso
Universidad Politécnica de Madrid, Spain

Genoveva López
Universidad Politécnica de Madrid, Spain

ABSTRACT

The Web is becoming in many respects a powerful tool for supporting business strategy as companies are quickly becoming more and more reliant on new Web-based technologies to capitalize on new business opportunities. However, this introduces additional managerial problems and risks that have to be taken into consideration, if they are not to be left behind. In this chapter we explore the Web's present and future potential in relation to information sharing, knowledge management, innovation management, and the automation of cross-organizational business transactions. The suggested approach will provide entrepreneurs, managers, and IT leaders with guidance on how to adopt the latest Web 2.0-based technologies in their everyday work with a view to setting up a business Web strategy. Specifically, Enterprise 2.0 is presented as a key enabler for businesses to expand their ecosystems and partnerships. Enterprise 2.0 also acts as a catalyst for improving innovation processes and knowledge work.

INTRODUCTION

There is no doubt that the Web is in many respects a powerful tool for supporting business strategy. Emerging Internet technologies continue to enable businesses to expand their ecosystems and partnerships. This, however, introduces additional managerial problems and risks that have to be taken into consideration to avoid being left behind.

This chapter explores the Internet's present and future potential in relation to information sharing, knowledge management, innovation management, and the automation of cross-organizational business transactions. It points out how a business Web strategy that takes into account this potential will help not only to improve the existing information sharing and knowledge management processes, but also to protect investments in technology that would otherwise have resulted in expensive failures and severe losses. The suggested approach is based on the emerging Web 2.0 vision and will help to minimize the risk of key information and knowledge being lost or simply not being available on time for the stakeholder, projects started and never finished, worse time-to-market, results not meeting expectations, failure of global, cross-organizational IT integration processes, or even incoherencies between technology and company strategy or structure and so on (Argyris, 1998, pp. 98-105). All managers, and particularly IT leaders, must be aware of this new potential and its implications in order to come up with innovative and effective answers to both known and new problems related to information sharing and knowledge management within their organizations (McAfee, 2006).

The chapter's contents are designed to guide entrepreneurs, managers, and IT leaders through the adoption of the latest Internet technologies, such as Web 2.0, Enterprise 2.0, and the global service oriented architecture (SOA), and their application to their everyday work with a view to setting up a business Web strategy. Musser and O'Reilly (2006) claim that by defining and following a set of architecture building blocks, architectural design decisions, and normative guidance, they can build flexible, extensible, and reusable solutions for exploiting the best features of the emerging Web 2.0 technology suite to achieve the best return on investment (ROI) by leveraging the upcoming Web of user-centered services.

BACKGROUND: THE ADVENT OF ENTERPRISE (WEB) 2.0

There are several different definitions of Web 2.0 (a.k.a. *social networking*) that mostly only describe certain aspects of the overall concept. Tim O'Reilly (2007), who originally coined the term, initially identified seven major characteristics inherent to the Web 2.0 concept. First, the Web is considered as a platform for building systems that do not necessarily have a specific owner and are "tied together by a set of protocols, open standards and agreements for cooperation." Harnessing Web users' collective intelligence represents the second major paradigm. This promotes architecture by participation and democracy and encourages users to add value to the application as they use it. The ownership of mission-critical data is regarded a further cornerstone of numerous Web 2.0 applications. Fourth, O'Reilly propagates the end of the software release cycle as another central paradigm. The use of lightweight programming models that allow for loosely coupled systems and applications, the provision of software above the level of a single device, and the realization of rich user experience represent the last major paradigms inherent to the Web 2.0 concept. Besides such analyses that properly describe parts of the super-ordinate concept, there are only very few comprehensive scientific definitions available. An in-depth investigation of numerous different, successful Web 2.0 applications conducted by Högg, Meckel, Stanoevska-Slabeva, and Martignoni (2006) condensed the respective characteristics into the following statement, which serves as underlying definition for this chapter: "Web 2.0 is defined as the philosophy of mutually maximizing collective intelligence and added value for each participant by formalized and dynamic information sharing and creation" (pp. 23-37).

The Enterprise (Web) 2.0 concept (henceforth referred to as *Enterprise 2.0*) is related to its big brother Web 2.0, because, to a certain extent, it can be viewed as many existing Web 2.0 consumer-oriented services maturing to include features that are important for enterprise users. Enterprise 2.0 represents on its own a new paradigm in which employees, regarded as *knowledge workers*, are coproducers of content, knowledge, applications, and services. Therefore, there is an imperious need to revisit and reconsider the very definition of knowledge worker during this chapter, because it is used extensively throughout the chapter from Davenport's (2005) viewpoint, that is, to refer to employees, partners, suppliers, customers and other possible stakeholders. Enterprise 2.0 provides enterprises with new models and tools for emergent collaboration and cocreation. Enterprise collaboration is thus being enhanced by virtual communities that leverage social linking and tagging tools (e.g., tools for social networking, social bookmarking, and social search), user-contributed content management platforms (e.g., enterprise wikis, blogs, and forums), tools that leverage user opinions (e.g., tools supporting comments and voting), subscription-based information distribution tools (e.g., Enterprise really simple syndication [RSS] feeds), user-centered services (e.g., mash-up- and pipe-based services), and so forth (Drive et al., 2004).

These digital platforms are already popular on the Internet, where they are collectively labeled Web 2.0 technologies (Laso, 2006). Now though, a number of Enterprise 2.0-based collaboration platforms are beginning to proliferate. These platforms are aimed at providing enterprises with specialized subsets of these out-of-the box capabilities (Coveyduc, Huang, Ostdiek, & Reif, 2006; Intel Corporation, 2006). These new collaboration platforms provide enterprises with an ecosystem of knowledge workers who collaborate to develop capabilities by collectively generating, sharing, and refining information, business knowledge, and services. Enterprise 2.0 collaboration enables firms to leverage desirable Web 2.0 attributes, including harnessing collective intelligence and architecture by participation.

The user's production is now primarily based on the customization, composition, remix, and reuse of existing material, such as sampling or mash-ups, by the employees themselves. Enterprise 2.0 technologies have the potential to usher in a new era by making both information sharing, service provision, and consumption, and knowledge work practices and their outputs more visible. At the same time, they provide support for the extraction and the emergence of both knowledge and organizational structure.

In representation of the Gartner Group analyst firm, Smith (2006) recently predicted that by 2008 the majority of Global 1,000 companies will adopt several technology-related aspects of Web 2.0 to advance their businesses. As companies quickly increase their reliance on new Web-based technologies to capitalize on new business opportunities, the industry is showing greater demand for technology experts who can build and manage Web 2.0 resources, including blogs, wikis, forums and user groups, and mash-up enabler tools to centralize the management of all of these resources, supporting technology and knowledge experts' work.

RESEARCH FRAMEWORK AND OBJECTIVES: FOSTERING INNOVATION THROUGH AN ENTERPRISE 2.0-BASED BUSINESS STRATEGY

As global market opportunities, competition, and availability of human resources increase, enterprises are assigning high strategic priority to speeding up innovation, even by modifying their own business and global presence strategies. Enterprises want to speed up innovation to improve their market and business opportunities in the fierce global competition by collaborating and cocreating with partners and users (Coveyduc et al., 2006). The desired outcome is improved time-to-market and increased value of more new products and services. The emergence of Enterprise 2.0 Web-based platforms provides enterprises with new models and tools for collaboration and cocreation.

Enterprise collaboration can be fostered by virtual communities that leverage user content sites (e.g., Wikipedia, Flickr), social linking, tagging, and searching tools (e.g., MySpace, del.icio.us), and sites that leverage opinions of all who participate (e.g., Amazon ratings). Enterprises already leveraging cocreation are eBay API and SalesForce.com AppExchange. These platforms, as Weill and Ross (2004) note, provide enterprises with an ecosystem of partners, suppliers, and customers collaborating to develop capabilities by integrating knowledge and services. Enterprises want a solution that delivers these capabilities out of the box.

The key idea behind Enterprise 2.0 vision, and the lesson many businesses must learn, is that next generation IT systems must be conceived to acquire the knowledge they operate on directly from who really has it, that is, the employees (seen as knowledge workers) and from the operation and communication processes employees enter into (Morris, Pohlmann, & Oliver, 2005). The knowledge of a business has less to do with the IT infrastructure than with the employees themselves. The IT infrastructure must be capable of extracting and managing that knowledge for it to evolve and adapt to the business processes. Any other means to model and exploit the business knowledge will never be flexible enough. If user knowledge changes (and it does change), the IT infrastructure must seamlessly adapt to such changes.

In any case, the design of both traditional and Enterprise 2.0-based solutions has focused primarily on creating a structure that supports common processes and stores information to assure that it is easy to find, reliably available, and backed up. They all have been conceived under the premise that teams need to focus on their core business rather than IT issues. The entire operating environment has therefore been traditionally subordinated to IT departments. Nevertheless, this approach has proved to have a number of collaboration-related drawbacks that slow down the pace of innovation. Knowledge workers are thoroughly acquainted with routine procedures and are capable of extracting automatic behavior, suggesting improvements on the IT systems they use through their operating environment and, more importantly, innovating new operating procedures. Operational innovation is an essential requirement in today's competitiveness-driven business markets, enhancing as far as possible collective intelligence-based knowledge work.

With this in mind, this chapter's main objective is to elaborate on the synergies the Web 2.0 concept and several IT technologies have with regard to the enterprise innovation. Web 2.0's focus on the inclusion of human beings and the exploitation of users' collective intelligence is considered a key enrichment of the knowledge emergency within enterprises. This research objective will be contrasted with relevant literature reviews. The remainder of the chapter is structured as follows. First of all we revisit the notion of knowledge worker and its duty, analyzing its features, needs, and problems they find in their daily innovation effort. In this section we elaborate on factors that can facilitate or instead inhibit a knowledge work process. We then present solutions for fostering enterprise innovation based on Enterprise Web 2.0 ideas and technologies, a pragmatic guideline recommending the alignment of the business strategy exploiting Enterprise 2.0 advantages and the application of this idea to other key Web strategy areas. Finally, the final section concludes this chapter and presents a brief outlook on future trends.

ISSUES, CONTROVERSIES AND PROBLEMS IN ENTERPRISE COLLABORATION AND KNOWLEDGE EMERGENCY

Knowledge Work Revisited: Novel Ways to Foster Innovation through Social Capital and Collective Intelligence

Collective intelligence has existed for at least as long as humans have. Ancient social groups, nations, and modern corporations all act collectively with varying degrees of intelligence. But this ancient phenomenon called innovation emergence is now occurring in dramatically new forms. With new communication technologies, and especially the Internet, huge numbers of people all over the planet can work together in previously unsuspected ways. For this reason, it is more important now than ever before to have an in-depth understanding of collective intelligence to be able to create and take advantage of the new possibilities. Our current definition of collective intelligence is "*groups of individuals doing things collectively that seem intelligent*" (Davenport, 2005).

In general, "collective intelligence" is a *perspective* that can be applied to many different kinds of phenomena. For instance, this perspective suggests another way of thinking about things like "firm productivity," "organizational effectiveness," "teamwork," "firm profitability," and "leadership." When people hear the term "collective intelligence," they tend to assume that it implies individuals giving up their individuality to be somehow subsumed in a group or team. This is *not* what we mean. Collective intelligence, as we understand and explore it, is not about false consensus, hive minds, cults, or groupthink. Collective intelligence relies upon and emerges from a synergy between the individual knowledge, creativity, and identity of its constituent parts (Brown & Duguid, 2000). In its highest forms, participating in collective intelligence processes can actually help people self-actualize while solving collective problems. This collective intelligence is developed within enterprises by the innovative engines called knowledge workers.

As we mentioned above, nowadays enterprises need to accelerate innovation to improve their market and business opportunities in global competition, and therefore it becomes essential to revisit features, profiles, and characteristics of key actors in every innovation process, hence, the knowledge workers. This will help in understanding of the relevance of Enterprise 2.0 technologies and models for both improving their work and expanding their productivity, as we will tackle in subsequent sections.

Knowledge Workers. Mission, Relevance and Novel Ways to Improve their Work

There is a range of ideas about what *knowledge workers* are and what characterizes them. Some examples are:

- "The term *knowledge worker* was coined by Peter Drucker some thirty years ago to describe someone who adds value by processing existing information to create new information which could be used to define and solve problems. Examples of knowledge workers include lawyers, doctors, diplomats, law makers, marketers, software developers, managers and bankers." (Fallows, 2005)

- "Knowledge workers use their intellect to convert their ideas into products, services, or processes" (Davenport, 2005).
- "Their main value to an organization is their ability to gather and analyze information and make decisions that will benefit the company. They are able to work collaboratively with and learn from each other; they are willing to take risks, expecting to learn from their mistakes rather than be criticized for them." (Davenport & Prusak, 1997)
- "Knowledge workers are continually learning, aware that knowledge has a limited shelf life" (Davenport & Prusak, 2000).

What then is a knowledge worker?

- A problem solver vs. a production worker
- A person who uses intellectual rather than manual skills to earn a living
- An individual who requires a high level of autonomy
- A manipulator of symbols; someone paid for quality of judgment rather than speed of work
- A worker who uses unique processes
- Someone who possesses uncodified knowledge which is difficult to duplicate
- A worker who sources between the ears
- Someone who uses knowledge and information to add to deeper knowledge and information

Fewer and fewer people are subordinates, even in fairly low-level jobs; increasingly they are knowledge workers. Knowledge workers cannot be managed as subordinates; they are associates. The very definition of a knowledge worker is one who knows more about his or her job than anyone else in the organization (Davenport & Harris, 2007).

According to Nonaka and Takeuch (1995), what motivates workers—especially knowledge workers—is what motivates volunteers. Volunteers have to get more satisfaction from their work than paid employees precisely because they do not get a paycheck. They need, above all, challenge. They need to know the organization's mission and to believe in it. They need continuous training. They need to see results. Implicit in this is that employees have to be managed as associates and/or partners, and not in name only. The definition of a partnership is that all partners are equal.

Davenport (2005) elaborates on this idea when stated that the productivity of the knowledge worker is still abysmally low. It has probably not improved in the past 100 or even 200 years for the simple reason that nobody has worked at improving the productivity. All our work on productivity has been on the productivity of the manual worker. The way one maximizes their performance is by capitalizing on their strengths and their knowledge rather than trying to force them into molds.

Types of Knowledge Workers

From the practical perspective outlined in this chapter, it can be very useful to consider three separate types of knowledge worker: "core knowledge workers," "*high-end* knowledge workers," and "everyone else."

- Core knowledge workers are those in specific "knowledge management" roles and enterprise duties. Examples of these kinds of roles include chief information/knowledge officers, librarians, knowledge managers, content managers, knowledge analysts, information officers, and so forth.
- "High-end" knowledge workers, or those with the highest degree of education and

expertise, would seem to be particularly important to enterprise innovation. They are the scientists who develop the new products, the professionals who plan and sell the big consulting or legal projects, or the hardware or software architects who envision and deliver the new product line. In the knowledge economy, these should be the horses that pull the plow (the people to whom we should look for the new ideas, products, and services that fuel revenue growth and ensure organizational longevity) (Davenport & Harris, 2007).
- Everyone else is all the other knowledge workers (e.g., dentists, doctors, nurses, managers, pharmacists, technicians, administrators, etc.). In short, everyone in the organization engaged in some form of "knowledge work."

Of course, there is not always a clear dividing line between these classes, but the distinction can be a helpful one at the start. It can be particularly useful for helping people to understand that everyone in a company is a knowledge worker to some degree, and knowledge work is everyone's responsibility, not just that of a few people with "knowledge" or "information" in their job title.

Features of HEKWs (High-End Knowledge Workers)

1. They control their own work structure; high-end knowledge work remains relatively unstructured and autonomous. No one generally tells these workers where to work, when to work, or what specific tasks to perform during work.
2. They are highly collaborative.
3. They work in multiple settings.
4. They do individual and group work.
5. They have high levels of passion, power and occupational mobility.

Knowledge Types Managed by Knowledge Worker

Here is one classification for different types of knowledge.

- **Logical:** There is knowledge that is the result of the understanding of how ideas relate to one another in a domain.
- Semantic: There is knowledge that is the result of learning the meaning of words or concepts. Knowledge of words is knowledge of definitions. Such definitions are set out in dictionaries. You can look this knowledge up.
- **Systemic:** There is knowledge of mathematics and geometry, for example, which is the result of learning a system of words, or symbols, and how they relate to one another and the rules for operating in that system. Any claims made that are consistent with those definitions and rules are called knowledge.
- **Empirical:** There is knowledge that comes through our senses. This is empirical knowledge. Science is the best example of a method for ascertaining the accuracy of such knowledge. Scientific knowledge is a result of practicing the scientific method, that is, observation, abduction of a hypothesis, careful observation, refinement of the hypothesis, deduction of a test for the hypothesis, testing and experimentation, and confirmation or falsification of the hypothesis.

In addition, knowledge can be viewed from another point of view as implicit and explicit knowledge.

Informative Channels used by Knowledge Workers

Traditional knowledge management programs attempt to manage the process of creation or identification, accumulation, and application of knowledge or **intellectual capital** across an organization.

By exploiting several informative channels, the guidelines of core knowledge workers can be made accessible for everyone else in the company. This approach looks to emerge and apply collaborative and social knowledge to create a **social capital** across the organization (Lin, Burt, & Cook, 2001).

The following is a list of informative channels used by core knowledge workers in Enterprise 2.0:

- e-mails
- chats
- blogs
- RSS feeds
- portal or Web content
- links and reverse references (links in blogs)
- wiki tools
- folksonomies
- bookmarks (tagged or not)
- documents of every kind, including files in ftp, printed papers, and so forth
- physical communication in person
- physical communication in distance (e.g., phone)
- common applications as enterprise mash-ups

Factors in Knowledge Work

Facilitating Factors

Bloom (2000, pp. 42-44) identifies the following five elements as causing a group to be intelligent (a "collective learning machine"):

1. **Conformity enforcers:** Mechanisms that cause consensus and similarities among most members of the group.
2. **Variety generators:** Mechanisms that cause some differences and discussion among members of the group.
3. **Inner judges:** Mechanisms that cause individual members of a group to reward themselves for successes and to punish themselves for failures, and cause everyone to evaluate a concept or idea, and validate it after their own experience-based verification.
4. **Resource shifters:** Mechanisms that shift resources (e.g., admiration, information, data, concepts, knowledge, money, or influence) to members of the group.
5. **Intergroup tournaments:** Competitions between subgroups or departments (such as games, corporate competitions, rivalry discussions, etc.)

Other authors, like Surowiecki (2005), say that there are three conditions for a group to be intelligent (for a "crowd to be wise"):

1. **Diversity:** The group includes members with a wide diversity of knowledge or abilities (and the ability to recognize successful and unsuccessful outcomes).
2. **Independence:** Group members use their own knowledge and abilities without being overly influenced by others. (When group members have too much influence over each other, various kinds of bad outcomes can result. See inhibitory factors section below.)
3. **A particular kind of decentralization:** Group members' actions are *aggregated* in a way that finds the right balance between: (a) "making individual knowledge globally and collectively useful," and (b) "still allowing it to remain resolutely specific and local."

Inhibitory Factors

Finally, there are several general factors that can inhibit collective intelligence, such as groupthink and informational cascades, social dilemmas, coordination failures, or failures in thinking itself (Malone, Jay, Legay, & Kosorukoff, 2006).

The first barrier to collective intelligence is called groupthink and social conformity, which was described by Janis (it is perhaps the best explored factor) and developed further in numerous experimental studies. The key point of this research is that people's tendencies to conform, imitate, and avoid conflict can reduce the effective diversity of opinions, and lead to judgments and decisions that are inaccurate, premature, systematically biased, and so forth. The analogue to this phenomenon in pragmatic distributed collective intelligence, as James Surowiecki points out in *The Wisdom of Crowds*, is the informational cascade, where imitation produces fads and conformity instead of individual decision making. In other words, the knowledge of the whole turns out to be less than the sum of the parts, because ***only some parts are actually contributing while everyone else conforms or imitates.*** For this reason, mechanisms that foster diversity and independence might improve collective intelligence. At the interpersonal level, this means practices and norms surrounding respect for individual ideas and contributions (as in the early stages of a brainstorming session). At a distributed level, this results in structural barriers in the physical, legal, or IT "code" (e.g., the walls around a voting booth).

An excellent example based on heuristic experiments of how early decisions by some group members can unduly bias the decisions of later group members is the study of online music ratings by Salganik, Dodds, and Watts. Here is a summary of that study from its online abstract:

Hit songs, books, and movies are many times more successful than average, suggesting that 'the best' alternatives are qualitatively different from 'the rest'; yet experts routinely fail to predict which products will succeed. We investigated this paradox experimentally, by creating an artificial 'music market' in which 14,341 participants downloaded previously unknown songs either with or without knowledge of previous participants' choices. Increasing the strength of social influence increased both inequality and unpredictability of success. Success was also only partly determined by quality: the best songs rarely did poorly, and the worst rarely did well, but any other result was possible. (Salganik, Dodds, & Duncan, 2006)

A second category of barriers includes prisoners' dilemmas, social loafing, and tragedies of the commons (Davenport & Prusak, 2000). These dilemmas or barriers, which involve disincentives for collective performance, are less explored, and described in less depth in economics. They apply at both the interpersonal level (e.g., social loafing in teams) and at the distributed level (low voter participation in democracies). The knowledge and intelligence of the whole turns out to be less than the sum of the parts because ***some parts contribute but others slack off.*** This tells us that it would be important to consider carefully structured incentives to reward individual participation as well as collective intelligence. This naturally occurs in futures markets and betting, but further application and innovation on such ideas is possible.

While the first two barriers involve mechanisms that suppress or delete individual contribution, a third category involves failures to integrate contributions when they are made adequately. Surowiecki offers the traffic jam as a simple example. Information overload on the Internet is another. The knowledge and intelligence of the whole turns out to be less than the sum of the parts because ***the parts' contributions interfere with or cancel each other.*** Solving this problem chiefly involves evolving structures and practices that coordinate individual and group contribution. At a distributed level, structures are highly visible, albeit incompletely studied. For example, there are congestion pricing for traffic systems, eBay's auction and interpersonal rating system, Amazon.com's collaborative filtering, Google's algorithm for search ranking, Wikipedia's review practices (as studied by Giles [2005, pp. 900-901]), and so forth. There is similar wealth on the interpersonal side, although many of the practices remain proprietary or tacitly in the hands of professional facilitators.

A final category of barriers to collective intelligence and innovation emerges from William Isaacs' work on dialogue. His theories build upon work by physicist David Bohm on "thought as a system," a new perspective in which all thinking and intelligence is understood to be collective. Within the system of collective thinking, Isaacs identifies four key pathologies that decrease collective intelligence. For each of pathology, he describes a principle that should be kept in mind and a dialogue practice for individuals to refine their own awareness and intelligence quality and contribute to fostering collective intelligence (Malone et al., 2006):

- **Abstraction/Fragmentation:** The tendency to hold oneself distant or separate from the world, for example, by abstracting or compartmentalizing it.
    - Siloing is a clear symptom of this kind of phenomenon: "That's an economics problem, not a psychology problem"; "That's a marketing problem, not a manufacturing problem"; "Not invented here."
    - Staying high on the so-called "ladder of inference" (Argyris, 1998), that is, arguing at the level of fragmented and reified ideas instead of about the flow of experience and data.
    - An example of this issue is: "This is a unique case" instead of "this is a symptom of how the whole thing is working."
    - Antidote: Listening (to data, to people, to the innovation emergence); that is, the principle of *holographic participation*, which is all things are whole, connected.
- **Idolatry of Memory:** The repetition of automatic answers, routines, stereotypes, and behavior patterns from memory.
    - "We solved that problem years ago"; "That's just the way we've always done things around here"; "We have a human resources department, therefore we're taking care of our people."
    - Antidote: Voicing what is actually new and emergent in one's understanding and experience; that is, the principle of *unfolding potential*, which is, the universe is always unfolding and producing the new.
- **Certainty:** The "knowledge" that one's view (often a manager or chief officer) is correct.
    - "That's impossible"; "There's no way that could be true."
    - Antidote: *Suspending* one's assumptions and prejudices for personal and collective reflection, that is, the principle of *proprioceptive awareness,* which is learning to see and feel how your assumptions are affecting your thinking and actions.
- **Violence:** The repression, disrespect, and destruction of alternative points of view in order to force acceptance or consensus of one's own understanding.
    - "No educated person could take that view"; "You're an idiot for believing that"; "That's all well and good, but..."
    - Antidote: Respecting diversity of opinion, style, and knowledge; that is, the principle of dif*ferentiation*, which is, diversity is natural and valuable, and

collective intelligence means fostering differentiation and integration.

Enterprise 2.0 ground rules are related to these strong and weak factors of collaborative intelligence (Davenport & Prusak, 1997) and tackle each one of them to adequately emerge social capital.

SOLUTIONS AND RECOMMENDATIONS TO FOSTER COLLABORATION AND KNOWLEDGE EMERGENCY THROUGH ENTERPRISE 2.0 IDEAS

Enterprise 2.0 Key Technologies and Models for Improving Knowledge Work

As mentioned above, Web 2.0 and its application in enterprises can be seen as the computer industry's business revolution caused by the move to the Internet as a platform, and an attempt to understand the rules for success on that new platform. The key rule is to build applications that harness network effects to improve as they are used by more and more people.

The concept of Web-as-participation-platform captures many of these characteristics from the viewpoint of the new software as a service (SaaS) paradigm. Bart Decrem, founder and former CEO of Flock, calls Web 2.0 the "participatory Web" and regards Web-as-information-source as Web 1.0 (O'Reilly, 2004).

The following sections deal with the different existing Enterprise 2.0 technologies from two different viewpoints: first, a service-oriented perspective that is paving the way for a user-centered Web of services, recently termed as global SOA (Schroth & Christ, 2007); and second, a user-centered content driven perspective, comprising enterprise blogs, wikis, RSS, and other business knowledge channels.

Service-Oriented Enterprise 2.0 Technologies

The number of enterprises that are bringing their business systems to the Web to automate cross-organizational business transactions is constantly growing. Porter (2001) says that benefits of performing such transactions electronically include extending market reach, saving time, cutting costs, and responding to customer queries more agilely. Renowned scientists such as Malone (2001) cite the relentless march of improvements in the cost-performance ratio of information technology as the main driver of this development. SOAs have attracted, as McAfee (2005) notes, a lot of interest during the last few years as they are expected to play a key role as enablers of seamless application-to-application integration, both within company boundaries and on a global, cross-organizational scale, required to build this scenario.

From a technological viewpoint, Web services (Alonso, Casati, Kuno & Machiraju, 2004) have been massively adopted as the technical foundation for the realization of SOAs. Even so, Web Ssrvices-based SOAs mostly only exist within company boundaries at present (Roman et al., 2005, pp. 77-106). The global provision and consumption of services over the Internet is still at an early stage and has not yet taken on a significant role in realizing cross-organizational collaboration in an *Internet of services.*

Several reasons, such as high technical complexity, implementation and maintenance costs, inflexibility, and the lack of widely accepted standards for defining service choreographies as well as message semantics, have been repeatedly identified as key factors that have prevented the emergence of a global mesh of interoperable Web services, as Hinchcliffe says (2007). Further hurdles on the path to a "global SOA" include the lack of global-scale service discovery, as well as platforms allowing for intuitive human-guided service interaction and composition. Recently, the emergence of the Web 2.0 phenomenon is expected to act as a facilitator of such a global SOA (McAfee, 2007). Novel Web 2.0 technologies and design principles are now about to experience increasing acceptance as they allow for reusing, customizing, interconnecting, composing, and finally exposing Web-based content or functionality again as new resources. They are, therefore, considered not as a substitute for, but as an enrichment of, SOA concepts and technologies (Schroth & Christ, 2007; Schroth & Janner, 2007).

User-Centered Global Service Oriented Architectures (Global SOA)

A number of Enterprise 2.0 collaboration platforms are beginning to proliferate. By leveraging desirable Web 2.0 attributes, these platforms provide enterprises with an ecosystem of employees, partners, suppliers, and customers who collaborate to develop capabilities by collectively generating, sharing, and refining business knowledge. Nevertheless, enterprise collaboration should evolve towards a new paradigm in which knowledge workers are considered as coproducers not only of information, but also of software services and applications that promote specific competitive advantages and/or meet their immediate needs, without involving IT departments. The Web 2.0-based approach to a global SOA empowers users to coproduce and share instant applications and thus represents a major step forward to evolving the above ecosystem into one in which all the stakeholders will also be able to collaboratively develop capabilities and innovate operating procedures by remixing and integrating already available services through the emerging ideas of Enterprise 2.0 mash-ups ("Mashing the Web," 2005).

Enterprise 2.0 Mash-Ups

Content-driven mash-up-oriented programming (a.k.a. situational programming or instant programming) (Smith, 2006) is a new agile application development paradigm[1] in which knowledge workers, who do not have previous coding skills but do have extensive domain expertise, visually assemble and combine off-the-shelf gadgets (a.k.a. widgets), that is, discrete self-contained domain data-oriented components, with both development (service and data binding and interconnection) and runtime rendering capabilities. These gadgets represent the basic building blocks for knowledge workers to assemble new services (e.g., SOAP or REST-based lightweight Web services), data sources (e.g., Atom/RSS feeds) and other gadgets, and to render them as necessary to develop the application they need in a very short time. The kind of hybrid application that results from applying this new paradigm is often called enterprise mash-up (a.k.a. situational application or instant application) (Hof, 2005).

A simple example would be a mash-up connecting three gadgets: a list of tasks involving customers, an agenda of customers, and a Google map. By attaching the three to each other, the agenda gadget will display the customer's details and the Google map gadget will display the customer's address on a map as you scroll the task list. This would be useful, for example, for an employee responsible for the task of geographically locating the customer. It is the knowledge worker who can develop this "service," and do it on the fly with the help of mash-up enablers. This way a business person could build a "dashboard" to see how weather is affecting sales at retail outlets. By aggregating information from public Web sites, such as mapping and weather services, the business person could assemble a very useful, albeit simple, content-driven application. Companies are trying to capitalize on these technologies (Smith, 2006) with software and services for relatively short-lived, quick-to-build applications.

Figure 1 depicts a real scenario extracted from a Telefónica-based mash-up which connects four gadgets: a list of tasks involving customer requests, a customer agenda, a Google map, and a network status map. Figure 1 shows how Telefónica's operational support systems' (OSS) knowledge workers create a fully functional environment on their own by visually attaching these gadgets to each other and to the enterprise backend; the agenda gadget will display customer details and have a customer/task selection option, the network map will represent the selected customer's network status, and the Google map gadget will display the selected customer's address on a map when a given task is selected from the list. This enterprise mash-up environment is useful for a user responsible for the task of testing the status of all systems used by a customer. In the event of a problem in the customer's local telecommunications infrastructure, customer geographical location is a big help for the technician to prepare the visit to the customer's home.

As mentioned above, traditional Web services are provided as functionality described by arbitrary (mostly WSDL compliant) interfaces that define input and output messages, as well as the supported service functions. These interfaces are not human-readable and do not facilitate the interaction of users with the underlying services. In the mash-up context resources no longer target technical experts in the corporate context but now go for the huge number of individuals (Anderson, 2006), of Internet users who require intuitive visual means for retrieving resources on the Web and for capturing their respective functionality.

This way, the Web 2.0-based approach to a global SOA delivers a mash-up-enabled infrastructure to help businesses share and collaborate with the business ecosystem and partners instantly. In doing so, enterprise collaboration architectures introduce the mash-up-oriented lightweight programming model as a means for knowledge workers to collaborate in solving an immediate, specific business problem by blending externalities with private business content and services.

The way services are discovered, used, and managed by knowledge workers is fundamental in terms of both the ICT technology and the cultural aspects involved in implementing this enterprise collaboration paradigm shift (Salganik et al. 2006, pp. 854-856). In this respect, user-service interaction must embrace a number of principles to ensure the widest acceptance by

Figure 1. Creation of an EzWeb platform-based enterprise mash-up

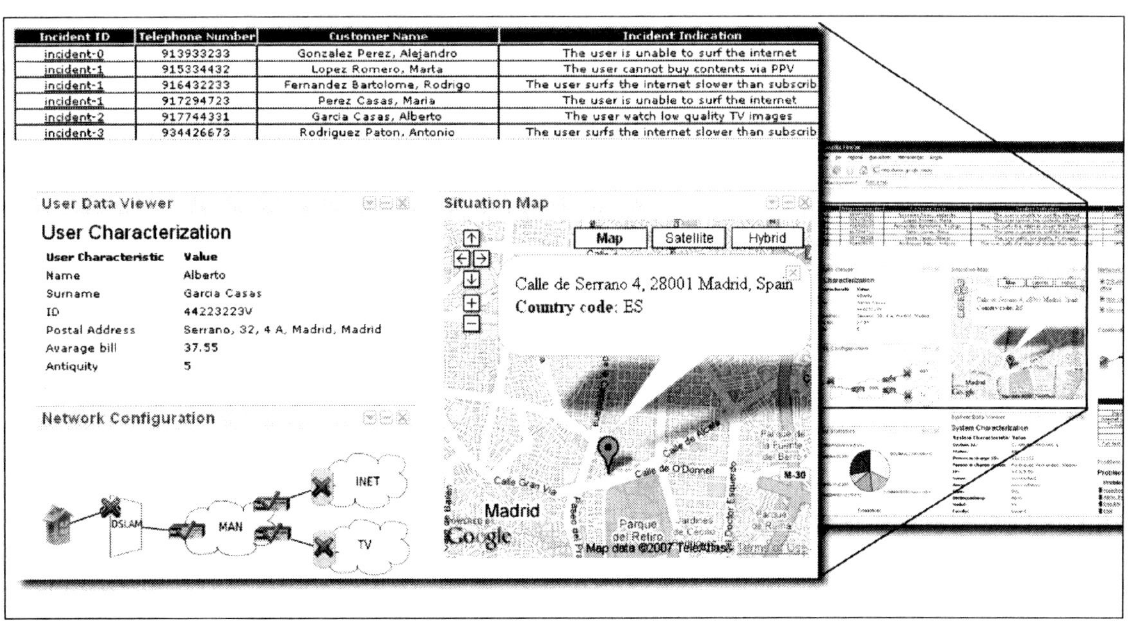

knowledge workers. The most important that have been identified are:

1. *Knowledge workers must feel fully empowered* and able to serve themselves from available resources that provide them with access to the content and services they can use to set up their own personalized operating environment in a highly flexible and dynamic way.
2. *Active user participation must be enabled.* Knowledge workers must be able to contribute new and improved versions of resources, as well as share further knowledge about these resources, their use, and their interrelationships.
3. *Community-based collaborations need to be fostered.* The introduction of a *share, reuse, and assembly culture of collaboration* will boost and speed up this process thanks to the network effect.

To exploit this approach to the maximum, IT departments will need to embrace the SaaS model as an effective software-delivery mechanism. This will change the department's focus from deploying and supporting applications to managing the services that those applications provide. Knowledge workers will now extend and improve these services in a collaborative fashion to exploit their extensive domain expertise and their thorough business knowledge.

Enterprise Mash-ups as a Means to Drastically Improve Time-to-Market

Internet technologies continue enabling businesses to expand their ecosystems and partnerships. This expansion process means, on the one hand, that the information technology effort focuses on work items related to integration, usually requiring a minimum of 6 months per request. On the other hand, partnerships change, and some business collaborations last less than 12 months. Consequently, there is a whole bunch of applications not being written today because they are not affordable due to time-to-market constraints and/or because there is no justification for IT investment. Knowledge workers' needs are typically of short duration (ranging from one week to several months), thereby limiting justification for IT investment too. It would be sufficient if they were provided with informal, just-in-time access to domain content, and were able to create their own short-lived ad hoc application for each individual need, without the need for IT investment.

Enterprise 2.0 mash-ups, and the associated application development paradigm, clearly fulfill these needs. Mash-ups will help businesses share and collaborate with the business ecosystem and partners instantly. This in turn will help evolving enterprise collaboration towards a new paradigm in which knowledge workers (without previous programming skills, but with thorough business knowledge) are considered as coproducers not only of information, but also of software services and applications that promote specific competitive advantages and/or meet their immediate needs, without involving IT departments, and share the solution with the remainder of the organization. This will lead at last to an ecosystem of knowledge workers collaborating to develop capabilities and innovate operating procedures by remixing and integrating available services, exploiting the collaborative intelligence. Community-based collaborations could speed up this process thanks to the network effect caused by broader community support and participation (architecture by participation). Additionally, the introduction of a reuse and assembly culture will boost this process, allowing strong business value synergy and linkages.

Current Impact of Enterprise 2.0 Mash-Up Technology

The ProgrammableWeb.com Web site acts as a major aggregator of numerous (currently over

1,600) references of assorted mash-ups existing all over the world. It provides statistics about used resources, a classification of mash-ups by categories, as well as user statistics to evaluate mash-up popularity. Therefore, it serves as a central source of empirical data to work with. Several facts can be inferred from these data:

- First, large companies such as Google, Amazon, and Microsoft started to successfully provide Web-based resources that were leveraged by numerous users to create mash-ups. At least 836 Web-based applications have already integrated the "Google Maps," a resource that offers geographical data.
- Second, there is a shift away from the professional corporate context towards a private, end-user driven field of applications. As opposed to the focus of traditional Web services, the resources used for building mash-ups target the long tail of Internet users and deal with media management, shopping functionality, entertainment, or desktop applications. Heavy-weight enterprise applications for automating business transactions can only rarely be found in this context.
- Third, besides the SOAP protocol, which is also used in the context of traditional Web services, "light-weight" protocols, like REST (Fielding, 2000) and RSS, are widely leveraged to allow for fast and seamless mashing of different resources.

Summing up, the Web can increasingly be considered as a comprehensive and global development platform containing numerous easily usable and mashable resources that are provided by large firms, as well as by SME and even individual end-users. As argued above, the provision of resources that use lightweight protocols based on uniform interfaces, such as REST or RSS, and also the focus on end-user requirements rather than business-to-business relationships represent core success factors for this new global SOA consisting of numerous mash-ups. Figure 2 shows the current most popular providers of resources leveraged for the creation of mash-ups.

User-Contributed Content-Driven Enterprise 2.0 Technologies

Of the existing categories of Enterprise 2.0 technologies now available, technologies for content management (enterprise *wiki/blogs*), contract management, project management, enterprise mash-up platforms, messaging and e-mail, signaling (feeds), listing services, social network analysis and analytics, social search, media, collaborative categorization (a.k.a. *folksonomy*), online Web storage, supporting infrastructure, organizing, social networking, consumer or workgroup *wikis*, massive collaboration, and business process management are worth mentioning (Redding, 2006). There will be significant differences in companies' abilities to exploit these technologies due to the challenges they bring with them. Because of the opportunities these technologies offer, these differences will matter a great deal. It is important to get an understanding of their real potential and drawbacks, as well as how to take advantage of them holistically.

Andrew McAfee (2006, pp. 21-28) first introduced the acronym SLATES to indicate the six components of Enterprise 2.0 technologies: **s**earch, **l**inks, **a**uthoring, **t**ags, **e**xtensions, and **s**ignals (SLATES). As technologists build Enterprise 2.0 technologies that incorporate the SLATES components, they seem to be following two intelligent ground rules following McAfee's vision of SLATES. First, they are making sure their offerings are easy to use. Second, Enterprise 2.0 technologists are trying hard not to impose any preconceived notions on users about how work should proceed or how output should be categorized or structured. Instead, they are building tools that enable these aspects of knowledge

Figure 2. Chart of most popular mash-up resource providers (according to ProgrammableWeb.com)

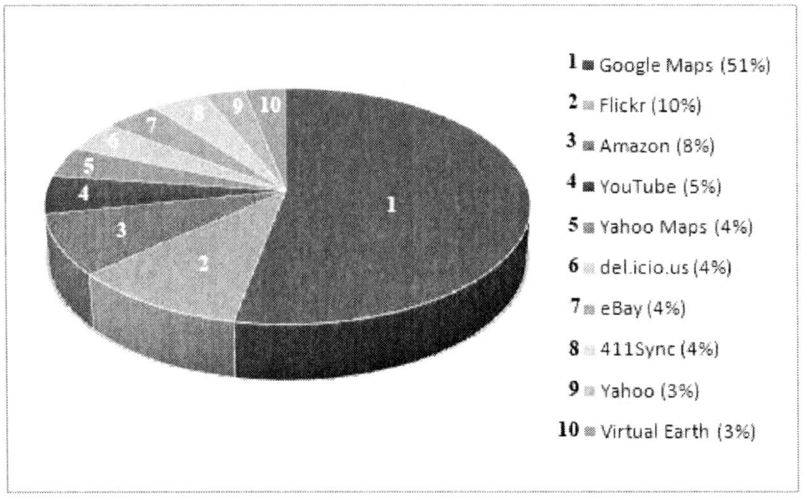

work to emerge. In the following sections we elaborate on the key Enterprise 2.0 technologies that will help a business to successfully exploit every SLATES component.

Enterprise Blogging

Blogging allows users to keep track of ideas and their authors, including concept redefinitions or business process information on a time line in a Web page, as if it were traditional Web content information. This information evolves like an approximating definition cycle based on dialogue and creative discussion, looking around for near ideas and issues to express enterprise knowledge.

There are several social networking functions in enterprise blogging that show the difference between simple blogging systems and systems used for triggering network effects across an organization. Some interesting insights about internal blogging in the *Enterprise Blogging in Practice* case study follow (Rand, 2004).

Michael Cot notes, *"At the department level, I wouldn't say that blogs have been a wide-reaching, raging success, primarily because people don't post to them as much as you'd hope. However, for the people who do post to and read the blogs, they've been very successful"* (Rand, 2004).

Among many other things, people usually post information at their company about their own stories and experiences, the status of tests, brainstorming-based ideas or issues, customer visit/phone call notes, comments that are only useful to bring attention to or track an employer with a project, requests for ideas or help, and even off-topic posts that can be useful for enforcing social networking relationships.

But the biggest problem is limited search capabilities. Without a Google-like quality search (i.e., near real time and full indexing, page-rank, quick search results, etc.) of the enterprise Intranet, it is very hard to find anything, let alone blog posts on relevant topics. People have been shown to get smarter a lot faster using social networking software (a lot of organizations of all sizes are using this Enterprise 2.0 application to trigger network effects and for innovation fostering). In addition to the enterprise blogging platform, it is interesting to stimulate several social networking system functions in Enterprise blogs (Wacka, 2005):

- Contact list (create informal groups or social networks) to maintain easy contact with community members
- Private messages
- Attach files (.doc, .pdf, .zip) and add tags and notes to them
- Advanced taxonomy with both structured and unstructured (free tagging) support
- Powerful ajax editor
- Tasks (to-do lists) for easy collaboration and project management
- Threaded comments for robust discussions
- Revision control
- Basic polls (advanced polling, surveying, and quizzes are available)
- Profiles
- Advanced search
- News aggregator (read RSS feeds and XML files)
- Syndication (generate RSS feeds and XML docs for content, profiles, tags, categories, etc.)

This way, it would be easier to make content and profiles precise. With structured and unstructured categories, it is easier to find people or content on the "long tail" of a curve (Kline & Burstein, 2005). In addition, it is possible to use tracking of individual profiles and posts to monitor user activity (i.e., to find out what others in the community are reading and writing).

Enterprise blogging should offer a useful way for connecting, creating, and collaborating on project management, help desks, finding and identifying experts (people aggregator), recruiting experts, talent, or ideas or innovation management, open innovation and its visibility, knowledge management, product development, and other off-topic tasks.

This platform would provide a comprehensive online ecosystem, a tightly integrated set of publishing, communication, and networking features that support and enable an online experience like never before. Users could engage, create, and share their content online (publicly or privately) in a multitude of ways to achieve greater performance. This technology will help to discover the purpose (i.e., connecting, creating, and collaborating) of an enterprise community in order to improve the company's business strategy. Knowing the purpose determines how the system is configured, how it flows, and how well it succeeds, and therefore the best strategy to carry out.

The following are some tips from Dion Hinchcliffe's (2007) "Nine Ideas for IT Managers Considering Enterprise 2.0," which are easily adaptable to enterprise blogging:

1. It is about ease-of-use, first and foremost.
2. Change requires motivation. Provide it.
3. Emergent does not mean a blank slate.
4. Discoverability is not an afterthought, it is the core.
5. It is okay to fear loss of control and misuse.
6. Dynamic, effective advocates are a key enabler.
7. The problems will be with the business culture, not the technology.
8. Triggering an Enterprise 2.0 ecosystem quickly is likely to be an early activity driver.
9. Allow the tools to access enterprise services

Enterprise Wiki: The 2.0 Approach of Content Management Systems (CMS)

A wiki can be defined as a piece of server software that allows users to freely create and edit Web page content using any Web browser. Wiki supports hyperlinks and has a simple text syntax for creating new pages and crosslinks between internal pages. Because it allows "everyday users to create and edit any website page, it is exciting in that it encourages democratic use of the Web and promotes content composition by non-technical users."

There are some fundamental Wiki design principles, but the most important principle that makes wiki different from any other Web site source is that it allows any enterprise readers to edit the page content as they see fit, if they feel the content is insufficient or poorly organized. If you come across any mistakes in the document as you read an information item, or you have more information that you would like to add to the item, just click on "Edit Text" and you, too, can change the content.

Because wiki is mainly designed to promote content composition by nontechnical users, the formatting rules for editing a wiki are fairly simple, and there are no complicated markup languages. Wiki content generally contains a lot of accurate information, because inaccurate information will be very quickly corrected by other readers.

The main idea of not having anyone to control the content in a centralized way, and/or of allowing *anyone* to edit and publish a document real-time is inconceivable for most people.

Therefore, a wiki can be the underlying technological support for creating a common sharing, emergence, and conceptualization of mash-up data, playing a similar role to ontologies in the Semantic Web vision. In fact, wiki software has demonstrated that ***it works well in a small community of like-minded people, like an enterprise community.*** Documents build up very fast as many people contribute small, manageable pieces. Some contribute contents and information, some contribute links, some correct grammar, while others fix the structure and formatting. Therefore, enterprise wiki allows the input of common descriptions and definitions of key business concepts, that is, everyone in the company can find a resource, bring experience to bear to evaluate and improve the resource. Wiki contents and data follow an iterative lifecycle, and their description undergoes constant improvements, refinements, and evolution.

There are several applications defined as enterprise wiki enablers, like Confluence, Social Text, or Twiki (*InfoWorld* proclaimed 2004 to be the "Year of the Enterprise Wiki"). Heightened interest comes in response to the increasing number of organizations like Google, Nokia, and Yahoo! who are turning to wikis as a way to improve internal efficiency.

Wikis and Content Management

Wikis fall conceptually under the broad concept of content management systems, and users could certainly use the existing CMS to create a *wiki-like* site. However, wikis have unique characteristics that differentiate them from traditional CMS (Choate, 2006).

Wikis emphasize ease of content creation. This simplicity has many sources: a wiki markup language that provides a short-hand way of linking documents and formatting text; the ability of users to edit and create pages independently and directly; a bottom-up approach to site navigation and structure; a very simple templating framework; and, finally, a conscious decision to eschew workflow or even simple approval steps.

Content Creation and Editing

Wiki software focuses on the empowerment of users to create and edit their own pages, but content management systems provide tools for creating and editing content, too. The difference is in their approach. When wikis first came out in 1995, there were not a lot of options for WYSIWYG editing from within a browser, so the wiki markup language (sometimes called "wikitext") provided a particularly valuable short-hand for formatting text that was much easier to learn than pure HTML (Heigl, Glaser, & Anja, 2006).

A good CMS will offer a WYSIWYG interface that makes writing content for the Web, like using a word processor. More wikis nowadays have WYSIWYG editing features, so the wiki markup language is a less interesting feature in terms of formatting, although it does provide the benefit of being supported by all browsers on all platforms, something that is not usually the case with rich-text editors. Many wikis support both wikitext and rich-text editors.

However, there is one area where wikitext still retains its power: linking resources and knowledge. Wiki software still provides a much easier way to link pages within the wiki to each other. Links are made based on the title of a page, so the author does not need to use, remember, or type long URLs in order to link one page to another.

Site Structure and Navigation

Contributors can create new knowledge, pages, contents, and can easily link one page and data to another (Venners, 2003); wikis offer a new unique approach to navigation and site structure.

Traditional information systems usually take a more formal approach to site structure and navigation through enterprise knowledge, with the site organized into a hierarchy by an information architect. User-created pages in a wiki mean that the hierarchy and structure of the site is created in an ad hoc way. Navigation tends to be simple, and the hierarchies are flat. For example, the Wikipedia online encyclopedia has hundreds of thousands of articles on a broad range of topics, but these topics are not arranged in any conceptual hierarchy. The entry for dogs serves as a good illustration. The URL for the article about dogs is: http://en.wikipedia.org/wiki/Dog

A pug is a kind of dog, and the URL for the pug entry is: http://en.wikipedia.org/wiki/Pug

Since a pug is a kind of dog, you might expect to see the following URL for pugs: http://en.wikipedia.org/wiki/Dog/Pug

But it is not there. Several wiki software solutions support more complex content categorization, but many are totally flat, like Wikipedia. Even if the software supports subconcepts, contributors are still allowed to create subpages in an ad hoc fashion and there is no systematic approach to the architecture of enterprise knowledge.

Content Repository and APIs

An experienced architect or administrator will ask of any content technology what the repository looks like. This is a good approach, because they are concerned about back-up, compatibility, performance, and a raft of similar issues.

Wikis have traditionally taken a very simple approach to information storage. Original wikis stored content in plain text files written with a wiki markup language. When a reader requested a page, the page was rendered. This was not speedy, but it worked. These days, wiki packages employ one of several different back-ends, with many housing their content in databases.

A transcendental consideration is whether the software supports automatic back-ups (commercial wiki applications often do). Another thing to think about is what this means in terms of integrating wiki content with content managed by other systems. For example, should an enterprise search system be able to index wiki content, and should the indexed content be raw wikitext, or rendered HTML pages?

This question leads on to the issue of wiki APIs, which, in fact, very few wikis have.

Templates

When a wikitext page is required, it is rendered as HTML in a two-part process. First, the wiki markup is converted to HTML, and links are created between pages. Then, this content is wrapped by a template that provides a consistent look to all the pages in the wiki.

Comparing wikis to a CMS, most wikis have template systems that are very simple, often only enabling one general template for the entire site. Wiki templates (and page rendering in general) are often not cached, so the page is rendered with each request. From an enterprise perspective, a lack of caching can obviously limit system scalability. On the other hand, there is no finicky caching mechanism to deal with.

Workflow

Wiki software completely changes the main idea of a workflow. Wikis are decentralized and typically lack the controlling mechanism of a workflow system with a formal approval process. Wikis workflow systems often lack sophisticated and dis-aggregated and approval processes is commonly considered a feature and not a fault of wikis. This is contrary to the basic idea of many CMS, which prioritize control over empowerment. Despite this decentralized approach, there is one important thing to remember: the main idea that anyone can edit content is only a general policy and not an inherent characteristic of wiki software.

Control vs. Flexibility

There is a traditional trade-off between control and flexibility in information systems software. Decision-making is centralized by some sort of editor that verifies and approves content prior to publishing in a traditional CMS. With a wiki, the writer writes then publishes without editorial oversight or approval. This direct channel to publication is what makes wikis so wonderful in scenarios that emphasize speed and flexibility.

An important issue to deal with is what possibilities there are if enterprises want to exercise at least some control. In the absence of workflow controls, content creation in a wiki is managed through change monitoring, automated spam prevention, and user access control. Let us look at each one in turn.

Change Monitoring

One simple defense mechanism is to monitor changes in the wiki and enable the rolling back to a previous version through versions control.

Recent changes can be monitored as follows:

- Most wikis have a "Recent Changes" page that lists all the pages that have been changed.
- E-mail notification of changes or support for RSS syndication.
- If more than one person has been tasked with monitoring changes, some wikis offer the capability to track whether a recently changed page has been checked yet, reducing the chances of the work being done twice.
- More sophisticated systems identify and differentiate "trivial" changes from more substantive ones.

In addition to the above ideas, it is important to consider that people make mistakes and sometimes deliberately do things badly. Therefore, the ability to roll back changes is a necessity. Features to look for include capabilities similar to what you would find in a CMS, like the ability to roll back changes to the previous version, to compare different versions side-by-side or the use of diffs between versions so that specific differences between them can be easily identified.

Spam Prevention

Another approach is to monitor the content of changes programmatically, that is, to manage spam prevention. This differs from user access control in the sense that it monitors wiki edits based on the content itself, or patterns of user behavior. Some systems can block access to IP addresses and URLs, or they can block the posting of individual changes based on restricting the use of certain words or phrases, using word lists or regular expressions, and blocking access based on excessive activity.

User Access Control

Enterprise wiki usually means that it has user access control. An increasing number of wiki projects offer sophisticated **more granular level** in user access control issues. Users and groups can be assigned rights to tasks such as reading, editing, writing to, and rolling back a resource to a previous version. There is a lot of variance among wiki packages in terms of how those rights are applied to the site. A less common but useful feature is the ability to restrict access to parts of resources. The most sophisticated enterprise wikis work with single sign-on security systems like Siteminder, or offer network and directory integration (e.g., LDAP and Active Directory) for user authentication and authorization.

Contrary to their reputation, wikis are CMS that can be managed efficiently. They simply take a different approach to content management by choosing to emphasize speed and flexibility rather than strict controls. In order to successfully implement a wiki software package you will need to look at workflow from a different perspective and be sure to select wiki software that provides the right level of content monitoring and access control for your organization.

Enterprise RSS

RSS is a family of Web feed formats used to publish frequently updated digital content, such as blogs, news feeds, or podcasts.

End-users that receive data from this technological channel use programs called feed "readers" or "aggregators", as follows. The user "subscribes" to a feed by supplying to their reader a link to the feed; the reader can then check the user's subscribed feeds to see if any of those feeds have new content since the last time it checked, and if so, retrieve that content and present it to the user.

The initials "RSS" are variously used to refer to the following standards:

- Really Simple Syndication (RSS 2.0)
- Rich Site Summary (RSS 0.91, RSS 1.0)
- RDF Site Summary (RSS 0.9 and 1.0)

RSS formats are specified in XML (a generic specification for data formats). RSS delivers its information as an XML file called an "RSS feed," "Webfeed," "RSS stream," or "RSS channel."

Essentially, Web 2.0 is fully centralized in its conception. Why are skype, del.icio.us, or Flickr Web sites instead of protocols (as foaf is)? The reuse of Web 2.0 data is limited only to the hostside and only with the help of feeds are data able to break out from centralized sites (Hammond, Hannay, & Lund, 2004, pp. 1082-9873). Therefore, feeds and RSS are the key to a new data-distributed model in the Web 2.0, where data are disaggregated on the Internet, and RSS allows data, information ,and remote events to be distributed to end-users through the Internet.

Content Tagging

Content tagging is a growing Internet trend that empowers users to add their own contextual tags to Web content, information, or resources. Typically, as Gruber (2005) affirms, this results in excellent content categorization in a way that is relevant to the needs of users.

Tags are, therefore, Web page and/or database descriptors (e.g., title, author, language, date, subject) that are assigned to knowledge (e.g., information, Web content, distributed resource, etc.). One of their main purposes is to help people find information. Tags can be assigned to document descriptions (e.g., card catalog cards in a library) or they can be assigned to the documents themselves or both.

Tags can be assigned by document authors, information professionals, editorial assistants, or even by computer programs. Artificial intelligence programs are a fast and easy (but not always the most accurate) method of tagging. Information professionals can produce highly accurate and effective tags that take into account all the nuances of language and subject matter, but there is a limited supply of people with these skills.

These tags are useful for creating an emerging user-centric categorization of content in a folksonomy (a user-generated taxonomy used to categorize and retrieve Web content). Folksonomic tagging is intended to make a body of information that is increasingly easy to search, discover, and navigate over time. A well-developed folksonomy is ideally accessible as a shared vocabulary that is both originated by, and familiar to, its primary users. Two widely cited examples of Web sites using folksonomic tagging are Flickr and del.icio.us.

Folksonomies are developed in Internet-mediated social environments. Therefore, knowledge workers can discover who has created a given tag for a concept, and see the other tags that this person created. In this way, folksonomy users often discover the tag sets of another user who tends to interpret and tag content in a way that makes sense to them. The result is often an immediate and rewarding gain in the user's capacity to find related content. Part of the appeal of folksonomy is its inherent subversiveness: when faced with the choice of the search tools that Web sites provide, folksonomies can be seen as a rejection of the search engine status quo in favor of tools that are created by the community.

Folksonomy creation and searching tools are not part of the underlying World Wide Web protocols. Basically, these folksonomies arise in Web-based communities where special provisions are made at site level for creating and using tags, as in del.icio.us. These communities are established to enable Web users to label and share user-generated content or to collaboratively label existing content. Since folksonomies are user-generated and therefore inexpensive to implement, advocates of folksonomy believe that it provides a useful low-cost alternative to more traditional, institutionally supported taxonomies or controlled vocabularies like enterprise IT solutions. An employee-generated folksonomy could therefore be seen as an "emergent enterprise taxonomy." Some folksonomy advocates believe that it is useful for facilitating workplace democracy and the distribution of management tasks among people actually doing the work.

As many authors note in blogs and articles, "workplace democracy is also seen as a utopian concept at odds with governing enterprise reality, the majority of which exist and thrive as hierarchically-structured corporations not especially aligned to democratically informed governance and decision-making." Also, the folksonomy may facilitate workflow, but it does not guarantee that the information worker will tag and, then, tag consistently, in an unbiased way, and without intentional malice directed at the enterprise.

Strategic Sensemaking

The increased importance of sensemaking will prove to be one of the central drivers for Enterprise 2.0 technologies adoption. The organizational theorist Karl Weick says that sensemaking is a central task in new organizations. Dan Russell at Creating Passionate Users provides a definition of sensemaking that will serve as a useful starting point: "Sensemaking is in many ways a search for the right organization or the right way to represent what you know about a topic. It's data collection, analysis, organization and performing the task" (Dervin, 1983).

Sensemaking can be a solution for constructing sensible accounts out of ambiguous, ambivalent, equivocal, and conflicting data in organizational settings for managers and leaders in the knowledge organization and management. In a world characterized by significant technology and strategic change, the problem of sensemaking becomes more acute.

One of the attractions of Enterprise 2.0 technologies is that they make business strategies more feasible and scalable. Most of the technologies depicted in this section take participation as far as what face-to-face methods can support. They make it possible to generate and organize more extensive raw materials and inputs to planning/

sensemaking processes. Wikis with good version tracking and refactoring capabilities make it both safer and easier to generate and work through alternative representations/sensemakings. Realizing this sensemaking potential will require brokering some introductions and partnerships. Those adept in the techniques are likely not to be versed in the ways that the technologies reduce or eliminate some of the key barriers to successfully using the techniques. Those who understand the technologies may not be aware that the techniques exist, much less that they could benefit from technological improvement. One starting point is to investigate the sensemaking planning techniques and practices and map points where the technologies enable, simplify, or improve the techniques for those promoting Enterprise 2.0 technologies.

Social Networking

Nohria and Eccles (1992) give a common definition of social network as "a social structure made of nodes (which are often organizations or individuals) tied by one or more specific types of relationships, such as values, visions, idea, financial exchange, friends, kinship, dislike, trade, web links, etc."

Social network analysis approach consists of relations in terms of *nodes* and *ties*. Nodes are actors within the networks, and ties are the relations between the actors. There can be many kinds of ties between the nodes. Research in a number of academic fields has shown that social networks operate on many levels, from families up to the level of nations, and play a critical role in determining the way problems are solved, organizations are run, and the degree to which individuals succeed in achieving their goals.

A social network is a map of all of the relevant ties between the nodes being studied. The network can also be used to determine the social capital of individual actors. These concepts are often illustrated by means of a social network diagram, where nodes are the points and ties are the lines. In traditional social network communities, an initial set of founders sends out messages inviting members of their own personal networks to join the site. New members repeat the process, adding to the total number of members and links in the network. Sites then offer features, such as automatic address book updates, viewable profiles, the ability to form new links through "introduction services," and other forms of online social connections. Social networks can also be organized around business connections, as in the case of LinkedIn.

The combination of networking is a new point of view to social networking that combines both off-line elements (face-to-face events) and online elements. The newest social networks on the Internet are becoming more focused on niches such as travel, art, and so forth. Other social networking sites focus on local communities, sharing local business, and entertainment reviews, news, event calendars, and happenings.

Traditional social networks on the Internet were public, and any user could participate. However, large enterprises and organizations also have access to private social networking applications often called enterprise social networking software. For example, Microsoft released an enterprise social networking application in the form of a free add-on for Microsoft Office SharePoint Server called Knowledge Network (currently in beta) in February 2007. Organizations install these applications on their own servers and enable employees to share their networks of contacts and relationships to outside people and companies.

Aligning the Business Strategy with Enterprise 2.0 Ground Rules, Diving Forces and Best Practices

Figure 3 shows a subset of the main Enterprise Web 2.0 technologies described throughout this chapter and what innovation promoting factors are fostered by each technology or ideology.

Figure 3. Enterprise 2.0 technologies: Innovation fostering framework

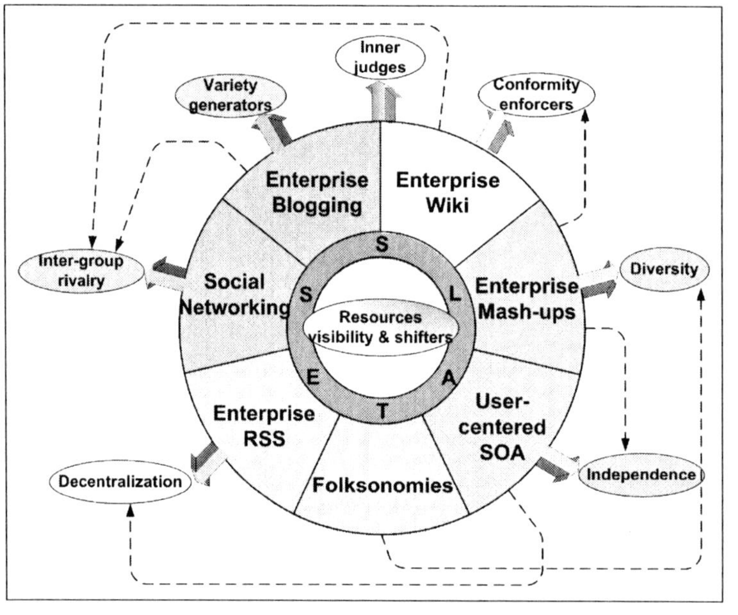

Generally, each Enterprise 2.0 approach follows the SLATES principle described by McAfee. In addition, one of the most important innovation factors described in high-level knowledge working, that is, resources and content visibility through the Web and its simple modification and reedition in a shifting of resources is fostered by all technologies in the framework.

- Enterprise Wiki is used to input common descriptions and definitions of main business concepts. Everyone in the company can bring his or her own experience and knowledge to bear to find, evaluate, and improve a content resource. This iterative process enforces common data and knowledge conformity across the enterprise. Each knowledge worker must review the wiki concepts and will use an inner judge of the content to do his or her best to improve the concepts. In addition, this phenomenon shows group and department knowledge outputs, allowing the management of constructive intergroup rivalry in order to improve the descriptions of concepts or knowledge about business processes.
- Enterprise blogging is useful for keeping a track on ideas, concept redefinitions, or business process information on a time line. This visibility stimulates each knowledge worker to discuss this information, generating a variety of ideas in an approximate definition cycle. Obviously, this technology, as the wiki approach, fosters the ideas of judgment and inter-group rivalry through the contribution of their own experiences. Therefore, wiki and blogging together can tackle the emerging knowledge, social capital, and collective enterprise intelligence created by the groups of knowledge workers.
- The social networking idea promotes the rivalry and communication between enterprise employers in a social net, using several communication channels to increase the dialogue in the enterprise and its departments.

Table 1.

	Conformity & imitation	Social loafing & slack off	Coordination failures	Abstraction/ Fragmentat.	Idolatry of memory	Certainty
Enterprise Wiki	✗	✓	✗		✓	✓
Enterprise RSS	?	?	✓	✓		✗
Folksonomy	✓	✓			✓	✓
Enterprise Blogging	✗	✓	✓		✗	✗
Enterprise Mash-ups	✗	✓	✗	✗	✓	✗
Global SOA	✓	✗	✗	✓	✓	
Social Networking	?	?	✓			✗

- Enterprise mash-ups deal with the diversity of solutions to a definite problem within the enterprise. Software solutions based on a heterogeneous merge of separate components communicated and parameterized by knowledge workers fosters diversity and originality in the enterprise, eliminating barriers to the innovation bloom. This approach also helps to enforce conformity in the enterprise about systematic knowledge, that is, what mash-up solution is the best to afford a particular solution for a business process. Users follow a "do-it-yourself" ideology that encourages the independence of end-users from the service providers and legacy.
- Global (user-centered) SOA mainly fosters the idea of the independence of end users and end-user innovation from the content and services providers, breaking down traditional innovation barriers in SOA approaches. In addition, it motivates a decentralization of resources across the Web, fostering client-computation and disaggregated data models and composition against the service traditional front-end.
- Enterprise RSS can manage the decentralization of data in Web 2.0 and Enterprise 2.0. Feeds and RSS are the keys of a new data-distributed model where data are disaggregated in Internet, and RSS is useful for distributing data, information, and remote events to end users through the Internet.
- Finally, folksonomies are related to the visibility of resources, and resources discovery and recommendation issues. The use of a relaxed taxonomy based on tagging by end users improves the diversity of knowledge related to these resources, and the collaborative intelligence present in enterprises can be better emerged.

Table 1 shows how the Enterprise 2.0 technologies are related to the factors that inhibit or slow down innovation.

These relationships are of three types: a technology can be appropriate for deleting a negative factor (shown as ✓), could cause this factor (shown as ×), or must be applied very carefully because it can cause or delete a factor depending on its use.

Next these specific relationships between technologies and inhibit factors are depicted in more detail:

- An enterprise Wiki is useful for collaboratively editing contents through a Web platform. In this technology, a new content or concept would commonly be inserted to the wiki, imitating the description, structure, and form used in other previous concepts, that is, it leads to a negative conformity and imitation process without concern about enterprise integration. On the other hand, it is very simple to edit a concept or contribute with one's own experience and knowledge to a wiki. Therefore, this idea reduces social loafing and the slack-off caused by traditional complex content management systems. The wiki uses an iterative description of concepts, that is, a new edition overwrites the previous one and could cause coordination failures. Finally, a wiki content is usually anonymous and iteratively improved. For this reason, it is easy to eradicate the idolatry of memory in enterprise knowledge, constantly improving the contents and anyone in the enterprise to properly discuss the manager's ideas.
- Enterprise RSS is a communication channel to manage the decentralization of data in Enterprise 2.0. Therefore, some factors like conformity or slack-off depend on the content transmitted through the feeds. The RSS architecture, based on client aggregators and feed channels, decreases the coordination failures during contents accessing in the enterprise. In addition, this data communication channel, split into several different disperse channels merged in the client, is a correct way to deal with data and their abstraction and fragmentation. The negative part of RSS is that the data origin is known, and this could foster the feeling that data created by heads or managers and distributed by RSS are certain.
- Folksonomies can create informal taxonomies based on tags (anonymous or not) in a very simple way using a Web platform. A new concept is very quick to tag. Therefore, this technique decreases the conformity in the conceptualization and the social loafing surrounding tagging contents and applying knowledge. Each knowledge worker will use his or her own experience, refining the tags used, even if these tags have been imposed by managers. For this reason, the knowledge is extrapolated and transformed, and therefore social capital emerges as obsolete enterprise conceptualizations are forgotten.
- Blogging keeps a track of ideas (and their authors), concept redefinitions, or business process information on a time line. This technology often causes a conformity feeling among the personnel, imitating structures, ideas, natural language descriptions, and schemas during a new track of knowledge in other enterprise blogs (in fact, wiki could cause a parallel effect). As with wiki editing, it is so simple, friendly, and quick to edit a new comment, or to refine an idea that social loafing is evidently decreased. However, each contribution is logged and stored in a blog, which has several consequences: coordination failures and overwrite issues are decreased but knowledge workers could be afraid of expressing their ideas in public or of arguing a traditional notion (provoking an idolatry of memory) or a leader's opinion.
- Enterprise mash-ups motivate software solutions based on a heterogeneous merge of separate components. A working mash-up often causes a conformity feeling in knowledge workers. Therefore, it is recommendable to force them to create their own solutions fitted to their own problems. In this same sense, nonprogrammer users should be offered a simple way to create mash-ups and reduce slack-offs. This approach may often cause coordination failures across departments, creating solutions to similar business problems. Likewise, this technique can foster a wrong abstraction level at solving

problems, creating partial software solutions without considering the whole problem dimension. This approach can improve previous mash-up-oriented solutions in a very fast and simple way, reducing the traditional idolatry of previous software or enterprise solutions so harmful in business strategies. Finally, it is very dangerous to publish strict mash-up compositions as unique software solutions created by managers or specialized departments, because this could provoke an innovation barrier to new ideas or improvements to these solutions.

- User-centered SOA mainly fosters the idea of independence of end users and end-user innovation from the content and services providers, focusing on reducing conformity and imitation issues caused by the traditional rigid SOA approach. This idea must be applied carefully because it could foster a slack-off in nonprogrammer users that have a poor perspective of Web services, or problems coordinating efforts in pragmatic developments. This issue can be easily improved using this approach and a mash-up orientation together. A strong point of this technique is the correct abstraction/fragmentation view of enterprise complex problems, coordinating or orchestrating user-centered services to tackle with whole problems through interface charts and storyboards linked to concrete workflows. One of the most important ideas is the high parameterization in enterprise services. This makes it easy to forget traditional memorized solutions, improving them in an adaptable way.
- The social networking is a general philosophy focused on improving effort coordination across a social group, fostering collective intelligence emergence and exploitation, reducing the possible coordination failures, and increasing outsourcing visibility. Like RSS, social networking can be considered a family of communication channels. Therefore, aspects like imitation, conformity, or social loafing depend directly on the content and the management of social information and knowledge. In this philosophy, it is harmful to introduce managers and manager-generated knowledge that can cause the social group concern and delimit its innovation process and evolution.

Application to Other Key Business Web Strategy Areas

This section looks at each of the implications of the explained Enterprise 2.0 vision for communication and information sharing, knowledge management business intelligence and business process management, and its application to key business web strategy evolution.

High-Performance Collaboration and Community-Building

More and more often organizations tend to behave like dynamically reconfigurable networked structures that carry out their tasks through collaboration and teamwork. Effective teamwork is an essential part of any nontrivial engineering process, and collaborative capabilities are an essential support for these teams.

Traditionally, collaboration has been a means for organizations to do their work. As illustrated throughout this chapter, however, the context in which they do this work is changing, especially in regards where the work is done, how the work is organized, and who does the work, and with this the characteristics of collaboration. Work teams face sizeable collaborative challenges, for which they have need of tools that they can use to communicate and coordinate their work efficiently. These challenges have been tackled traditionally by profuse research in *computer supported collaborative work* (CSCW). CSCW has a great deal of drawbacks can be dealt with under the Web 2.0 vision. Web 2.0 has taken a step forward in this respect with the emergence of *social networking* and *communities*, where the emphasis is on *open source communities*.

Open source communities are one of the most successful—and least well understood— examples of high-performance collaboration and community-building on the Internet today. *Open source communities* began as loosely organized, ad hoc communities of contributors from all over the world who shared an interest in meeting a common need. However, the organization of these communities has proven to be very flexible and capable of carrying out all kinds of developments, ranging from minor projects to huge developments. Businesses following the *Enterprise 2.0* vision can benefit enormously by learning what open source communities are and how they work. It is important to remember how the use of the *Enterprise 2.0*-based IT infrastructure will transform today's Intranets into virtual spaces where all project stakeholders, possibly distributed in time and space, can negotiate, brainstorm, discuss, share knowledge and resources, and, generally, work together to carry out some task. The vision presented in previous sections will help to definitively change Tom Allen's well-known "30-meter" rule, stating that two scientists or engineers whose desks are more than 30 meters apart have a communication frequency of almost zero.

Collaborative Knowledge Emergence and Management

The concept of knowledge management introduced previously has been an elusive chimera to corporations since the mid-1990s. Ever since employees came to be seen as knowledge workers, companies have been searching for ways to capture and disseminate the stuff inside their heads. Knowledge management systems have traditionally tried to do this by both relying on distributed production and providing high commonality. This way, they have sought to elicit tacit knowledge, best practices, and relevant experience from people throughout a company and put this information in a widely available database.

Nevertheless, a corporation's knowledge is scattered across a multitude of communication media channels, including e-mail threads of conversation, instant messaging talks, and communication media platforms or Intranets, corporate Web sites, and information portals. Production in the first group is distributed (that is, knowledge workers can create and distribute digital information free of charge), and many of them leave communication traces (e.g., instant messaging talks or e-mail threads of conversation). However, the degree of commonality of this information is low (e.g., only the participants in an e-mail exchange have access to the knowledge held in the thread of conversation). In the second group, commonality is high, but production is centralized and visits to platforms leave no traces. Both the "low commonality" factor in current channels and the "centralized production" and "lack of traces" factors in current platforms imply that most knowledge work practices and output are invisible to most people in most companies. For this reason, it is very important to understand the presented vision and especially how *Enterprise 2.0*-based IT introduces new channels and platforms that enable distributed production, communication tracing, and high commonality of information and services simultaneously to improve user productivity in the way explained in the framework of Enterprises 2.0 technologies and their application to innovation, knowledge emergence, and content visibility.

On the other hand, current knowledge-work-specific technologies, like highly structured knowledge management systems using complex taxonomies and/or ontologies are not doing a good job at capturing, sharing, and applying their knowledge, which is typically highly unstructured and textual. In this respect, a recent study (Morris, 2005) has shown that only 44% of respondents agreed that it was easy to find what they were looking for on their Intranet. The channels and platforms in traditional use are not much good at providing answers to questions like who is working on a similar problem right now. Or what is the right way to approach this analysis? The presented practical Enterprise 2.0-based collaborative (and social) approach can catalog and search knowledge so that employees can easily leverage it throughout the firm. Briefly, the application of the broad spectrum of Enterprise 2.0 of technologies to business Web strategy should be considered.

Finally, most current platforms, such as knowledge management systems, information portals, Intranets, business process management (BPM), business activity monitoring (BAM), and workflow applications are highly structured from the start, and users have little opportunity to influence their structure or to customize their functionality and their interfaces. Emerging platforms, like wiki, blogging, or folksonomies (explained throughout this chapter), for generating, sharing, and refining information under the Enterprise 2.0 vision umbrella focus not on capturing actual knowledge, but rather on knowledge workers' practices and output.

In conclusion, the Enterprise 2.0 vision is significant in this respect because it can potentially knit together an enterprise and facilitate knowledge work in ways that were out of the question before. Putting it simply, Enterprise 2.0 technologies have the potential to make the knowledge management infrastructure of a corporation what the Internet already is, that is, an online platform with a constantly changing, searchable structure built by distributed, autonomous, and largely self-interested peers. Technologies like blogs, wikis, and labeling systems capable of emerging folksonomies make a decisive contribution to the elicitation of knowledge, best practices, and relevant experience that is scattered across the corporation and make this information trustworthy, searchable, and accessible to people throughout a company, at the same time as creating a cooperative and helpful culture capable of boosting knowledge production and guaranteeing convergence and quality through highly egalitarian collaboration.

Social Network Analysis and Business Intelligence

Howard Dresner, a Research Fellow at Gartner Group, popularized the term business intelligence as an umbrella term to describe a set of concepts and methods to improve business decision making using fact-based support systems. This discipline aims to describe how end users could access and analyze information stored on their company systems in order to provide a better understanding of the business and its customer. To do this, they used a broad category of applications and technologies for gathering, providing access to, and analyzing data from the earlier business activities for the purpose of helping enterprise users make better business decisions.

Much of the early research in this field took place before the widespread use of the Internet and even corporate e-mail. We now need to understand the implications of Enterprise 2.0 technologies in the process of sharing business intelligence (Barquin, 2006). Web 2.0 can be seen (as it has been explained previously) as the ability to communicate and share knowledge with other participants that have similar interests, resulting in a key means of producing, communicating, and sharing business intelligence. In this line, this chapter focused on describing how Enterprise 2.0 can be used to spread, publish, and manage data from previous business activities through a new breadth of collaborative social knowledge networking tools like blogs, wikis, messaging, e-mail channels, and so forth, improving the added value outlined in enterprise business intelligence.

In previous sections, we showed how the Enterprise 2.0 vision introduces new channels and platforms that enable distributed production, communication tracing, and high commonality of information and services simultaneously. Apart from making information and knowledge work practices and outputs far more visible to the entire company, another key advantage pointed out during the chapter is that it can be used to carry out *social network analysis* of *business intelligence*. This is primarily concerned with the paths that information exchanges take between individuals and the fact that some individuals act as key nodes in the network and become critical factors in the successful communication and sharing of business intelligence.

Social network analysis allows us to identify the pathways that *business intelligence* must travel if it is to be used, integrated, enriched, and applied by individuals within enterprises. It will help to identify which the critical nodes are. Star nodes usually represent key points in the routing at which important intelligence can be either effectively tunneled to other network members that share the same interests or are working on the same type of problems or tragically choked, as we have seen up to now. Star nodes have been identified in almost any work environment grouping knowledge workers. Social network analysis is a powerful tool for improving a company's intelligence capabilities. Therefore, business intelligence analysis should be carried out not only on the content of communications, but preferentially on the structure of the social network, its topology, communication patterns, and links to identify the key nodes.

Business Process Management Revisited

Finally, BPM represents another key area in which Enterprise 2.0 vision will help to make great improvements. It is generally accepted that the current orchestration, BPM and workflow systems and technologies have failed to achieve a seamless automation and integration of business processes. Although there are currently a lot of approaches and standards in this respect, none of them seems to offer a sound solution to real enterprise needs. Most of the problem lies in how they handle the knowledge acquisition process for automating business operations. As we have seen, *Enterprise 2.0* provides a new approach to this problem that integrates flexibility, human interaction, and modeling facilities (different from the flowchart-based or the algorithmic approaches, and now based on networks of resources and interrelationships). Following the exposition carried out it is simple to find out the need to apply the Enterprise 2.0 approach to improve and evolve this kind of systems.

CONCLUSION AND FUTURE TRENDS

In this chapter we have explored emerging Internet technologies, highlighting their potential for supporting business Web strategy as companies' reliance on new Web-based technologies to capitalize on new business opportunities increase quickly. Specifically, Enterprise 2.0 has been presented as a key enabler for businesses to expand their ecosystems and partnerships, as well as acting as a catalyst for improving innovation processes and knowledge work in general. On the one hand, we have elaborated on the concept of Enterprise 2.0 mash-up as the main technological enabler of a global, user-centered SOA that spans company boundaries. On the other hand, we have discussed the most relevant Enterprise 2.0 models and tools suitable for fostering emergent collaboration and cocreation, thus enabling firms to leverage desirable attributes, including harnessing collective intelligence and organization by participation.

The key idea behind the Enterprise 2.0-based approach to a business Web strategy, and the lesson many businesses must learn, is that next generation IT systems must be conceived to acquire the knowledge they operate on directly from who really has it, that is, the employees (seen as knowledge workers) and from the operation and communication processes employees enter into. The knowledge of a business has less to do with the IT infrastructure than with the employees themselves. The IT infrastructure must be capable of extracting and managing that knowledge in order to evolve and adapt it to the business processes, and finally to the business strategy. Any other means to model and exploit the business knowledge will never be flexible enough. If user knowledge changes (and it does change), both the IT infrastructure and the business strategy must seamlessly adapt to such changes.

Future work would concentrate on evolving practical Enterprise 2.0 frameworks, as open source packages of technologies, tools, and platforms that build on all the key technical and theoretical enablers described above and on the proposed model of collaboration and enterprise knowledge emergency. We expect this framework and its pragmatic application could have great research opportunities within the domain of the topic.

REFERENCES

Alonso, G., Casati, F., Kuno, H., & Machiraju, V. (2004). *Web services concepts, architectures and applications*. Springer.

Anderson, C. (2006). *The long tail: Why the future of business is selling less of more*. Hyperion Books.

Argyris, C. (1998, May-June). Empowerment: The emperor's new clothes. *Harvard Business Review, 76*, 98-105.

Barquin, R. C. (2006, July 18). *Social network analysis and business intelligence*. Retrieved January 8, 2006, from http://www.b-eye-network.com/view/3088

Bloom, H. (2000). *Global brain: The evolution of mass mind from the big bang to the 21st century* (pp. 42-44). Wiley.

Brown, J. S., & Duguid, P. (2000). *The social life of information*. Boston: Harvard Business School Press.

Choate, M. (2006, April 28). What makes an enterprise wiki? *cmswatch*. Retrieved at April 15, 2007, from http://www.cmswatch.com/Feature/145-Wikis-in-the-enterprise

Coveyduc, J., Huang, C., Ostdiek, L., & Reif, J. (2006, October). IBM innovation factory. An integrated solution for accelerating innovation. *IBM*. Retrieved July 3, 2008, from http://ibm.com/websphere/developer/zones/hipods

Davenport, T. H. (2005). *Thinking for a living: How to get better performance and results from knowledge workers*. Boston: Harvard Business School Press.

Davenport, T. H., & Harris, J. G. (2007). *Competing on analytics: The new science of winning*. Boston: Harvard Business School Press

Davenport, T. H., & Prusak, L. (1997). *Information ecology*. Boston: Harvard Business School Press.

Davenport, T. H., & Prusak, L. (2000). *Working knowledge: How organizations manage what they know*. Boston: Harvard Business School Press.

Dervin, B. (1983, May). *An overview of sense-making research: Concepts, methods, and results to date*. Paper presented at the annual meeting of the International Communication Association, Dallas.

Drive, E. et al. (2004, August 2). *Road map to an enterprise collaboration strategy*. Forrester Research.

Fallows, D. (2005, January). *Pew Internet & American Life Project. Search engine users*. Retrieved January 1, 2006 from http://www.pewinternet.org/PPF/r/146/report_ display.asp

Fielding, R. T. (2000). *Architectural styles and the design of network-based software architec-*

tures. Unpublished doctoral thesis, University of California, Irvine.

Giles, J. (2005, December). Internet encyclopaedias go head to head. *Nature, 438*(7070), 900-901.

Gruber, T. (2005). *Ontology of folksonomy*. Paper presented at the First on-Line Conference on Metadata and Semantics Research (MTSR'05).

Hammond, T., Hannay, T., & Lund, B. (2004). The role of RSS in science publishing. *D-Lib Magazine, 10*, 1082-9873.

Heigl, R., Glaser, M., & Anja, E. (2006). *Wiki: Web collaboration*. Springer. ISBN 3540229396.

Hinchcliffe, D. (2007). i-Technology viewpoint: Is Web 2.0 the global SOA? *SOA Web Services Journal*. Retrieved February 23, 2007 from http://webservices.sys-con.com/read/164532.htm

Hof, R. D. (2005). Mix, match, and mutate. *Business Week Magazine*. Retrieved February 23, 2007 from http://www.businessweek.com/@@76IH*ocQ34AvyQMA/magazine/content/05_30/b3944108_mz063.htm

Högg, R., Meckel, M., Stanoevska-Slabeva, K., & Martignoni R. (2006). Overview of business models for Web 2.0 communities. In *Proceedings of GeNeMe 2006* (pp. 23-37). Dresden.

Intel Suite Two. *The Enterprise 2.0 Suite, powered by Intel*. Retrieved July 3, 2008, from http://www.suitetwo.com.

Kline, D., & Burstein, D. (2005). *Blog!: How the newest media revolution is changing politics, business, and culture*. Squibnocket Partners, L.L.C. ISBN 1-59315-141-1.

Laso, I. (2006, October 18). *A Web 2.0 research perspective. DG information society and media*. European Commission Representative at the 6th Internet Next Generation Workshop 2006, Madrid, Spain.

Lin, N., Burt, R. S., & Cook, K. (Eds.). (2001). *Social capital: Theory and research*. New York: Aldine de Gruyter.

Malone, T. (2001). The future of e-business. *Sloan Management Review, 43*(1), 104.

Malone, T. W., Jay, J., Legay, L., & Kosorukoff, A. (2006). *Handbook of collective intelligence*. Retrieved at January 15, 2007, from http://www.eu.socialtext.net/mit-cci-hci/index.cgi?handbook_of_collective_intelligence

Mashing the Web. (2005). *The Economist, 376*(8444), 4.

McAfee, A. P. (2005). Will Web services really transform collaboration. *MIT Sloan Management Review, 46*(2).

McAfee, A. P. (2006). Enterprise 2.0: The dawn of emergent collaboration. *MIT Sloan Management Review, 47*(3), 21-28.

McAfee, A. P. (2006). *The impact of information technology (IT) on businesses and their leaders*. Retrieved January 1, 2006, from http://blog.hbs.edu/faculty/amcafee

McAfee, A. P. (2007). Enterprise 2.0 vs. SOA. *Harvard Business School faculty blog*. Retrieved June 8, 2007, from http://blog.hbs.edu/faculty/amcafee/index.php/faculty_amcafee_v3/enterprise_20_vs_soa/

Morris, M., Pohlmann, T., & Oliver, Y. G. (2005, April 8). *How do users feel about technology? Business technographics United States*. Forrester Research.

Musser, J., & O'Reilly, T. (2006, November). *Web 2.0 principles and best practices*. O'Reilly Radar.

Nohria, N., & Eccles, R. (1992). *Networks in organizations* (2nd ed.). Boston: Harvard Business Press.

Nonaka, I., & Takeuchi, H. (1995). *The knowledge-creating company: How Japanese companies create the dynamics of innovation.* Oxford University Press.

O'Reilly, D. (2004, 7 May). Web-User satisfaction on the upswing. *PC World.* Retrieved January 1, 2006, from http://www.pcworld.com/news/article/0,aid,116060,00.asp

O'Reilly, T. (2007) *What is Web 2.0.* Retrieved June 9, 2007, from http://www.oreillynet.com/pub/a/oreilly/tim/news/2005/09/30/what-is-web-20.html

Porter, M. (2001). Strategy and the Internet. *Harvard Business Review.*

Rand, M. (2004, December 13). Best of the Web: Extreme blogging. *Forbes.* Retrieved January 1, 2006, from http://www.forbes.com/best/2004/1213/bow001.html

Reding, V. (2006, December 3). *The disruptive force of Web 2.0: How the new generation will define the future* (European Commission press release). Hong Kong: Youth Forum, ITU Telecom World.

Roman, D. et al. (2005). Web service modeling ontology. *Applied Ontology, 1*(1), 77-106.

Salganik, M. J., Dodds, P. S., & Duncan, J. (2006, February). *Experimental study of inequality and unpredictability in an artificial cultural market.* Watts, Science 10, 311(5762), 854-856.

Schroth, C., & Christ, O. (2007). Brave new Web: Emerging design principles and technologies as enablers of a global SOA. In *Proceedings of the 2007 IEEE International Conference on Services Computing (SCC 2007),* Salt Lake City.

Schroth, C., & Janner, T. (2007). Web 2.0 and SOA: Converging concepts enabling the Internet of services. *IT Professional, 9*(3), 36-41.

Smith, D. M. (2006, May 16). *Web 2.0 offers many opportunities for growth, but few enterprises will immediately adopt all aspects necessary for significant business impact.* Paper presented at the Gartner Symposium/ITxpo, San Francisco.

Smith, R. (2006, June 14-16). *Enterprise mashups: An industry case study.* Paper presented at the New York PHP Conference & Expo 2006, The New Yorker Hotel, Manhattan.

Surowiecki, J. (2005). *The wisdom of crowds* (pp. xviii – xix, Chapters 2-4). Anchor Books.

Venners, B. (2003, October 20). Exploring with wiki: A conversation with Ward Cunningham, part I. *Artima developer.* Retrieved January 1, 2006, from http://www.artima.com/intv/wiki.html

Wacka, F. (2005). Enterprise blogging in practice. *webpronews.* Retrieved at January 1, 2007, from http://www.webpronews.com/topnews/2005/01/24/enterprise-blogging-in-practice

Weill, P., & Ross, J. W. (2004, May). *IT governance: How top performers manage IT decision rights for superior results.* Boston: Harvard Business School Press.

ENDNOTE

1 Note that we are not talking here about things like agile development or eXtreme programming, because the target audience is the knowledge workers not a development team.

Chapter V
Customer Relationship Management (CRM):
An In-Depth Analysis

Mahesh Raisinghani
TWU School of Management, USA

Abdu Albur
Ministry of Education in the Kingdom of Saudi Arabia, Dhahran, Eastern Province

Sue Leferink
Montana Department of Commerce, USA

Thomas Lyle
PNC, USA

Stephen Proctor
CSC, USA

ABSTRACT

This chapter discusses customer relationship management (CRM) as a customer-focused business strategy enhanced by technology that automates and enhances business processes to proactively manage profitable and long-term customer relationships. CRM solutions span a continuum of implementations from a narrow tactical implementation of a specific technical solution to a broad strategic implementation of a customer centric solution. Furthermore, the authors hope that understanding the underlying assumptions and theoretical constructs through the use of CRM will not only inform researchers of a better CRM design for studying e-commerce and Internet marketing, but also assist in the understanding of intricate relationships between different factors.

INTRODUCTION

In the transition from the raw industrial economy of the last century to the pure knowledge-based economy of the Internet generation today, businesses have seen many evolutionary changes in how they conduct business and market products and services. It is no longer sufficient to expect that good products will sell themselves or that creative advertising and selling campaigns will overcome consumer resistance and convince them to buy from them rather than a competitor. Instead, we are in an era where the consumer is king, where providing service before, during, and after making a purchase can determine a business's level of success.

Investment in technology over the last two decades has created legacy systems defined by organizational boundaries containing silos of business information within functional areas. The consumer information collected by each functional area of the corporate value chain (such as production, distribution, sales, and marketing departments) remains buried in these silos. This prevents it from being shared or leveraged across the enterprise, which could improve the decision-making processes of all departments (Chan, 2005).

Back-end systems managing suppliers, production, inventory, and order fulfillment that remain disconnected from front end systems managing sales, marketing, and order processing can cause functional and process disparities that can adversely affect the consumer experience. For example, sales strategies that do not leverage marketing intelligence and marketing campaigns that conversely do not leverage previous sales data often result in consumers receiving many promotional offers without regard to their previous purchases (Chan, 2005).

With these inward facing information systems, the focus has primarily been product- and process-oriented. Now with the evolutionary changes brought on by the Internet, a shift has been made towards outward facing information systems where the focus has become consumer centric. "These systems promote seamless interactions between businesses and their customers to build strong customer relationships" (Shah & Murtaza, 2005). This represents a paradigm shift from the traditional concept of marketing as a simple exchange between buyer and seller, to the concept of relationship marketing that promotes the development, growth, and maintenance of long-term cost-effective relationships with individual consumers, suppliers, employees, and other partners for the mutual benefit of all (Chan, 2005).

An enterprise model which provides the framework for an information and process sharing architecture can integrate the relationship marketing concept and create a unified view amongst disparate systems throughout the entire enterprise value chain. This concept has led to the creation of customer relationship management strategies and technologies.

Customer Relationship Management (CRM)

CRM is a customer-focused business strategy enhanced by technology that automates and enhances business processes to proactively manage profitable and long-term customer relationships (Seeman & O'Hara, 2006). CRM solutions span a continuum of implementations from a narrow tactical implementation of a specific technical solution, to a broad strategic implementation of a customer centric solution (Payne & Frow, 2005).

The architecture of these solutions encompasses three CRM functional areas. The first area, referred to as the "operational" area, covers the front end business processes managing sales, marketing, order entry, and customer service. The operational area's function is focused on the collection of consumer data from any and all of the customer touch points or contact points (Alexandrou, 2007). Constructing a customer database is a necessary first step, thus the foundation for any complete CRM solution. With the advent of the Internet, constructing a customer database is a relatively easy task, as customer transactions are gathered as a natural part of the interaction with customers (Winer, 2001). The database should contain the following information: registration details, prior transaction histories, customer contacts, and responses collected to marketing stimuli.

The second functional area is the "analytical" area. The analytical area contains the back-end business processes as well as the analysis of consumer data collected by the operational area. The function of the analytical area is focused on identifying market segmentation and expanding it through cross-selling and up-selling opportunities (Alexandrou, 2007). More recently the focus has shifted from general segmentation to targeting each consumer in the database; that is, targeting each customer individually. Depending on the nature of the service or the product, the company can address its customers' needs individually or by market segment.

The third functional area, referred to as the "collaborative" area, facilitates, coordinates, and supports customer interactions throughout the enterprise. The function of the collaborative area focuses on bringing people, processes, and data together so that businesses can better serve and retain their customers (Alexandrou, 2007). Regardless of the implementation employed, the application of the CRM architecture results in a single version of the truth about customers created by a seamless internetworking of people, processes, and technology, all working together to achieve a business objective. The CRM objective provides an ongoing one-on-one relationship with consumers by integrating the use of organizational knowledge and knowledge management technologies to facilitate the decision-making processes regarding product offerings, marketing strategies, order processing and fulfillment, customer service and support, and customer retention (Cunningham, Song, & Chen, 2006).

BUSINESS CHALLENGES

Many organizations attempt to solve universal business problems by purchasing and deploying CRM applications. The business problems listed in Box 1 represent some of the pain points organizations want to reduce or eliminate. These problems affect an organization's people, processes, and data, spanning the entire value chain from front office systems to back office systems. Box 1 illustrates business problems; a detailed analysis of each business problem follows.

Data Fragmentation

Typically, organizations gather information from many different sources, whether it is directly from customer touch points like physical store locations, telephone, e-mail, or Web sites, or indirectly from the analysis of data collect from sales and marketing campaigns, order processing and fulfillment reports, or other useful areas of information (Seeman & O'Hara, 2006). Data captured from these different sources can end up

Box 1. Business problems

- Data Fragmentation
- Data Quality
- Data Redundancy
- Customer Service
- Customer Acquisition and Retention
- Operational Costs
- E-Commerce Capability
- Increased Business Velocity
- Regulatory Compliance
- Change Management
- Customer Dissatisfaction

in different databases spread across departments, or worse, fragmented across internal functional data silos serviced by a variety of legacy systems. Access to customer information collected from all areas of an organization can provide a complete 360-view of the consumer that can help an organization's employees service customers more efficiently (Raisinghani, Tan, Untama, Weiershaus et al, 2005). Fragmented consumer data prevents organizations from having a clear and accurate picture of their customers. Gartner estimates that more than 75% of organizations looking at CRM solutions cannot create a comprehensive view of their customers due to fragmented data sources (Reid & Catterall, 2005).

Data Quality

The quality of an organization's data indicates how well they can accurately report findings about real world customers at any given time and moment. "Poor data quality can have a severe impact on the overall effectiveness of an organization" (Reid & Catterall, 2005). Recent studies suggest that losses of 10 to 25% of an organization's revenues can be attributed to poor data quality amounting to over $600 billion per year in the U.S. alone, where less than one third of organizations feel confident in the quality of data they collect (Reid & Catterall, 2005). Sources of poor data quality range from lack of industry data coding standards to poor data entry practices.

Data Redundancy

An inefficient nonoptimized value chain is challenging for most organizations especially when an organization's systems are not integrated with other enterprise systems in the value chain. This lack of integration between organizations leads to data redundancy problems. In this case the consumer information ends up being duplicated across vertical data silos. The data redundancy problem compounds the data fragmentation and quality problems. It becomes problematic to link together fragmented data when there are no standards between silos. It is also difficult to determine which version of duplicated data is the authoritative version. Integrating consumer data sets is extremely challenging when attempting to gain a single view of each customer's value and need (Reid & Catterall, 2005). With multiple versions of the truth, it is difficult to be sure that the "true" customer is being reflected in the database.

Customer Service Issues

In today's global marketplace it is essential that organizations maintain good customer relationships if they wish to remain profitable. Customer service processes must be able to effectively interact with all customer touch points at every opportunity whether they are initiated at a physical store location, by telephone, via e-mail, or from a Web site. The data collected from these touch points can be a problem if they are inaccessible to employees who could use and benefit from them during future customer interactions (Shah & Murtaza, 2005). Customer satisfaction diminishes quickly when customers must repeatedly provide the same information every time they interact with the organization, due to lack of stored real time customer data at all touch points.

Customer Acquisition and Retention

In marketing circles it is an established belief that it is more expensive to acquire new customers than it is to retain existing customers using cross-selling and up-selling techniques. It is estimated that repeat customers can generate more than twice the income that new customers can generate (Chan, 2005). The essence of a good customer strategy is to identify profitable customers. Not all existing customers are equally profitable. It is estimated that less than 20% of an organization's customers are responsible for producing over 80% of its profits (Roberts, Liu,

& Hazard, 2005). Businesses need to develop and manage their customer relationships by serving the consumer's unique and individual needs in a way that makes these relationships long term and profitable (Chan, 2005). Simply identifying how often and how much customers are purchasing is not a sufficient analysis to grow a business' sales volume. If organizations cannot identify the most successful methods for obtaining new customers and cannot identify what makes current customers loyal, it is much more difficult to maintain a solid customer base.

Operational Costs

Organizations with inward facing information where the focus has been primarily product and process oriented, not only have a data redundancy problem but also have process and people redundancy problems. These activities directly impact operational costs. The front end business processes designed to collect and store information in conjunction with the back-end business processes designed to massage this information are duplicated wherever data silos exist. As the volume of data stored in databases increases, the cost to store and manage this information becomes a huge challenge (Reid & Catterall, 2005). With each disparate data silo, more effort is required to import and export data between systems. In addition, valuable labor resources are spent entering the same data into multiple systems.

E-Commerce Capability

With the evolution of the information age and the introduction of the Internet, organizations have had to scramble to keep up with emerging technologies. New e-commerce technologies like e-mail, Web browsing, and online shopping have dramatically changed how business is conducted today. In particular, the business-to-consumer (B2C) business model has fundamentally changed. Today there are new types of products to sell like downloadable software, music, and video and new ways to pay for these purchases like e-payments and PayPal. There are also new ways to sell products like online auctions as well as new ways to create and execute sales and marketing campaigns online. Businesses need to be able to adjust and adapt to advances in e-commerce technologies in order to remain competitive and stay ahead of the competition. Established "brick and mortar" businesses must engage in e-commerce or risk being overtaken by online competitors. Engaging in such activities can result in additional silos of information across an organization, further dampening collaboration efforts. Business processes may require alterations for adaptation to the Internet environment.

Increased Business Velocity

With the evolution of the Internet, businesses have been able to move their sales and marketing operations online to vastly expand their product and service offerings in the global marketplace. As capabilities evolve, customers expect faster service and product delivery. In the past, financial transactions routinely took up to 5 days to be completed. These transactions now occur in seconds over the Internet and there are more alternatives for paying via the Internet. It no longer takes days to book airline tickets and hotel reservations through a travel agency. Instead, travel arrangements can be made online in real time by the consumer. With e-commerce, more demanding consumer behaviors have emerged where consumers now have the information at their fingertips to help them make more informed purchasing decisions. They know exactly what they want, how they want it, and when they want it. This change in consumer behavior has changed how businesses compete today. Most organizations today are being squeezed with time pressures.

Regulatory Compliance

Government regulations passed as part of the Sarbanes-Oxley, HIPPA, Gramm-Leach-Bliley, and CAN-SPAM legislation are presenting businesses with new compliance-related problems including, but not limited to, executive accountability, business transparency, industry oversight, privacy of health records, consumer rights of privacy, spam, e-mail opt-out notification, and real-time reporting and accounting requirements. All these new regulations can hinder or even incapacitate an organization. Attempting to understand the legal requirements is difficult. Integrating these legal requirements into current business processes can be overwhelming.

Change Management

Today's increased business velocity has been supported by major advances in technology developed since the origin of the commercial Internet. Advances in hardware and software have been evolutionary in nature and will continue to develop over time. Legacy hardware and software systems of yesterday with their vertical data silos and monolithic applications are slowly being upgraded or replaced with integrated networked and distributed systems. This cycle of change presents problems that not only affect a business' technology but also its processes and employees. Changes in one system can have a ripple effect throughout an organization. Worse yet, by the time a new technology can be integrated, it may become obsolete.

Customer Dissatisfaction

Customer satisfaction rates are decreasing, while complaints, boycotts, and other expressions of consumer discontent rise (Biba, 2005). As customer expectations increase, more demands are placed on the organization's customer service representatives. This mounting wave of pressure affects employee morale and performance.

BUSINESS OPPORTUNITIES

The business opportunities listed in Box 2 represent the objectives organizations are attempting to achieve. Box 2 illustrates business benefits; a detailed analysis of each business benefit follows.

Cost Reduction

Many of the problems solved by CRM implementations can produce cost reductions. CRM can help address data fragmentation problems by focusing on the acquisition, development, and retention of customer relationships through the collection and sharing of consumer information across an entire organization (Raisinghani et al., 2005). CRM solutions that incorporate the creation of centralized data warehouses eliminate excess operational costs caused by redundant staff positions, processes, and data. Replacing disparate vertical silos with centralized integrated systems streamlines processes and technologies and can achieve a positive ROI very quickly. The creation of a centralized data warehouse that enables a 360-degree view of each customer can drastically reduce customer service costs and

Box 2. Business benefits

- Cost Reduction
- Improved Customer Service
- Customer Loyalty
- Customized Products and Services
- Improved Efficiency
- Better Integration of Systems
- Speed of Processes
- Customer Management

provide the ability to create unique sales and marketing campaigns designed to cross-sell and up-sell to profitable customers as well as convert unprofitable customers into loyal, profitable, and retainable customers (Eisenfeld & Hagemeyer, 2004). In addition, CRM solutions can help create the processing framework necessary to meet regulatory requirements, thus reducing compliance costs.

Improved Customer Service

A quality CRM system offers comprehensive summary reports on critical business information essential for making informed decisions. Home sales agents, for example, can utilize CRM to help them provide better service to their customers with less effort. CRM makes it easier for them to increase profits and acquire repeat business by managing and analyzing homebuyer databases, launching targeted e-mail campaigns, and administering purchasing agreements (Clements, 2006). Offering consumers multiple interaction touch points scores convenience points especially when the information entered from one location is quickly retrievable from all other locations. CRM solutions are designed to provide these types of customer-centric benefits.

Customer Loyalty

CRM solutions can have several positive effects on customer satisfaction and loyalty. First, CRM applications enable organizations to customize their offerings for each customer. Accumulating information from previous customer interactions and analyzing this information to discover hidden patterns helps organizations personalize their offerings to suit the individual tastes of their customers. Second, in addition to enhancing the perceived quality of the offering, CRM applications enable organizations to improve the reliability of consumer experiences by facilitating the timely and accurate processing of customer orders, requests for information, and the ongoing management of customer accounts. Third, CRM applications can help organizations manage customer relationships more effectively across the stages of relationship initiation, maintenance, and termination. Effective management of the customer relationship is the key to managing customer satisfaction and customer loyalty (Mithas, Krishnan, & Fornell, 2005). Customer-centric CRM solutions also build loyalty by creating customer appreciation processes and programs.

Customized Products and Services

The ability to target profitable customers in the pursuit of a customer relationship strategy is a primary benefit realized in the initial stages of a CRM implementation. Specifically, an organization's value equity can be improved by targeting customers that are more likely to find their products and services attractive. This improves customers' perceptions of the price, quality, and convenience of the products and services offered by the selling organization leading to increased customer equity (CE).

With increased CE, consumer interaction with CRM technology can positively impact the creation of customized products and services through the soliciting of specific customer desires providing the organization with a competitive advantage. Brand equity can be positively impacted as well by CRM technology as current customers' attitudes toward the brand are enhanced through their experience with these customized products and services. Relationship equity can also be enhanced as consumers become aware of the special recognition and treatment that results from their input in customizing products and services. This special treatment serves as "glue" to cement the relationship. Thus a carefully crafted customer relationship strategy implemented via CRM processes benefits an organization by facilitating the creation of popular customized products and services.

Improved Efficiency

CRM technology allows organizations to gain a better understanding of customers' implicit and explicit needs and wants. With improved information, prices can be efficiently set to cover costs, deliver value, and extend profits. As consumers perceive a greater value in pricing, their perceptions of what they are getting for the money, also known as consumer value equity, will improve.

The combination of improved customer perceptions of convenience and greater customer retention by using CRM can boost an organization's value equity. In addition to increasing value equity, providing customers with multiple points of access to products and information increases exposure to the seller's brand, improving the ability to up-sell and cross-sell and increasing brand equity. The marketing and sales force operations also benefit from improved efficiency and effectiveness as a result of increased values and brand equity.

Utilizing CRM strategies can help reduce redundant data by integrating different data stores into a single repository where a complete customer knowledgebase is always available (Shah & Murtaza, 2005). Since data only has to be entered one time, employees are free to work on other business objectives. This benefit also creates one version of the truth, so less time is spent trying to reconcile conflicting data. Data quality is a sign of a good CRM strategy and can go a long way to improving an organization's effectiveness and competitive advantage (Reid & Catterall, 2005).

Better Integration of Systems

E-commerce CRM solutions can help organizations reengineer their infrastructures to take advantage of the latest online technologies like Web services as well as other future advances. Well-architected CRM solutions integrate easily with existing systems and can support future expansion and changes in business requirements.

Speed of Processes

CRM's ability to tie front end processes to an organization's back-end processes can dramatically increase the speed of business processes demanded by e-commerce environments. With increased processing speeds, organizations can increase the number of customer transactions handled at any given moment. Transactions consist of consumer purchases as well as customer service requests transmitted over the Internet. Customer service requests can be made faster, better, and cheaper by employing various Internet tools integrated into CRM solutions such as e-mail, Web chat, and order tracking. E-mail automated responses provide a highly personalized interaction with customers who are always expecting instant feedback (Singh, 2002). Web chat provides customers an instant messaging service with real time personalized access to customer service agents, and order tracking allows customers to see the status of their orders without the need to contact customer service (Hamid, 2005).

Customer Management

A CRM solution can help analyze and create differentiated customer strategies that not only target customers with high customer value but also target segments of unprofitable customers in order to develop ways to make them loyal, satisfied, and profitable customers, ultimately increasing the organization's overall customer retention (Roberts et al., 2005).

Information sharing that extends beyond an organization's boundaries to both its suppliers and customers can be a key factor for achieving a sustainable competitive advantage (Shah & Murtaza, 2005). The implementation of a CRM solution can extend these boundaries and enable the traditional customer's physical proximity to be substituted by digital proximity (Kennedy, 2006). By helping to create a single, centralized repository for all customer information collected both offline and online, a CRM solution can provide that 360-degree view of its customers desired by today's customer service departments.

BANK ONE CASE STUDY

Bank One is the sixth largest bank in the United States (Kolsky, 2002). It is based in Chicago, with total assets of more than $260 billion (Kolsky, 2002). Bank One services a broad customer base, offering commercial and private banking products along with lending and credit card services. In 2000, Bank One wanted to provide Internet access to its customers' data using workflow capabilities that could span its multiple customer channels to provide real-time data modeling, decision making, and personalization capabilities in order to become more customer-centric (Kolsky, 2002).

Business Problems

Bank One services a large number of customers spread across many different market segments which in turn are traditionally serviced by different divisions within the bank. Each division maintained its own data. This information did not flow between departments and divisions. Therefore, Bank One was not able to create a 360-degree view of its customers across products and market segments, especially critical for identifying and catering to its best customers. Bank One wanted to implement an initiative to give personalized Internet access to its commercial customers so that they could perform simple account management by entering and retrieving data (Kolsky, 2002). Additionally, Bank One wanted to create better marketing profiles for its commercial customers to improve their marketing programs (Kolsky, 2002). Unfortunately, Bank One's legacy systems were not integrated and were unable to provide a consolidated profile of its commercial customers.

Business Benefits

Bank One was able to provide additional interactive channels over the Web. This Web site was tied to their enterprise legacy systems thus providing a consistent, positive experience and a single image of Bank One to its customers (Kolsky, 2002). This enabled Bank One to create a single version of the truth about customers. Creating a 360-degree view of its customers provided a mechanism to systematically gather and analyze its customer information. Bank One was able to reduce service costs by leveraging the Internet in place of more expensive private networking channels between offices (Kolsky, 2002). Increased customer satisfaction through personalization has resulted in more profitable customer relationships across Bank One's entire value chain. The deployed CRM solution also provides the capability to easily add new customized CRM modules. This agility is designed to improve other departmental processes and connect to additional legacy systems within Bank One as the business strategy evolves.

In light of today's intense competition to attract and retain customers, it has become necessary for organizations to adopt strategies that employ an enterprise approach to understanding and influencing customer behavior (Payne & Frow, 2005). CRM enterprise solutions can provide an information and process sharing architecture that creates a single version of the truth about customers that is current, consistent, and complete across disparate systems along the enterprise value chain (Shah & Murtaza, 2005).

APPLICATION DEFINITIONS

The Internet has changed the way enterprises do business as well as the way they deal with customers. With the evolution of the new technologies and the Web-based customer mentality, meeting customers' requirements is far more complex than simple advertisement. CRM has become a necessity for many companies to gain customers' confidence and loyalty. The speed of processing and size of computer storage, developments in client-server technology, databases and data warehouses, communications technology, and the rapid spread of computer literacy, Internet connectivity, and e-commerce into the consumer field have contributed to the widespread use of CRM in the enterprise culture and increased the reach of CRM applications. Such developments make it possible to gather vast amounts of customer data and to analyze, interpret, and utilize it constructively. A 2005 survey by the United States Society for Information Management showed that CRM was ranked 11th by CIOs and 7th by other IT executives regarding its importance (Hart, 2006). It is estimated that approximately 40% of businesses have made investments in CRM communications technology (Jukić, Jukić, Meamber, & Nezlek, 2002). According to Gartner (as cited by Payne and Frow [2006]) the worldwide global market for CRM systems and consultancy was estimated to grow to $US 47 billion by 2006.

There is a wide range of CRM application software, packaged differently by different vendors, with varying degrees of integration. According to Pan and Lee (as cited by Hart [2006]) "a high level CRM application is classified into information integration, customer analysis, campaign management, real time

decision-making and personalized messaging." CRM has become a central business strategy for many companies. This strategy must take in consideration all phases of the customer buying process, from prepurchase to purchase to post-purchase and after-sales service, and the types of interactions required at each stage (Jukić et al., 2002). However, this does not mean that such benefits will automatically be achieved by purchasing CRM software solutions, as illustrated by many examples of CRM failures. If the benefits of CRM are to be realized, a more integrated approach is needed for both the formulation and implementation of CRM strategy (Payne & Frow, 2006).

Application Space

There are two types of CRM solutions: on-premise and on demand. In the on-premise solution everything from hardware and IT support is managed in house. On the other hand, in the on-demand solution, Web-based systems provide an instant external resource, readily accessible by all who need it, at a per-user flat rate. And even though some analysts are skeptical about an on-demand solution, it is quickly evolving into a more strategic, more capable CRM technology option. There are three types of on-demand solutions: shared on-demand, private on-demand, and on-premise and on-demand (Carlson Marketing, 2007).

Application Definition

CRM is the art of dealing with customers. Although two artists may not share the same prospective about a painting, hopefully, they will agree about the beauty of the painting. Defining CRM is similar to defining a painting. There are many definitions, even by contemporary specialists in CRM. In its basic definition, CRM entails every aspect of interaction a company has with its customers, whether it is sales or service related. According to Thompson, Dunne, Radcliffe, Herschel, Prentice, Mertz et al. (2007), CRM is a business strategy with outcomes that optimize profitability, revenue, and customer satisfaction by organizing around customer segments, fostering customer-satisfying behaviors, and implementing customer-centric processes. Zablah (as cited by Hart [2006]) defines CRM as an ongoing process that involves the development and leveraging of market intelligence for the purpose of building and maintaining a profit-maximizing portfolio of customer relationships. In simple terms, CRM involves the intelligent deployment of database technology for the management of all customer interactions with the company (Jukić et al., 2002). Payne and Frow (2004) define five important cross-functional processes related to CRM:

strategy development, value creation, multichannel integration, information management, and performance assessment.

Recently there has been some increased clarity in the definition and the field of CRM has now begun to converge into a common definition: CRM relates to strategy, managing the dual-creation or value, the intelligent use of data and technology, the acquisition to customer knowledge and the diffusion of this knowledge to the appropriate stakeholders, the development of appropriate (long-term) relationships with specific customers and/or customer groups, and the integration of processes across the main areas of the firm and across the network of firms that collaborate to generate customer value (Payne & Frow, 2006).

Application History

CRM was first coined as a term in 1983 by the academic community and adopted by strategy and management consultants in the early 1990s and by applications technology vendors by the mid-1990s. In 2001, after a peak in technology spending, CRM gained an ugly reputation for large-scale, technology-driven projects that failed to meet expectations. This resulted in a CRM *winter* from 2001 to 2003, when wholesale cutbacks were made in CRM initiatives. Companies that continued to invest in CRM technology justified these expenditures primarily through cost savings. With CEOs' renewed focus on organic revenue growth at the start of 2004, CRM consultancies began growing in 2004 and CRM application technology spending also began to grow (Thompson et al., 2006).

CRM has emerged in recent years as the convergence of a number of factors that contributed to its evolution. Berry (1983) coined the term "relationship marketing," which encouraged a new movement towards customer relationships rather than customer transactions. Peppers and Rogers (1993) promoted the concept of one-to-one marketing and of mass customization. Reichheld (1996) further motivated companies with his research on loyalty and empirical evidence, demonstrating of the profitability of customer retention. Customer lifetime value (CLV) has become a key element of CRM (Hart, 2006).

The concepts of database marketing (DBM) and sales force automation (SFA) were introduced in 1985. These concepts have laid the foundation for marketing on the basis of knowledge of the needs and behavior of the individual customer. These advanced systems require real time data integration which was initially achieved using proprietary application programming interfaces (APIs). From a technical viewpoint, CRM systems can be considered an advancement of the earlier sales force automation (SFA) systems. SFA systems primarily support the sales process of companies whereas CRM systems integrate support for marketing, sales, and service processes (Messner, 2005).

Despite the steady growth in number of worldwide installations and sales, not all is perfect in the world of CRM applications. Industry studies suggest that approximately 60% of CRM software installations fail (Jukić et al., 2002).

Application Features and Functionality

At the end of the 20th century, capital and work have lost their importance as the driving force for generating value, and enterprises became an integrator that try to offer as much information, and as many products and services as possible, based on deep knowledge about the customer. This is considered the driving force for the move from transaction-based marketing to relationship marketing. Furthermore, the one-to-one marketing approach, in which the company approaches every customer at the perfect time with a customized offering, provides a further individualization of products, which are customized and bundled to match the needs of the individual customer (Jukić et al., 2002; Messner, 2005). Buyers of CRM products and services face a nontrivial challenge when evaluating and selecting the appropriate technologies and service providers to maximize project success. The importance of picking and managing a service provider is becoming greater; given the higher proportional spending, understanding how to differentiate and choose the right product or service provider requires a pragmatic, balanced approach incorporating short- and long-term enterprise objectives. These are some practices that insure a successful implementation of a CRM solution (Thompson et al., 2006):

1. CRM strategy and implementation best practices
2. Building a business case and measuring the success of CRM
3. The process of selecting a CRM applications vendor
4. The evolution of CRM technology architectures
5. Improving the customer experience
6. Creating a single view of the customer
7. Creating improved customer insight through analytics
8. Implementing CRM in a resource-constrained organization
9. The wants and needs of buyers of CRM applications
10. The size of the market and demand for CRM applications

Application Design/How it Works

CRM must take into consideration all phases of the customer buying process, from prepurchase to purchase to postpurchase and after-sales service, as well as the types of interactions required at each stage. The challenge is to communicate with customers at the right time, in the correct manner, and on the correct topic in order to manage single-customer, multiple product relationships (Jukić et al., 2002).

The main scope of CRM is marketing, sales, and customer support. But other factors such as system management capability, interface support management, and knowledge management have important roles. In order to maintain the smooth management of business operations, adoption of data mining and data warehouse techniques are needed to construct business intelligence (BI). At the same time, a continuous adoption of an enterprise application integration (EAI) system connects the back-office applications (i.e., ERP) and front-office applications (i.e., CRM). By those actions, the enterprise can develop an API, thus connecting call centers to provide smooth after-selling services (Chang, Hsiung, & Tsai, 2006).

CRM application providers frequently force businesses to adapt their business processes to fit applications rather than adapting the applications to fit their business processes. Standardization of the technology makes it challenging for businesses to customize their systems to meet their needs. When their needs change, either in terms of data management or with any change to their business processes, maintaining customized applications can be costly. Businesses with complex information systems or businesses with customers having complicated information needs may not be able to alter their business processes to fit the applications. In addition, the adaptations may require a significant investment of corporate resources. In fact, successful CRM applications must exhibit relatively volatile data content and be much more readily manageable through the use of well-crafted CRM techniques (Jukić et al., 2002)

COMPARATIVE VENDOR/ PRODUCT MATRIX

In the 1990s, relationship marketing became the chosen method for marketing products (Janjicek, n.d.). This tactic was based on forming a relationship with a customer in hopes of building loyalty.

After all, studies have shown that it takes six times less money to keep a customer happy than to acquire a new one (Janjicek, n.d.). With the evolutionary changes brought on by the Internet, many businesses have lost sight of the real meaning of CRM. Engaging customers through an online environment is more impersonal then direct face-to-face contact. CRM is about refocusing on customers at a personal level. This is accomplished by utilizing automated customer management modules to acquire and retain customers, improve customer loyalty, gain customer insight, and implement customer-focused strategies (SAP, 2007). CRM tools have great potential to once again build intimacy with customers regardless of the size of the business.

CRM Business Requirements

CRM tools allow businesses (regardless of size) to understand and cater to the needs of their customers in an automated fashion. A typical CRM tool provides marketing, sales, service, and analytic functions. These primary functions provide the following benefits (Janjicek, n.d.):

- Increased customer loyalty
- Improved effective marketing campaigns at a lower cost
- Customized products and pricing to meet specific needs for each customer segment
- A single "true picture" of the customer throughout the company
- Cross-selling opportunities

The key to implementing a successful CRM solution is to understand the process and then apply it to a specific company. In general, CRM is valuable for identifying customer's needs, differentiating between the customer segments, customizing product lines to meet these needs, and then interacting with customers to convince them to purchase these products. Next the company solicits feedback on past purchases and uses this feedback to begin the cycle again.

CRM VENDOR SELECTION PROCESS

When selecting a CRM vendor, it is vital to find the right set of tools to provide a seamless interaction between the business and its customers. These tools also promote the development, growth, and maintenance of long-term cost-effective relationships between consumers, employees, and other business partners for the mutual benefit of all (Chan, 2005).

With the wide selection of CRM solutions available today, it is important to select a product that meets the specific business needs of the company. Every business that requires a CRM solution will need to ensure it encompasses the three main functional areas of CRM. First, this includes the "operational" area that deals with the front end business processes which manage the collection of data from the sales and marketing efforts as well as the order entry and customer service interactions (Alexandrou, 2007). Second, this includes the "analytical" area that deals with the back-end business processes which manage the analysis of consumer data collected by the "operational" area for purposes of identifying market segmentation for cross-selling and upselling opportunities (Alexandrou, 2007). Third, this includes the "collaborative" area that focuses on bringing people, processes, and data together by facilitating customer interactions throughout the entire enterprise (Alexandrou, 2007). The CRM implementation selected should provide the enterprise a seamless internetworking of people, processes, and technology, all working together to create a single version of the truth about its customers.

Criteria Categories Recommended for Evaluation

Table 1 lists the CRM criteria categories that were rated during the vendor evaluation process. An itemized list of the criteria used in each category is included for references purposes.

Vendor and product selections were determined based on ratings allocated by the Gartner Magic Quadrants listed:

- Customer Service Contact Center
- Sales Force Automation
- Marketing Resource Management
- Multichannel Campaign Management
- Field Service Management
- E-Commerce
- Enterprise Marketing Management

An attempt was made to select vendors that fit into all four magic quadrants so that niche

Table 1. CRM criteria categories recommended for evaluation

Category	*Details*
Vendor Information	Contact Information Financial Stability Vision Gartner Magic Quadrant Rating Value Chain/Strategic Partners
Functionality	Automation, ease of use, personalization, and language options. Integrated communications, data integration, and ERP integration. Business intelligence, search capabilities, and analytics.
Marketing Features	Campaign, lead, marketing resource, segment and list management.
Sales Features	Sales force, quotation, order, contact, and territory management. Transactional e-commerce and forecasting support.
Partner Channel Management Features	Partner relationship management.
Service Features	Automated service, service contract, complaints, and returns management. Web self service.
General Features	Privacy and knowledge management.
Technology	Service Oriented Architecture (SOA), Web services, open standards, maintenance, and compatible platforms. Scalability, security, and reliability of service.
Costs	Deployment, training, and maintenance costs.
Other Items of Consideration	Technical support, training, and consulting services.

players, challengers, as well as visionaries would be covered by the analysis. As a result, the vendors chosen for this evaluation model included: Microsoft (Dynamics), RightNow Technologies (RightNow 8), Salesforce.com (Salesforce CRM), SAP (mySAP CRM), and Oracle (Seibel CRM). Each rated criteria was assigned a priority based on a scale of 1 to 5 (1 being low priority and 5 being high). Each product was then evaluated, based on the objective criteria selected, and scored on a scale of 1 to 5 (1 being unsatisfactory and 5 being excellent). The scores were then multiplied by the assigned priority to attain the weighted scores and totaled to provide a comparative summary for each product evaluated.

Each CRM product has strengths and weaknesses associated with it. The key in evaluating products is to match the product strengths with the company's core business needs. This can be achieved by altering the priorities assigned in the corresponding comparative vendor/product matrix, titled ECOMGroup1ARPVendorMatrix.xls, to more heavily weigh those criteria that are most important. For the purposes of this overview, priorities were based on a group consensus determined by general business needs.

CRM solutions are in the innovation stage of the product lifecycle (Desisto, 2006). As the competition heats up, CRM vendors are beginning to merge (Ragsdale, 2007). Vendors are providing two types of solutions, that is, on-demand and on-premise. On-demand solutions are vendor hosted applications where the vendor maintains all the hardware and software. On-premises solutions are a more traditional solution where the business purchases hardware, software, and deploys the solution. On-demand solutions are becoming a more attractive option for businesses (Ragsdale, 2007).

Analysis and Synthesis

In evaluating the selected vendors, it was clear that some products had advantages over others in certain areas. In comparing products, Microsoft Dynamics primarily provides various editions of the software to target the on-premise market, while RightNow and Salesforce use on-demand modules accessible through the Internet to target their markets (Fitzgerald, 2007). Oracle and SAP provided both on-demand and on-premise robust solutions.

Looking at the vendors evaluated, each has built brand recognition for their products in the CRM field. Oracle is clearly the leader in market penetration with over 30% on average across all industries, while Microsoft Dynamics is just entering the CRM market (Thompson & Goldman, 2006). The big players, Microsoft, SAP, and Oracle, have been in business for at least 30 years. They show financial stability, have many strong allies, and a large impressive customer base. Both RightNow and Salesforce have been operating less than 10 years, but while their sales growth is very strong, their net income growth is negative. This is typical for companies just starting out. While it is not cause for great concern, there is some risk that these newer companies could fail or be assimilated. All vendors with the exception of Salesforce place great importance on research and development. Overall, all the vendors demonstrate that they are currently reliable and focused in their strategy.

In terms of functionality, all products provide some type of automated workflow engine that allows businesses to integrate custom business rules that can be dynamically changed. They are all easy to use and incorporate intuitive interfaces. All products offer some sort of integrated communications tool to meet the needs of all customers. Each one also has strong analysis and reporting tools.

Seibel (now merged with Oracle) has a more mature feature rich product since it has been involved with CRM since its inception (Fitzgerald, 2007). SAP is similar in its offering and provides robust integration features with other third party products. Microsoft's strong suit is its "common look and feel" that is familiar to Microsoft office users (Fitzgerald, 2007). This allows end users to adapt quickly to their software with little or no training. RightNow's strongest feature is the self-learning automated knowledge management tool. It is exceptional for understanding and catering to specific customers (Fleischer, 2006). Salesforce concentrates primarily on sales force automation to provide staff with the automated tools needed to sell products (Fitzgerald, 2007).

On the flip side, Microsoft is just entering the CRM market and its products are not yet feature rich. They are still lacking some tools in their arsenal, such as a sales management commission tool (Fitzgerald, 2007). Oracle has some complications, such as being able to provide real-time information (Fitzgerald, 2007). SAP's complexity and lengthy implementation schedules remain one of its drawbacks (Band, 2007). RightNow requires more IT staff to maintain the application that small to medium businesses can afford (Fitzgerald, 2007). Salesforce is weak in marketing and service automation (Fitzgerald, 2007).

Each product is based on open standards and provides integration tools. Most products require a proprietary utility to interface with third party applications. They all have a good security and privacy component. Each product offers scalability and flexibility. In terms of technology, they are all sufficient. Comparing costs can be the most difficult part because each company's pricing structure is different. When purchasing Dynamics from Microsoft, the software is then owned by the business, so even though this cost per user is greater, the business has a tangible asset. In contrast, RightNow and Salesforce use a subscription-based pricing so while there would be no hardware to purchase and maintain, there is also no tangible asset received. Siebel and SAP have mix strong robust products allowing them to charge more for their solutions.

Recommendations

The 47 rated criterions resulted in: Microsoft with a total score of 897; RightNow with a total score of 891; Salesforce.com with a total score of 892; SAP with a total score of 991; and Seibel with a total score of 1005. Oracle's Seibel CRM and SAP's mySAP CRM products are the clear leaders across industry segments. They offer both broad functional capabilities and industry specialization with the scalability to support global organizations. They are unequivocally the industry leaders. At the same time, however, their leadership is being challenged by niche players like RightNow and Salesforce.com who offer software as a service (SaaS) deployment and low upfront costs at the expense of some limited functionality (Ragsdale, 2007). Meanwhile, challenger Microsoft Dynamics provides businesses the opportunity to leverage the Microsoft platform to lower the total cost of ownership. Microsoft's reputation and previous experience in developing widely accepted applications is allowing it to quickly catch-up to the competition as it continues to incorporate new functionality into their CRM product offerings.

Based on the findings, large organizations that can afford the robust solutions would benefit the most from a SAP or Oracle solution. Businesses that already have IT support for other hardware and software solutions would lean towards Microsoft's solution which would integrate with software or hardware already in place. Smaller firms that cannot afford the upfront cost of a good CRM solution would benefit from a RightNow (customer service focused) or Salesforce (sales focused) solution.

CRM is about acquiring and retaining customers, improving customer loyalty, gaining customer insight, and implementing customer-focused strategies that help businesses meet the needs of today's consumer demands. This fluid, agile customer service model allows businesses to grow, maintain competitive agility, and attain operational excellence (Forrester, 2007). Finding the right CRM tool at the right price can really improve customer service by integrating various data silos within a company to depict one "true picture" of the customer. This allows companies to "know thy customer" intimately and use this knowledge to cross-sell and up-sell products as well as alter product offerings quickly as customer's needs change. An effective CRM tool has the potential to provide great customer service, thus building the desired customer loyalty.

CRM IMPLEMENTATION CHALLENGES

From an IT standpoint, the traditional definition of "system implementation" includes such items as loading and configuring software, integration with existing software packages, performing the initial data loads, and developing operational procedures such as task scheduling and backups (Rainer, 2005). While these are all sizable and worthy tasks, the real challenge in CRM implementation is not the technical integration of computing system, but rather the successful integration into the day-to-day operation of a business enterprise. Rob Bois, research director at AMR Research has stated, "Integration to the back office isn't the biggest challenge right now. Companies are already doing this. Integrating sales, marketing, and customer service is still the biggest dilemma" (Beasty, 2007).

CRM implementations are big business for software vendors. In 2005, $4 billion was spent on CRM software licenses (Marchand, 2006). Yet many CRM implementations have failed to deliver the anticipated return on investment. McKinsey and Company have reported that two-thirds of companies surveyed classify their CRM investments as disappointments and in a 2004 study, IBM found an astounding 85% of survey respondent were dissatisfied with their CRM implementations (Roberts et al., 2005).

Virtually all IT implementations require business process and organizational changes. CRM implementations, however, may have a higher dependence on successful business process and organizational change than other ERP elements. A firm's entry into CRM may be its first attempt at true customer-focused marketing: blending sales, marketing, and customer service into a seamless set of transactions and interactions (Chalmeta, 2006).

The value of a CRM system, like any other business system, is not in the technology but in how the technology is used. Technology is simply an enabler for driving business solutions. Implementation teams must recognize that every part of the enterprise had some sort of CRM system beforehand, such as, the loose collection of spreadsheets, Rolodexes, and personal files that represent the "way we have always done things." A well designed and executed implementation then, goes well beyond racking and stacking servers. Business processes must be reengineered, organizational alignment needs to be considered, and most importantly, the project must have wide support.

PREVENTING FAILURES

There are many schools of thought on ensuring successful CRM implementations. Common features found in a survey of the literature include developing acceptance and support, a disciplined change management plan, initial and ongoing training, and communicating value through metrics.

Gaining Acceptance and Support

Buy-in across the enterprise is crucial to a successful implementation. As the project moves through development and implementation, it is important that the early momentum and acceptance is maintained. The deployment team should develop a complete communication plan early in the project lifecycle. The communication plan must include the "entire definitive vision of the project" (Chalmeta, 2006), and outline where the organization is and where it needs to go. Frequent and consistent communiqués from the implementation team will help maintain interest, thus creating wider acceptance of the CRM solution throughout the organization upon project completion. Of course, management support is crucial. A firm's senior management must support the project and create an environment that can maintain the focus and momentum necessary for success (Roberts et al., 2005).

Develop a Change Management Plan

As Benjamin Disraeli said, "Change is inevitable. Change is constant." Business drivers and requirements will invariably change during the project development and implementation. Without a solid change management plan the project faces an uphill climb. However, change management cannot be simply the realm of IT. New requirements must be quickly translated into system modifications. A successful CRM implementation requires a change management plan that can support constant change while simultaneously maintaining the software baseline configuration, and providing rigorous testing and rapid deployment.

Training

Ultimately, it is the order takers, sales representatives, service technicians, and marketing professionals that are the key to a CRM implementation success or failure. Employees must understand the project in detail and embrace the change. They need opportunities to provide feedback early in the development phase. This feedback should then be incorporated to establish some ownership of the new CRM system and increase acceptance among key staff upon deployment. Most importantly, employees must see the holistic value of the system. In many instances, they will have to be trained in not only the technology, but the new customer service philosophy that often accompanies an investment in CRM. Without training in both aspects, users will invariably fall back into their old, comfortable ways. This risk is very real. Brenner, Fontana, and Godbout (2003) state, "While there is often a period of user involvement, after 6, 12 or 18 months, the intended users...go back to their own spreadsheets and personal files." The implementation team cannot simply turn the new system on and walk away. Training and communication must continue for the lifecycle of the project.

Demonstrate Value through Metrics

The CRM implementation team will need to do some internal marketing to show the value of a new system to the organization. No matter how great and wonderful the system is, it will not be used if no one knows about it. Employees must know about the system and understand the benefits (Chalmeta, 2006). A very effective tool for demonstrating the value of any project is regular publication of simple, business driven metrics. Gartner's Eisnefeld writes, "The unsatisfactory or inadequate results often associated with CRM come mainly from enterprises' inability to demonstrate value" (Eisenfeld, Apfel, & Smith, 2004). Eisenfeld lists some excellent heuristics for defining useful metrics, summarized in Table 2.

Business users need different metrics to measure their success. While improvements in sales, marketing, and service are each critical components of a successful CRM implementation, the measurements of success are different. Some suggested business value driven metrics for each of these components are listed in Table 3.

Table 2. Characteristics of good metrics

	Characteristic of good metrics
1.	No more than seven metrics should be used at any given time
2.	Metrics should be collectively exhaustive and mutually exclusive
3.	Metrics should be tied to executive and middle management activities and compensation
4.	CRM metrics should foster collaboration across the processes in an enterprise's sales, marketing, and service functions
5.	Metrics should capture cause and effect relationships between business functions
6.	Metrics should develop over time

Source: (Eisenfeld et al., 2004)

Table 3. Business value driven metrics

Sales Metrics	Marketing metrics	Service metrics
Prospects and new customers	Number of campaigns	Calls per agent
Retained customers	Customer retention rates	Average time to resolution
Open opportunities	Purchases and revenue by campaign	Average number of service defects
Up-sells and cross-sells	New customers and leads by campaign	SLA compliance
Renewal rate	Referrals	Repair time

Source: (Eisenfeld et al., 2004)

Inappropriate use of metrics can lead to short-term benefits and long-term failure. An example is L.L. Bean's over reliance on customer retention as a measure of success. By focusing on customer retention, profitability initially improved. Eventually, however, profitability began to slip. L.L. Bean adjusted their strategy to blend both customer retention and new customer acquisition metrics as part of their overall performance targets (Boulding, Staelin, Ehret, & Johnston, 2005).

Although metrics and measurement are important tools, they are no excuse for poor management and the complexity of the real world. Albert Einstein once said, "Not everything that counts can be counted, and not everything that can be counted counts." Likewise, managers must view metrics as one of many wrenches in the toolbox.

CASE STUDY: CIGNA HEALTHCARE

Many case studies documenting successful CRM implementations have been published by software vendors, systems integrators, and academics. To understand what defines success however, it may be more helpful to look at a failure instead. Cigna Healthcare is one of the largest health insurers in the United States, with more than 9 million subscribers enrolled in a variety of health plans including PPO, HMO, Point of Service, indemnity, dental, vision, pharmacy, and behavior health (Hoovers, 2007). The company is headquartered in Philadelphia, Pennsylvania and traces its roots in the U.S. back to 1792 (Yahoo! Finance, 2007). In 2006, Cigna reported a net income of $1.1

billion on sales of $16.5 billion (Cigna, 2007), a big improvement from 4 years earlier when the company reported a net loss of $398 million on sales of $19.34 billion (ICFAI, 2007). Many analysts blame Cigna's earlier problems on a failed IT restructuring, including problems associated with a massive new CRM system.

Great Expectations

Cigna's CRM project involved migrating 3.5 million users from a collection of legacy systems to an integrated solution. The new system would provide customer service representatives a 360-degree view of its member's data. When a user called with a question, the reps would be able to see a complete history of all interactions (Bass, 2003). Cigna had three critical business drivers pushing this deployment:

1. Cigna and other major U.S. insurers were being sued by thousands of doctors for slow payment.
2. Cigna's sales team had promised large accounts that the new system would improve customer satisfaction and supplied a go-live date of early 2002.
3. Cigna's management believed the system would allow the company to consolidate 20 service centers into nine and replace 3,000 senior employees with 1,000 new hires (Bass, 2003).

Cigna pushed ahead with its layoff and consolidation plans, paying $33 million in severance and $32 million to build the new centers. In January 2002, 3.5 million of Cigna's customers were migrated to the new system (Bass, 2003). Immediately, problems began surfacing: member healthcare coverage for customers could not be confirmed; erroneous identification cards were issued; and prescriptions could not be filled (Bass, 2003). As a result, Cigna lost 6% of its subscribers in 2002, almost 900,000 individuals (ICFAI, 2007).

What Went Wrong?

In a nutshell, the system was deployed before it was ready. Wrapped up in a billion dollar IT upgrade, the CRM system did not receive the attention it required. Business requirements were not adequately defined and tested. Most importantly, the company terminated its most senior customer service specialists just when they were needed the most. Cigna CIO Andrea Anania stated, "You can have the best system in the world, but if you have people with relatively little tenure, you're not going to get the best service" (Bass, 2003). Cigna has since rehired many of these representatives.

Back On Track

Cigna slowed its implementation. By July 2002, Cigna was able to restart migration to the new system. In January 2003, Cigna turned a corner, successfully launching an online self-help portal for its customers. Customer satisfaction rates improved and Cigna was finally able to achieve the workforce reductions it needed, reducing its sales and case management team by 3,900 positions (Bass, 2003).

CONCLUSION

Customers today have come to expect the high levels of service a quality CRM implementation delivers. It has become a required part of doing business. Implementing a CRM system touches every part of an enterprise, from sales entry to the C-level suite. Likewise, a successful implementation requires the support of virtually every part of the organization. Like any business process reengineering, communication, training, and executive support are mandatory. Organizations considering a CRM solution should create a comprehensive implementation plan that takes into account integration with all CRM modules (e.g., sales, marketing, and service) as well as other enterprise resource planning systems. This plan should incorporate current business strategy and generate "buy in" by all stakeholders. These individuals should be actively engaged throughout the entire process. A CRM solution should not be rushed, rather it should be slowly adopted, in phases if necessary, to allow for thorough testing and system improvements early on to maximize user acceptance and customer satisfaction.

An automated CRM system is only as good as the underlying business process methods and personnel skills used to develop it (Allman, 2004). Continued senior level management involvement providing vision, funding, and cross-organizational collaboration is critical to a successful implementation and ongoing support of a CRM system (Payne & Frow, 2006, p. 153). Managers and teams should not over promise and under deliver a CRM solution. Communication should be focused on what stakeholders want to know, not what the sponsor or project manager thinks they should know. Stakeholders must understand their roles, where to access project information, and how to provide feedback (Brenner et al. 2003). Proper planning and team assignments ensure the long term success of CRM implementations (Mitra, 2002). It is important to choose the right person for the job or task at hand, clearly outline the tasks and expected deadlines, monitor and address stakeholders concerns immediately, and evaluate each phase of the project.

REFERENCES

Alexandrou, M. (2007). *CRM definition*. Retrieved February 1, 2007, from http://www.mariosalexandrou.com/definition/crm.asp

Allman, D. (2004). *Taking CRM to a higher level of interaction*. Retrieved February 18, 2007, from http://www.destinationcrm.com/articles/default.asp?ArticleID=3849

Anand, S., DuLaney, W., & Melling, W. (2004). *Smart advice: Consider long-range vision for IT when upgrading ERP systems*. Retrieved February 18, 2007, from http://www.informationweek.com/security/showArticle.jhtml?articleID=47902493&pgno=1&queryText=Definition+ERP

Band, W. (2007a). mySAP CRM is a leader in the enterprise CRM suites market. *Forrester research*. Retrieved April 1, 2007, from http://www.forrester.com/Research/Document/Excerpt/0,7211,41054,00.html

Band, W. (2007b). The Forrester Wave™: enterprise CRM suites, Q1 2007. *Forrester research*. Retrieved April 1, 2007, from http://www.forrester.com/Research/Document/Excerpt/0,7211,39975,00.html

Bass, A. (2003, March 15). *Integration management: Cigna's self-inflicted wounds*. Retrieved April 12, 2007, from http://cio.com

Beasty, C. (2007). The enlightened enterprise: Customer relationship management. *Medford, 11*(3), 26-40. Retrieved April 8, 2007, from ABI/INFORM Global database.

Biba, E. (2005). *Shoppers find less satisfaction online*. Retrieved March 3, 2007, from http://www.pcworld.com/article/id,119778-page,1/article.html?RSS=RSS

Boulding, W., Staelin, R., Ehret, M., & Johnston, W. (2005). A customer relationship management roadmap: What is known, potential pitfalls, and where to go? *Journal of Marketing, 69*(4), 155-166. Retrieved January 31, 2007, from Business Source Premier Database.

Brenner, M., Fontana, C., & Godbout, N. (2003). *Change management.* Retrieved February 18, 2007, from http://www.destinationcrm.com/articles/default.asp?ArticleID=3656+&TopicID=9

Carlson Marketing. (2007). *No more limits: On demand CRM goes strategic.* Retrieved March 8, 2007, from http://research.pcpro.co.uk/detail/RES/11634386511_120.html?src=pcpro

Chalmeta, R. (2006). Methodology for customer relationship management. *Journal of Systems & Software, 79*(7), 1015-1024. Retrieved January 31, 2007, from Business Source Premier Database.

Chan, J. (2005). Toward a unified view of customer relationship management. *Journal of American Academy of Business, 6*(1), 32-39. Retrieved January 30, 2007, from ABI/INFORM Global database.

Chang, H., Hsiung, C., & Tsai, R. (2006). The design and analyses of customer relationship management using ARIS technique. *International Journal of Management & Enterprise Development, 3*(1/2), 1. Retrieved March 8, 2007 from Business Source Premier database.

Cigna. (2007). *Cigna reports 4th quarter and full year 2006 results* [News Release]. Retrieved April 14, 2007, from http://cigna.com/about_us/investor_relations/disclosures_archived.html

Clements, D. (2006). The benefits of customer relationship management software. *Real Estate Finance, 23*(1), 15-17. Retrieved January 31, 2007, from Business Source Premier database.

Cunningham, C., Song, I., & Chen, P. P. (2006). Data warehouse design to support customer relationship management analyses. *Journal of Database Management, 17*(2), 62-85. Retrieved January 30, 2007, from ProQuest ABI/INFORM Global database.

Desisto, R. P. (2006). *Magic quadrant for sales force automation, 1H06* [Gartner Research Rep. No.G00141438]. Retrieved March 27, 2007, from http://www.gartner.com/mq/index.jsp?qid=141438

Dyche, J. (2002). Managing your CRM project: The CRM development team. *Informit.com.* Retrieved April 19, 2007, from http://www.informit.com/articles/article.asp?p=25181&seqNum=3&rl=1

Eisenfeld, B., Apfel, A., & Smith, M. (2004, June 2). *Choosing metrics for customer relationship management* (Gartner Research Issue Brief No. DF-22-7541). Stamford, CT: Gartner, Inc.

Fitzgerald, M. (2007). CRM made simple new software to supercharge your sales. *Inc. Boston, 29*(1), 46-47. Retrieved March 26, 2007, from Business -ABI/INFORM Global database.

Fleischer, J. (2006). RightNow technologies' two-sided approach to CRM. *Call Center Magazine.* Retrieved April 6, 2007, from http://www.callcentermagazine.com/showArticle.jhtml?articleID=193500843

Hart, M. L. (2006). Customer relationship management: Are software applications aligned with business objectives? *South African Journal of Business Management, 37*(2), 17-32. Retrieved January 31, 2007, from Business Source Premier database.

Hoovers. (2007). *CIGNA Corporation.* Retrieved April 14, 2007, from http://hoovers.com/cigna/--ID__10259--/free-co-factsheet.xhtml

ICFAI. (2007). *CRM implementation failure at Cigna Corporation.* Retrieved April 14, 2007, from http://www.icmr.icfai.org

Jukić, N., Jukić, B., Meamber, L., & Nezlek, G. (2002). Employing a multilevel secure approach in CRM systems. *The Journal of Information Technology Theory and Application (JITTA), 4*(2), 17-31. Retrieved March 8, 2007 from Business Source Premier database.

Kellen, V. (2002). CRM measurement frameworks. *DePaul University.* Retrieved February 18, 2007, from http://www.kellen.net/crmmeas.htm

Kennedy, A. (2006). Electronic customer relationship management (eCRM): Opportunities and challenges in a digital world. *Irish Marketing Review, 18*(1/2), 58-69. Retrieved January 30, 2006, from ProQuest ABI/INFORM Global database.

Kolsky, E. (2002). *CRM at Bank One: Becoming customer-centric.* Retrieved February 19, 2007, from Gartner database.

Marchand, D. (2006).). Customer relationship management challenging the myth: Focus on people, not the technology. *Perspectives for Managers,* (131), 1-5. Retrieved January 30, 2007, from ProQuest ABI/INFORM Global database.

Messner, W. (2005). Customer relationship management technology. *Business Information Review, 22*(4), 253-262. Retrieved January 31, 2007, from Business Source Premier database.

Mithas, S., Krishnan, M. S., & Fornell, C. (2005). Why do customer relationship management applications affect customer satisfaction? *Journal of Marketing, 69*(4), 201-209. Retrieved January 31, 2007, from Business Source Premier database.

Mitra, C. (2002). A strong IT team is essential for CRM success. *Cybermedia India Online Limited (CIOL).* Retrieved February 18, 2007, from http://www.ciol.com/content/search/showarticle.asp?arid=35658&way=search

Payne, A., & Frow, P. (2006). Customer relationship management: From strategy to implementation. *Journal of Marketing Management, 22*(1/2), 135-168. Retrieved January 31, 2007, from Business Source Premier:

Ragsdale, J. (2007). 2007 service and support technology trends. *SSPA News.* Retrieved April 6, 2007, from http://www.thesspa.com/sspanews/_07Jan/article1.asp

Rainer, E. (2005). New trends & techniques in managing CRM & SAP projects. *ITtoolbox.* Retrieved February 1, 2007, from http://blogs.ittoolbox.com/crm/trends/archives/erp-packages-feature-comparison-4309

Raisinghani, M. S., Tan, E., Untama, J. A., Weiershaus, H., et al. (2005). CRM systems in German hospitals: Illustrations of issues & trends. *Journal of Cases on Information Technology, 7*(4), 1-26. Retrieved January 30, 2006, from ProQuest ABI/INFORM Global database.

Reid, A., & Catterall, M. (2005). Invisible data quality issues in a CRM implementation. *Journal of Database Marketing & Customer Strategy Management, 12*(4), 305-315. Retrieved January 30, 2006, from ProQuest ABI/INFORM Global database.

Richards, K. A., & Jones, E. (2006). Customer relationship management: Finding value drivers. *Industrial Marketing Management.* Retrieved February 28, 2007, from ScienceDirect database.

Roberts, M. L., Liu, R. R., & Hazard, K. (2005). Strategy, technology and organizational alignment: Key components of CRM success. *Journal of Database Marketing & Customer Strategy Management, 12*(4), 315-327. Retrieved January 30, 2006, from ProQuest ABI/INFORM Global database.

SAP. (2007). *mySAP CRM: Software for enabling customer relationship management (CRM) without compromise.* Retrieved April 1, 2007, from http://www.sap.com/solutions/business-suite/crm/index.epx

Seeman, E. D., & O'Hara, M. (2006). Customer relationship management in higher education: Using information systems to improve the student-school relationship. *Campus - Wide Information Systems, 23*(1), 24-35. Retrieved January 30, 2006, from ProQuest ABI/INFORM Global database.

Shah, J. R., & Murtaza, M. B. (2005). Effective customer relationship management through Web services. *The Journal of Computer Information Systems, 46*(1), 98-110. Retrieved January 30, 2006, from http://proquest.umi.com.ezproxy.umuc.edu/pqdweb?index=33&did=914802221&SrchMode=1&sid=3&Fmt=4&VInst=PROD&VType=PQD&RQT=309&VName=PQD&TS=1170210789&clientId=8724

Thompson, E, Dunne, M., Radcliffe, J., Herschel, G., Prentice, B., Mertz, S. A., et al. (2006). *Key issues for CRM strategy and implementation* (Gartner Research Issue Brief No. G00137783). Stamford, CT: Gartner, Inc. Retrieved March 8, 2007 from Gartner database.

Thompson, E., & Goldman, M. (2006). *The industries in which CRM consultants and systems integrators deploy Oracle-Siebel and SAP* (Gartner Research Rep. No.G00143194). Retrieved March 27, 2006, from Gartner database.

Webopedia. (2007). *CRM.* Retrieved March 8, 2007 from http://www.webopedia.com/TERM/C/CRM.html

Winer, R. S. (2001). A framework for customer relationship management. *California Management Review, 43*(4). Retrieved February 1, 2007 from http://groups.haas.berkeley.edu/fcsuit/PDF-papers/CRM%20paper.pdf

Yahoo! Finance. (2007). *Yahoo! finance - business summary.* Retrieved April 14, 2007, from http://finance.yahoo.com/q/pr?s=CI

Chapter VI
Different Web Strategies for Different E-Marketplaces

L. Geppert
Catholic University of Milan, Italy

ABSTRACT

This chapter presents two possible models of electronic marketplaces put in place at the beginning of this century, which, after their introduction, the first incoming wave of connected economy-based paradigms was ended. Both the two models show a particular use of Web-based information technology in order to exploit their mission and represent meaningful cases of application of well-defined Web strategies. Even though, at the moment, the initially built Web sites supporting those related business have been closed and merged with other Web sites, they may introduce examples of a Web strategy approach having a relevant historical meaning that may be still redefined in practical implementations once revised and adequately updated. The mentioned cases described in this chapter are usteel.com and up2gold.com, two examples of Web-based business in two well-defined supply chains, that is, the "steel" chain and the "gold and silver" chain.

INTRODUCTION

After the first incoming wave of the connected economy ended, investors, business managers, consultants, and specialists began to review their plans, actions, and methodologies. Many e-business models, both business-to-business (B2B) and business-to-consumer (B2C), are currently under consideration in order to assess and verify their real effectiveness (Kalakota & Whinston, 1996). The models applied so far have generally shown that, with some exceptions related to e-procurement implementations of large enterprises (Timmers, 1999), vertical marketplaces owners must look for different business paradigms in order to economically survive and to be actually able to create value to market industries chains they address (McDonald, 1997). In the following pages, two real cases are described: one trading marketplace addressing the steel industry as an example of "first wave" e-marketplace and one supply chain-like Web portal built for the Arezzo's (Italy) gold manufacturing supply chain. The first case relates to a "traditional Web strategy"-based e-marketplace with a standard approach and contents (Shapiro & Varian, 1998) and the second shows an innovative strategy and a new business model (Porter, 1985). Even though, at the moment, both of the abovementioned e-marketplaces have merged their activities and brands in other larger e-commerce initiatives, at the time they were presented (2001), they represented important and innovative examples of an e-marketplace-based business model.

BACKGROUND

As some analyst observes, "the first high-visibility Internet businesses conceived to attract significant investor attention were, at the beginning, in the business-to-consumer (B2C) relationships, providing facilities for consumers to choose and purchase goods online. It seemed logical to think, as a related consequence, a development also of B2B, based on the "buying and selling" part of the relationship between businesses.

Thus, starting in 1999, and accelerating rapidly through 2000, the focus of B2B attention was very firmly on those initiatives that seemed to be capable of exploiting information and communication technology to alter the balance of power in trading relationships. These initiatives, which covered a wide spectrum of business and technical services, gained the general title of "marketplaces," and, throughout 2000, the focus of B2B activities was on developing marketplace interaction strategies and examining procurement processes.

The focus on the transactional elements of B2B relationships will come to be seen as

an aberrant commencement to the development of the connected economy that will emerge in the 21st century. Businesses interact with other businesses in hundreds of ways. It is true that the interactions are generally intended to result in a commercial transaction. However, for many business relationships, the crucial performance-enhancing interactions take place before and after the transactions, and B2B will expand to occupy these broader horizons. (Kyte, 2001)

Business managers' and investors' expectations were very high at the end of the last century. They understood the strategic importance of B2B electronic model in particular and invested their money and their commitment, demanding rapid deployment of innovative business systems, which the IT community was increasingly hard-pressed to deliver. Many of these systems were new breed of applications, designed to change the external relations in the B2B world.

These systems work fine on the drawing board. However, actually developing externally focused applications as robust, reliable, secure systems with high availability and very high adaptability was and still currently is a challenge that only few IT specialists know how to address.

Figure 1.

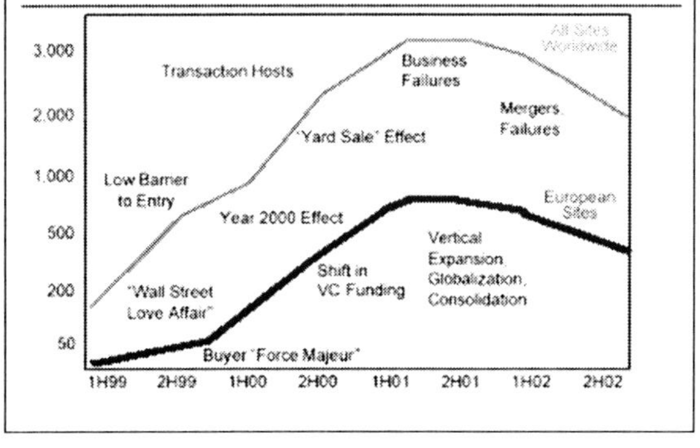

Figure 2.

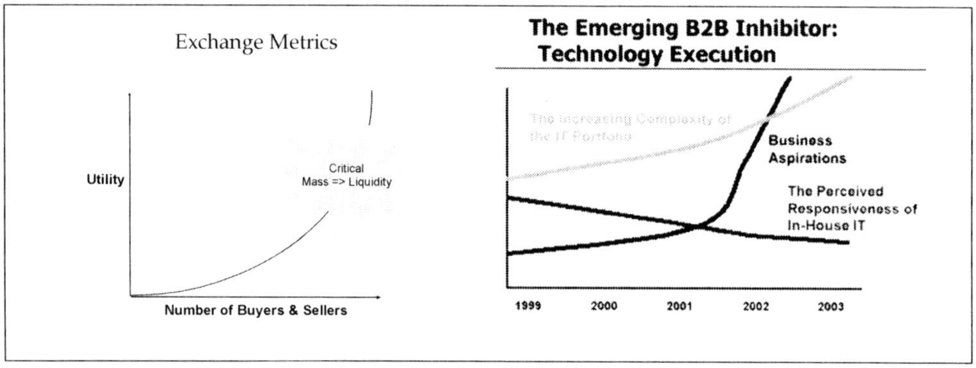

Furthermore, models of marketplaces so far implemented put the stress and focus of their business model on the transactional aspect. This model, which can act and succeed—or, better, survive—in all those case where the liquidity is reached in a sensible spell of time, does not take into account a natural evolution of electronic ecosystems where the complexity of relationships is becoming more and more vital.

However, the connected economy and B2B marketplace models with it have just started their evolution and growth, which will not be stopped; they just need some adaptation mechanism to consolidate the new paradigm.

Thus, after that, a first pioneer period went by, where marketplace models have attracted business managers and, consequently, finance investors, showing the possible strategic value of Web technology, that is, the ability to extend the enterprise business model and to improve process efficiencies; now that paybacks have not been respected and investors begin to retire their support, the marketplace model must be readapted and reengineered (Bollier, 1996).

In order to better respect investors' and business managers' constraints and expectations, that is, to present a more realistic behaviour of the promised return of investment, it is now crucial for marketplace business models to take into account the following aspects:

- Membership and brand is more important than transaction fee (Dibb, Simkin, Pride, & Ferrel, 1991).
- Quality and quantity of services can aggregate other partners, which can bring additional financial resources and customers, in order to give more value to the industry they address.
- Processes begin before transactions and continue after it; support must be given to a wider range of customer tasks.

In the next pages two significant real implementation cases have been considered. The first concerns the possible evolution of a typical vertical marketplace positioned in the steel supply chain. The second shows how a service center for a geographical industrial district populated by many fragmented small enterprises can give value to the related supply chain and presents a valid paradigm to overcome possible "digital divide"-based approaches.

MAIN THRUST OF THE CHAPTER

Case Study 1: u-steel.com[1]

General Aspects

U-Steel is the first next generation e-marketplace, neutral, independent, and completely online, to be dedicated to the commerce of steel and related services. Initially launched in September 2000 by the strong sponsorship of a group of European partners such as COFACE (F), Falck Group (I), Mastr@ (I), Iperbusiness (I), Sopaf (I) and Mittel Generale Investimenti (I), U-Steel's aim is to bring the most qualified players of the steel industry online, increase the consumers purchasing power, and expand the markets.

From an operational perspective, U-Steel is an e-marketplace integrated with a large number of support services. Strategically, however, it is not just a virtual channel between buyer and seller but also an important tool to allow all supply chain participants to become a part, easily and safely, of the steel's industry future dynamics.

All effective members of the marketplace who purchase steel through U-Steel are certified and assigned a credit rating by COFACE, a worldwide leader in credit certification. Sellers who require total guarantee of their credits towards other U-Steel members can take out a credit insurance policy offered by COFACE exclusively to the portal's subscribers at extremely favourable terms.

When in 1999, as a group of friends, the founders came up with the idea of starting a company with the purpose of building an e-marketplace for steel, they immediately understood the huge importance of the project. Shifting to "virtual" steel meant entering a dimension completely new to the steel industry, traditionally tied to the business rules of the so-called "*old economy*."

In order to be successful, this new initiative required three elements: high level stockholders, capable of ensuring neutrality and independence to the marketplace; managerial staff knowledgeable in both the steel and the information technology industries; and finally, the most advanced technology available (Ansoff 1987; Hammer, 1996).

The idea was presented to a number of potential partners, all of whom accepted the project with great enthusiasm, starting from Iperbusiness of the TC Sistema Group, who became the technology partner, taking responsibility for the coordination of the project and the system integration requirements.

U-Steel has chosen a business model mainly based on the concept of *membership*, despite other initiatives based solely on the transaction mechanism. In an industry like steel-working, which only in Europe claims a turnover of approximately 285 million Euros, U-Steel's further objectives is to bring together partners who have a technologically innovative offering, placing particular emphasis on the technological brand.

The marketplace starts off as a platform on which a number of additional services can be included; the objective is to increase the offering based on the efficiency of the cycle with other services ranging from insurance, credit, and logistics, all the way to integrating the partner's IT infrastructures with a technology across platforms.

Figure 3.

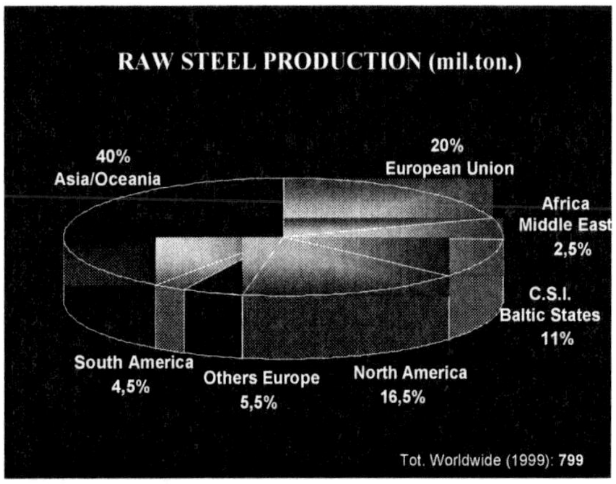

Figure 4.

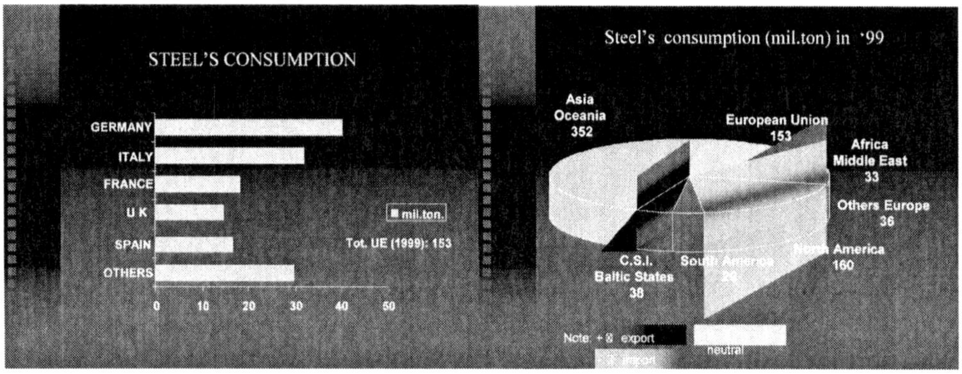

U-Steel's transactional system is globally among the most advanced, with a variety of exchangeable products including plate, coil, rod, wire, pipe, and special irons and steels. Additionally, there is the possibility of marketing excess lots or scrap lots for the plants.

When a purchaser wishes to find a potential seller, they choose the product to buy, select the supplier, and can contact them through the marketplace to establish prices and terms by means of an interactive chat.

This way U-Steel becomes the point of contact between purchaser and seller, the location where the negotiation is conducted and finalized, regardless of where in the supply chain the negotiation takes place. The model allows considerable savings in terms of time, cost, and back-office infrastructure.

But the portal is not only about buying and selling; this is just the initial step of an online process to which it will soon be possible to add additional services such as invoicing, logistics, warehousing, management and handling, accounting, and credit management.

Overall, a system capable of handling all processes not strictly connected to the steel's core business.

Who are the players interested in accessing the marketplace? The system potentially involves all components of the cycle, from the manufacturers

Figure 5.

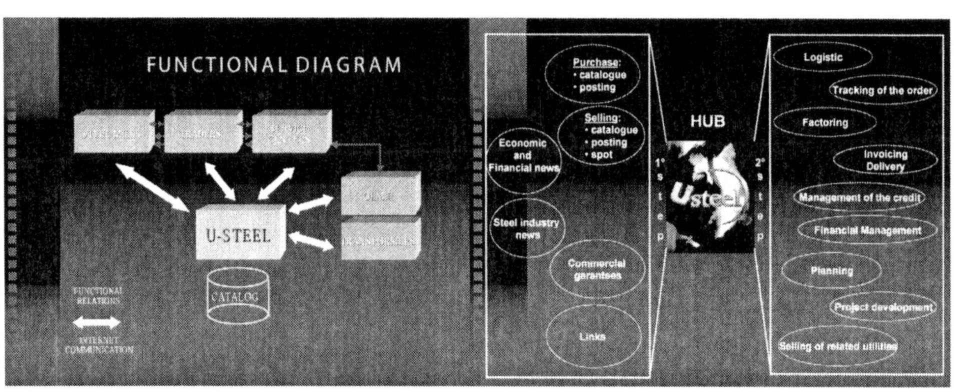

to the retailers, from service centers who purchase large amounts and resell smaller lots of semifinished product, to the retailer or storekeeper who distribute the product to end users.

U-Steel does not promote disintermediation. On the contrary, it encourages the repositioning of sales figures who previously benefited from the inefficiencies of the supply chain and for whom it is now possible to become value added consultants. A trading tool only adds another level of competition within the procurement process.

A seller approaching U-Steel is given the opportunity of publishing the seller's catalogue directly on the marketplace, complete with prices and current available stock, or with the option of showing only the products without any further information. In both cases, the subscriber is given the possibility of integrating the U-Steel catalogue with the seller's own information system so that each transaction results in a warehouse operation, a production cycle advancement, or a purchase order.

Alternatively, the manufacturer may simply replicate a portion of their product catalogue, put it online, and choose when to synchronize the two systems.

The technological components was chosen, developed, and implemented by Iperbusiness on the BroadVision application. Hewlett Packard was chosen as outsourcer of all the systems management services, including the environment infrastructure (i.e., premises, power supplies and surge suppressors/generators, security systems, fire prevention, local area networks, etc.) and the required hardware platform housed in the HP Data Centre located in Bergamo (I).

The system, designed and commissioned in just 4 months time, is based on a UNIX HP UX operating system, is configured on two HP9000 A400 clustered Web servers, two HP9000 A500 to run the BroadVision applications, and an HP9000 A500 box acting as DB server with the Oracle database. The information is stored on two HP SureStore ESC10 clustered devices with an HP SureStore E 2/20 backup library.

The network is based on two Cisco 2900 routers, a Cisco Local Director 416 firewall, and a Cisco PIX 525 in failover. The entire system was designed to provide the highest level of availability, and with doubled security and load balancing systems, it was built to handle an initial volume of 1.200 users in its first months.

Furthermore, HP specialists handle all the system resource administration activities ensuring the users' seamless and transparent functionality.

Figure 6.

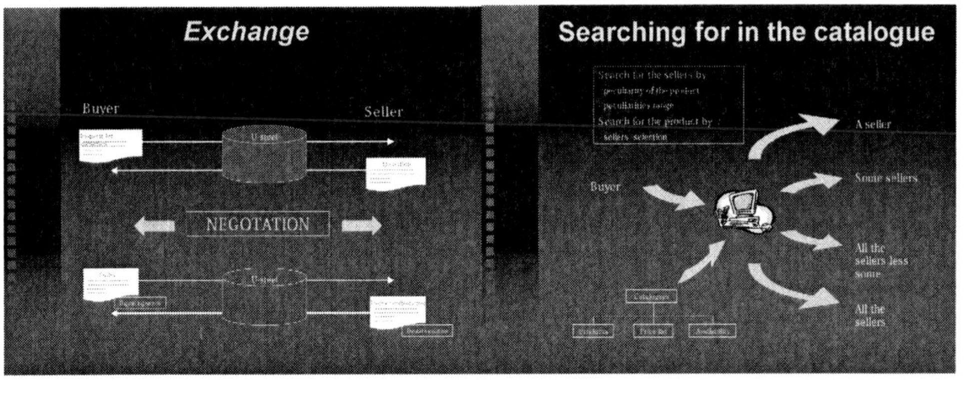

The E-Business Model

The business model of *u-steel.com* is based the on typical characteristics of the European steel supply chain and on the model of a B2B marketplace.

The European steel supply chain is formed by a very limited number of large steel producers and brokers, which are positioned at the top links of the chain, and by a great deal of other smaller players, grouped in dealers, steel service centers (SSC), and final companies which transform and use steel components in their own production. Even though this supply chain includes the presence of large producers, brokers, and even buyers (e.g., large automotive industry enterprises), the typical power play is well balanced, in that, medium and small companies can usually buy steel also oversee, that is, in America and Asia.

Thus, u-steel.com, which includes among its partners ex-business managers of the steel chain who know very well the market and the industry behaviour, is positioned as a vertical, *neutral*, *independent*, *aggregating*, and *collaborative* electronic market maker for steel trading exchanges and other services, in order to satisfy the needs of all the players and to support all the competitive forces of the steel supply chain.

Thus, first of all, u-steel.com is a vertical marketplace, which implements typical processes of a well identified market and industry where trading activities are accomplished in a particular context.

It is *neutral*, in that it is neither "seller" or "buyer" advocated. It offers to both the groups of players an electronic trading place where it enables secure commercial transactions and other services. Buyers and sellers can access the site and operate it using the typical trading approaches: catalogue searching and sourcing, online and even interactive exchange, and offer-posting processes.

The expected macro-benefits produced through these approaches can be grouped in the following main classes:

- Improving efficiency of intercompany operational processes by reducing the typical time usually needed to manually accomplish the same tasks.

Figure 7.

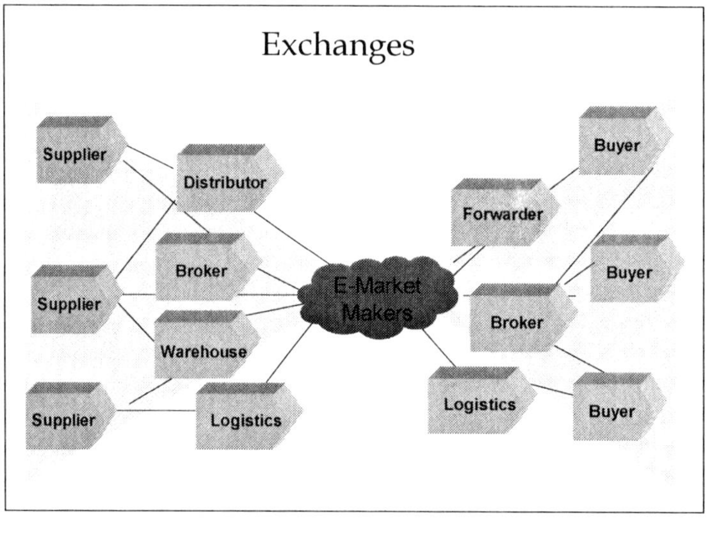

- Enhancing commercial visibility and exploitation of resources, particularly for Steel Service Centres and small dealers, which can reach a larger audience of final transformers.
- Reducing buying cost, in term of final price, for small and medium transformers that can increase their ability to compete.
- Reducing stocking and work-in-progress costs of the whole supply chain through a better synchronization of production, distribution, and delivery.

Furthermore, taking into account the peculiarity of the European steel supply chain, U-Steel will offer in the immediate future a very sophisticated auction-based system, which will give the opportunity to large producers (i.e., steel mills a brokers) to launch forward actions on the European market and which, at the same time, will provide to small dealers and transformers with a mechanism of reverse actions to buy at better conditions also on oversee marketplaces.

U-steel is also an *independent* electronic market maker, in that it lies in a well-balanced power play position. Its claimed mission is "acting as informediary on the steel market," with no particular preference for any side.

The historical role of shareholder partners ensures both the knowledge of the steel supply chain, through the Falck Group currently involved in the utilities industry, and the absolute transparency and independence, being the others well known players of other industries.

But U-Steel is an *aggregating* and *collaborative* market maker as well. These issues represent the innovative aspects of the U-Steel business model.

In fact, after the "hype cycle" of the connected economy has shown failures and constraints of an expected "no limits" growth, U-Steel proposes a further possible direction in a marketplace model evolution. U-steel offers other services by aggregating other business partners and delivering new functionalities which create new value for the steel business community.

There are two types of services that U-Steel delivers: business services and technological services.

Business services include

- Credit insurance (already delivered at the moment)
- Logistics delivery

Technological services include

- Process integration
- Platform outsourcing

Credit insurance gives the opportunity to sellers to insure their credit, following a credit rating mechanism classified and tuned by Coface (F), one of the worldwide insurance market leader and one of the U-Steel shareholder. Through these functionality, the seller can decide to insuring the seller's credit and thus electronically activate the operation with an automatic interaction with the Coface credit system, in order to assess and possibly accept economic conditions, as proposed by Coface. The seller pays a further transaction fee for that.

Logistics delivery, which is not currently available, but it will be in the future, will give

Figure 8.

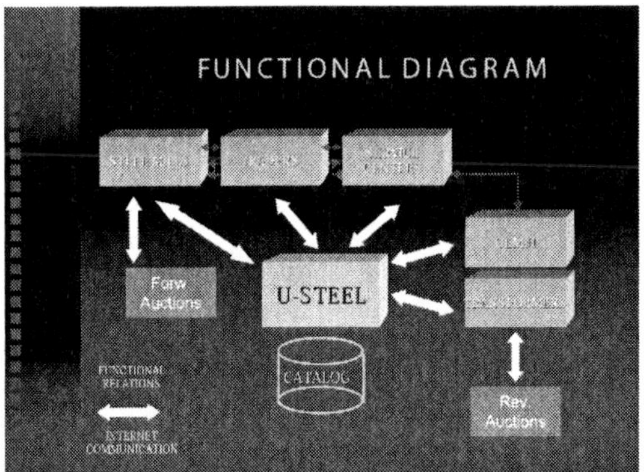

the possibility to the buyer to choose a possible logistics and transportation player to be electronically involved in the process in order to deliver goods in a specified spell of time.

Process integration means the availability of integrating, when needed, the information system of both seller and buyer, in order to automatically accomplish several collaborating processes, such as procurement, fulfilment, billing, and invoicing through an XML interface. It means also the possibility to be integrated with other marketplaces supporting the same feature in order to extend the catalogue interaction to other "ecosystems."

The most innovative *collaborative* aspect of U-Steel is the availability of a technological platform, which can be rented by supply chain players. The technological platform on which U-Steel is based is a very flexible architecture provided by BroadVision and Oracle "best of breed" components. It ensures the possibility to build separate and independent environments where different players can implement their corners. In this way, a steel mill company or a broker, for example, can also deliver their own extranet, outsourced to U-Steel, avoiding the investment of a huge amount of money in a capital expenditure which would ask them for large financial resources and would force them to track a control on the return of investment.

In this model, on the contrary, U-Steel has the possibility to reduce this payback time and thus to shrink its ROI, while other players receive the value to have their own environment, classified as a variable cost in their profit and loss account. In other words, U-Steel can reduce the initial trade-off between revenues and costs and the other players using the feature, use the platform as a general expenditure, and do not have to directly involve themselves in technology, which is not their core business.

In this sense, the U-Steel model appears as a vertical model for trading and commercial operations, and, at the same time, as a horizontal model for distributing technological skills and services; a sort of a hybrid, new model, where U-Steel can act as a business service provider as well. A model that, taking into account the current evolution of other vertical marketplaces in other industries, can be considered a valid evolution and a possible successful paradigm.

The Revenue Lines

The U-Steel model is built on the following economic revenue lines:

- Membership fee, which is an annual fee to be paid by sellers and buyer in order to operate on the marketplace.
- Transaction fees on goods, as a percentage of the total amount.
- Transaction fees on services delivered by other partners (credit and logistics).
- Annual fee for ERP integration, following defined technological standards.
- Annual BSP fee, which is a yearly based fee for using the platform hosted by u-steel and adequately customized on customer needs.
- Advertising.

U-Steel will involve more then 6,000 European users after 3 years and will trade more then 5 millions of steel tons in the same period (3% of the whole European market). Expected revenues will be over 40 Millions Euro at the third year.

The www.u-steel.com site, which is currently available for test purposes to a limited community of "family and friends" players of the steel supply chain, will be launched and delivered worldwide on June 20, 2001.

Case Study 2: up2gold.com[2]

General Aspects

Up2Gold.com is a service center conceived to deliver consulting and technological services to the supply chain of gold and silver manufacturing located in Arezzo (I). Arezzo is one of the most important industrial sites for gold and silver manufacturing, a typical Italian industrial district which geographically groups many SMEs belonging to the same supply chain.

At the starting point of the gold (and silver) supply chain there are the raw material suppliers, that is, larger companies which are in a limited number, while production plants are divided into a plethora of small and very small companies linked together by complex multidirectional channels.

The distribution layer is, again, a "long chain" where various intermediaries and dealers are involved before reaching the end user.

These large companies control the supply links and schedule main production flows, while large dealers mainly govern the sales channel and, in particular, the export market.

As mentioned before, the productive part of the supply chain is very fragmented and powdered in a myriad of SME, involved in production subphases mainly as third parties. As it is usual in these cases, they do not participate to the research and development process; they just execute and manage orders received from other supply chain links.

These enterprises are often familiar companies, involving even less than 10 persons and unable to base their activities on some information system, which would be useless yet.

The final production output is finished goods like gold chains, rings, and ear-rings. The product is considered to be medium quality goods, usually competing with Indian and Far East equivalent productions.

Such a supply chain, despite its pervasive presence in many other sectors, is obviously inefficient, for the following reasons:

- The main obstacle to a more efficient chain is, as said before, the dimension of the company involved, the number of companies, and the related fragmentation.
- Moreover, the efficiency of the chain seems to be influenced by the flow of information regarding stock or availability of goods or raw material. In fact, while the commercial side of the chain (the end) does not know the availability, prices, quality, and other information, the producers do not know the real needs of their markets in order to be able to adequate their production volumes, stiles, and so forth.
- The third type of inefficiency is linked to the sales forecast or production schedule, which are extremely difficult to govern, because of a very fragmented catalogue of products; is very common the case of a catalogue with more than 10,000 articles, and practically every company has a catalogue of thousands of articles. Moreover, many articles are subject to season waves, other are linked to styles and stylists.

The result of these inefficiencies are, of course, poor service at the end user level (e.g., unavailability of products in certain period of the year) or, which is worst due to the cost of raw gold, obsolescence of the entire stock with the consequent inefficiency at the producer level. This also causes lack of competitive presence in those markets where the newcomers are getting more and more aggressive due to a lower production cost and a more flexible supply chain.

In order to better understand the subject, the Arezzo's gold and silver supply chain is described in Table 1.

The total Italian production export of gold and silver goods is shown in Table 2.

Arezzo's supply chain shares the 20% of that global export, that is, more than 1.500 billion ITL. The total Arezzo's gold and silver production is more than 6.000 billion ITL.

The aim of up2gold.com is to offer a class of services to overcome the inefficiencies described above through the deployment of the a service center able to deliver:

Table 1.

Site	1-9 workers	10-49 workers	50-99 workers	100-999 workers	Total
Arezzo					
Companies	991	269	7	4	1271
Workers	3366	4547	423	1279	9615

Table 2.

Region	Production export 1989 (billions ITL)	Production export 1999 (billions ITL)
Europe	1173 (28.15 %)	2323 (28.76 %)
North America	1684 (40.41 %)	2694 (33.36 %)
Latin America	175 (4.20 %)	634 (7.85 %)
Middle East	468 (11.23 %)	747 (9.24 %)
Far East	379 (9.10 %)	860 (10.65 %)
Others	288 (6.91 %)	818 (10.13 %)
Total	4167 (100 %)	8076 (100 %)

- Consultancy services to help managing personnel, legal, and work trends issues;
- ERP services, with particular stress on an accounting and payroll application.
- A B2B supply chain marketplace to exchange raw material, semifinished goods, and human resources work availability and to interconnect along the chain goods' availability and stocks.
- A B2C marketplace, sell side-oriented, in order to increase sales of medium size companies.

The shareholders of the Up2Gold are:

- Arezzo's and National artisans associations (CAN).
- TC Sistema SpA and Datamat SpA, two famous IT players of the Italian market;
- FlyNet, a local ISP.
- Banca dell'Etruria e del Lazio, which is a local bank.

General concepts and strategic issues of the center have been designed in the following terms:

- The marketplace model appears innovative and suitable to fit chain needs and SMEs' expectations.
- It can be considered as a new paradigm for supply chain local systems and exported and disseminated to other SMEs districts; a real B2B best practice.

The E-Business Model

When considering the business model of up2gold.com, it is necessary to underline that the company is a service center which delivers a set of services supporting a supply chain. It means that the center offers traditional B2B services along with "connected economy"-like functionalities.

Yet, even though the business model of the center is very articulated and all of its components should be considered and analyzed in a broader perspective, for the purposes of this chapter we would like to draw the attention to the marketplace part, which is an innovative B2B model and includes elements of as the supply chain management approach.

When addressing this marketplace, it is necessary to take into account typicality and behaviours of the supply chain to which it belongs. As already mentioned before, the Arezzo's gold chain is a very fragmented network of small suppliers all working for a few larger enterprises, which act as raw material suppliers and distribution channels (Fariselli, Oughton, Picory, & Sugden, 1997). These companies control the chain and play a strong power role: they make the research and development, that is, design gold and silver final products, schedule large scale production (distribution and sale channels, in particular), and control the raw material procurement process (raw material suppliers). They usually involve the other links of the chain as third parties, requesting them to produce finished or semifinished goods by a well defined delivery date (make-to-order).

This links, on their turn, involve other links of the supply chain, ordering subcomponents and other semifinished goods, scheduling delivery dates, and so on along the whole supply chain. All the subproducers links must generate by and large the following tasks:

- Procuring raw material, buying it in general from large raw material suppliers.
- Procuring human resources or other smaller production companies.
- Procuring semifinished goods, if any, from other chain players.
- Producing components as ordered.
- Delivering goods.

The value of information in accomplishing these activities is very high; if existing, it would reflect in a larger economy of scale, and in a better stock and delivery efficiency. Sometimes it happens that sale channels must refuse oversee orders, in that they are not able to declare on time a right delivery date and final prices to customers, for a lack of updated information!

These activities are not supported by an integrated information system and this fact, as already mentioned above, generates inefficiency to the whole network chain.

Figure 9.

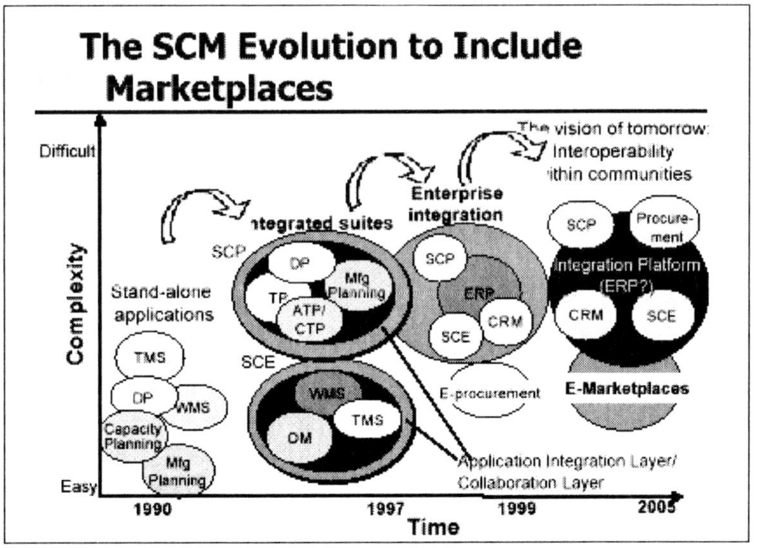

Up2gold.com is building a marketplace which offers two important services to the companies:

- A typical exchange marketplace in order to accomplish the first three activities as described above: procuring raw material, semifinished goods, and human resources. This marketplace gives raw material suppliers the ability to exchange gold and silver and give other enterprises the possibility to buy possible semifinished goods from others' stocks and/or human working resources availability. An additional auctions and reverse auctions system is included in order to sell stocks or buy components at a more convenient price.
- A collaborative system to support a supply chain management marketplace. Through this remote system, accessed via Web, any company, even small enterprises, can give their availability-to-promise, by claiming their actual stock availability, production capacity and, of course, prices. This system can give the sale channel an immediate picture of both the order progress status and the possible make-to-order availability of the whole supply chain.

The model that up2gold.com implements is not exactly that of a supply chain management marketplace, in that production orders scheduling is not taken into account. It is a sort of warehouse and stock availability integration, useful to catch up-to-date information before and after the order is assigned. "Before the order" means that the sale channel can assess and verify possible online availability along the whole supply chain and possible delivery dates to communicate to customers. "After the order" means that the sale channel and other links of the chain can assess online the real progress status of orders, controlling the actual delivery date.

This model is very simple in that no information system is required for SMEs; they are only requested to keep their order profile updated by entering simple information at the up2gold.com site.

Up2gold.com conceived this marketplace model in order to match actual requirements of SMEs-based supply chains, that is, the typical Italian industrial chains. Up2gold.com named this model *partner chain management* (PCM).

Figure 10.

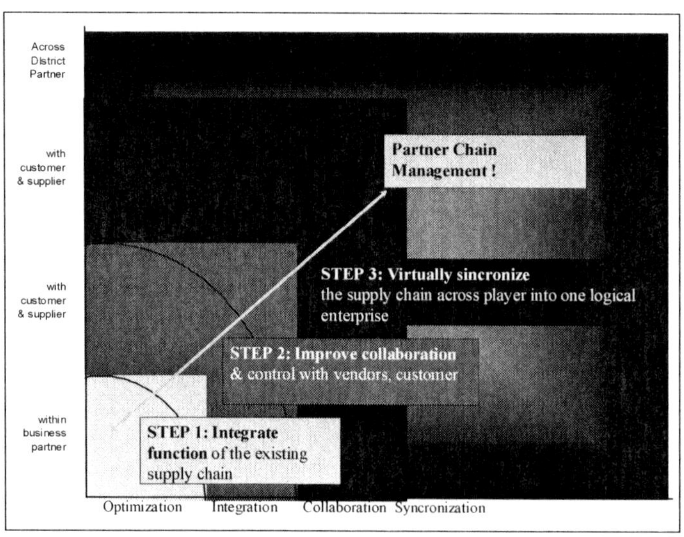

The Revenue Lines

The Up2Gold model is built on the following economic revenue lines:

- Membership fee, which is an annual fee to be paid by any company in order to access the center services.
- Transaction fees on goods, as a percentage of the total amount.
- Annual fee for consulting and ERP function access.
- Advertising.

Up2Gold will involve more then 1.200 district companies after 3 years; the value of transaction involved will be more then 1.000 billion ITL.

The current site is only for presentation and advertising purposes. The service center will release its services progressively, starting at the beginning of 2002. The marketplace in particular will be launched and delivered by the first quarter of 2002.

FUTURE TRENDS AND CONCLUSION

The B2B e-marketplaces will represent in the future the new efficient paradigm that will be used to accompany all trading process concerning goods, commodities in particular. They will be the "low cost" arenas where large companies will invite their provider to submit proposals at the best prices. But they will also be the "low cost" arenas where SMEs will present their products and their brands in order to be known and contacted by a larger and larger audience. E-marketplaces will more likely act per industry, instead of being "cross-industry, general purpose," and will show the two natures as previously mentioned, that is:

- Organized and driven by a large enterprise that will use the e-marketplace in order to attract a great number of suppliers, globally widespread, and able to enhance their good value proposition. This will ensure a strong competition and low supply prices.
- Organized and driven by communities (associations) particularly composed by small enterprises that will group their efforts in order to make their brand and their products more globally known as possible. Through this mechanism, SMEs will also implement collaboration tools and instruments in a sort of "partners' chain network," where different facilities, such as marketing services, information systems, administrative, and financial services, can be shared at lower and more competitive internal prices by creating at the same time both efficiency and new dimensions of effectiveness.

REFERENCES

Ansoff, H. I. (1987). *Corporate strategy* (2nd ed.). Harmondsworth: Penguin.

Bollier, D. (1996). *The future of electronic commerce.* Queenstown, MD: The Aspen Institute.

Dibb, S., Simkin, L., Pride, W., & Ferrel, O. (1991). *Marketing, concepts and strategies.* Boston: Houghton Mifflin.

Fariselli, P., Oughton, C., Picory, C., & Sugden, R. (1997). *Electronic commerce and the future for SMEs in a global marketplace: Opportunities and public policy requirements.* Brussels: CEC.

Geppert, L. (2001, December). *E-Marketplaces: New models.* Paper presented at the First Policy Group Meeting of DEEDS Project, Brussels

Hammer, M. (1996). *Beyond reengineering.* London: Harper Collins Business.

Kalakota, R., & Whinston, A. (1996). *Frontiers of electronic commerce.* Reading, MA: Addison-Wesley.

Kyte, A. (2001, February). Paper presented at the iEB Europe Gartner Conference 2001, London.

McDonald, M. H. B. (1997). *Marketing plans: How to prepare them, how to use them.* Oxford: Butterwoth-Heinemann.

Porter, M. (1985). *Competitive advantage.* New York: Free Press.

Shapiro, C., & Varian, H. R. (1998). *Information rules: A strategic guide to the network economy.* Boston: Harvard Business School Press.

Timmers, P. (1999). *Electronic commerce: Strategies and models for B2b trading.* New York: Wiley.

ENDNOTES

1 At the moment, u.steel is not longer directly operating.

2 At the moment, Up2gold.com is not longer directly operating. It has been included in another larger initiative.

Section II
Aligning Web Strategy to Corporate Strategy

The second section is dedicated to the subject of the Web strategy alignment with the corporate strategy. The six articles hosted in this section discuss this theme focusing their attention not only on the aspects related to the development of the public presence on the Web (Memmola and Tzannis; Frigerio), but also through an analytical perspective which can either be a technique (Yen; Wen-Chen Hu, Yanjun Zuo, Lei Chen and Chyuan-Huei Thomas Yang), or integrate other technologies (Axel C. Schwickert, Bernard Ostheimer) or integrate cost measurement and performance management processes (Massimo Memmola, Giovanna Palumbo and Mauro Rossini).

Chapter VII
Trends of Web Services Adoption:
A Synthesis

Vincent C. Yen
Wright State University, USA

ABSTRACT

The technology of Web services (WS) has been a hot area in the software industry for many years. Many organizations in the past 5 years have conducted surveys designed to get a profile of the state of Web services adoption in various subject areas. Some of those survey results are available free from the Internet. Since conducting a large scale Web services survey takes time and significant financial commitment, the research conducted in this chapter is a synthesis from published free survey results. All sources of surveys indicate Web services are being adopted more or less in all mid-size to large organizations because of realized benefits, and are anticipated to become a viable component of information systems infrastructure. Some of the current issues in Web services adoption and implementation are standards, training, and security.

INTRODUCTION

For software components to be reusable across different architectural environments new standards of integration and interoperability have been proposed and developed. The maturation of the Internet and the World Wide Web accelerates the idea for the global distributed computing. In order to make a large number of heterogeneous application systems on the Internet interoperable, many standards have been produced and practiced, for example, CORBA, COM, DCOM, and Java/RMI initiatives. *Common object request broker architecture* (CORBA) is a specification defined by the Object Management Group, DCOM is an extended version of COM of Microsoft's distributed common object model, and Java/RMI is the remote method invocation mechanism. However, these technologies are not compatible and are difficult to use. The success of these standards is rated as marginal (Chung, Lin, & Mathieu, 2003).

A recent approach to tackle the interoperability problem is XML-based Web services, or simply Web services (WS) (Alonso, Casati, Kuno, & Machiraju, 2004). The definition of Web services as offered by W3C (w3.org) is: "A Web service is a software application identified by a URI, whose interfaces and binding are capable of being defined, described and discovered by XML artifacts and supports direct interactions with other software applications using XML based messages via internet-based protocols." This approach uses Web standards of HTTP, URLs, and XML as the lingua franca for information and data encoding for platform independence. Three XML-based protocols, one for communication, one for service description, and one for service discovery have become de facto standards. They are:

- The simple object access protocol (SOAP) provides a message format for communication among Web services.
- The Web services description language (WSDL) describes how to access Web services.
- The universal description, discovery, and integration (UDDI) provides a registry of Web services descriptions.

Additional standards that are essential for applications of Web services have been developed. Two major standards under the category of "Web services composition" are

the business process execution language for Web services (BPEL4WS) (Fischer, 2002), later called business process execution language (BPEL), and another competing standard called the business process modeling language (BPML) developed by the Business Process Management Initiative (BPMI, www.bpmi.org). Programming tools are now available for creating and composing Web services. For example, BPEL4WS has been incorporated in Microsoft's ASP.Net and BPML has been incorporated in Java.

Solid Foundations in Web Services

It is obvious that Web services technology could be the catalyst for a potential revolution in providing "services" within a company and on the Internet, and its impact might be paramount. Web services are not only a key development area of the software industry in languages, tools, and standards, but also are very active in research both in industry and academic institutions (Zhao & Cheng, 2005). Because the platform of Web services is designed to allow complex composition of a new service from arbitrary (at least in theory) number of services, thus an important challenge is how to correctly describe, compose, and verify them. Many popular modeling languages for Web services such as BPEL4WS and BPML have theoretical underpinnings using Petri Nets (Petri, 1962) and Pi calculus (Milner, 1999). For example, Smith and Fingar (2003) claim that the conceptual representation and execution of business processes of BPML is based on Pi calculus.

Web Services and SOA

The concept and framework of Web services bring about a new level of abstraction over object- and component-oriented software development. This new level of abstraction is referred to as service oriented architecture (SOA) which is not tied to a specific technology. The Organization for the Advancement of Structured Information Standards (OASIS, www.oasis-open.org) defines SOA as "a paradigm for organizing and utilizing distributed capabilities that may be under the control of different ownership domains. It provides a uniform means to offer, discover, interact with and use capabilities to produce desired effects consistent with measurable preconditions and expectations." A simpler definition of SOA is: "A standards-based approach to managing services made available by different software packages for reuse and reconfiguration" (Austvold & Carter, 2005). Any SOA architecture requires defined standards for services to be interoperable on the Internet or any other network independent of the underlying platform and programming language (Hamid, Nezhad, Benatallah, Casati, & Toumani, 2006). Lewis, Morris, Simanta, and Wrage (2007) point out that "SOA is an architectural pattern/ style/paradigm and not the architecture of the system itself; SOA-based systems are about more than just technology; and SOA cannot be bought off-the shelf." SOA can be implemented using Web services technologies, but it does not require the use of Web services technologies. Currently, Web services are becoming the technology of choice for realizing SOA. Since SOA is a broader concept than Web services, recent surveys have extended the core concerns in Web services to SOA planning and development.

Business Processes and SOA

Business process management (BPM) is a very popular topic in the software industry, business sector, and research communities today. Its importance is highlighted by the "reengineering" era of the 90s and the book *Reengineering the Corporation* by Michael Hammer and James Champy. They call for conducting analyses of all critical business processes of an organization because of potential existence of opportunities for making dramatic (not minor) improvements in operational effectiveness and efficiencies. Consequently, analysis, design, and construction of business process tools are needed. Since core business processes have strategic implications, being able to continuously make changes, designs, modifications, or even create new ones in short notice is a very desirable feature of the IT environment. Orchestration and choreograph languages for Web services and SOA are well suited to such an agile requirement.

Under the framework of Web services, the software industry has developed companion products using a graphical modeling approach for business process designs. For example, BPMI (www.bpmi.org) has developed a standard business process modeling notation (BPMN). It consists of a small set of readily understandable graphical notations for the designing of business processes and is convertible to BPML and BPEL4WS. Such products could raise the productivity and agility of IT and business analysts to a higher level.

Source of Survey Results and Interpretations

From trade magazines to professional journals, one could sense that the IT industry and end users have immense interest in learning the paces of adoption, benefits, and drawbacks of Web services and SOA. Unfortunately, the authors are unable to find an academic journal grade survey research article in this area. The research is based on survey results published on the Internet and aims to understand what the market trend looks like and reasons for adopting or not adopting such technologies. From the year 2001 to 2004, surveys conducted are limited to the subject of Web services, but beginning 2005, many surveys shift from Web services to a broader context of SOA. A reason could be that Web services adoption has reached to the mainstream among medium to large companies and the IT industry wants to emphasize more broader and fundamental issues underlying service oriented infrastructure. In analyzing free survey results, two difficulties among many others surfaced: 1) lack of clear definition of terms used and 2) numerical values are not additive. Due to the vastness of the Internet, it is inevitable that the research has overlooked some published surveys and therefore the article does not represent a total complete investigation. Nevertheless, the conclusions drawn from the synthesis of collected publications should shed new light in the use of publicly available surveys.

OBJECTIVES OF STUDY

The study intends to find some evidence of how companies/organizations motivate, adapt, apply, and benefit under the new paradigm of WS/SOA with an expectation that such findings would be useful addition to their own experience. The objectives can be simply explained in the diagram below. There are four primary target areas of interest:

1. Business applications - to get an understanding of the role of WS/SOA applications development.
2. People – to collect data on WS/SOA skill levels, acceptance, and support.
3. Technology – to identify what programming languages, tools, and standards used.
4. Organization – to learn the basic demographic organizational data, management commitment, policies, and so forth.

In order to find information in the four areas stated above, it is important to define what information is of interest for collection before conducting the study. Below are some examples of questions of interests.

- Find out whether companies are ready to adopt Web services/SOA, and if so, with what timeframe.
- Find out what kind of Web Ssrvices/SOA will emerge, and how they will be distributed (payable or not).
- Find out about choices in terms of technologies and solutions.
- Identify the main technological drivers and threats.
- Identify attitudes and concerns of developers in their development efforts in the Web services.
- Where would Web services/SOA be generating values for business?
- How much interest existed in applying WS/SOA to the supply chain operations?
- Readiness in using WS/SOA. What steps are taken to build a company's WS/SOA capability?
- Internet-based procurement
- About future uses of and barriers to Internet-based e-commerce activities.

Due to the limitations of available data, it is difficult to find adequate answers to some of these questions, with the "organization" category

Figure 1.

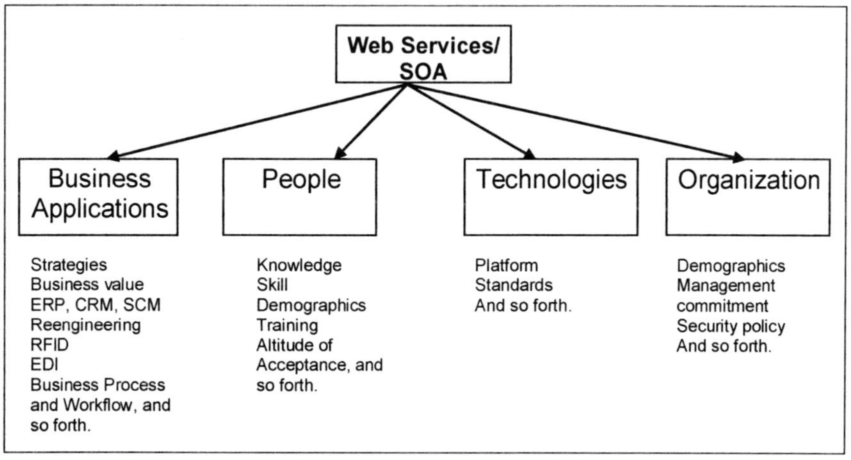

in particular. Nevertheless, published data do provide information to certain vital questions of interest. These areas are presented in the following sections.

REVIEW OF PUBLISHED SURVEYS

The name of Web services was first mentioned by Bill Gates in July, 2000 (Levitt, 2001). Beginning in the year 2001, surveys conducted reveal some aspects of Web services adoption characteristics. In this chapter, the concept of "adoption" is defined as any combination of the actions such as using, building, testing, implementing, and/or deploying. For each year an estimated mean adoption rate is calculated which is based on the average of the adoption rates found from that year's published surveys.

It is important to keep in mind that the study does not use all surveys conducted in the past because they are either not easily available or they require huge fees for access. This study uses data primarily published on the Internet and accessible to the public. The form of the free survey results that are available are often contained in a briefing or in a summery style of news release, not a complete report. This form of data is usually the result of data analysis, not the raw data, thus has limitations for interpretation. However, the data do carry useful information and have value when they are used collectively.

Some WS survey data are available in the year 2000; for example, Evens Data Corporation reports 42% of the large companies surveyed started working on WS standards. For what follows, this report will compile survey results annually beginning from the year 2001. Unfortunately, the context of each survey is not the same from year to year, thus it causes follow-up difficulties. At any rate, an assessment of the mean WS adoption rate is determined using the average adoption rate reported in that year.

2001 Surveys

Two well-documented survey reports available from the Internet are published by TechMetrix Research (2001, 2002). In 2001, TechMetrix Research conducted a survey titled "Web Services Adoption & Technology Choices" among the subscribers of TechMetrix/SQLI's TrendMarkers newsletters. A key finding is that "28% of companies claim to have already started using/building Web Services". The responses come from various countries of the world. The survey contains information such as company size, geographical distribution, job title, and industry type. The aims of the study are to find out the readiness of Web services adoption, what kind of Web services will emerge, the main drivers and threats, and technology choices and solutions. This chapter only uses portions of data from these surveys. A survey by Evans Data Corporation (EDC) finds 80% are already incorporating the leading Web services standards (i.e., XML, WSDL, SOAP, and UDDI) into applications. The key point learned from the survey is that 2001 is a year of companies showing evidence of significant interest in experimenting and believing in the potential of the new paradigm of Web services.

2002 Surveys

An interesting finding is that WS adoption rate may be very much dependent upon the software vendor's product. An example is a survey conducted among the users of Borland Software during its user conference which indicates that an unprecedented 80% of respondents are either currently using Web services or are planning to use them in the very near future. Borland customers are using Web services across many industries, but of those surveyed, a surprising 24% are in healthcare, 14% are in finance, and 14% are in government. Unfortunately the report cannot differentiate the percentage of the adopted from those of the planned. In another report, IDC has estimated that just 5% of U.S. businesses in 2002 had completed a Web services project. This 5% will not be used for the average adoption rate because it only concerns completed Web services. Evans Data Corp. survey reveals in its summer 2002 Enterprise Development Management Issues Survey that 75% of the companies are incorporating some level of WS functionality into their applications.

Fontana (2002) quotes the result from the Hurwitz Group's Web Services Primary Research Opportunity Study, which is based on a polling of 300 IT professionals, that 45% of companies are implementing Web services, while another 36% are testing the technology. When the two percentages are added it equals 81%. The results also show that 47% are using Web services for internal integration projects and 25% for external integration projects.

From the same survey conducted by TechMetrix Research in 2001, which was rerun in 2002/2003, the percentage of already started Web services projects stays around 27%. A survey of CIOs by BEA Systems (Hayday, 2002) in Europe shows 54% of European companies adopted Web services, and 59% expect benefits.

2003 Surveys

According to Mimoso (2004a), of 273 Global 1000 companies surveyed recently by Westbridge Technology, 37% are currently using Web services in production, and 26% plan either to deploy a Web service within 6 months or to complete a proof of concept (a combined adoption rate of 63%). Of those using Web services, 70% are using them internally, while 48% are exposing them to the Internet for business-to-business transactions. Gartner Dataquest (2003) released the results of several surveys on North American enterprises' adoption of Web services and the analysis showed that virtually all the respondents were using some form of Web services or plan to do so within the next 12 months.

In a survey conducted by IDC (2004), it shows about 61% of government organizations (central/local) are already using Web services. The estimated average WS adoption rate (of 63% and 61%) in this period is 62%.

2004 Surveys

A Gartner survey of 110 companies indicates that 54% are already working on Web services projects or plan to begin such projects this year (Leavitt, 2004). It is interesting that 34% of developers indicate that they are deploying Web services applications to their users (Evens Data Corporation, 2004), which is a 7% increase from fall of 2003. In a survey of 280 large North American firms by Forrester Research (Leavitt, 2004) 66 (23%) respondents said their Web services were either in production or in development. Yankee Group surveyed more than 400 enterprise IT decision makers in 2004 Enterprise Web Services Survey 48% of those surveyed have Web services in production, 39% say plan to join them within the next 12 months. If it is assumed that by the year end half of the 39% has successfully put WS in production then Yankee's survey could mean a 67.5% WS adoption rate.

The Radicati Group said in a report entitled "Web Services Market 2004-2008" that 52% of Web services deployments have occurred in the United States, with the rest of the world accounting for the remaining 48%; Europe accounts for 39% of all deployments this year, followed by Asia Pacific with 6% and the rest of the world with 3%.

2005 Surveys

WebMethods (2005) reported that SOAs and Web services are catching on with global 2000 companies. From the responses of 480 survey respondents, more than 80% of respondents deploy Web services within their enterprises. In one IDG study, 28% of respondents stated that their companies are implementing SOA. A Forrest Research survey of large North American companies reported that more than 70% of respondents have already implemented SOA. Austvold and Carter (2005) of AMR Research state that 21% of their survey respondents have deployed SOA and 33% plan to implement it within a year. Using the AR survey data it is projected that the adoption rate for SOA is around 36.5% (sum of 21% and 16.5%). The survey by AMR is from 134 companies with at least 1,000 employees. The main anticipated benefit of adopting SOA from the respondents is "faster and more flexible reconfiguration of business process."

The *Information Week* (Babcock, 2005) survey reveals reasons for adoption and difficulties during the adoption in two questions:

1. What is your company's business case for adopting an SOA or Web services standards? Top five responses are: standardization (60%), business process automation (59%), business flexibility (52%), operational savings (48%), and better integration with business partners (43%).
2. If your adoption has fallen short of expectations, which of the following reasons best describes why? Top three responses are: Legacy systems could not be cost-effectively integrated (50%), cost more than expected (33%), and introduced more complexity into IT system (36%).

Vahidy's (2005) survey report finds how most components of SOA were used in actual deployments; two items relevant to this chapter are: Web services (71%) and BPM framework (14%). The survey also finds most companies (57%) have standardized on Web services as an SOA component.

Predicting Investments in SOA

As discussed in the beginning, SOA is getting more attention as it is more general than WS. But Web services are one of the major implementations of SOA today, which means the two are closely related. Surveys for the future direction of SOAs conducted by Yakee Group of Boston in 2004 and 2005 reveal the following:

1. Projections in 2005 (Mimoso, 2004b). Seventy-five percent plan on investing in the technology and staffing necessary to enable a service-oriented architecture. By industry, the greatest investments in SOA are coming from the wireless telecom and manufacturing markets (78%), financial services (77%), and healthcare (71%).
2. Projections in 2006 (Stansberry, 2005). The surge of SOA implementation in 2006 reaches saturation in many verticals: wireless (93%), retail (92%), financial (89%), manufacturing (76%), and government (75%).

However, a 2005 IDG Research Services Group survey reveals that SOA is not a well understood concept by the IT professionals because the survey finds an almost even split between people who claim some familiarity with SOA (52%) and those who admit they have not a clue (48%). In that same IDG study, 28% of respondents stated that their companies are implementing SOA, with about half of those SOA implementers merely conducting pilot projects.

2006 Surveys

A surprisingly good piece of complete SOA survey results conducted by eBizQ (2006) is published free on the Internet. The survey was conducted online on the eBizQ Web site and received responses from 313 companies in 21 industries. The top-five industries were technology, financial/banking, public sector/government, insurance, and healthcare/pharmaceutical. Key findings of the SOA survey are as follows:

1. In the four stages of SOA adoption the respondents are distributed as: exploration (34%), planning (17%), pilot (21%), and deployed (28%). I exploration rate is excluded, the adoption rate is 66%.
2. The distribution of number of Web services in production is: 0-10 (49%), 10-50 (31%), 50-100 (8%), and more than 100 (12%);
3. The majority of services (59%) deployed are for internal use. The industries offers of higher external availability of services are technology, financial/banking, and public sector/government. The survey also contains information on SOA governance issues.

In Gartner's (2006) survey of telecommunication carriers on SOA the adoption rate by that industry is at 90%.

A survey conducted by CodeFutures (2006) from 5000 U.S. and European Java developers reveals the following, based on a 10% response rate:

1. 76% of enterprises using Java are using or SOA.
2. Over 30% have scalability problems and 56% have performance problems due to bottlenecks.
3. The vast majority of Java developers surveyed believe they have high performance requirements.
4. About one third of the Java developers surveyed had groups using C++ as well, and Web services technology is still relevant for C++ development.

Surveys sponsored by Mindreef Inc. (2006) in September and October of 2006, conducted by Hurwitz & Associates, found 99 IT executives from companies in North America and the UK with a size greater than 250 employees who had

Table 1.

Source	2001 (WS)	2002 (WS)	2003 (WS)	2004 (WS)	2005 (SOA)	2006 (SOA)
TechMetrix	28%	27%				
EDC	80%	75%				
Borland		80%				
Hurwitz Group.		81%				
Westbridge Technology			63%			
IDC			61%			
Gartner				54%		90%
Forrester Research				23%	70%	
Yankee Group				67.5%		
WebMethods					80%	
AMR Research					36.5%	
eBizQ						66%
CodeFutures						76%
Mindreef Inc.						66%
Average (rounded)	54%	66%	62%	48%	62%	75%

expressed an interest in SOA or Web services. Approximately two-thirds of respondents surveyed (66%) had begun their SOA journey.

Following is a table summarizing the WS/SOA adoption rates from year 2001 to 2006. A mean score is calculated for each year as the estimated adoption rate. Note that the last 2 years adoption rates are focused on the SOA.

BENEFITS AND THREATS

Published surveys also have revealed information on motivations and reasons for nonadoption of Web services and SOA. Their findings seem quite consistent across different surveys. They are summarized in two categories: benefits and technical drivers, and threats.

Benefits and Technical Drivers

- Standards compliance and interoperability.
- Business agility and business process optimization.
- Scalability.
- Tools for development productivity.
- Tools for administration.
- Reuse services.
- Lower integration costs.
- Faster delivery of products.
- Making application development more flexible.
- Increase customer satisfaction and revenue.
- Important to business goals.
- Mainframe modernization.
- They reduce the burden of internal and external integration.
- They allow for true reusability.
- They are a platform-independent facilitator, enabling data to flow across applications and systems.
- They break down internal silos by providing information across traditional technological barriers.
- Web services extend the life of legacy systems by extracting specific business processes, such as licensing and appointment and quoting, and making them available in new forms.

Threats

- Security and authentication issues.
- Interoperability issues (e.g., noncompliant SOAP implementations).
- Lack of standard business schemas.
- Service level agreement of WS providers.
- Lack of awareness in the business.
- Developing effective ROI cases.
- Standards compliance and interoperability.
- Scalability.
- Tools for development productivity.
- Tools for administration.
- General knowledge of SOA within their enterprise.
- Governing development standards within their enterprise.
- Determining return on investment for SOA development and deployment efforts.
- Getting organizational buy-in.

CONCLUSION

Publicly available and free survey results on the adoption of WS and SOA are available in limited ways in that they either are contained in news releases or in reports. Even though the data are limited they still contain valuable information. One can conclude that Web services/SOA technology is now becoming critical and has widespread acceptance, even though the technology development is not entirely complete. WS and SOA are being recognized as generating benefits because the technology provides a platform for making the business process agile and optimized. A simple example of using the mean yearly WS adoption rate reported above can show trends. In the years 2001 to 2006, the percentages of major companies or organizations started projects of some kind involving WS/SOA are 54%, 66%, 62%, 48%, 62%, and 75%. The first four percentages are WS adoption rates and the latter two are SOA adoption rates. Roughly between 2001 and 2005, about half of the U.S. businesses were using WS. Apparently the adoption rate is on the rise in WS and SOA. In fact, the same phenomenon is true in Europe. Beginning in 2006, more companies have entered into SOA, and SOA is the center of attention among IT developers. It is safe to conclude that WS/SOA are in the mainstream or a pacing technology of the technology lifecycle. Another conclusion could be drawn is that Web services are primarily used internally within an organization; external use of Web services is rather limited. Further, more the data reviewed have information on "what motivates or deters the users from using Web services." That information is useful not only to managers and end users but also to IT vendors. Language-wise, Java and Microsoft.net are used by an equal number of SOA developers. Other information contained in the survey but not extracted here are technologies used, governance policies, industry group differences, and so forth.

In reading the published surveys, some questionnaire designs and interpretations are vague. For example, Austvold and Carter (2005) of AMR Research report does not make it clear whether WS is considered as a part or subset of SOA. Another issue is when a company may be declared as an adopter of WS/SOA. It is obvious that this research must be continued in order to answer the research objectives set forth in the early part of this chapter.

REFERENCES

Alonso, G., Casati, F., Kuno, H., & Machiraju, V. (2004). *Web services*. Springer-Verlag/Berlin Heidelberg.

Austvold, E., & Carter, K. (2005). *AMR research: Service-oriented architectures: Survey findings on deployment and plans for the future*. Retrieved November 2, 2007, ftp://ftp.software.ibm.com/software/soa/pdf/amr_files.pdf

Babcock, C. (2005). *InformationWeek*. Retrieved October 31, 2007, from http://www.informationweek.com;jsessionid=35XHAV0I1RVSWQSNDBCSKH0CJUMEKJVN

Broenen, J. (2006). Evans Data Corporation. *CodeFutures*. Retrieved June 19, 2008, from www.codefutures.com/weblog/corporate/archives/2006/03/java_developer_survey.html

ebizQ. (2006). Retrieved June 19, 2008, from http://www.actional.com/resources/whitepapers/SOA-Governance-Adoption/SOA-Survey.html

Evens Data Corporation. (2004, Spring). *North American development survey*. Retrieved June 19, 2008, from www.evensdata.com

Fontana, J. (2002, May 17). *Network world fusion*.

Gartner. (2006). *User survey report: Carriers' service-oriented architecture and Web services*. Author.

Hamid, R., Nezhad, M., Benatallah, B., Casati, F., & Toumani, F. (2006). Web services

interoperability specifications. *Computer*. IEEE Computer Society.

Hayday, G. (2002). *CIOs slowly turning to Web services*. Retrieved June 19, 2008, from http://zdnet.com.com/2100-1106-960985.html

IDC. (2004). *Western Europe, government sector, Web services and new technologies, levels of adoption and investment plans: An IDC survey.* Retrieved June 19, 2008, from www.idc.com

Leavitt, N. (2004, November). Are Web services finally ready to deliver? *IEEE Computer.*

Levitt, J. (2001, October 1). From EDI to XML and UDDI: A brief history of Web services. *Information Week.*

Lewis, G., Morris, E., Simanta, S., & Wrage, L. (2007). *SOA common misconceptions.* Retrieved June 19, 2008, from www.iccbss.org/2007/more-info/SOA-Common-Misconceptions.pdf

Microsoft. (2006). Retrieved June 19, 2008, from http://www.microsoft.com/biztalk/solutions/soa/overview.mspx

Milner, R. (1999). *Communicating and mobile systems: The Pi calculus.* Cambridge: University Press.

Mindreef, Inc. (2006). Retrieved June 19, 2008, from www.mindreef.com/report

Mimoso, M. S. (2004a). *More enterprises exposing Web services to Net.* Retrieved June 19, 2008, from www.SearchWebServices.com

Mimoso, M. S. (2004b). SOA prominent on 2005 budgets. *Yankee Group of Boston.* Retrieved June 19, 2008, from www.SearchWebServices.com

Petri, C. A. (1962). *Kommunikation mit Automaten.* Unpublished doctoral thesis, Institut f¨ur instrumentelle Mathematik, Bonn.

Smith, H., & Fingar, P. (2003a). *Workflow is just a Pi process.* Retrieved June 19, 2008, from www.bpm3.com/picalculus

Stansberry, M. (2005). *Yankee group: SOA everywhere by 2006.* Retrieved June 19, 2008, from www.SearchDataCenter.com.

TechMetrx Research. (2001). *Web services adoption & technology choices report.* Retrieved June 19, 2008, from http://www.techmetrix.com/products

TechMetrx Research. (2003, February). *Adoption of Web services & technology choices, version 2 – February 2003 analysis of survey results.*

Vahidy, T. (2005). *The SOA revolution.* Retrieved June 19, 2008, from www.line56.com/print/default.asp?ArticleID=7099

Webmethods. (2005). *Survey of 480 global 2000 companies.* Retrieved June 19, 2008, from www.Webmethods.com

Zhao, J. L., & Cheng, H. K. (2005). Web services and process management: A union of convenience or a new area of research? *Decision Support Systems, 40,* 1-8.

Chapter VIII
Web & RFId Technology: New Frontiers in Costing and Process Management for Rehabilitation Medicine[1]

Massimo Memmola
Catholic University, Italy

Giovanna Palumbo
Ospedale Valduce, Italy

Mauro Rossini
Ospedale Valduce, Italy

ABSTRACT

Radio frequency identification (RFId) has recently begun to receive increased interest from practitioners and academics. This type of technology has been widely used in healthcare organizations for different purposes, like to localize patients, devices, and medical instruments. This chapter presents the results of a study in which we used RFId technology and modern systems of cost management methodologies (e.g., activity-based costing, activity-based management, and process management) in a "proof of application" aimed at defining some specific data on care needs of a person with a disability, costs of the main activities performed during the person's rehabilitation process, and level of performance which could be reached in order to improve the "disability management" process, from a clinical as well as a managerial perspective.

INTRODUCTION

In recent times there has been a sustained trend of growth of healthcare and rehabilitation services demand coming from people affected by different types of psycho-physical disabilities related even to genetic characteristics and/or different type of traumatic events (e.g., stroke, traumatic brain injury, spinal cord injuries, anoxia, etc.).

In order to have an idea about the "dimension" of this phenomenon, it is useful to refer to the World Health Organization (WHO), who has estimated a total of around 600 million people with some kind of disability all over the world.

Some reasons for this rapid growth may be found in the interaction among different variables:

- Demographic, referred to population growth and/or progressive aging.
- Social, mainly related to the higher probability of traumatic events related to traffic crashes, domestic and/or sportive accidents, work accidents, war-related, and/or violence-related event injuries.
- Medical, related to a higher surviving rate after critical events due to the use of new trauma management techniques mainly in the acute phase and in general, to the higher survival rate of subjects with some kind of disability coming from congenital and/or acquired malformation of the central nervous system.

As an answer to such a growing demand, rehabilitation medicine has became a specific discipline aimed at the integral recovery of physical, psychical, social, and functional capacities of people with disabilities.

These change-drivers have an important impact on the organization of activities in rehabilitation which, inevitably, follow different criteria if compared with acute care services. In rehabilitative medicine it is necessary to deal with "long term complexity" coming from the different potential development of a disability itself as time passes by. This characteristic, which is not present in acute care, is usually concentrated on short-term care. Furthermore this long-term feature has promoted a progressive trend toward home care vs. the option of in-hospital stay. This trend has been supported by the development of technological devices able to guarantee an appropriate quality of life for the person with disability while staying at home.

Beside these issues, which are mainly referred to clinical factors, there are others mainly related to managerial aspects such as the way the healthcare service is organized, how much it costs, and how the rehabilitation process may be evaluated. In fact, rehabilitative medicine, as a fundamental aspect of disability treatment and even more than acute care field, poses some difficult questions about the measurement of type and quality of service offered and/or performed. Up to now those questions have received only partial answers through experimentation. Furthermore, experimentation in rehabilitation management has been considered a nonclear and nonprecise experimental field.

Obviously, it is not just a trivial problem of working-load identification. Rather, its objective is to know if the care service performed is suitable to the diverse needs a person with a disability presents. This includes in-hospital as well as home care, in a continuous-care perspective. In rehabilitative medicine it is necessary to define what to do, how many people should be involved, which kind of professional must be involved, how long they will be involved, during which part of the day-time service will be provided, and which type of knowledge should the caregiver have in order to be able to assist the person with a disability. Therefore, the critical points are: Which and how much care service should be provided, who should be responsible for each activity, how it must be performed, and, finally, how much it costs.

The answer to such questions depends on the possibility to measure costs through a structure that permits taking into consideration the specific nature of chronic care clinical paths. Those clinical paths may be considered as very complex "production processes," which are usually performed in medium/long term periods (maybe during years) and which results present serious difficulties for being evaluated, if they could be evaluated at all. It must be considered that a rehabilitation process, different from an acute care process, does not mean the complete overcoming of a disability itself. Moreover, they usually are very "labour intensive" processes which produce a high working load to all healthcare operators involved, such as medical doctors, nurses, or physiotherapists.

It is therefore, necessary, to define a system able to register clinical performance, (i.e., improvement of the degree of functional independence of the person with disability) as well as managerial performance (i.e., costs of rehabilitative clinical paths, optimization of working load assignment, and maximization of the use of available resources)

As in other managerial fields, in rehabilitative medicine information and communication technology (ICT) may play an important role. This chapter presents an exploratory case study (Yin, 2003) aimed at clarify the potentiality of using new technologies of active and passive geographical localization (i.e., radio frequency identification [RFId]) in order to solve some of the problems mentioned above. RFId technology is nowadays widely used in different sectors, from logistics to massive distribution and production systems. Using it in the healthcare field is not a new trend, since literature refers to plenty of experiences where this technology has been used to trace, for instance, the movement of patients, devices, and other clinical equipment in different areas.

This research has been performed considering the use of RFId technology and modern systems of cost management methodologies (i.e., activity-based costing, activity-based management, and process management) in a "proof of application" aimed at defining some specific data on care needs of a person with disability, costs of the main activities performed during a person's rehabilitation process, and level of performance which could be reached in order to improve the "disability management" process, from a clinical as well as a managerial perspective.

THE LIMITS OF A TRADITIONAL APPROACH TO ACTIVITY-BASED COSTING AND IMPLICATIONS OF TIME-DRIVEN APPROACH

Activity-based costing (ABC) is a particular methodology used for the determination of the costs of a product or a service. It is widely known in literature (Brimson, 1990; Cooper, 1988a, 1988b; Cooper, 1989a, 1989b; Cooper & Kaplan, 1988; Johnson & Kaplan, 1987; Shank & Govindarajan, 1992) as well as in the current managerial practice.

This methodology was developed during the 80s as an answer to the changes that have, since then, characterized the organizational management and the corporation governance. Changes like the globalization of the market place, the rapid growth of automation applied to productive processes, the development of new products, and an increasing complexity of corporate structures were the framework in which ABC intended to go over the typical limitations and inefficiencies of the traditional systems of cost-control.

In fact, the absence of alignment between the new structures and the traditional accounting methodologies introduced some dangerous distortions into the information related to costs. Among them:

- A wrong evaluation of the costs related to corporation complexity

- Attention concentrated exclusively on economic/financial values
- Goals oriented mainly toward short-term results with special emphasis on budget vs. execution control
- Low consideration to production-supporting activities
- A rigid corporate structure based on functional units with no attention paid to accountability centres
- Incapacity to see the trans-functional dimension of corporate processes
- The ability to highlight the symptoms of problems but not their true causes
- The absence of a strategic conception, able to conceive corporation as an open system in continuous interaction with costumers, providers, and competitors.

The essence of the ABC approach is that it is very difficult to observe the direct relationships between the quantitative measurements of production factors and the measurements related to costs and prices of final products.

The possibility to determine the cost of a particular product or service from an ABC perspective involves the performance of three main tasks:

- Phase 1: identification of activities involved in each process
- Phase 2: identification of the amount of resources used in each activity
- Phase 3: identification of the contribution each activities gives to the final object of cost

However, after two decades have passed since the first contributions appeared in literature, which had provided the general framework and the definition of the model, there is a kind of "feeling" that activity-based costing has not maintained its promises (Kaplan & Anderson, 2004). Contrary to expectations, traditional approaches have not been abandoned, ABC has not shown the expected generalized diffusion, and, above all, some implementation projects have been unsuccessful.

The reasons for such failures may be brought back to the complexity of the calculation model, which makes the task particularly difficult and complex. Inevitably, the costs of designing and maintaining the system itself have proven to be higher than those related to traditional systems. Moreover, it is important to mention the negative impact of such a complex and difficult system on the staff satisfaction level.

However such a complexity is linked to the companies' policy when organizing an activity-based costing system. After having outlined an *activity map* they usually proceed to quantify how much time in percentage every employee dedicates to each activity compared with the total time at disposal. Such a percentage enables one to obtain a parameter (*resource driver*) through which you can determine the cost of the resources used for every activity. These estimated values are based upon the employee's perception and are obtained through interviews or surveys carried out within the company.

So far there is really not much to add since the approach undertaken is quite normal, and, at least from a theoretical point of view, it does not raise any other consideration. From a practical point of view, instead, the matter changes.

Experience has shown that this kind of approach, where detailed processes, activities, and resources must be identified, may be easily applied to small and medium companies, or even in big corporations, as far as it is used just in some specific unit and/or on experimental basis.

Difficulties emerge when the system is developed on a large-scale basis, mainly when a highly complex structure is present. And this is the case of a hospital structure.

In such a context you have to organize thousands of employees under different professional categories. Most of all you have to handle a particularly complicated *activity map* due to the numerous treatments to undertake according to the various pathologies which may affect a patient. It is quite easy to imagine how difficult it can be to quantify how much time clinicians and nursing staff may dedicate to the various activities linked to different treatments.

Furthermore, even if you succeed in quantifying time after long and laborious work (very expensive too), it would be ever so difficult to be maintained. The course of a process in a hospital is often subject to modifications to comply with the innovations, which, as time goes on, are brought to the diagnostic-therapeutic protocols relevant to a specific pathology.

In practical contexts the activity-based costing fails to achieve one of the targets for which it had been created, that is, adequately capturing and processing the organization complexity. Besides, it has another limit: It has no perception of the complexity implied in carrying out each activity.

Rehabilitation medicine can produce many examples of this. The activities undertaken, especially when they require more interaction with the patient and more assistance, mostly depend on the type of disability of the patient. Although being caused by the same pathology, disabilities are a very personal matter which can vary according to sex, general conditions, psychological attitudes, and to previous rehabilitation treatments performed either in a hospital or/and at home.

It could be useful to define different types of activities which would enable one to detect the level of disability of the patient at a certain time[2]. The complexity of the *activity map* reveals that this solution can only be theoretical.

As a final consideration on the limits of a traditional approach for implementing an ABC system, we could enquire with the hospital personnel on how they cope with the various activities; this may lead to the definition of a resource driver erroneously based upon a vision of full saturation of the availability of production capacities. It is necessary, therefore, to make a distinction between theoretic and real production capacities. In the first case it corresponds to the number of hours personnel is present within the structure (e.g., 8 hours a day for 6 days a week); the second case derives from the first but you must deduct a 15-20% of time which personnel dedicates to other nonspecific productive activities such as arrivals, departures, breaks, communication with other colleagues, formation courses, and so forth. A similar thing may involve machinery, for maintenance, repair and scheduling which reduces the availability of production capacities.

If resource drivers have to show resources (such as personnel, medical material, and facilities) consumed by each activity, parameters obviously must consider real production capacities or must quantify the real time each person has dedicated to the various activities.

This is necessary not only to improve efficiency levels, but mostly to enable a correct handling of the work load created by patients. In fact, according to the number of patients in hospital, assistance will be directly proportioned to the disabilities displayed at a certain time, since this value can change time to time.

So the questions are: Must the activity-based costing be left aside, reiterating old cost accounting systems? Has the calculation of activity-based costs revealed it to be a failure?

It is the personal opinion of the authors, supported by literature, that the ABC system has shown to be, during these past years, a wonderful instrument to improve company efficiency, to rationalize production systems through the elimination of activities with no value added, and to identify opportunities for continuous improvement even where economical-financial aspects are not directly involved.

A possible solution to the problems discussed above is to implement the activity-based costing system proposed by Kaplan and Anderson (2004), known as "time driven activity-based costing." The solution proposed is quite simple and may be achieved by totally changing the way the matter has been faced up until now. According to their proposal, managers should proceed to measure:

a. The effective capacity of production as a quota of the total capacity theoretically available.
b. The total overhead divided by effective capacity of production measured in time units, in order to obtain the cost per time-unit.
c. The effective time used by each person for the performance of a particular task.
d. Quantification of overhead weight on each activity.

The two authors wish to point out that it is not crucial to quantify in detail the time required by an employee to carry out a specific activity; it is just sufficient to make an accurate and credible evaluation. The thing to emphasize is that it is no longer necessary to carry out far-ranging surveys to calculate in percentage the time personnel dedicate to the various activities, but it becomes essential to know *how long it takes to complete a certain activity*. This estimation must be carried out by the managers, who can in any case involve personnel through interviews, meetings, and so forth.

It is in the authors' opinion that even this approach will not completely eliminate limitations and problems as aforementioned.

The weak point of Kaplan and Anderson's ABC system is when it has to quantify time required by employees to carry out each activity. In the opinion of the two authors, managers should determine this parameter by cooperating with personnel and searching for a credible quantification which does not have to necessarily be detailed.

Although, by acting this way the quantification of a resource driver will not be objective enough since it could be somehow affected by personal perceptions. Managers, in fact, do not always have a complete knowledge of each process and operation phase relevant to every activity. This is also due to their different professional formation. For example, the financial manager of a hospital may only have a broad idea of the various clinical activities. In this case, the cooperation of personnel directly involved in these activities becomes fundamental. The danger is that they may give wrong evaluations not necessarily for personal interests but due to consolidated habits which somehow may ignore situations of inefficiency.

Besides, quantification would express average values which would not enable one to detect the differences due to different professional formation or experience; as a matter of fact, quite likely time dedicated to some of the activities could be reduced if carried out by more competent people.

Another further consideration is that Kaplan and Anderson state that the cost of an activity is subject to modification in the following two cases: a) when costs change for personnel needs (e.g., wage increase, operational cost increase when acquiring new instrumental goods); and b) when there is a change in the resource driver, in other words, in the quantity of resource employed for each activity due to either improvement or worsening of production efficiency. In both cases a manual updating of the parameters of reference is necessary to allow the costing system to record the changes. This is a common requirement for traditional accounts systems, but it becomes quite difficult to handle for an ABC system when it has to face a huge activity map.

In the end, in order to handle the complexity of activities strictly linked to the level of disability of a patient, in the case of rehabilitation medicine, Kaplan and Anderson propose an *if-clauses* structure which leads, however, to a considerable increase of complexity and maintenance of the system[3].

Through the specific design used in the experimentation reported in this chapter, we have tried to get through the problems discussed above, using the potentialities of a combination of new technologies (i.e., RFId, the Web, and WiFi) in order to obtain relevant data able to sustain an ABC system on real basis.

The technological infrastructure, indeed, facilitates the registration of the working load generated by each person with a disability being assisted inside the structure in a given moment, on a real-time basis. This means that it is possible to have, on real time, the information about the consumption of resources (i.e., staff time, sanitary materials, and/or instruments) associated to the activities performed in order to fulfil a specific clinical path.

In other words, it is possible to know in each moment, maintaining the privacy of patients and operators[4], how much and which kind of assistance is performed, given a particular "mix" of pathologies, which is (if any) the saturation degree of resources and which is the cost of the service.

RADIO FREQUENCY IDENTIFICATION (RFID) TECHNOLOGY

The RFId technology is, among all emergent technologies, the one that presents the highest potentiality of development and application (Nextinnovator, 2004; Teradata, 2004). It consists "of small integrated-circuit 'tags' that can store information and announce their existence passively (or actively, NOA) through wireless radio communication to a networks of RFId readers. RFId potentially can track physical items–medical equipment, patient charts, even patients- in real time" (Janz, Pitts, & Otondo, 2005).

Literature, even if it is still rather scarce, refers to some applications of this technology in a healthcare context. Until this moment it has been fundamentally used to localize devices and/or instrumental goods in order to keep track of them and make them available whenever they are needed (Becker, 2004; Glabman, 2004). There are also some applications used to localize patients inside the hospital structure (Becker, 2004) or to evaluate the conditions of under and/or subuse of some particular equipment (McGee, 2004).

The benefits obtained from the use of RFId technology in a healthcare environment are not restricted, however, to the improvement of supply chain efficiency, but they could also be applied to the improvement of outcomes in a given clinical path (Wicks, Visich, & Suhong, 2006). In this sense some experiences show the use of RFId tags aimed at verifying the correct administration of medicines, to the proper patient and in the proper dosage (McGee, 2004), or else to avoid that some clinical equipment and/or surgical instruments may be "forgotten" inside the patient's body, or that some instruments may be not correctly sterilized. In the future, RFId technology may be used to control the access of patients, staff, and/or visitors to some restricted areas of hospital structure.

Reviewing the literature, it seems to be clear that there have not been reported experiments in which RFId technology is used to support costing systems and/or process management systems. This is the challenge of this chapter.

THE ADVANCED HEALTH PROCESS MANAGEMENT (AHPM) PROJECT

The AHPM project is aimed at developing an exploratory case-study (Yin, 2003), performed with the objective of evaluating the usability of RFId technology and combined with the potentialities of Web and WiFi systems, in order to define an ABC system along with a process management system able to get through the typical inefficiencies coming from traditional accountability approaches.

This case study was performed in a specialized rehabilitation centre, that is, "Centro di Riabilitazione Villa Beretta - Costa Masnaga (Lecco)", Italy. Villa Beretta, a secondary structure of Valduce Hospital, is an Italian excellence centre dedicated to patients with traumatic brain injury, stroke, and other cognitive and movement disorders. It is a 81-bed hospital, with 154 personnel staff, organized as follows: 10% physicians, 18% therapists, 24% nurses, 26% health support operators, 14% administrative staff, and 8% consultants. In this centre is a functioning WiFi network since 2006.

Research Objective

This study is aimed at determining the possibility of using RFId and Web technologies in order to define the cost of a person with a disability in a rehabilitation context using working load and activity-based costing as the main methodological approach.

Research Methodology

The project has been performed in three main phases. The first one was technologically-oriented, and it involved the application of technical devices to the measurement of contact time among service users and healthcare operators.

The second phase, analytical, was dedicated to the development of an activity-based costing system able to combine organizational strategic goals with operational aspects. This phase was developed in five steps:

- Preliminary analysis, which consists of the selection of specific clinical paths and the organizational units involved.
- Activity and process mapping, which is the identification of macro- and micro-activities and their relationship in selected clinical paths.
- Hospital costing-based system, which is the study of costs related to selected clinical paths.

The third phase refers to the evaluation of the organizational impact of RFId/Web technology in a rehabilitation hospital (RH).

Execution Steps of the Project

Given the complexity of the project from a technological as well as an organizational point of view, the first part has been defined as an experimental phase which will enable researchers to identify:

- The reliability of current technology;
- Design, test, and tuning of software; and
- Identification of potential organizational barriers.

In this sense, in order to obtain such preliminary assessments, the experimental phase has been performed in the following steps:

A. Identification of processes/activities to be studied
B. Implementation of technology and devices (main criteria used was to choose a standard technology available in the market place)
C. Software development (description of macro- and micro- activities)
D. Identification of patients
E. Data collecting (organizational aspects)
F. Data analysis

A. Identification of Processes to be Studied

In order to simplify the experiment, it has been decided that just some activities will be followed, which are the normal sequence of actions performed by operators during a standard in-hospital day.

This includes: general clinical control, medical visit, physiotherapy, personal hygiene, medicines administration, meals administration, and preparation of patient for night time.

B. Implementation of Technology and Devices

The technological structure was defined in three steps:

- Selection of standard technologies and providers
- Testing of technology
- Start-up

Some important criteria were considered,in order to select the most suitable technology:

- First of all, environmental and technological limits and conditions were evaluated in order to define a proper research strategy.
- Based on this information, a basic infrastructure was defined, taking into account that patients' activities were usually performed in different places; therefore, technological infrastructure should allow the "following" of patients wherever the activities were being performed.
- Each activity could involve more than one actor (i.e., patient, nurse, medical doctor, objects, and so on) which must be properly identified in order to avoid ambiguity during data collecting and analysis.
- Technology must be transparent to all participants and must be as minimally personal-dependant as possible.

This set of criteria has evolved into a decision of using a technology able to establish a specific "zone" around the patient which will allow the identification of roles and activities on a real-time basis. In order to strictly follow those criteria, a semiactive RFId technology was chosen; meanwhile, architectural design is patient-centered

Figure 1. Three-layered architecture

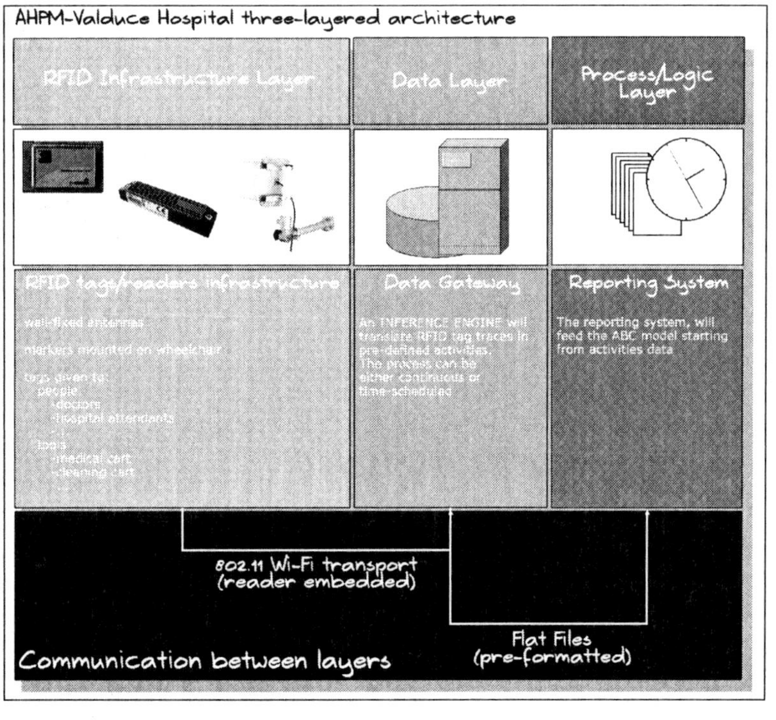

and it is based on three technological levels plus a fourth one to facilitate communication among them. Each one of these levels is independent from each other from a technological as well as functional perspective (Figure 1).

Level 1: RFId Structure

The first level includes the RFId structure. This level is dedicated to physical data collecting and is performed through specific technological components defined as RFId tags, antennas, and markers.

The RFId structure included following components:

- Four activity recording stations placed at strategic points determined by the previous environmental study. Each station includes a PC and a RFId antenna (Figure 2).
- Five markers: three of them were placed under patients' wheelchair, and the other two were placed near patients' beds (Figure 3). Each marker was battery supplied.
- Thirty-six RFId tags with univocal identification code numbers associated with different actors and/or objects.

The devices used were:

- Active UHF Tag i-B2 (*Identec Solution*) Intelligent Long Range® beacon tag. Static data are written on these tags and are continuously sent at a programmable interval (ping rate) without being requested by a reader.
- Position Marker i-MARK 2 (*Identec Solutions*). This device is used to determine exact locations of the i-B2 tag. As the tag passes the induction loop connected to this position marker, it is "woken up" and identification data are downloaded onto the tag. This device was adapted in order to allow the constant following of the patient.
- UHF Reader i-CARD R2 (*Identec Solutions*). It is ILR® beacon tag reader in a PCMCIA card format. It allows real-time information form tags.
- UHF Read/Write Antenna i-A9185 (*Identec Solutions*). It is an elliptically polarised RFID antenna specially designed for the identification of numerous tags at the same time in a large read zone. Since the antenna's polarization is elliptical, the direction of the tag relative to the antenna does not matter.
- CV60 Computer (*Intermec Technology Corporation*). These computers with full PC functionality provide the connections between UHF antenna through UHF reader cards and data storage servers. The connection was realized using Cisco® Compatible 802.11b/g radio standard wireless LAN.

Each patient was provided with a position marker and an active tag. Operators (i.e., medical doctors, nurses, and physiotherapists) were provided with active tags. Other active tags were attached to lunch tables, medication trolleys, and hygiene trolleys. When a tag passes near by position marker it is "woken up" and its identification code is registered. In this way the system is able to recognize the association between patient and

Figure 2. Timekeeping station

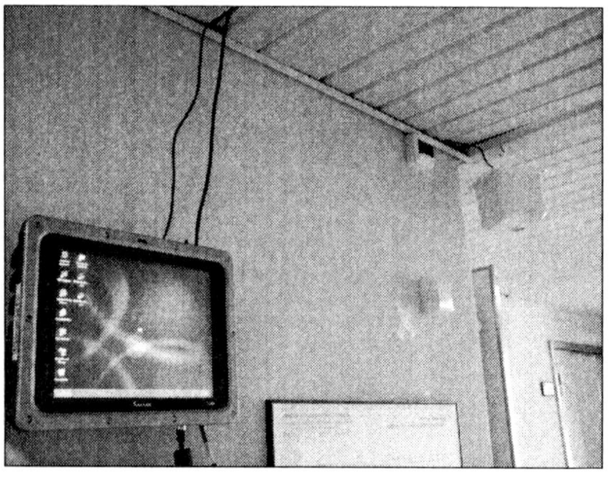

Figure 3. Example of meal administration scheme

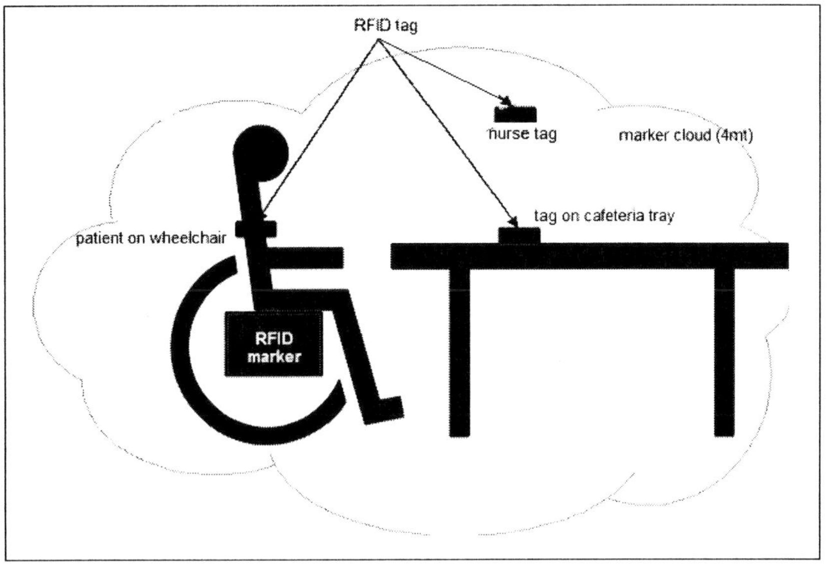

activities. All these information are collected by the RFId antennas and acquired by an activity recording station. Using wireless Wi-Fi network this information is transmitted and stored into a server.

Figure 3 shows an example of meal administration activity, where it is possible to observe the relationship among one marker reader and three RFId tags, that is patient, nurse, and cafeteria tray. Once the marker "wakes up" the tags it is possible to identify that a meal administration activity is being performed for the specific patient.

Level 2: Data Level

This level is mainly referred to the point where all information coming from tags arrives. In this point information is filtered and consolidated in different activities.

The activity monitor is the software component responsible for the registration of activities among patients, healthcare operators, and hospital assets. The activity monitor role is to receive signals from the RFId/WiFi infrastructure, which has already been described. After that, using a particular predefined system of rules, a specific activity is associated to a special combination of signals[5].

The activity monitor is a Web WAMP[6]-based application. This platform has been chosen for two main reasons: a) to reduce time related to technology and go-live development, and b) to reduce project costs.

Figure 4 shows the activity monitor functioning. The data collecting stations send information

Figure 4. The activity monitor

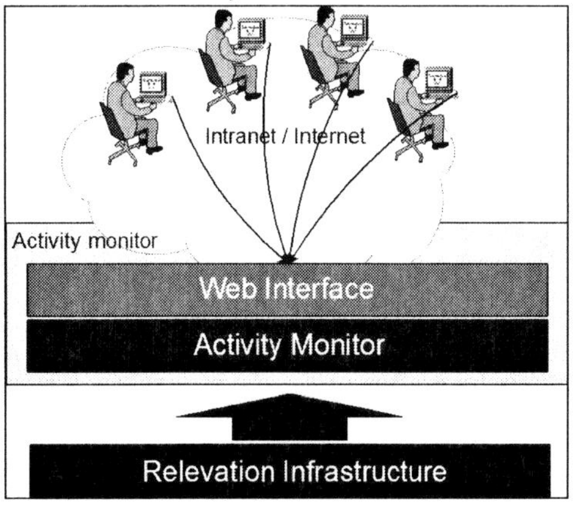

Figure 5. Activity monitor functioning scheme

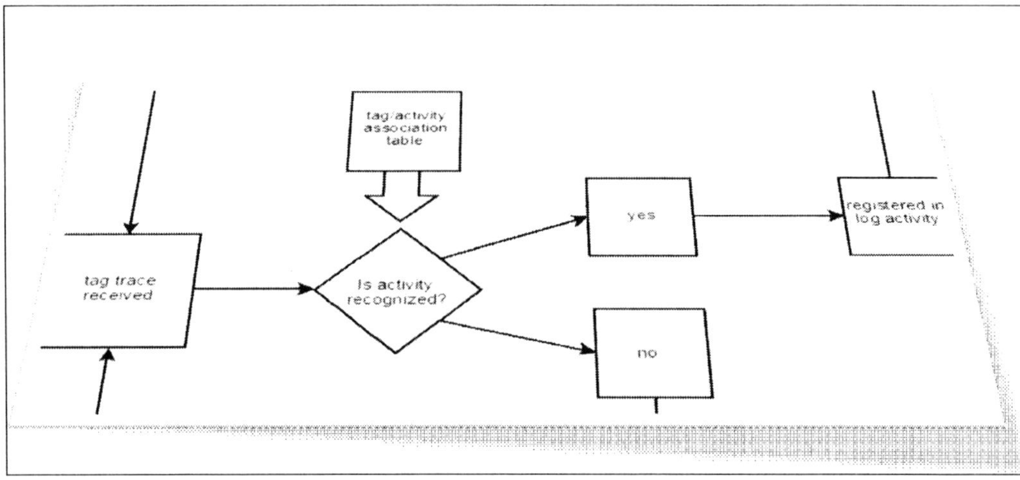

Figure 6. Patients activities dashboard (Activity Monitor main screen)

patient status: last update Wed Aug 22 13:47:27 UTC+0200 2007 — Refreshing every 5 seconds

Patient ID — Activity ID

pid	aid	Descrizione	Istante Inizio attività	Durata (s)
701	7	Preparazione Serale O	2007-07-21 17:17:19	65
702	7	Preparazione Serale O	2007-07-18 17:02:32	55
703	13	Igiene Personale O	2007-07-21 15:19:51	53
704	5	Somministrazione Pasti I	2007-07-21 18:02:22	536
705	4	fisioterapia	2007-07-17 19:39:01	45363

to the activity monitor, which is the back-end of the main application. This application includes a database and a routine which from time to time brings out the activities from the raw data (Figure 5).

The Web interface application allows operators to continuously control current activities with observed patients. Given the characteristics of the application, it has been chosen a Web application as architectural paradigm.

Using a simple PC with an Internet/Intranet connection (or even an Extranet) and a standard browser (e.g., Microsoft Internet Explorer, Mozilla Firefox, etc.) any operator may monitor the activities during its performance and the operator may also act in case of abnormal functioning[7].

Furthermore, the application allows one to obtain additional information on current activities through a simple selection of the activity's name, which may be observed on the third column of the Figure 6. In this way it is possible to observe all information related to involved operators as well as general information about patients' hospitalization.

The activity monitor is an example of how Web applications may change the business application in a substantial way; in fact, it is possible to go from the old concept of installation of applications on each user's PC to a new logic client-server based on a Web interface.

The benefits of this approach are not only the facility to access applications, but also include other critical issues like:

- Simplification and optimization of maintenance activities of the application.
- It is possible to develop a business intelligence application based on the data coming from the activity monitor in a completely transparent and automatic way.
- Application is available anytime, anywhere.

Level 3: Logic/Process Level

This is the analytic level where all detected activities are associated with processes and costs. Its development depends on software design.

C. Software Development

The software was developed in such a way as to allow the identification of specific activities through the relationship between the different tags indicating the duration of each contact. For instance, whenever a position marker simultaneously reads a medicine trolley, a patient, and a nurse tag, software identifies this association as medicine administration activity.

The information is available to researchers on real-time basis and may be observed directly on PCs screen and/or through a Web-based interface (Figure 6). All activities detected by each position marker were registered into a proper log file on the server.

D. Identification of Patients

Six patients were asked to participate to the project and all of them accepted to be followed using RFId/Web-based technology. Their selection was done according to three main criteria: adults, whose estimated hospitalization time was at least two weeks and who had a relatively high programmed working-load. The hypothesis on their potential working-load was defined based on their evaluation on the new Barthel index (NBI)[8], which determines the level of autonomy in daily life activities (DLA). In fact all patients except one presented a very low NBI, consequently there was a very high potential working-load for healthcare operators (see Table 1).

E. Data Collecting

In order to proceed to data collecting, some organizational arrangements were done. First of all, it was necessary to guarantee the consensus and

Table 1. Patients characteristics

Patient ID.	Sex	Age	Diagnosis	New Barthel Index
701	M	55	TBI (Traumatic Brain Injury)	0
702	M	64	TBI (Traumatic Brain Injury)	4
703	M	61	Ischemic Stroke with non responsive period	4
704	F	44	Hemorrhagic Stroke with non responsive period	0
705	M	26	TBI (Traumatic Brain Injury)	0
706	M	46	TBI (Traumatic Brain Injury)	10

compliance of operators as well as patients and care-givers. For the reliability of data, we asked for some pre-experimental actions such as training, technical trials, and monitoring.

Training. Special training was provided to healthcare operators, patients, and caregivers. They were told about how technology worked, how to used it, which kind of implication could results have, and how data obtained would be used. Additionally, IT providers were trained on rehabilitation clinical path, healthcare management and policies, and organizational potential impact of technology.

Technical trials. Three experimental trials were performed in order to determine the ability to use devices and the effective technological capacity to collect data using a RFId/Web-based system.

Monitoring. A continuous process of monitoring was performed in order to verify technology functioning as well as operators' and patients' compliance. Two main problems were solved using monitoring.

The first one was the maintenance of data collecting continuity, since some difficulties arose from crashes of the software during beta test and faults of WiFi service due to maintenance operation on the infrastructure not communicated to project manager.

The second problem was related to technical granularity for activity identification. Having a range of action predetermined by the "marker" in a 3 meter diameter sphere, whenever more than one patient's tag was read along with one single operator's tag, it could be difficult to identify the right patient/operator relationship. In order to avoid potential interferences, operators were asked to keep themselves away at, at least, 1.5 meters from any other patient they were not working with. This requirement put a strong stress on operators, given rehabilitation room dimension and layout.

Data was collected during a period of 8 days.

Data Analysis

Tables 2 and 3 show the data of the activities carried out first during the preliminary experimental phase then during the adjustment phase of the system.

As mentioned previously, the research group has decided to carry out an experimental preliminary session on the technological services in order to:

- Go ahead with a fine-tuning of the various hardware and software components;
- Evaluate what impact would this technology have on hospital personnel;
- Understand up to which extent technology could be bearable or invasive to patients, care-givers, and generally to the rehabilitation treatment; and
- To verify the reliability of information given by the system to establish whether a cost accounting activity-based system can be implemented.

Let us focus on the last point, leaving comments to points two and three to next paragraph.

The estimation of the technological efficiency as to the targets of the project (defining an activity-based costing system based upon real-time and objective measuring of time hospital personnel dedicate to various activities) has been the most important aspect of the research. This is the reason why during the experimental phase one person of the team accurately followed either operators or patients in order to check whether the system would correctly identify the different types of activities and properly measure time of each activity.

Facts reveal that this effort has not been in vain. This experimental phase has pointed out some technological discontinuities already mentioned before, which must be contained, otherwise they could jeopardize the reliability of the system. We also had the chance to notice that the system

Table 2. Preliminary data collecting: activity information about Patients no. 701, 702 and 703

Activity	Time (minutes)	Total Counter Activity	Average of operator per activity
Patient 701			
General Clinical Control	420	9	1
Medical Visit	0	0	0
Medicines administration	0	0	0
Physiotherapy	0	0	0
Meals administration	5201	9	1
Personal hygiene	0	0	0
Preparation of patient for night time	934	7	1,8
Other residual interactions	0	0	0
Total	**6555**		
Patient 702			
General Clinical Control	201	2	1
Medical Visit	119	3	1
Medicines administration	10	1	1
Physiotherapy	3413	5	1
Meals administration	2100	5	1
Personal hygiene	50	6	1
Preparation of patient for night time	25413	44	1,5
Other residual interactions	0	0	0
Total	**31306**		
Patient 703			
General Clinical Control	5	1	1
Medical Visit	224	1	1
Medicines administration	0	1	1
Physiotherapy	110	1	1
Meals administration	0	0	1
Personal hygiene	15	1	1
Preparation of patient for night time	270	5	1,4
Other residual interactions	0	0	0
Total	**624**		

records very brief interactions between patients and clinicians, which are very difficult to identify as a specific activity. We decided to measure and quantify, during the definite survey, this interaction (see Table 3) which could depend on the following two factors:

- One could be technological due to the unevenness of the technology for identifying the activities as explained above. The marker creates a sphere around the patient of a diameter of 3 meters; within this sphere other operators may be present while carrying out activities with other patients. This creates interference which may be only partly eliminated;

- Another could be organizational, regarding assistance and performance of rehabilitation treatments. During research we decided to focus only on some activities of the process, especially concerning the patients' stay in the hospital, which were quite time-intensive in regards to medical performance, excluding the rest. We believe that the interactions difficult to record refer to these latter activities

Table 3. Data collecting: activity information about Patients no. 704, 705 and 706

Activity	Time (minutes)	Total Counter Activity	Average of operator per activity
Patient 704			
General Clinical Control	120	6	1
Medical Visit	84	8	1
Medicines administration	234	3	1
Physiotherapy	805	6	1
Meals administration	863	27	1
Personal hygiene	1278	12	1,7
Preparation of patient for night time	917	13	2
Other residual interactions	3875	-	-
Total	**8176**		
Patient 705			
General Clinical Control	128	11	1
Medical Visit	95	11	1,2
Medicines administration	244	6	1
Physiotherapy	1692	13	1
Meals administration	1116	8	1,3
Personal hygiene	1151	12	1,7
Preparation of patient for night time	987	10	1,8
Other residual interactions	3340	-	-
Total	**8753**		
Patient 706			
General Clinical Control	126	6	1
Medical Visit	45	12	1
Medicines administration	264	6	1
Physiotherapy	1892	13	1
Meals administration	73	9	1
Personal hygiene	231	7	1
Preparation of patient for night time	25	9	1,5
Other residual interactions	3104	-	-
Total	**5760**		

or to brief contacts between patient and clinicians which do not include any professional service.

Tables 2 and 3 have been structured in order to show in the first column the type of activity, in the second column the total time (expressed in minutes) for each activity, in the third column the number of times the specific activity was recorded in the established period, and the last column the average number of operators present during the performance of the activity.

The following are a few remarks to better understand the information on Tables 2 and 3. The first experimental survey was carried out in July 2007 referring to patients identified with codes 702, 702, and 703. It lasted 3 weeks, but shows inevitable discontinuities due to the reasons aforementioned. After this test session, we carried on with a second survey in September 2007 for a week on patients identified with codes 704, 705, and 706. As shown in Table 1, the two groups are made of different patients with different disabilities or pathologies which caused the disability.

The value carried in column 2 in Tables 2 and 3 is relevant to the quantity of each activity performed by the patient during the survey period. It refers to a partial measuring carried out during this period (one week) and not during the whole time of treatment. Column 3 shows the activity driver, that is, the number of times when a specific activity is carried out by a patient during the time of the survey. In this case, it is necessary to explain that value 12 for activity "medical examination" carried out for patient 706 does not mean that the patient personally went through this examination 12 times during one week. It means that the system recorded the activation of this activity 12 times during survey time. For instance, it could happen that during the medical examination, the doctor may get in and out from the sphere of influence of the marker various times.

Regarding information carried on Table 3, we may consider the following:

- Patient 704 and 705, although they have a similar disability (according to the Barthel scale), they carry out a different number of activities probably due to the different pathologies which caused the disabilities.
- Patient 706 confirms the fact that fewer activities are necessary for disabilities of minor seriousness.
- If you get close to a specific treatment, by studying each relevant activity, you may notice that: a) the most time-intensive activities are the ones linked to physiotherapy, catering, personal hygiene, and preparing patients with serious disabilities for the night (Patients 704 and 705); b) patients with minor disabilities (patient 706) mainly require only physiotherapy, although personal hygiene and medicine dispensing remain always considerable activities; c) generally speaking physiotherapy is carried out in proportion to the level of disability; and d) the number of people involved in a treatment is also proportioned to the level of disability (activities concerning personal hygiene, night preparation, and catering to a certain extent are very often carried out by two operators at the same time, in the case of Patients 704 and 705).

DISCUSSION

Results reflect some critical points in three main areas, that is, technology design and use, relationship between technology and task, and, finally, relationship between data obtained and ABC model.

Technology Design and Use

The decision to use standard existing technology revealed some difficulties due to the need stated by the experiment design to follow up the patient at any moment and in any place inside the hospital. In fact, the patient-centred architectural design chosen presented the need to convert a standard fix device (position marker) into a mobile device able to follow the patient.

This adaptation revealed three main difficulties. The first one was that a market-proved stable technology became an unstable technology. The second one was that since the device was provided with a battery which needed to be charged daily in order to guarantee movement during all day long, it showed to be too dependent on operator collaboration. This put a limit on the autonomy of the whole system and, therefore, the device loses its intrinsic autonomy.

Relationship between Technology and Task

The design of data collecting procedure imposed to each operator the use of an active tag during each operator's working shift. This characteristic revealed some critical points.

First of all, being a procedure to much dependent on operators' collaboration and/or compliance, it happened some times that some of them forgot to take the tag and/or it was difficult for them to carry the tag during day-long activities, due to the tag dimension. In fact, it was perceived as a heavy device.

Furthermore, it was perceived as a way to control their daily activities. Trying to understand this last finding, and in order to perform the experiment, it was necessary to do some interviews in which other organizational issues were detected as relevant variables. For instance, it was clear to researchers that giving a consensus did not necessary mean being compliant and it was taking into account a variable already identified by literature as a clue variable, which is that operators did not perceive a positive relationship between their individual performance of a particular task and the result expected from the use of that specific technology. Moreover, in some cases, they expressed a general perception of being controlled while performing their normal task. This, of course, meant that a perception of a potential threat to their right of privacy was present.

The appearance of these serendipity variables confirm the fit between individuals, task and technology (FITT) framework recently presented by Ammenwerth, Iller, and Mahler (2006), which invites researchers and managers to pay attention to organizational relationships as well as to technological issues.

Relationship between Data Obtained and ABC Model

The final aim, as the essence of the Advanced Heath Process Management (AHPM) project was to demonstrate the feasibility of using modern technology such as Web, WiFi, and RFId for creating a system of activity-based costing/activity-based management which could enable an objective measuring in real time of the consumption of professional resources (e.g., clinicians, nursing staff, and physiotherapists) dedicated to various activities during a treatment.

A system which would not only be able to overcome the limits of traditional approaches to the development of an activity-based system as mentioned in the previous chapter, but also make available a complete set of information for measuring the process performances both from a clinical and managerial point of view.

As a matter of fact, details on how different professional resources are used in performing the activities of a treatment at an established time or period may implement a system balanced by performance indicators which can be structured

Figure 7. Evaluation perspectives of the managerial performances according to the AHPM project

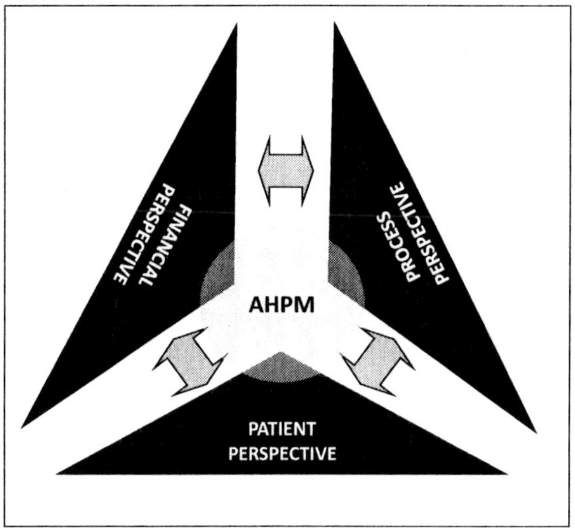

upon three main perspectives (Figure 7): financial perspective, patient perspective, and process perspective.

For example, from a financial perspective, you can either detect the impact the cost of each activity has on the total cost of the treatment or establish the economical feasibility through a comparison with the reimbursement system according to national standards. From a process perspective you can either quantify the saturation point of the different professional resources or, during both budget and final balance, estimate the impact a treatment may have on the personnel available. In the end, from a patient perspective, it could be either possible to determine the duration of a treatment or record any improvements of the disability as a result of a mix of assistance/rehabilitation activities carried out.

Any consideration regarding the above would not be much different from what years ago led to the creation of measuring systems of the multidimensional performances not any longer based upon a traditional (and unique) economical-financial dimension (Kaplan & Norton, 2004), but which widen the analysis towards a qualitative dimension (regarding process, client satisfaction, growth, and learning) allowing to better link daily work to the achievement of the company strategy.

In comparison with the systems of objectives, requirements, and working assumptions as previously defined, it could look as if the results obtained during research and outlined in these pages confirm the initial intuitions that modern technologies can bear both a cost accounting activity-based system and relevant multidimensional performance measuring system. Although, we ought to establish limits to this result since it is linked to research which is still at the experimental stage. We shall deal with these limits in the next paragraph.

Tables 4 and 5 show information which has determined the cost for each activity relevant to Patient 706. Table 4 shows in particular details regarding the consuming of different professional resources for each activity expressed in hours. Table 5 shows the cost per activity as the sum of the cost for professional resources involved in each activity.

Obviously, this is a partial cost measuring which takes into account—and we are aware of this—only the cost for the working resource. Although, we do believe this does not belittle the meaning of the results obtained, due to the following:

- In hospitals, especially where rehabilitation is performed, the cost for working resource is the highest; besides any estimation which can be made, no doubt that at the "Centro di Riabilitazione Villa Beretta" cost of work has a 70% impact on the total costs.
- It is quite simple to link the cost of each activity to another direct cost such as the medical and nonmedical material consumed. This information is available on the digital case history of the patient, therefore, a pos-

Table 4. Operators involved for each activity (analysis on Patient 706)

Activity	Time (hour)	Human Resources (time)			
		Medical Doctors	Nurses	Nurse Assistants	Physio-thera-pists
Patient 706					
General Clinical Control	2,10	-	1,90	0,20	-
Medical Visit	0,75	0,60	0,15	-	-
Medicines administration	4,40	-	4,40	-	-
Physiotherapy	31,53	-	-	-	31,53
Meals administration	1,22	-	-	1,22	-
Personal hygiene	3,85	-	0,95	2,90	-
Preparation of patient for night time	0,42	-	-	0,42	-
Other residual interactions	51,73	-	-	-	-
Total	**6807,2**	**0,60**	**13,85**	**15,74**	**31,53**

Table 5. Costs for each activity and professional involved for Patient 706

Activity	Costs per activity (euro)	Human Resources (euros)			
		Medical Doctors	Nurses	Nurse Assistants	Physio-thera-pists
Patient 706					
General Clinical Control	34,37	-	31,60	2,77	-
Medical Visit	31,59	29,10	2,49	-	-
Medicines administration	73,17	-	73,17	-	-
Physiotherapy	577,94	-	-	-	577,94
Meals administration	16,88	-	-	16,88	-
Personal hygien	34,37	-	15,80	40,14	-
Preparation of patient for night time	5,81	-	-	5,81	-
Other residual interactions	-				-
Total	**774,14**	**29,10**	**123,06**	**65,60**	**577,94**

sible improvement could be achieved by connecting the two databases.

- Company overheads have a 20% impact on the total costs. A simple and effective solution could be to reload at same percentage the cost of personnel for each activity.

In the end, Drawing 8 shows the saturation point of the different professional resources during the period of survey. It is a virtual measurement, not a real one, because this survey was carried out only upon three patients against many others who were there in hospital at the same time.

Figure 8. Saturation points of the various professional resources

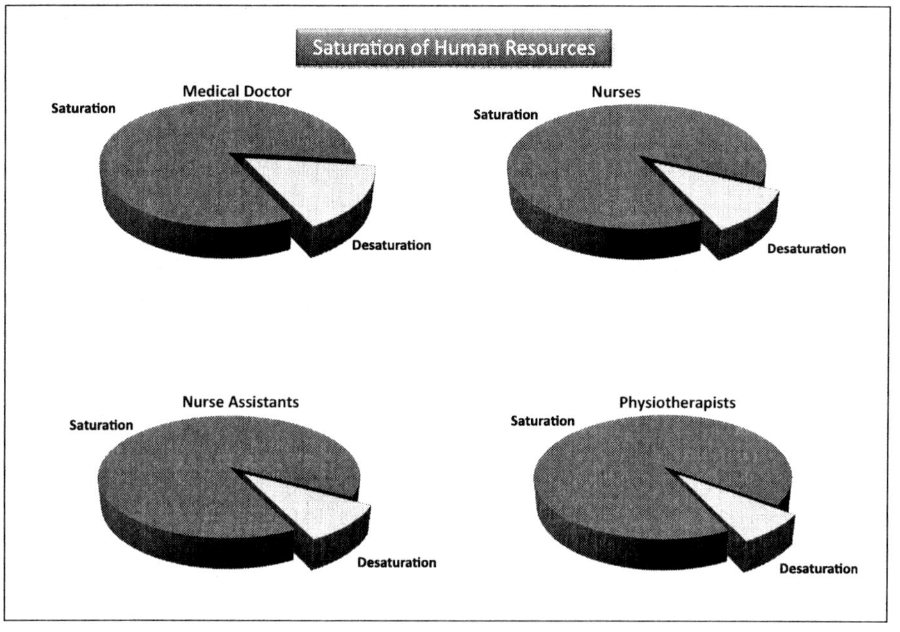

The way the survey was carried out (towards three patients for one week) did not really give a sense to the value of the key performance indicators (KPI), which had been assumed during the multidimensional measuring of the performances. Therefore we think it would not be much useful to outline and discuss the results now, postponing the whole matter once improvements have been made.

LIMITATIONS AND FURTHER RESEARCH AREAS

Instead of talking about the limits of this research, we believe it is much more useful to highlight in this paragraph what we have experienced and learned during this project development carried out at the "Centro di Riabilitazione Villa Beretta," since we are quite convinced that any proof of concept needs further proof of concept until the ideas, models, considerations, and technological structures presented in these pages are validated.

Obviously there are many elements to ponder upon. Starting from the technology aspect we must admit that the solution proposed is not really ready to start with, since:

- It basically needs a WiFi net, which is not commonly used in hospitals yet, especially here in Italy.
- The RFId technology, or some of its elements, are still experimental, especially with regards to the configuration we used during our research which was based upon a net with an antenna, a marker, and a series of active tags; in particular, the feeding system of the marker was bulky and invasive both for personnel and patients. For sure the availability of a RFId technology with WiFi, with a less invasive marker and by becoming functional for medical personnel (for instance, an active RFId on a palmtop

computer with Internet) could definitely be an improvement.

- The software used for capturing, storing, processing, and reporting data is another tessera of fundamental importance. This is another way to improve the quality and reliability of data by removing disturbances and wrong surveys.

Although miracles do not happen by only installing technology, it is notorious that the most advanced technologies and the best projects may fail if the human factor is not taken into consideration. Therefore the following become necessary:

- A **multidisciplinary approach** in all phases in projects of this kind where medical, managerial, technological and project management competences alternate.
- A **serious involvement of clinicians** explaining to them the real objective and functioning modalities of the technology; we have many times remarked that a project of this kind can be misunderstood and seen as a sort of "big brother" controlling each movement of doctors and staff, but the aim is not this at all! But thanks to a more advance technology, and to detailed information we can feed a cost accounting activity-based system and better handle processes linked to an activity-based management (e.g., estimation of work-loads, determining how many disabled patients can be borne, measuring performance, and linking company strategies to operational attitudes).
- **Formation activities** regarding functioning modalities and the logic of the system.

According to certain logic of the process, the following is essential:

- To **correctly map the company processes**, such as assistance and rehabilitation treatments which are different according to the specific disability of the patient. A multidisciplinary approach becomes essential as explained above.
- To **carefully calibrate the technology** and test its efficiency. The preliminary test we carried out (on Patients 701, 702, and 703) before a definite survey (on Patients 704, 705 and 706) enabled us to carry on with some necessary adjustments mainly required by the layout of the hospital building.

CONCLUSION

The research has demonstrated that it is possible to use the potentiality of RFId, the Web, and WiFi technologies to design a tailor-made system for activity-based cost and performance management that may overcome the limitation and inefficiency observed in traditional approaches and referred by literature.

This experimentation was performed in a highly complex context, that is, the one of rehabilitation medicine which emphasizes some of the previously discussed restrictions. Therefore, it becomes an optimal testing in the healthcare context.

Some of the critical elements arising from this experiment are:

- The identification and development of technological infrastructure. In order to be able to use this experience, the technology and devices used are easily available on the market place even if they must be reengineered to fulfil the specific goals and to reduce investments. These operational criteria have had a strong impact on the perception of patients and staff about the experimentation process.
- The organizational and behavioural dimensions of involved staff. The highest risk that this technology has shown is that it may be

perceived as a kind of "big brother" which controls every action and/or attitude performed by the clinical staff. As this experiment has demonstrated, this is not the actual goal of a project like this. On the contrary, it is oriented toward the identification of potential overuse of production capacity which could be a consequence of the presence, inside clinical structure, of patients with different disability characteristics. Nonetheless, the track-records of relevant information will permit to have a more effective clinical plan.
- Patients and personnel privacy protection. All information was anonymously recorded and treated in order to guarantee the maximum level of privacy protection.

This experimentation has demonstrated that current technology is able to produce valuable information to achieve the established goals. Probably, technology may be improved, reengineered, and industrialized. But, it is clear that just technology is not the complete answer. *It is necessary to foresee the organizational aspects and impacts as well as the methodologies and analytical tools which will enable the healthcare manager to obtain not only information, but actual results.*

ACKNOWLEDGMENT

We thank for the collaboration to the project the following partners: IBM, Cisco, Intermec, and Softwork.

REFERENCES

Ammenwerth, E., Iller, C., & Mahler, C. (2006, January 9). IT-adoption and the interaction of task, technology and individuals: A fit framework and a case study. *BMC Med Inform Decis Mak,* 6-3.

Becker, C. (2004). A new game of leapfrog? *Modern Healthcare, 34*, 28-38.

Brimson, J. A. (1991). *Activity-accounting: An activity-based costing approach.* New York: John Wiley and Sons, Inc.

Cooper, R. (1988a, Summer). The rise of ABC (part I): What is an ABC system. *Journal of Cost Management*, 45-54.

Cooper, R. (1988b, Fall). The rise of ABC (part II): When do I need an ABC system? *Journal of Cost Management*, 41-48.

Cooper, R. (1989a, Winter). The rise of ABC (part III): How many cost drivers do you need, and how do you do select them? *Journal of Cost Management*, 34-46.

Cooper, R. (1989b, Spring). The rise of ABC (Part IV): What do ABC systems look like? *Journal of Cost Management*, 38-49.

Cooper, R. (1990, Spring). Implementing an activity-based cost system. *Journal of Management Accounting*, 33-42.

Cooper, R., & Kaplan, R. S. (1988). Measure costs right: Make the right decision. *Harvard Business Review, 66*(5), 96-103.

Davis, H. E., & Luehlfing, M. S. (2004, November). Radio frequency identification: The wave of the future. *Journal of Accountancy*, 43-49.

Fanberg, H. (2004, Fall). The RFID revolution: Healthcare is ready to embrace this new technology. *Marketing Health Services*, 43-44.

Glabman, M. (2004, May). Room for tracking: RFID technology finds the way. *Materials Management in Health Care*, 26-38.

Janz, B. D., Pitts, M. G., & Otondo, R. F. (2005). Information systems and health care II: Back to the future with RFID: Lessons learned – some old, some new. *Communications of the Association for the Information Systems, 15*, 132-148.

Johnson, T. H., & Kaplan, R. S. (1987). *Relevance lost. The rise and fall of management accounting.* Boston: Harvard Business School Press.

Kaplan, R. S., & Anderson, S. R. (2004, November). Time driven activity-based costing. *Harvard Business Review*, 131-138.

Kaplan, R. S., & Norton, D. P. (2004). *Strategy maps*. Boston: Harvard Business School Press.

Mahoney, F. I., & Barthel, D. (1965). Functional evaluation: The Barthel index. *Maryland State Medical Journal, 14*, 56-61.

McGee, M. (2004). Health-care I.T. has a new face. *Information Week, 988,* 16-26.

NextInnovator. (2004). *Top ten technologies for 2005.* Retrieved June 20, 2008, from http://technologyreports.net/nextinnovator/?articleID=3004

Shank, J. K., & Govindarajan, V. (1992, Winter). Strategic cost management and the value chain. *Journal of Cost Management*, 5-21.

Taghaboni-Dutta, F., & Velthouse, B. (2006, November). RFId technology is revolutionary: Who should be involved in this game of tag. *Academy of Management Perspective*, 65-78.

Teradata. (2004). *Survival of the fastest.* Retrieved June 20, 2008, from http://www.teradata.com/t/page/126753/

Wicks, A. M., Visich, J. K., & Suhong, L. (2006, Summer). Radio frequency identification applications in hospital environments. *Hospital Topics: Research and Perspective on Healthcare, 84*(3), 3-8.

Yin, R. K. (2003). *Applications of case study research* (2nd ed.). Thousand Oaks, CA: Sage Publication.

ENDNOTES

1 This chapter is the result of joined research work done by all authors, nevertheless, Massimo Memmola is the author of section I, II and III; all the authors contributed in equal proportion in section IV, discussion and conclusion.

2 There are different scales for evaluating disabilities based upon daily activities which can be carried out autonomously by the patient (e.g., the *Barthel Scale*).

3 "The key insight is that although transactions can easily become complicated, managers can usually identify what makes them complicated. The variables that affect most such activities can often be precisely specified and are typically already recorded in a company's information systems. To take an example, let's assume a manager is looking at the process of packaging a chemical for shipment. In this situation, complexity arises from the potential need for special packaging and the additional demands of air as opposed to ground transportation.
Let's say that if the chemical is already packaged in a way that meets standard requirements, it should take 0.5 minutes to prepare it for shipment If the item requires a new package, however, the manager estimates, either from experience or from making several observations, that an additional 6.5 minutes will be required to supply the new packaging. And if the item is to be shipped by air, he or she knows (or can quickly determine) that it will take about 2 minutes to put the package in an air-worthy container.
This information allows the manager to estimate the time required for the packaging process;

Packaging Time = 0.5 + 6.5 [if special packaging required] + 2.0 [if shipping by air]" (Kaplan & Anderson, 2004).

4 Moreover, this measurement may express average values which allow neither the identification of individual operator nor the differences due to the experience each operator shows mainly related to his/her expertise and time-experience.

5 See example on Figure 3 about meal administration activity.

6 Windows, Apache, MySQL, and PHP

7 For instance, whenever a specific activity is not performed during a predefine period, the operator may ask for performance of the responsible operator.

8 Mahoney and Barthel (1965)

Chapter IX
The Web Strategy Development in the Automotive Sector[1]

Massimo Memmola
Catholic University, Italy

Alessandra Tzannis
Catholic University, Italy

ABSTRACT

Especially in recent years, a transformation is ongoing: the Web, besides being a means of information sharing (internal-external), becomes a powerful tool for saving costs, reducing the distribution structure, initiating distance transactions, and ever more, becomes a mechanism of integration with the external environment and a catalyst of experiences for all stakeholder. Starting from the identification of the key elements, potentialities, and of the impact of the Internet on firms' performance, competitiveness, effectiveness, and efficiency, this chapter is focused on the changes in the automotive sector due to the integration between business strategy and Web strategy. Therefore, starting from the consideration of a clear identification and subsequent sharing need of strategic goals, a research work will be presented exploring, on the basis of an interpretative model, the Internet potential in the automotive sector in order to achieve the identification of an optimal path definition and development of Web strategy. This objective will be developed through a desk analysis focused on the strategic positioning of the current businesses in the automotive sector (i.e., complexity evaluation of the presence on the Internet, strategic architecture, quality, and effectiveness of this presence).

INTRODUCTION

I need a new car!

Centy, as I affectionately call my old car, is gasping its last breaths. The "poor thing" has really had it, but then, it is really old; 10 years have passed from that happy day it made its entry in our family.

However, I am not so sure I like the idea of searching for a worthy successor. I am not an automobile fan, and my knowledge of them is rather limited. I am, however, sensitive to environmental problems, and I try my best to reduce the impact of my own person on the ecosystem as much as I can, considering its balance has already been seriously damaged.

What bothers me most is the idea of the long search ahead of me. I do not have any particularly difficult requests, but I want to consider such an important purchase very carefully, avoiding hurried decisions. I need a sensible family vehicle, with a good balance between performance, consumption, and above all cost. I need to acquire information, but just the idea of going from one dealer to another puts me in a bad mood. I can not stand having to waste time on this, and above all, having to pollute for a car I do not even own yet!

Also, interaction with the dealers is more often than not rather uncomfortable. At times—especially the more prestigious brand dealers—they treat you with a presumptuousness as if you were a nuisance, a matter that needs to be settled as quickly as possible. Other times, for the more sporty brands, they give you so many figures and acronyms, that they make you feel like "you did not do your homework" as in your school days. Lastly, with some there is feeling of a lack of transparency, something like "tell 'm everything but not really the truth."

Of course I do not want to demonize all car dealers. These are only impressions, and on top of that, only personal ones. But I do believe that, at least once, we have all felt the uneasiness I described above, either completely or in part.

I am holding the yellow pages in my hands, ready to copy my city's car dealers addresses on a piece of paper, but then a question just comes to mind, popping up almost unconsciously: "Why not use the Internet?"

I do most of my work nowadays through the Web: I keep in touch with my friends scattered around the world, I keep my bank account, book my holidays, buy music, and lately I have also taken care of my physical well-being using the telemedicine services offered by my city's health department. But then I say to myself: "But for a car it is different!" But is it really? Why would the Internet not have brought about the same changes in life style, in the way of thinking, of buying, in the automotive sector as it did in other areas?

I start my Web search by typing the word "car" into the search engine. Promptly a long list appears with sites specialized in online car sales. I am starting to question my beliefs. Some sites only offer general information, deferring the sale to a moment of real interaction. There are some sites though that are true virtual car dealerships, and allow you to get through the whole buying procedure on their site, with even a home delivery service of the newly purchased car. So after all I would not spend too much time, as I did with Centy.

However, I do not feel ready for an online purchase. I have identified the model that seems right for me, but I need some extra information. Through the faithful search engine, I reach the Web site of the car manufacturer. I am welcomed into a very sophisticated ambient. I click on "Product Range" and I find myself in a virtual car dealer showroom.

The models are well-presented, with clear 3D images. The technical features provided are exhaustive for each model. I discover the "car configurator," a very interesting tool that allows you to configure your car by choosing the color, the interior, the optional features, and, once the final price has been determined, to have access to another service that allows one to request personalized financial options.

It is also possible to print out a customized brochure showing the specific car model with the chosen color, interior, and optional extras. I find out that the site offers me the possibility to not only get to know my local dealer, but also book a test-drive at my pleasure.

And there is more! The site offers a whole series of services that I would never have imagined that go well beyond the usual general information about the manufacturer. I realize that the contents vary according to the phase of the purchasing process: when the customer is in the process of choosing a car, the customer can benefit from tools that compare the models and optional extras, and that allow people to create the car that is closest to their expectations. Afterwards, some gadgets that make the car look closer and more real, can be downloaded (pictures and videos). Once the purchase has been completed, the customer can monitor the delivery process to see when it will arrive at the customer's door and after that, book the assistance and maintenance services, activating an SMS reminder service.

Often the Web becomes an occasion to create a community of customers of the same brand. Various services have been created to this end, from the normal forum to the more sophisticated blog, that allow an almost one-to-one interaction with the manufacturer, offering the possibility to interact in various ways and degrees with the managers, and to know the reasons behind some project choices, or to be informed about new models before they are launched on the market.

I also find an online magazine that I can virtually read and that, very coincidentally, not only contains car related articles, but "talks" about environment, style, and sports. I realize I am understanding, really entering the brand's life style. Once I own the car, through the Web site, I can request a loyalty card that, apart from working as a credit card on the main international circuits, allows me to obtain discounts not only for the car's assistance (i.e., maintenance, repairs etc.), but also in many shops in line with the brand's style and features.

At this point I am ready to leave the virtual world to get back to the real one. I can calmly think about my visit to the dealer where I booked my test drive. I have got all the information I need and I have made my choice. I could even make the purchase online. I have seen that some sites offer this possibility, but entering the dealer's showroom will allow me to satisfy the emotional value of the purchase, that is, to hold tight the steering wheel of the long-desired vehicle, breathe in the "smell of new," touch the dashboard. The Internet is not able to satisfy these senses, at least not yet!

The lines above are obviously fantasy, but they are not science fiction, as they describe a situation that is very likely in the light of the information and service contents that we have found on the Web analyzing the companies' Web sites of the automotive sector during our research work. We had to use our creativity simply because we imagined an ideal Web site that should hold all the solutions, information, and services that in reality the car manufacturing companies activate only partially in their portals.

It has rightly been observed that the Web represents "a new space, a new territory, made of computers, connections, software and above all, information...inhabited by individuals, companies, and by organizations; a place made of games, commerce and exchange" (Porter, 2001).

Comparing the results of our research with similar ones in other branches, such as banking (Frigerio, in this book), tourism (Yaobin, Zhaohua & Bin, 2007), retail business (Duke, Chul, Sang-Il & Soung, 2006), and the health sector (Baraldi & Memmola, 2007), it seems that the automotive sector is, generally speaking, running far behind the others. The feeling that we get is that the companies belonging to this branch, even though they have been developing a "public presence" on the Web for quite a time, have hardly exploited its potential, limiting themselves in various cases to a low profile Web strategy, structured as a source of information rather than as a service. The Internet has therefore played a role up to now that is not dissimilar from any other media that allows a company to develop a one-way communication with the customers, and more generally, with the company's stakeholders.

This delay and lack of sensitivity toward "evoluted usage forms" of the Internet's potentialities can be partly justified by the strong emotional value linked to the purchase of a car, leading the customer to the dealer's showroom to touch the product that the customer is going to buy, to verify its qualities, thereby trusting tactile senses to guide and reassure. And it must be said that past attempts to develop forms of e-commerce in the automotive field, although they requested the purchase to be finalized at the dealer's, did not have much success.

Over the past years however, this trend has been changing dramatically. According to a research carried out by McKinsey (2003), during the next 10 years we will see the third revolution in the automotive sector, after the creation of the Ford mass production factory and the lean production of the Toyota Production System. Customers of the automotive branch, according to the research, will be expecting "hightest performance" for the same price, generating strong pressure on cost reduction and innovation capacity in the automotive companies. These factors will, without a doubt, lead to a whole new set-up of these companies' supply chain.

In fact, various companies are starting to grasp the idea that the Web can become an essential component in the so-called "low cost car strategy" to which many companies (e.g., Renault, Tata, General Motors, Fiat, Volkswagen, and Toyota) are decidedly heading. Through the Web the distribution network, which currently represents an important part of a car's final price, can be removed or reduced to a minimum.

Not only! The Web becomes the privileged instrument of interaction with the so-called "iPod generation," who through the company's portal or a specific product site, can contribute to the various steps of the product's project or image definition (e.g., accessories, internal and external graphic features, etc.). In this sense Fiat's experience with its Web site www.500wantsyou.com is surely significant.

Finally the Web can play a vital role in supporting customer relationship management (CRM) policies, or in the broadening of the product system with a series of complementary services linked to the automotive world (e.g., assistance, insurance, loans, etc.).

This chapter presents the results of a complex research aiming to define in what measure and with which modes the companies of the automotive sector are using the potentialities of the Internet to create value for their main stakeholders.

POTENTIALS OF THE INTERNET FOR THE AUTOMOTIVE SECTOR

Why does the Internet continue to receive so much attention? Why are the Internet and its potentialities always being discussed? What makes the Internet a better technology than so many others that preceded it? Does the Internet really change the way "to do business"? Does the Internet create new business models or does it change the existing ones?

Quite a few years after the boom of the New Economy and the subsequent enlivenment of the managerial literature at the turn of the century, perhaps the question will receive more sensible answers and above all, be sustained by a longer experience. The Internet surely is a surprising technology through which it is possible to communicate, interact, sell or distribute products and services, and above all, create a powerful tool to bind customers through communities in the form of forums, chat groups, blogs, user groups, and so forth.

What mainly distinguishes the Internet from previous technologies is that it sums up all their features and all their potentials in one low-cost standard, that is, through the Web people can watch TV, listen to the radio, and talk over the telephone. Therefore it may make more sense to talk about Media strategy rather than Web strategy: to underline the necessity, from a company's viewpoint, to govern the various interactive channels, on and off the Web, in a harmonious and coordinated way.

The Internet, therefore, changes or can change the way to do business. It has rightly been observed that the brick and mortar companies, that is, which existed before or which were not founded in function of the Internet, have to develop their own business model in order to seize opportunities, taking into account the limits (but also the strong points) of the current situation.

The most important thing is not to succumb to the charms of the Internet, which is only a tool after all, a dependent variable of the model of analysis of the company's governance. It is therefore necessary to align the choices that have been made within this scope with those which have determined the business strategy, selecting the Internet's best potentialities that better fit the purpose, thereby carefully studying realization time and modes.

The Internet as a Media and a Geographical Distance Reduction Tool

The Web facilitates exchange based relations, overcoming the obstacles of time and space. More than 10 years since the diffusion of the Internet, this may seem very obvious. According to the authors, this is the main reason behind its success though. The Internet is certainly not the only media available: the alternatives are the press, television, and radio. As said above, the Internet is extraordinary because it summarizes them all; at the same time developing interaction and the possibility to realize a communication process that is not just one way.

As in other areas, also in the automotive sector this potential allows re-examination of the development logic of the supply chain (e-procurement), the distribution processes (relations with the dealers), and of the relationship with and assistance to the customer. Therefore, it does not matter if the dealer is located far from the manufacturer; through the company's Intranet[2] the dealer will be able to see if a particular model is available or, alternatively, how long the production will take. A small size, niche manufacturer will be able to offer cars for sale to potential customers all over the world, even if the manufacturer has no dealerships at all.

The Internet as a Means to Rationalize Time

The Internet enables efficient time use, as a result of its power to reduce or dilate it. The first distinctive feature appears when looking for information about a particular car, the Web offers the possibility to find it quite quickly. In the same way the Internet allows a time dilation realization if services, streamlining the transaction, are implemented (e.g., booking of test-drive, of assistance services, etc.).

Web Externalities at the Base of the Internet

Technologies or products present Web externalities if their value is proportional to the amount of users, that is, the more the users, the higher the value (Katz & Shapiro, 1986). In order to grasp the meaning of these words, think of the usefulness of a telephone that can be used to talk to only one person. The value of such a phone would most certainly be lower than that of a phone which has a potential for worldwide communication. Clearly, the more people connected to the telephone system, the higher its value will be for its users. The Internet represents, in this sense, clear proof of this: the higher the number of people connected, the higher is the value of the Net.

The authors have had a clear demonstration of this potential, participating at a discussion forum about a new small Italian car manufacturing company. The continuously increasing number of participants and their interactions is progressively creating a pressure group aimed at demanding the satisfaction of the customers' wishes on the manufacturer's part.

The Internet as a Distribution Channel

The Internet, in general, operates as a distribution channel for all products with high information content that can be digitized (e.g., software, music, videos, images, plane tickets, various services related to banking, insurances, and even healthcare).

In the automotive sector, the Internet could be an interesting substitute for the existing distribution channels. Through the Web, low-cost cars could be sold, making it necessary to reduce the cost of the value creation channel as much as possible. And there is more! The Web could act as the main distribution channel for all those companies that have difficulties with traditional approaches, for example, small size or niche manufacturers or companies of a collateral sector such as car tuning.

The Internet as a Tool to Reduce Asymmetrical Information

Information is asymmetrical when, in a transaction, one party holds information that is relevant to the same transaction that the counterparty does not have. A classic example, printed in all economics books, is that of the transaction of a used car: here the asymmetry lies in the fact that the buyer does not only ignore the real condition of the car, but often does not even have the means to know if the asking price is congruous with the market value, or to evaluate the technical features. The Web offers such information, for example, the prices asked by different dealers for similar models, or consumers' associations that certify the quality of the product one intends to buy.

The reduction of these asymmetries can work both ways. Through the Web a car manufacturer can get to know about the expectations, tastes, and opinions of its customers. A very good example is Fiat's blog *quellichebravo* and it's Web site www.500wantsyou.com, gathering information essential for the development of new products.

The Internet as a Transaction Cost Reduction Tool

In the automotive sector too, the Internet could reduce the costs of the transaction process, as a direct consequence of the above-analyzed elements. Often in fact, companies have to do research work to find suppliers that offer the raw materials needed for the requested quality standards; the customers have to acquire information on the products' features, on the prices, and, in general, on the alternatives offered by the market; the dealers have to acquire information on the customer's financial situation, or interact with the headquarters to evaluate availability and delivery terms of each model. In the above activities, surely the majority of the information necessary to the transaction's completion can be easily obtained on the Web.

The Internet as a Tool to Support Corporate Governance and Personnel Training Courses

Finally, the Internet's technology is revealed as an irreplaceable tool sustaining the following activities:

- The company's analysis, control, and governance (business intelligence systems, ERP systems that are ever more oriented towards Web-based applications).
- Distance training of operators, with huge cost savings and significant benefits regarding the service level.
- Management of the company's information service and integration (through Intranet and management networks) of the company's divisions, organizational units, production sites, and dealerships.

OBJECTIVES AND METHODOLOGY OF THE RESEARCH

Objectives of the Research

This research work, carried out through a general survey of the Web sites of the main companies of the automotive sector, aimed at verifying how and to what extent these companies are currently exulting the potentials offered by the Internet to create value for the major stakeholders. In particular, the aim was to collect data that would allow us to evaluate:

- The strategic positioning—obviously on the Internet—of the companies of the automotive sector.
- What results have been achieved, or rather how much of the potential offered by the technology, is being used by the companies to define their own public presence on the Web.
- If and to what degree the strategic initiatives are successful in terms of visibility and registered traffic.

The Methodology of the Research

R. M. Grant (1998) defines the main objective of a company's strategy as the aim of "guiding the management's decisions to excellent results through the search of a competitive advantage; it is simultaneously a means of communication as of co-ordination inside companies."

In a correlated way, the Web strategy could be defined as that systemic set of decisions and actions aimed at determining in what measure and through which modes the company's positioning on the Internet can constitute an advantage in the pursuit of the company's strategy.

It is evident that business and Web strategies need to undergo an integrated process of definition, development, and implementation. On the other hand, if through the business strategy the company's management can understand what to use the web for, through the Web strategy it should be able to understand how to use it. The colonization of the virtual space offered by the Web should then be realized through a progressive alignment and a systematical evaluation (Memmola, 2007) of:

- The company's main strategic orientations.
- The degree of acceptance and awareness of the Internet's potentials on the part of the company's internal and external stakeholders (i.e., customers, personnel, suppliers, dealers, etc.).
- The possible impact on the organizational structure and internal processing.
- The company's general usage logic of other information and communication technologies sustaining its processing.
- The costs and benefits linked to the service or information contents activated one by one on the Web.
- The performance measurement mechanisms enabling the evaluation of the project's success.

However, the creation of a strategy (business strategy or Web strategy) consistent with the internal and external environment in which the company operates is not sufficient. Porter (2001) observes that it is essential to also "act upon" the strategic positioning, that is, on the ability to do business in a different way than the competitors, in order to create value for the customers, that is, not imitable. Which means using the Internet and ICT solutions in order to apply an univocal characterization of the product system to the image, the quality, and to complementary services rather than to the distribution logistics in such a way that it will be impossible, or at least very difficult, for the competitors to copy.

It is subsequently necessary to define the strategic profile of the Internet presence, in order to be able to continue with the positioning of the single company compared to the direct competitors, and to do the necessary categorizations and possible benchmarking actions.

According to Vittori (2004) and Buttignon (2001), the evaluation of a company's approach to the Web is normally based on the technological dimension (i.e., show-case or static site, interactive site), the recipients of the communication process (i.e., B2B, B2C, suppliers, customers, etc.), or on the goal(s) of the site (i.e., e-commerce, customer service, branding, database creation, etc.).

Such orientations are not consistent with the objectives and the set-up that we have put into practice in our research work. In order to proceed with the evaluation of the strategic positioning of the companies of the automotive sector, we have used the model created by Angehrn (1997) that goes under the acronym of ICDT, which stands for *information, communication, distribution, transaction*. This approach explains in a simple, but effective way, the sort of strategic approach used by companies for the definition of their own Web site, evaluating how and to what degree they are using the potentials offered by the Internet.

The model's name (Figure 1) derives from the segmentation of the Internet's virtual space into four main spaces: the *virtual information space* (VIS), the *virtual communication space* (VCS), the *virtual distribution space* (VDS), and the *virtual transaction space* (VTS). This segmentation highlights the fact that "the Internet has extended the traditional market space by providing new spaces in which economic agents can interact by

Figure 1. The ICDT model (Angehrn, 1997). © 1997. Albert A. Angehrn. Used with permission.

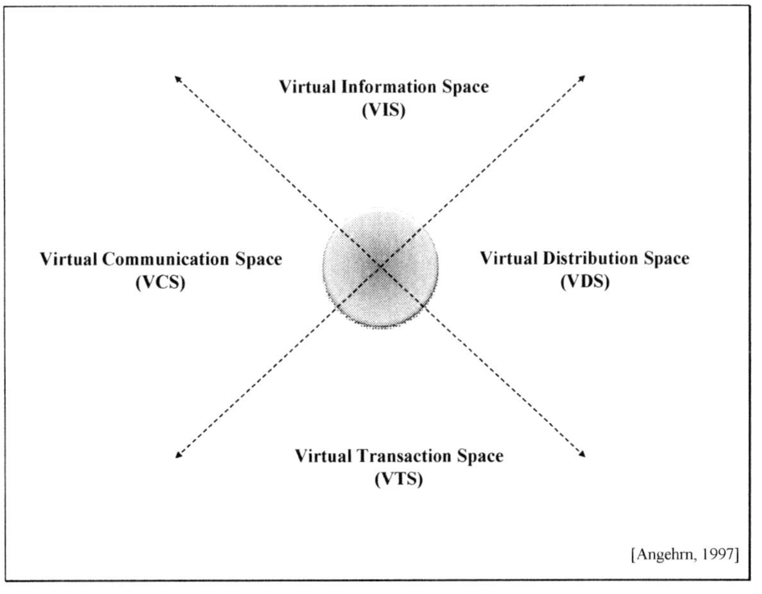

exchanging information, communicating, distributing different types of products and services and initiating formal business transaction" (Angehrn, 1997). In particular:

- VIS: The space that companies use to introduce themselves to customers or other stakeholders; in essence, it is a "one way," low-cost communication channel that the companies of the automotive sector use to present their organizational structure, mission, commitment to the environment, and linked services (e.g., assistance, maintenance) or complementary products (e.g., secure driving courses etc.).
- VCS: Offers the possibility to create a space on the Web to build relationships and exchange ideas and opinions; in this case, the aim of automotive companies is that of customer retention through the creation of virtual communities such as a forum, a chat, or a blog where customers and fans may exchange information or express their views on products, brand, services, and so forth.
- VDS: The Internet offers a new, efficient, and inexpensive distribution channel, suitable for a variety of services and products. All companies offering goods or services that can be digitized (e.g., e-books, videos, music, etc.), but also companies wanting to offer services to support their traditional products (e.g., online assistance, training, etc.) can benefit. Obviously, the automotive companies belong to the latter category and they can, for instance, offer the customer the possibility to download their car instructions manual, to receive reminder services for car maintenance, and so forth.
- VTS: The space allows one to realize remote transactions with different interlocutors (e.g., customers, suppliers, personnel, etc.); this Internet scope has traditionally been identified as e-commerce, related to the online sale of goods or a series of contents which in any case "assists" the transaction

Figure 2. Creation of the map of the Internet presence

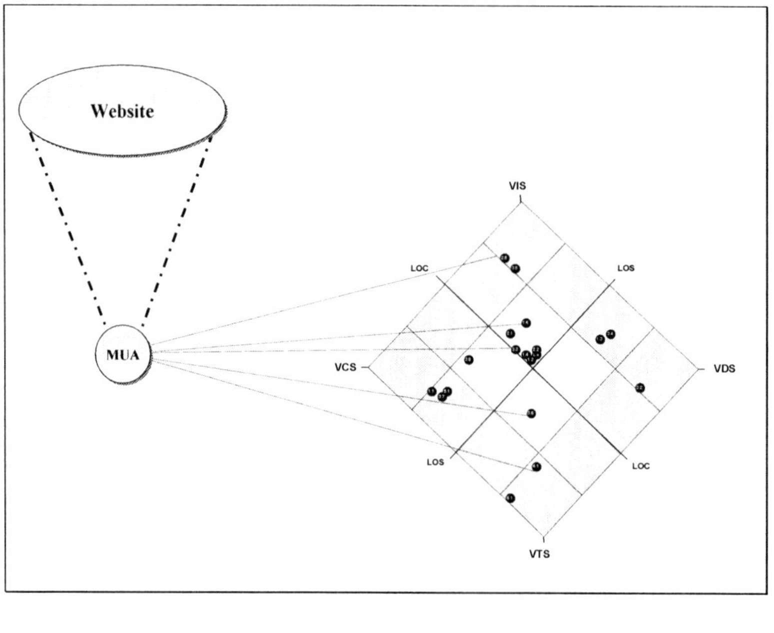

process (e.g., booking of test-drives, maintenance and repair services, job offers, etc.).

With Angehrn's model, a "map" of the Internet presence can be drawn up, aiming at offering a synthetic and prompt view of the Web strategy pursued by the company (Figure 2).

To this end the Web site is ideally divided into a series of minimum units of analysis (MUAs). A MUA represents an area inside the site characterized by the homogeneousness of the contents (informative or service) representing a precise "occupation" area of the Net's virtual space. As such, it does not necessarily coincide with the single Web page, but can be spread over more pages and can share the same page with other MUAs. In the VIS area the MUAs which can be activated are mainly informative ones, for example, those informing about the company's history, mission, product range and, so forth. In the VDS area instead, a MUA does normally have the task to give a content service, such as a reminder service for the car maintenance. Each single MUA will then be placed within its virtual correspondent area (i.e., VIS, VDS, VCS, VTS). The more external the MUA, the higher the degree of technological sophistication (LOS) and the level of content customization (LOC) measured along the orthogonal axes in the figure.

The Internet presence map allows us to evaluate and compare how and to what degree the automotive companies are using the Web to pursue their business strategy. It is therefore possible to work out a taxonomy of the strategic approaches to the Internet that we have divided into four main scenarios (Figure 3):

- The *low profile strategy*: In this scenario (represented by typology X in Figure 3) the company, although present on the Net, is (almost) unable to exploit its opportunities. Probably, the company's top management does not consider the Internet's technology as a means through which the company can pursue its business strategies, or the company has not clearly defined what the Web should be used for, or even, it may simply represent the company's answer to their stakeholders' low degree of acceptance and awareness of the Internet.
- The *medium profile strategy*: The scenario (typology Y in Figure 3) defines an operative context in which the company seems only interested in exploiting a few opportunities that the Web is able to offer. Very likely, in this situation the company's top management is still not fully aware of the opportunities that the Web can offer, or in any case it does not want to make substantial investments in this technology, due to the fact that the stakeholders do not value the Internet very highly.
- The *high profile strategy*: In this case (typology Z in Figure 3) the Web strategy starts to actively support the pursuit of the institutional strategy. The company, although it started exploiting the Web as a mere low-cost communication tool, is now starting to significantly broaden and qualify its presence on the Web, "aggregating" other types of virtual space (e.g., VTS, VDS, VCS).
- The *very high profile strategy*: In a context as the one represented by Company K in Figure 3, the top management considers the Web strategy an essential part of the business strategy and the stakeholders accept and are aware of the opportunities offered by the Internet. The occupation of the virtual space offered by the Internet is homogeneous and characterized by high levels of technological sophistication and of content customization; the company also "lives" through, and on the Web. An essential part of the relationship with the customers (or the other stakeholders) is lived through this means of interaction.

The "colonization" of these virtual spaces can obviously be developed according to different

Figure 3. The taxonomy of the strategical approaches to the Internet

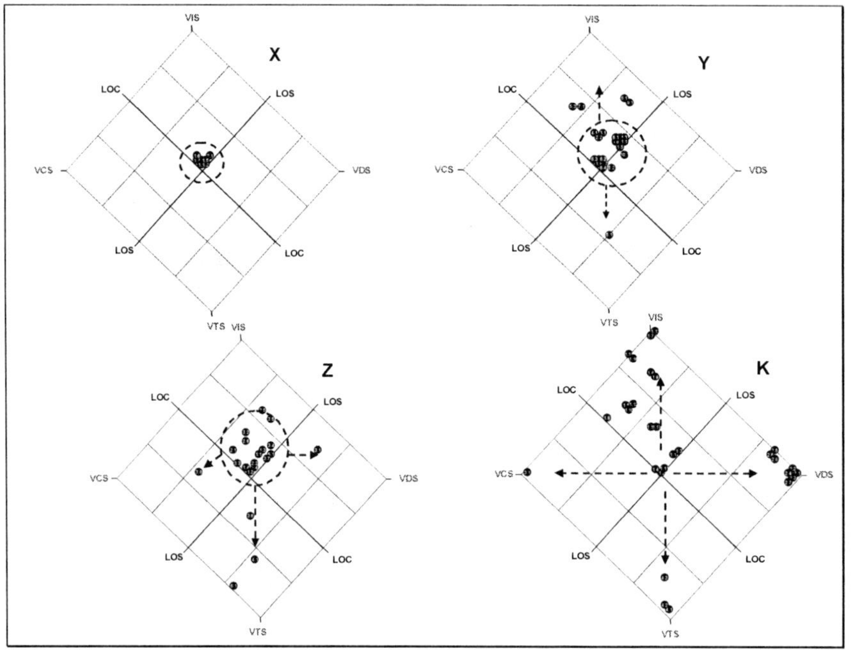

approaches and modes. Subsequently, the evaluation of the strategic positioning on the Internet of the automotive companies has been carried out through the consideration of four analytical viewpoints:

- The **complexity of the Internet presence**, or the intensity with which a company uses the Internet to sustain its institutional activities. An intensive and extensive usage of this technology will very likely be realized through a Web site that has got "plenty of" content able to give information about the company (VIS), furnish goods/services (VDS), create virtual places of communication (VCS) and carry out transactions (VTS).
- The **strategic architecture of the Internet presence**, or the more or less balanced modes through which the company has continued with the colonization of the different virtual spaces made available by the Internet, privileging some stakeholder typologies (e.g., private customers, suppliers, dealers, rental companies, business customers, etc.) and, perhaps, neglecting other typologies.
- The **quality of the Internet presence**, measured by the LOS and by the LOC of the contents proposed by the Web site.
- The **effectiveness of the Internet presence**, in terms of visibility (based on the main search engines) and success (visitors' count) of the company's Web site.

The Complexity of the Internet Presence

The aim of this analytical perspective is to measure the consistence of the Internet presence in the automotive sector, not so much in terms of the Web site's number of pages, but content richness serving the purposes of:

Table 1. Methodology for analysing the Internet presence

Step 1: Internet presence complexity	• examine Web site contents and classify them in MUAs • assess the existence of other structural features (e.g., foreign language version, native language search engine, Web site map, Extranet)
Step 2: Internet presence strategic architecture	• position retrieved MUAs in corresponding virtual space areas (VIS, VDS, VCS, VTS) • verify Web site use of Internet potential
Step 3: Internet presence quality	• assign the LOS of each MUA • assign the LOC of each MUA
Step 4 : Internet presence effectiveness	• verify Web site visibility by means of the most common search engines • measure average Web site traffic

- Providing information about the VIS area.
- Distributing services in the VDS area.
- Creating spaces for virtual communication in the VCS area.
- Realizing transactions in the VTS area.

The higher the number of contents offered in the various areas, the higher the degree of complexity that can be attributed to the site, which is measured through two indicators:

- The amount of MUAs counted in the site.
- The content usability index (CUI).

The concept of MUA has been explained above, whereas the CUI allows one to measure, in complex terms, the existence of a series of "structural characteristics" that have been activated within the site to improve content usability. It is infact logical to suppose that when the amount of content increases, for a Web site user looking for information, the degree of complexity increases accordingly; the presence of an internal search engine or a site map can be very helpful to this end. We also wanted to consider the site's usability by a particular category of stakeholders, that is, people with poor eyesight, verifying the possibility of easy modification of character size, background color, or the existence of a voice describing the contents. The CUI is thus the outcome of the ratio between the number of structural characteristics deemed noteworthy according to the methodology of the research, that is:

- Site map.
- Search engine.
- Site accessibility by poor sighted people (character size and background modification, etc.).
- Reserved area/Extranet.

An Extranet area can be defined as a company's nonpublic presence on the Internet. Whereas the use of an Intranet is totally private, within the company and by personnel only, an Extranet may give information and contents outside the company's scope in a reserved way. In fact, customers can use it to share information or services (the so-called customer clubs). Therefore, although the Extranet is not a tool aimed at improving the legibility of the site's contents, it actually increases the degree of complexity of the Internet presence.

The Strategic Architecture of the Internet Presence

Building a Web presence allows a company to use virtual space, which areas (i.e., VIS, VDS, VCS, VTS) can be used in various ways and to various extents, depending on the company's Web strategy.

This analytical perspective allows one to highlight the Web strategy followed by the company in the "colonization" of such space in terms of:

- Number of occupied virtual spaces, in order to evaluate the size of the presence built on the Internet.
- Exploitation degree of the Net's potential in global terms and of each typology of virtual space (VIS, VDS, VTS, VCS).
- Stakeholder typologies that the company is privileging (or, possibly, neglecting).

The indicators used to evaluate these aspects are respectively:

- The degree of "colonization," resulting from the simple enumeration of the Internet's virtual space areas occupied by the company (VIS, VCS, VDS, VTS).
- The *coverage index*, or the degree of coverage of the virtual space considered as a whole and of each single area.
- The *coverage index* in respect to the different stakeholder categories, that is, the degree of coverage of the demand for information and services expressed by the site's users (i.e., customers, dealers, suppliers, etc.), who for whatever reason are interested in the company.

Defining the degree of colonization of the Internet's virtual space is a simple but effective way to measure the evolutionary phase of a company's Web strategy. However, since an area can also be "colonized" through only one MUA, it is clear that this indicator furnishes a very approximate indication that needs to be integrated by the study of the mean coverage and the single area coverage index. If the degree of colonization indicates the company's presence in one of the Internet's virtual areas, through the coverage index it is possible to assess how much the company is benefitting, through its site, from the opportunities offered by the Net in that specific area.

In particular, the coverage index for each area (VIS, VDS, VCS, VTS) is fixed according to the ratio between:

- The number of MUAs found in the company's site for each area.
- The total number of MUAs found in the same area during the research.

The average coverage index is subsequently calculated through the mean average of the values calculated in each single area and supplies a synthetic, though approximate measure, of how much the company is exploiting the advantages offered by the Internet in general terms.

Finally, the coverage index regarding the different stakeholder categories is the outcome of the relation between:

- The number of MUAs found in the site that relate to a particular stakeholder category.
- The total number of MUAs that relate to the same stakeholder category found during the research.

The Quality of the Internet Presence

With regard to the qualitative dimension of the Web contents, the objective of the research work aims at using a measurement process that would be able to take into account the technical aspects as well as the traceability of the contents within the site, and that would respond as much as possible to people's informative or service needs.

Among the many alternatives, the drivers that have been adopted by the research methodology to evaluate the Internet presence are the LOS and the LOC, presented with their mean value through the MUAs found in the sites of the analyzed companies.

The evaluation of the level of sophistication (in the broadest sense of the word) is linked to the technology used. A progressive variable scale from 1 to 15 has been used to represent the various intermediate situations that can occur through:

- A MUA based on static pages realized with programs that were not specifically ideated for the Internet (e.g., MS Word, Powerpoint, etc.).
- A MUA based on real pages with the most modern tools of multimedial presentation (e.g., Flash View animations, audios, videos, images) and with a high degree of interactivity.

The level of customization is also based on the possibility to "cut" the offered site contents, according to the user's specific customization or service needs. In this case too, a progressive evaluation scale from 1 to 5 has been used. These values represent the extremes of a continuum of intermediate solutions that can turn up between:

- A MUA based on static pages, without any possibility of customizing the content, and without any additional services.
- A MUA based on pages that allow the creation of dynamic, customized contents with accessory services, for example, to build your own customized home page.

The Effectiveness of the Internet Presence

We wanted to complete the evaluation process of the Internet presence of the companies of the automotive sector through an analysis that would express the degree of "success" obtained by the Web sites. This intention implies, nevertheless, the solution of two methodological matters:

- The first concerns the need to translate in actual terms the very concept of "a Web site's success."
- The second is closely linked to the first and concerns the finding of variables able to give a measurement as objectively as possible.

There are basically two sorts of goals a company can have when setting up a Web site:

- Maximizing its visibility on the Web and therefore be reached by the highest number of potential users.
- Maximizing the use of the Web site by all the stakeholders.

The higher the degree of the above goal accomplishment, the more effective (and successful) the Internet presence will be, thereby also solving the second critical point, that is, the effectiveness level measurement of a company's Internet presence. The measurement has been obtained through the combined consideration of:

- A traceability index resulting from the major Italian and international search engines.
- A ranking of the Web site's registered traffic supplied by "Alexa.com," which draws up an international "hit parade" based on the registered traffic.

The first indicator tells how easy it is to "reach" the Web site through the major search engines (e.g., Yahoo.com, Google.com, Msn.com, Altavista.com, Virgilio.com, Arianna.com). As keywords, we have used, with an incremental approach, the company's name, the name of a product of the company, and lastly, a series of other information (i.e., the dealers and the promotions related to the company, a product's accessories line).

The index is equal to the sum of the mean of the positive results (search engines with positive outcome compared to the total number of search engines) obtained for each level of measurement (i.e., company, product, and other information). A result is positive when the search engine has included the searched information in the first seven proposals (normally visualized on the first page of the search results).

Alexa.com is a site that makes a classification of all Web sites worldwide, giving them points according to the number of registered visitors. Alexa.com uses an evaluation model that takes into account both the number of accesses to the site, and the number of pages visited within the site. The number of accesses quantifies the number of visitors that have accessed the site on a particular day. The number of pages visited is calculated by the amount of internal URLs requested by the same visitor. Obviously the repeated request of the same URL, on the same day, is counted as a single page. The site with the highest combination of accesses and visited pages will be given a value equal to 1.

The evaluations furnished by Alexa.com relate to all the active sites on the Web. Therefore, a data normalization process had to be carried out in order to obtain a classification specifically related to the analyzed companies.

MAIN FINDINGS

The Research Reference

This project was developed to focus on the Italian customer point of view. We took into account all the car-maker operating in Italy at the time of the research[3] from the databases of the National Association of Italian Automobile Manufacturers (ANFIA) and National Union Foreign Car Dealers (UNRAE). The study took into consideration the Italian version of the Web site, or when this was not available, original language Web sites. This choice was made following two criteria:

- Automotive companies, except for a few cases, are global in nature. They develop a different approach in terms of products and services offered, reflecting the specific characteristics and the importance of the market, depending on each geographical area (state or continent). We decided, therefore, to evaluate their strategic positioning in a specific market area: Italy, in our case.
- The international sites of larger sized companies (singled out by domain name_company.com) are simply "entry systems" through which they are readdressed to the site of the country or language (singled out, for example, by domain name_company.it or name_company.uk).

Obviously, this decision may have an impact on the results of the research, which are in fact influenced by the strategic choice which the company decides to follow in that specific market. Nevertheless, research does not imply any change or adjustment to the framework which remains the same, regardless of the geographic area considered.

The research sample examines 56 automotive companies that roughly fall into two macro categories: generalist companies and niche companies. The first category includes those automotive companies with differentiated offer (in terms of price level and type/model of car and service features), multiple targets, and different strategic areas of market covered; while the second category includes automotive companies that have a strong brand and image, often part of a group, offer a limited range of luxury cars and focus on a segment, target type, or special use. These tend to present a higher level price range with many customized options (even if the variable is not discriminatory) (see Table 2).

For each dimension of analysis (e.g., complexity, strategic architecture, quality, effectiveness) a positive and negative benchmark was created, defining the average value of 5% of companies with the best and worst performances, respectively.

Analysis of the Internet Presence Complexity

The complexity of the Internet presence was evaluated through the combined consideration of the following indicators:

- Number of MUA activated within the Web site.
- CUI, calculated on presence or absence, within the site, of fixed structural characteristics such as: a) site map, b) internal search engine, c) accessibility to site for poor sighted people, and d) reserved area (Extranet).

The research identifies a total of 56 MUAs variously distributed in the different areas where it is possible to articulate the virtual space of Internet (VIS, VDS, VCS, VTS). Three distinct clusters (see Table 3) stand out from the Web sites of theAnalyzed companies:

- **Cluster A**: Companies, using the Web, with a high number of informative or service contents, with over 30 activated MUAs.
- **Cluster B**: Companies with an average profile approach and between 25 and 30 activated MUAs

Table 2. Macro categories of reference of the analyzed companies sample

Generalist Manufacurers	Niche Manufacturers
• Alfa Romeo	• Abarth
• Audi	• Aston Martin
• BMW	• Bentley
• Cadillac	• Corvette
• Chevrolet	• Daimler
• Chrysler	• Dr
• Citroen	• Ferrari
• Dacia	• Hummer
• Daihatsu	• Lada
• Dodge	• Lamborghini
• Fiat	• Lotus
• Ford	• Maserati
• Honda	• Mayback
• Hyundai	• Mini
• Jaguar	• Pagani
• Jeep	• Porsche
• Kia	• Rolls-Royce
• Lancia	• Santana
• Land Rover	• Smart
• Lexus	
• Mazda	
• Mercedes	
• Mitsubishi	
• Nissan	
• Opel	
• Peugeot	
• Renault	
• Saab	
• Seat	
• Skoda	
• Ssangyong	
• Subaru	
• Suzuki	
• Tata	
• Toyota	
• Volkswagen	
• Volvo	

- **Cluster C**: Companies with a low profile approach and less than 25 MUAs.

The simple enumeration of MUAs must not be misleading. We are not looking at performance criteria and stating that a company that activates a higher number of MUAs inside its site is better than one that activates just a few. A company's Internet strategy is nothing more than a "translation" of its business strategy on the Web. This explains how the activation of some MUAs, for example, those supplying informative content or provide a booking service for their company museum, is typical of companies that focus on tradition and historical authenticity of their brands as an added value element (e.g., Alfa Romeo, Ferrari, Lamborghini, Maserati, etc.)

From the above considerations, it results that the number of activated MUAs is not a satisfactory indicator of the complexity of the Internet presence, which gives way to the introduction of the second indicator, the CUI as previously defined.

Overlapping the two parameters used to evaluate the complexity of the Internet presence, certain inconsistencies come to light: one would expect a high number of CUIs to result from a high number of MUAs, due to greater usability of the numerous contents offered. In fact, in several situations this hypothesis is denied by sites which, faced with a rather large number of activated MUAs, present a quite low CUI index (Alfa Romeo, Volkswagen, Volvo, Saab).

Table 3. Cluster of 56 analyzed companies

Companies with less than 25 MUA	Companies with between 25 and 30 MUA	Companies with over 30 MUA
Daimler	Honda	Citroen
Lada	Suzuki	Ford
Dr	Bmw	Jaguar
Abarth	Hyundai	Mazda
Pagani	Kia	Saab
Rolls-Royce	Seat	Volkswagen
Mayback	Chevrolet	Volvo
Santana	Ferrari	Lotus
Jeep	Smart	Porsche
Chrysler	Subaru	Fiat
Dodge	Audi	Skoda
Lamborghini	Cadillac	Lancia
Tata	Nissan	Alfa Romeo
Dacia	Land Rover	Mercedes
Hummer	Mini	Renault
Daihatsu	Mitsubishi	Maserati
Toyota	Opel	
Bentley	Peugeot	
Ssangyong		
Lexus		
Aston Martin		
Corvette		

The good positioning of Maserati, Renault, Mercedes, and Lancia stands out. Alfa Romeo's situation is worth mentioning since it presents the greatest number of activated MUAs with a poor usability level due to the absence of basic navigational instruments (i.e., site map, search engine, accessibility for poor sighted people) on which the CUI index was developed. It is usually the general manufacturers who offer the greatest number of contents, compared with the niche manufacturers. Lotus is an exception, which even if it belongs to the second category, shows up in the higher part of the second quadrant.

Analysis of the Strategic Architecture and Internet Presence

This perspective of analysis recognizes the strategic profile of the Internet presence of companies more in the automotive sector than in other areas of research. A company pursues a high profile Web strategy the moment it achieves a presence that is:

- "Widespread" in each of the areas in which the use of the Internet virtual space has been developed (degree of colonization).
- Quantitatively important in terms of number of MUAs activated compared to those potentially activeable (coverage index).

The results of the research demonstrate a generally elevated degree of colonization (average value 3.57) as a consequence of the fact that the automotive companies have predisposed Web sites with at least one MUA in each virtual space. Nevertheless, the potential of the Internet is still only used to a limited degree, that is, less than half of its potential. The average value of the coverage index is just 37%. The fact that many companies still only have a weak presence in the areas of greater value content such as VDS, VCS, and VTS (the coverage indices specific to these areas are generally very low) contributes to this result.

Nearly all of the analyzed companies, however, adopt an Internet strategy which concentrates on the information space (VIS), in which the

Figure 4. The complexity of the Internet presence

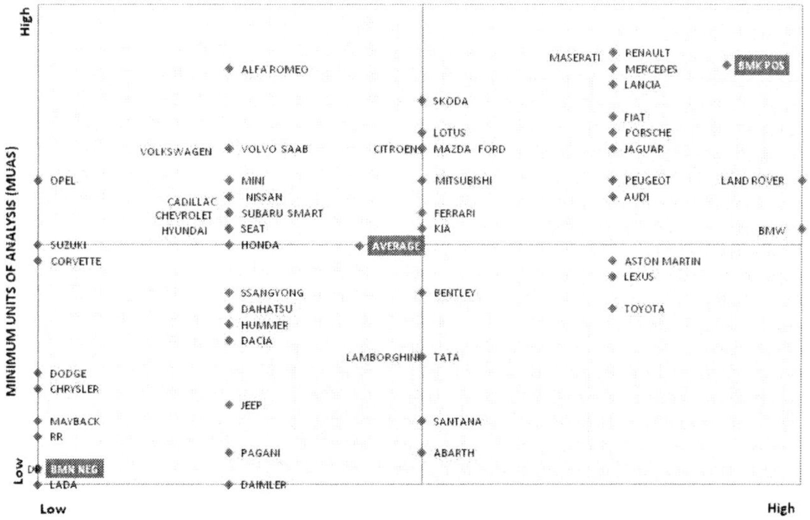

contents are expressed in information regarding the company structure, the products and services offered by the company, the prices of the products or services, contacts, and so forth.

There are five distinct clusters of indexing (Figure 5):

- **Cluster A:** Companies with a Web strategy of medium/high profile that develop a widespread presence in all the areas and with an over average coverage index. Companies nearest to the positive benchmark like Alfa Romeo, Maserati, Fiat, Volkswagen, Renault, Lotus, but also Cadillac, Jaguar, Smart, Land Rover, and Audi, who show lower degrees of exploitation of the Internet opportunities, appear in this cluster.
- **Cluster B:** Companies with a medium profile Web strategy which develop a widespread presence in all the areas and with a coverage index value in line with or lower than average, for example, Honda, Seat, Chevrolet, Corvette, Hummer, Abarth, and so forth.
- **Cluster C:** Companies with a Web strategy of average profile which develop a presence limited to three areas, but with an average coverage index value (e.g., Ford, Saab, Opel, Nissan, Aston Martin, and Toyota).
- **Cluster D:** Companies with an average/low Web strategy profile which develop a presence limited to three areas, but with a consistently lower than average coverage index value (e.g., Chrysler, Jeep, Pagani, DR, and Lada).
- **Cluster E:** Companies with a low profile Web strategy which develop a presence limited to only two areas and with a lower than average coverage index value (e.g., Tata, Santana, Rolls Royce, and Daimler Jaguar).

Another evaluation parameter of strategic architecture of the Internet presence concerns the evaluation of informative or service contents activated for each individual stakeholder category.

Figure 6 shows the performance of the average coverage index of the automotive sector with

Figure 5. The strategic architecture of Internet presence

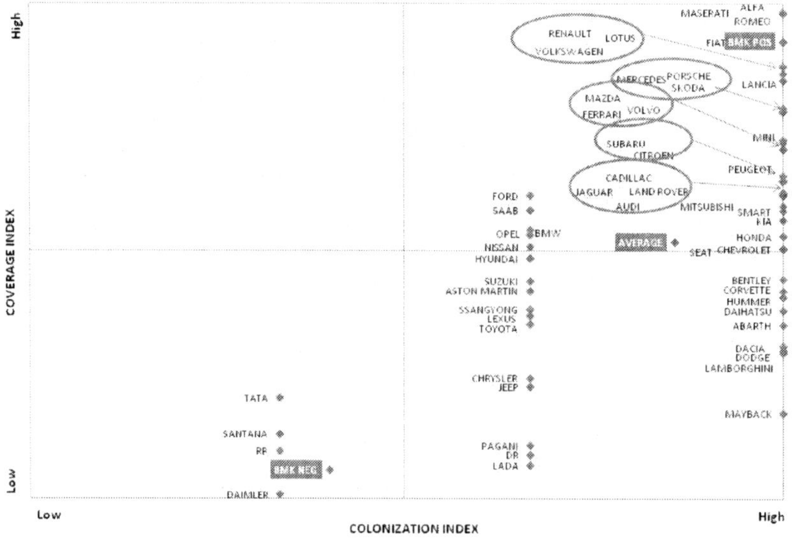

regard to the Italian market compared with the informative or service contents activated for each stakeholder; it is possible to develop the following considerations:

- It is confirmed the trend highlighted in the previous lines underlining that the companies in this sector are not yet able to fully exploit the potential of the Internet. On average, the index of coverage for each category of stakeholder varies its range from a low of 41% to a maximum of 54%.
- Certain categories of stakeholders (i.e., suppliers, business customers, rental companies, public companies, new employees) are more privileged than others (e.g., partners, employees, protected categories).
- The relatively low coverage for private customers shows how the potential of the Internet, compared with those who may be classified as the institutional recipients of the Web site remains high and largely unexplored; this result is the consequence of a large number of MUAs related to this type of stakeholders (see Figure 7) a still relatively modest coverage especially in more "advanced" areas such as the VDS, the VCS, and the VTS.

Analysis of the Quality of the Internet Presence

The quality of the Internet presence is intended as the level of technological sophistication of the

Table 4. Stakeholder considered in the research

Stakeholder	Description
Providers	Companies providing semifinished goods (e.g.m alloy wheels, plastic wheel cover, piston, connecting-rod, etc) or providing services (financial companies, consulting, etc.) for the analyzed company.
Partner	Companies that, participating in the design or development of the product, may be considered a strategic partner of the analyzed company (e.g., Microsoft developing in collaboration with Fiat the Blue & Me service or Brembo, a world leader in braking, involved with a decisive role in the design of a new model).
New Employee	Those who are interested in starting a working relationship with the company analyzed (e.g., work with us...).
Employee	Those who have an employment relationship with the company analyzed.
Dealers	Companies that are holders of a concession of sales of the company analyzed (includes the car dealers and the so called "authorized" and "retailers").
Car Rental Companies	The car rental societies.
Private Customers	Actual customers or prospects of the analyzed company.
Business Customers	This category is different from the previous one because it does not refers to private customers in a B2C logic, but to companies in a B2B logic. Expectations at a content or service level are obviously different from the previous stakeholder identified.
Persons With Disability	Those who are actual or prospect customers of the analyzed company, carrying a form of disability.
Public Administration	The public corporations (e.g., state, region, province, municipality) which may be interested in the territorial development policy implemented by the analyzed company.

distributed contents and the possibility to enter in a personalized or assisted way. The following parameters of evaluation have been used to develop such a perspective analysis:

- Level of technological sophistication (LOS)
- Level of customization (LOC)
- measured on a scale from 1 to 5.

As a first approximation, Figure 7 shows the indexing of automotive companies regarding the qualitative dimension of their presence on the Internet. Although in this case the significant differences noted in other analysis perspectives were not evident. Three distinct clusters however arise:

- **Cluster A:** The first group basically shows the presence of companies like Mini, Ferrari, Maserati, and Alfa Romeo and assumes a "pioneering" position regarding the qualitative characterization of their presence on the Net. The more sophisticated the technology, the higher the possibility to allow customization of the Web site content; this is far superior compared to the average value of the two parameters.
- **Cluster B:** The second group in the position of "chasers" is particularly noteworthy considering it combines the majority of the companies present on the market with performances that are not far or around the average measured value.
- **Cluster C:** A final group of "latecomers" who present a considerable weakness in respect to this dimension of analysis, consisting of those companies that have values close to the negative benchmark, such as Aston Martin, Santana, Tata, Lada, and so forth.

A further evaluation takes into consideration the balance between LOS and speed of access to the contents in each individual Web site. Most of the sites (41%) present a "strong preference" for LOC rather than LOS (with higher values of the first index than the second one). So the companies tend to favor highly customized contents over technology, which ends up being a little sacrificed. Only 11 of the 56 analyzed companies, in fact, consider the level of sophistication more important than the level of content customization.

A final consideration concerns the alignment between business strategy and Internet strategy.

Figure 6. The stakeholder coverage index

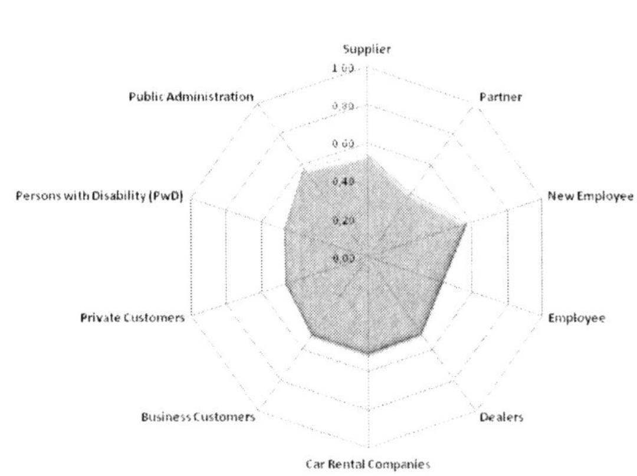

Figure 7. MUAs per stakeholder

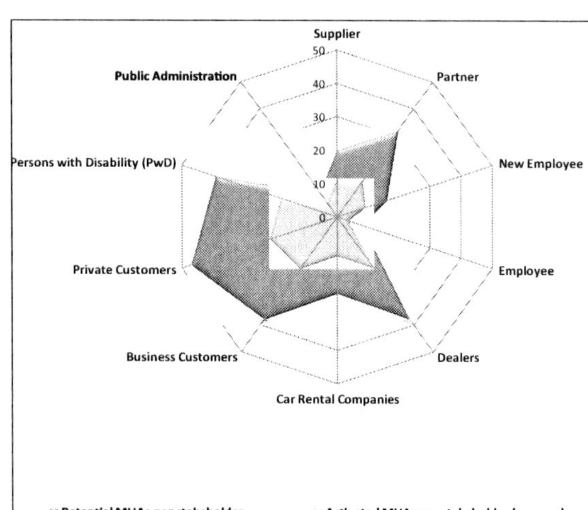

Companies with greater evocative power, with a middle to high position in terms of market level or which target the young or a niche, tend to present sites that focus on aspects of emotional involvement, full of images or interactive contents. The aim, in this case, is to create a bond with the brand rather than the mere gratification of informative demands.

In this sense, it can be noted: how Mini stands out thanks to a technologically advanced site and to highly customized contents; how Fiat has a good position, thanks, above all, to new ideas introduced for the launching of the new Fiat 500; how Ferrari has one of the richest sites regarding service contents and high path diversification; and how Abarth presents an extremely distinguished image from a technological point of view, offering the possibility to customize the site structure, its aspect, colors, and music, almost achieving the type of tuning which is at the basis of this brand's philosophy.

Analysis of the Effectiveness of the Internet Presence

The measure of success of the strategic choices by companies in the automotive sector regarding Web positioning was achieved through the combined consideration of the following parameters:

- A tracing index of the site through a panel of search engines.
- A ranking, created by Alexa.com, which supplies an indication of the traffic registered by the Web site.

Generally speaking, the Web sites of automotive companies enjoy good visibility: 63% of them presents a high traceability index (more than 2), 23% presents average visibility (with an index between 1.60 and 2), and only 8 out of the 56 companies presented a decidedly low visibility.

Concerning registered average traffic, the results of the research offer two considerations:

Figure 8. The quality of the Internet presence

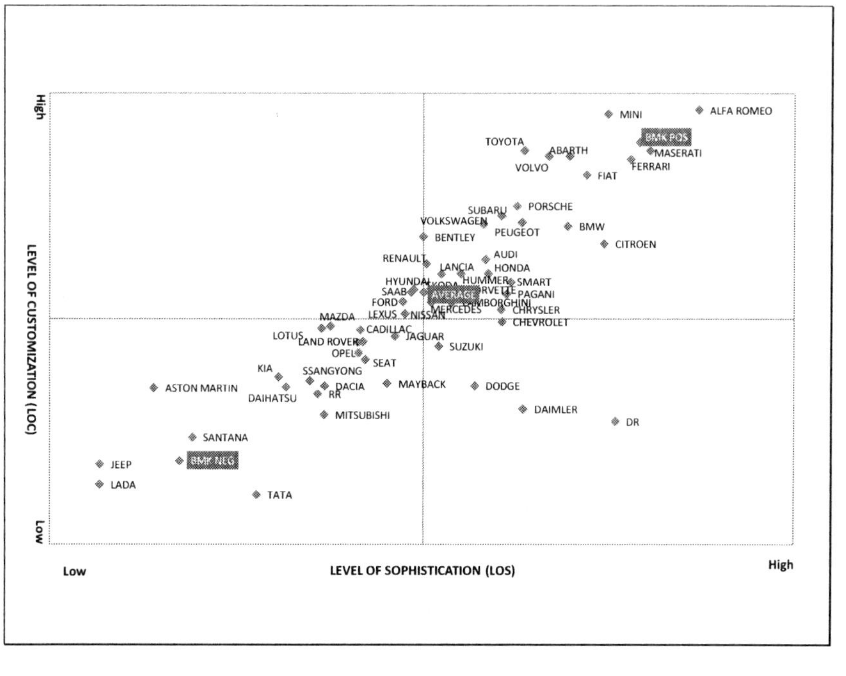

Figure 9. The effectiveness of the Internet presence

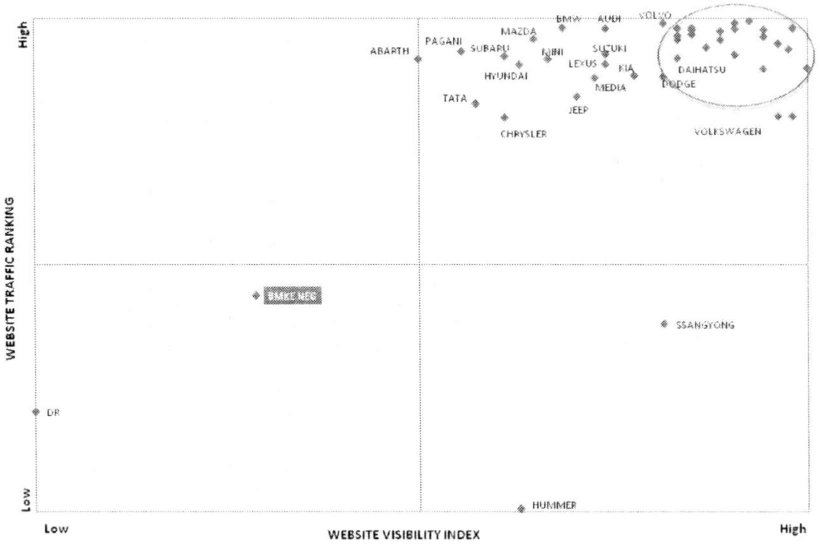

- A substantial aggregation of performance values corresponding to the positive benchmark; with the exception of a few companies (e.g., Fiat, Nissan, Volvo) great differences in terms of traffic registered do not exist in the surveyed Web sites.
- The consistent distance which exists between the average reference values and the companies which present a performance close to the negative benchmark (Figure 7). The companies in question (i.e., DR, Hummer, Ssangyong) are relatively new on the Italian market, or have an extremely focused offer. In our view, these could be the main reasons for such different results.

Focusing the attention on the companies present in the quadrant II, it is possible to note the existence of four clusters:

- **Cluster A:** Companies that have an effective presence on the Web with a high level of visibility and traffic registered. Companies like Fiat, Nissan, Volvo, Alfa Romeo, Ford, Citroen, Mercedes, and so forth, which are closest to the positive benchmark, are included in this first area.
- **Cluster B:** Companies that record high levels of traffic, but show a lower visibility compared with the search engines we used (e.g., BMW, Mazda, Audi, Suzuki, Subaru, Mini, etc.).
- **Cluster C:** Companies which present high visibility levels, but at the same time achieve a relatively worse performance in terms of recorded traffic (e.g., Volkswagen or Jaguar) who therefore present atypical behavior.
- **Cluster D:** "Follower" companies with visibility levels and recorded traffic lower than the average (e.g., Jeep, Tata, and Chrysler).

The previous arguments refer to the dynamics of competitive interaction among companies in the automotive sector specific to the Italian market. As in every other research area, also with regard to this perspective of analysis, the reference was the Italian language Web sites of Italian automotive companies, whose contents, in certain cases, are not managed directly, but by a branch or the Italian importer. It is evident that the results obtained in terms of visibility, but above all recorded traffic inside the Web site, are influenced by the competitive positioning that the company achieves in the real world. In other words, the particularly high performance rate of Fiat, Alfa Romeo, and Lancia is also a consequence of the fact that the study deals with well-known national brands with high values of brand awareness. On the contrary, the results of Chrysler or Jeep can be fit into the logic of a minor diffusion of these brands, having almost a niche presence on the Italian market.

We have therefore tried to repeat the evaluation of this perspective of analysis using original language sites of the different companies (singled out by the extension .com or from the extension of the country of origin as, for example, Lotus [www.lotuscars.co.uk]). The results shown in Figure 8 are, evidently, more consistent compared to the impact that different automotive companies, or rather their brand names, have on the international scene; Toyota, Mercedes, Chevrolet, Ford, and Volkswagen are the players who achieve the highest performance levels. Italian brands undergo a decisive reduction, with the exception of Ferrari which, although it is a niche producer, has an international standing. Lancia's case is noteworthy; the brand name is present on the Italian market, but not particularly diffused on the international scene, which in this new evaluation presents performance levels that are a lot lower than the average.

LIMITATIONS AND FURTHER RESEARCH AREAS

In order to present a complete view of the phenomenon, the methodology used to analyze the presence of automotive companies on the Web inevitably experiences a few structural limits which may be clarified as follows:

- The Internet is an ongoing reality. The research project photographs the situation as it was at the time of the analysis of the Web sites, which took place in the months of March and April 2007.
- The aim of this research is to provide an objective evaluation of the presence on the Internet. Even though a precise scale of evaluation of the LOS and of the possibility to proceed to a customization of the contents offered by the Web site (LOC) had been defined, there are inevitable elements of subjectivity present in this phase which cannot be eliminated.
- The research examines the Italian user's point of view. We have already discussed the reasons for this choice which, however, do not prejudice the possibility of applying this methodology to any other geographical context. It was in our interest, furthermore, to evaluate the dynamics of the competitive positioning of the companies in the automotive sector in a specific geographical area.

The research presented in this chapter is the first phase of a complex project, still being carried out, which also includes the development of an in-depth analysis through a series of case histories which help to fully explain the possible definition paths of the Web strategy of an automotive company.

The intent is to move from an analysis which takes place "outside" of the company, which is resolved in the evaluation of the "finished product" of the Web strategy, through the site to an analysis

Figure 10. The effectiveness of the Internet presence (focus on 2nd quadrant)

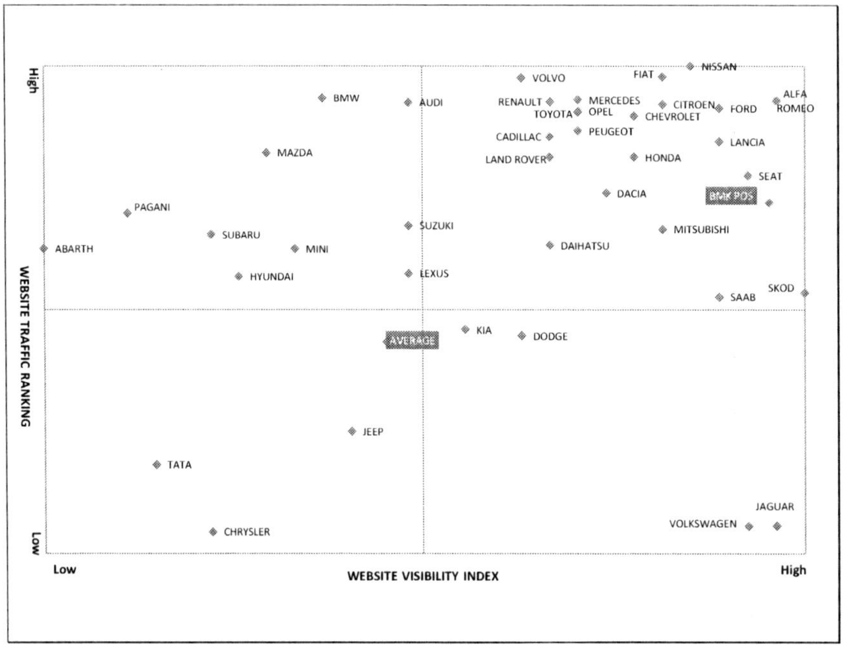

Figure 11. The effectiveness of the Internet presence (international positioning)

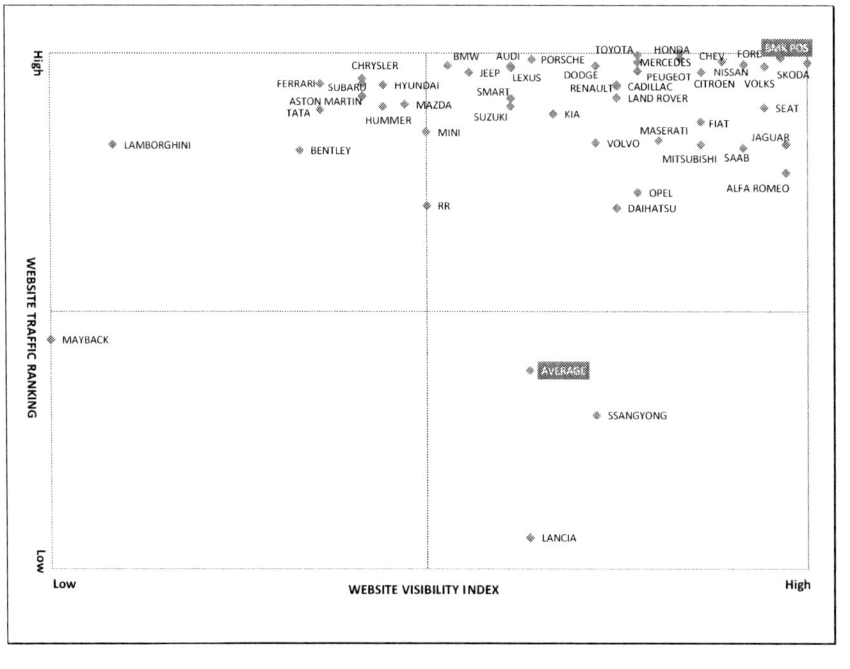

"within" the company. The aim is to have a complete vision of the phenomenon object of the survey, studying the path of preparation, definition, and successive implementation of the Internet strategy, the critical aspects and the duration of development, the impact on the organization, and the expectations and requirements of the persons either directly or indirectly involved.

CONCLUSION

This research would not have had much sense 15 years ago. More to the point, it would have been impossible to carry out. The raw materials on which our study is based would have been missing, or, anyway, the object phenomenon of the survey would have had totally different levels and dimensions.

It is in this simple and obvious reflection, but absolutely corresponding to the truth, that the extraordinariness of Internet must be found. It is almost banal to say that Internet has overturned the logics and paradigms of our daily life. This same book has been written through the Internet! Its authors have interacted virtually over the Net, without ever having met physically.

Inevitably, Internet also changes the company logics and paradigms. Thanks to (or because of) this technology, the way to run a company has changed. The Web "opens" new opportunities. And again, not even these are new concepts. Very concisely it is possible to affirm that through a redesigning of the company proceedings, the Net paves the way for improved performance in terms of lower costs, higher quality, and greater capacity of innovation. The Internet is a remarkable instrument for interaction with the company stakeholders, while other media provide only one-way channels of communication. Through the Net it is possible to improve logistics and interaction. Parallel considerations can be made for other phases of the company's value chain, such as the distribution and the relationships with the car dealers in the case of automotive companies. Finally, the Web could become the preferred instrument for interaction with customers (not just the private ones, but also business-to-business), in order to fully understand their expectations and requirements, gain their loyalty, and even involve them in the design and launching of new products (as Fiat did with the site 500wantsyou.com concerning the new Fiat 500).

The feeling which emerges from the results of this research is that companies in the automotive sector are not fully exploiting the opportunities.

Frequently, a limited use of the Web is witnessed, focused on the less complex areas, such as the allocation of contents usually of the informative sort and low technological level. Although it cannot be said that automotive company sites are merely digital transpositions of the brochures or product catalogues, wider areas of improvement could be made, both in terms of services offered and greater personalization of the proposed navigation routes. In general terms, it can be stated that the focus is still basically on the product and on the image associated with the product.

A further consideration examines the approach to the Net, which does not always seem to occur strategically, but often with a logic tied to the improvisation or an image of the Internet as a mere communication tool. In several cases, the impression is that the Web site "frees" itself from the logics of strategic positioning that the company follows in the real world in order to follow merely commercial or promotional logics, on one hand, or to present the product and the company sometimes too quickly and superficially. Companies rarely demonstrate awareness that the investment in this technology must come through an appropriate alignment with the institutional strategy of the company (Minard, 2001), with a correct review of the main operative mechanisms and company governance.

If we were to retrace the results of our analysis to different steps of the purchasing process, it can be stated that the initial stages of seeking basic information and presentation of product alternatives are definitely well covered, with often complete contents and of a multimedia sort. The phase of the actual transaction, and above all, the one linked to the management practices of post sales services (e.g., assistance, maintenance, etc.) tends to be poorly managed, with a few exceptions. The "pioneeristic" practices allow the online management of accessory services such as financing, purchase or ordering, and possible payment for maintenance/repairs, or the activation of a reminder service which advises the client when to proceed with maintenance.

The contents allowing companies to gain customer loyalty both through playful components (e.g., screen savers, images, logos, wallpaper but also real videogames) and through mechanisms of online community (e.g., forums, chats, blogs and also personalized credit cards) are beginning to catch on. At times contents are increased without adequate attention to their usability, causing confusion and a frequent absence of a profiling logic of the user, of the user's interests and needs.

The road towards a mature use of the Web by the automobile companies is still a long one, at least judging from the results we have gleaned from our research. The question, or rather, the questions that must be asked at this point are: What is it worth? Is it worth investing in the Web? How can the Web, through either its more traditional ways or from its more evolved declinations (like the various blogs, Wikipedia, YouTube, and Facebook) help to accomplish the corporate strategy of a company?

These are important questions, we realize, that could generate a specific trend of research. As far as we are concerned, having studied the Internet phenomenon for several years, we have become more aware of the fact that it is not a question of how much to invest, but rather how to invest. The success and choice effectiveness of the strategic positioning on the Web are influenced by the company's brand, but they also depend heavily on the grade of originality of the informative contents or proposed services.

We would like to conclude with a sentence from one of our many "specialists" we met during the planning phases of our research work: "The point is that on the Internet we are all guests... and as guests, we have to bring something!" What is required, therefore, is to gain the attention, "bringing something which is not normally present in the home of our host, or rather, on the Net," that is, contents that are new and original. But, in order to avoid that, this becomes an exercise for its own sake, it is necessary "to understand what type of invitation we have received and how we intend to behave accordingly." We must know what the expectations of our stakeholders are and how we can achieve them on the Web. By doing this, the corporate strategy of our company will be realized.

REFERENCES

AA. VV. (2002). *Stanford-Makovsky Web credibility study 2002: Investigating what makes Web sites credible today.* Retrieved June 23, 2008, from www.webcredibility.org

Alemi, F. (2000, Fall). Management matters: Technology succeeds when management innovates. *Frontiers of Health Service Management, 17*(1), 17-30.

Angehrn, A. (1997, August). Designing mature Internet business Strategies: The ICDT model. *European Management Journal, 15*(4), 361-369.

Angehrn, A., & Meyer, J. F. (1997, Summer). Developing mature internet strategies (cover story). *Information Systems Management, 14*(3), 109-140.

Baraldi, S., & Memmola, M. (2006). How healthcare organisations actually use the internet's virtual space: A field study. *International Journal of Healthcare Technology and Management, 7*(3/4), 187-207.

Buttignon, F. (2001). *Strategia e valore nella net-economy.* Milan, Italy: Il Sole 24 Ore.

Duke, H. C., Chul, M. K., Sang-II, K., & Soung, H. K. (2006, December). Customer loyalty and disloyalty in Internet retail stores: Its antecedents and its effect on customer price sensitivity. *International Journal of Management, 23*(4), 925-942.

Elango, B. (2000). Do you have an Internet strategy? *Information Strategy: The Executive's Journal, 16*(3), 32-38.

Glaser, J. P. (2002). *The strategic application of information technology in health care* organizations (2nd ed.). San Francisco: Jossey-Bass.

Grant, R. M. (2005). *Contemporary strategy analysis 5/e.* Oxford, UK: Blackwell.

Kaplan, R. S., & Norton, D. P. (2004). *Strategy maps.* Boston: Harvard Business School Press.

Katz, M. L., & Shapiro, C. (1986). Technology adoption in the presence of network externalities. *Journal of Political Economy, 94*, 822-841.

Malcolm, C. (2001, February). Making a health-care Web site a sound investment. *Healthcare Financial Management*, 74-79.

Malec, B., & Friday, A. (2001). *The Internet and the physician productivity: UK & USA perspectives.* Paper presented at the EHMA Congress, Granada.

Memmola, M. (2007). The development of a Web strategy in a healthcare organization: A case history. In L. Al-Hakim (Ed.), *Web mobile-based applications for healthcare management* (pp. 1-35). Hershey PA: Information Science Publishing.

Minard, B. (2002). CIO longevity and IT project leadership. *IT Health Care Strategist, 4*(1), 3-7.

Porter, M. E. (2001, March). Strategy and the Internet. *Harvard Business Review, 79*(3), 62-79.

Robert, M., & Racine, B. (2001). *E-strategy.* Milan, Italia: McGraw-Hill.

Shapiro, C., & Varian, H. R. (1999). *Information rules.* Boston: Harvard Business School Press.

Vittori, R. (2004). *Web strategy.* Milano, Italia: Franco Angeli editore.

Yaobin, L., Zhaohua, D., & Bin, W. (2007, June). Tourism and travel electronic commerce in China. *Electronic Markets, 17*(2), 101-112.

ENDNOTES

1. This chapter is the result of joint research work done by all authors. Nevertheless, Massimo Memmola is the author of section I, II and III; all other authors have equally contributed to section IV, V and VI, the discussion and the conclusion.
2. The research was drawn up in February, 2007; The analysis of the Web site took place in March and April, 2007.
3. A net that can be accessed only by users of a particular company/organization.

Chapter X
Adaptive Mobile Web Browsing Using Web Mining Technologies

Wen-Chen Hu
University of North Dakota, USA

Yanjun Zuo
University of North Dakota, USA

Lei Chen
Sam Houston State University, USA

Chyuan-Huei Thomas Yang
Hsuan Chuang University, Taiwan

ABSTRACT

Using mobile handheld devices such as smart cellular phones and personal digital assistants (PDAs) to browse the mobile Internet is a trend of Web browsing. However, the small screens of handheld devices and slow mobile data transmission make the mobile Web browsing awkward. This research applies Web usage mining technologies to adaptive Web viewing for handheld devices. Web usage mining is the application of data mining techniques to the usage logs of large Web data repositories in order to produce results that can be applied to many practical subjects, such as improving Web sites/pages. A Web usage mining system must be able to perform five major functions: (i) usage data gathering, (ii) data preparation, (iii) navigation pattern discovery, (iv) pattern analysis and visualization, and (v) pattern applications. This approach improves the readability and download speed of mobile Web pages.

INTRODUCTION

With the introduction of the World Wide Web, electronic commerce has revolutionized traditional commerce and boosted sales and exchanges of merchandise and information. Recently, the emergence of wireless and mobile networks has made possible the extension of electronic commerce to a new application and research area, that is, mobile commerce, which is defined as the exchange or buying and selling of commodities, services, or information on the Internet through the use of mobile handheld devices. In just a few years, mobile commerce has emerged from nowhere to become the hottest new trend in business transactions. The future of mobile commerce is bright according to the following predictions:

- Figure 1 shows the growth in demand for smart mobile devices including handhelds, wireless handhelds, and smart cellular phones through 2006, as estimated by the research firm Canalys (2004a, 2004b, 2004c, 2005a, 2005b, 2005c, 2005d, 2006).
- Cumulative sales of smartphones will reach 1 billion units by the first quarter of 2011 according to IDC, a market research company (Symbian Limited, 2006).
- According to various reports, the estimated worldwide shipments of the following three equipments in 2006 are
    - *PDAs and smartphones*: 84 million (Gartner, Inc., 2006),
    - *Celular phones*: 986 million (cellular-news.com, 2006), and
    - *PCs*: 250 million (*Silicon Valley Daily*, 2006).

The worldwide shipments of PDAs and smartphones in 2006 had a 57% increase from the same period last year, according to Gartner, Inc. Smartphone shipments bolstered the market, growing 75.5% to reach 34.7 million units, more than four times the size of the PDA market. PDA shipments increased by 5.7%, totaling 7.4 million units. Though the unit sales are less than one tenth of the worldwide mobile phone sales in 2006, they are not too far away from the worldwide PC sales in 2006.

Regardless of the bright future of mobile commerce, its prosperity and popularity will be assured only if information can be properly, speedily displayed. Traditional Web pages are designed for desktop or notebook computers and loading and visualizing large text documents on handheld devices are often arduous tasks. They usually do not suit the devices well because the pages, especially the large text files, cannot be properly, speedily displayed on the microbrowsers due to the limit of screen size, narrow network bandwidth, small memory capacity, and low computing power. This research investigates a method summariz-

Figure 1. Worldwide total smart mobile device market (Canalys, 2004a, 2004b, 2004c, 2005a, 2005b, 2005c, 2005d, 2006)

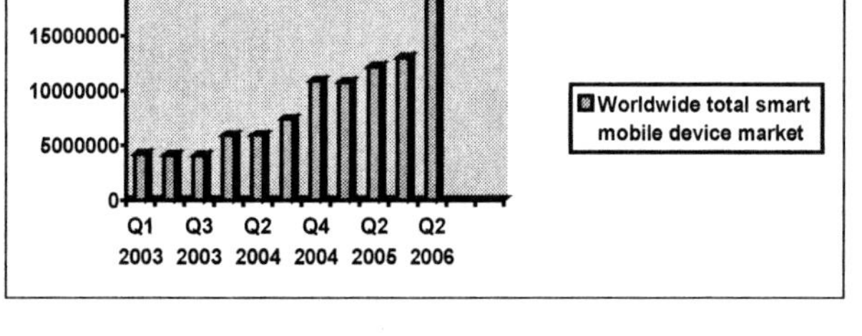

ing large text for mobile handheld devices. This method applies a Web usage mining technology, that is, Web navigation pattern discovery and applications, to adaptive mobile Web browsing. An experimental example is given. Other related issues such as mobile handheld devices will also be discussed in this research.

BACKGROUND

Mobile users interact with mobile commerce applications by using small wireless Internet-enabled devices, which come with several aliases such as handhelds, palms, PDAs, pocket PCs, and smartphones. To avoid any ambiguity, a general term, mobile handheld devices, is used in this chapter. A mobile handheld device is small enough to be held in one hand and is a general-purpose, programmable, battery-powered computer, but it is different from a desktop PC or notebook due to the following three major special features:

- *Limited network bandwidth*: This limitation prevents the display of most multimedia on a microbrowser. Though the WiFi and third generation (3G) networks go some way toward addressing this problem, the wireless bandwidth is always far below the bandwidth of wired networks.
- *Small screen/body size*: This feature restricts most handheld devices to using a stylus for input.
- *Mobility*: High mobility of handheld devices is an obvious feature that separates handheld devices from PCs. This feature also creates many new applications such as mobile recommendations that normally cannot be done by PCs.

Short battery life and limited memory, processing power, and functionality are additional features, but these problems are gradually being solved as the technologies improve and new methods are constantly being introduced. Figure 2 shows a typical system structure of handheld devices, which includes six major components: (i) a mobile operating system, (ii) a mobile central processing unit, (iii) a microbrowser, (iv) input/output devices, (v) memory, and (vi) batteries (Hu, Yeh, Chu, & Lee, 2005b). Brief descriptions of the components follow:

1. *Mobile operating systems*: They are the core software of handheld devices. Mobile operating systems are different from those

Figure 2. A system structure of mobile handheld devices (Hu et al., 2005b)

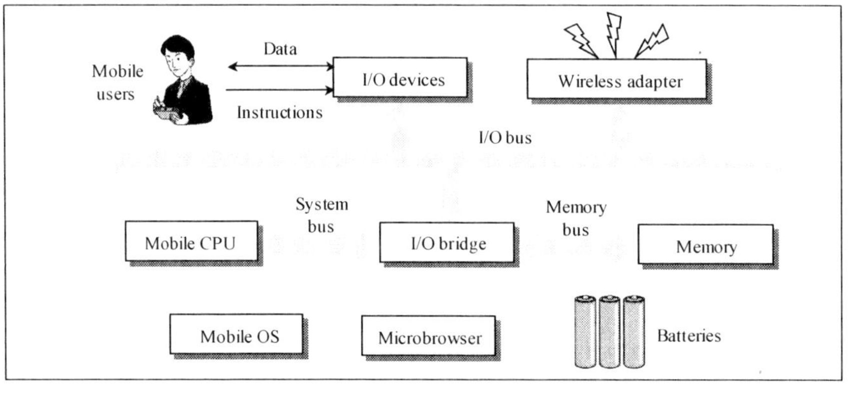

in desktop computers as they include the following additional features: (i) power management to prolong the battery life, (ii) real-time capability for time-critical operations such as voice communication, and (iii) wireless infrastructure for wireless communication.

2. *Mobile central processing units*: Mobile CPUs are the core hardware of mobile handheld devices and the performance and functionality of the devices are heavily dependent on the capabilities of the processors.
3. *Microbrowsers*: Microbrowsers are Internet browsers specifically designed for use in mobile handheld devices. They differ from desktop browsers in several ways, specifically the languages they use, security, footprint, and smaller windows. The last feature, smaller windows, enables them to provide simplified interfaces, although it also eliminates much of the desktop browser's multimedia functionality, such as streaming audio and video support.
4. *Input/output devices*: There is only one major output device, the screen, but there are several popular input devices; in particular, keyboards and touch screens/writing areas that require the use of a stylus.
5. *Memory*: Three types of memory are usually employed by handheld devices: (i) random access memory, (ii) read-only memory, and (iii) flash memory.
6. *Batteries*: Rechargeable Lithium Ion batteries are the most common batteries used by handheld devices.

MAIN THRUST OF THE CHAPTER

HTML Web pages usually do not suit Internet-enabled mobile handheld devices well because the pages cannot be properly, speedily displayed on the microbrowsers of the devices due to:

- Small screen size.
- Narrow network bandwidth.
- Low memory capacity.
- Limited computing power and resources.

To relieve this problem, this research applies Web usage mining technologies to Web page summarization for handheld devices.

Web Usage Mining

World Wide Web data mining includes content mining, hyperlink structure mining, and usage mining (Cooley, 2000). All three approaches attempt to extract knowledge from the Web, produce some useful results from the knowledge extracted, and apply the results to certain real-world problems. The first two apply the data mining techniques to Web page contents and hyperlink structures, respectively. The third approach, Web usage mining, the method used by this research, is the application of data mining techniques to the usage logs of large Web data repositories in order to produce results that can be applied to many practical subjects, such as improving Web sites/pages, making additional topic or product recommendations, user/customer behavior studies, and so forth.

Figure 3. A Web usage mining system structure

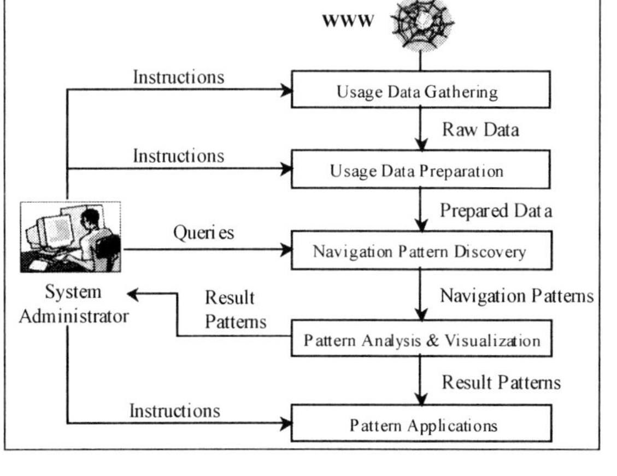

A variety of implementations and realizations is employed by Web usage mining systems. Figure 3 shows a generalized structure of a Web usage mining system (Hu, Yang, Lee, & Yeh, 2005a). A Web usage mining system performs the following five major tasks:

1. *Usage data gathering*: Web logs, which record user activities on Web sites, provide the most comprehensive, detailed Web usage data. A log file can be located in three different places: (i) Web servers, (ii) Web proxy servers, and (iii) client browsers, as shown in Figure 4.
2. *Usage data preparation*: Log data are normally too raw to be used by mining algorithms. This task restores the users' activities that are recorded in the Web server logs in a reliable and consistent way. This phase should at a minimum achieve the following four major tasks: (i) removing undesirable entries, (ii) distinguishing among users, (iii) building sessions, and (iv) restoring the contents of a session.
3. *Navigation pattern discovery*: This part of a usage mining system looks for interesting usage patterns contained in the log data. Most algorithms use the method of sequential pattern generation, while the remaining methods tend to be rather ad hoc.
4. *Pattern analysis and visualization*: Navigation patterns show the facts of Web usage, but these require further interpretation and analysis before they can be applied to obtain useful results.
5. *Pattern applications*: The navigation patterns discovered can be applied to the following major areas, among others: (i) improving the page/site design, (ii) making additional product or topic recommendations, and (iii) Web personalization. Learning user/customer behavior and Web caching, less important applications for navigation patterns, are also worth studying.

Figure 4. Three Web log file locations

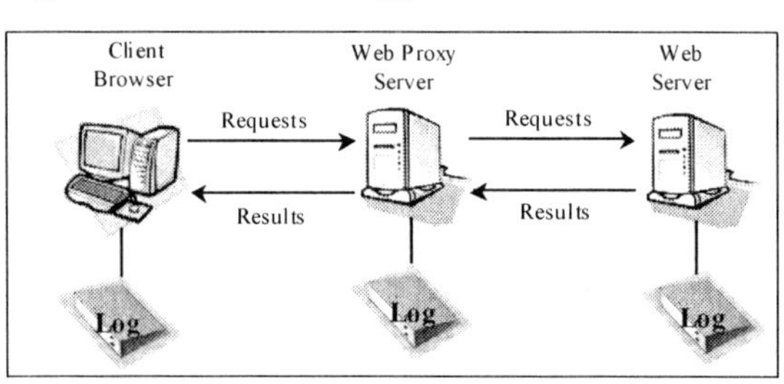

An Example

This example shows how to apply Web usage mining technologies to adaptive Web browsing for handheld devices. Figure 5 shows an HTML page—the World Health Organization of the United Nations—which is too large to be properly displayed on a microbrowser of a mobile handheld device. An HTML page can provide the following information:

- *Content*: Web page content provides the most accurate and full-text information. However, it is also the least-used information for a search engine since content extraction is far less practical than other methods.
- *Descriptions*: Web page descriptions can either be constructed from the metatags or submitted by Webmasters or reviewers. A metatag is an HTML tag that provides information (e.g., author, expiration date, a list of keywords, etc.) about a Web page. Page descriptions are either from the metatags or submitted by Web masters or reviewers.
- *Hyperlinked text*: Hyperlink text is normally a title or brief summary of the target page.
- *Hyperlinks*: Hyperlinks contain high-quality semantic clues to a page's topic (Chakrabarti, Dom, Kumar, Raghavan, Rajagopalan, Tomkins et al., 1999). A hyperlink to a Web page represents an implicit endorsement of the

Figure 5. The about page of the World Health Organization

page being pointed to. However, exploiting this link information is challenging because it is highly noisy.

- *Keywords*: Keywords can be extracted from full-text documents or metatags. Before obtaining keywords from a full-text document, some filtering operations are applied to the document. Typical operations would include the removal of common words using a list of stopwords, the transformation of uppercase letters to lowercase letters, word ranking, and so forth (Korfhage, 1997).
- *Page structure*: An HTML page is usually organized as a tree, in which top-level nodes are normally more important than lower-lever nodes.
- *Page title*: The title tag, which is only valid in a head section, defines the title of an HTML document. A title is usually chosen that makes sense with no context.
- *Text with different fonts*: Emphasized text is usually given a different font to highlight its importance.
- *The first sentence*: The first sentence of a Web document is likely to give crucial information related to the document, as it is usually an introduction or an abstract.

From the above HTML page information and the Web usage mining technologies, the following three major steps are used to achieve adaptive Web browsing for handheld devices:

1. The first sentence of each paragraph is kept because the sentence usually contains important information or an introduction about the paragraph.
2. The rest of the paragraph is ignored. By doing this, the page size can be reduced significantly.
3. The hyperlinks are sorted based on popularity from the Web usage mining. The proposed system installs a counter on each Web page. Each page visit increases the counter by one. A popular page therefore has a higher number in its counter.

Figure 6 shows three microbrowser screen dumps (one page), which are from the results

Figure 6. Screen dumps of the results after applying the proposed methods to the page in Figure 5

of applying the proposed methods to the page at Figure 5. Note that the proposed methods are still in the very early stage of development. This is probably the best example the methods could get so far. The major reason of this—ineffectively summarizing a Web page for handheld devices—is the high complexity of Web pages, many of which are not written by humans. Instead, they are generated from software such as Adobe Dreamweaver or include a variety of scripts such as JavaScript and active server pages (ASP).

FUTURE TRENDS

Many methods are proposed for rendering Web pages for microbrowsers. This research uses Web usage mining technologies to summarize Web pages and then displays the results on microbrowsers. Other promising methods include:

- *A block importance model*: This method extracts and presents more condensed Web search results to mobile users by using a block importance model (Xie, Miao, Song, Wen, & Ma, 2005), which assigns importance values to different segments of a Web page.
- *Link analysis*: Yin and Lee (2004) use a ranking algorithm similar to Google's PageRank algorithm to rank the content objects within a Web page. The PageRank model suggests the reputation of a page on a topic is proportional to the sum of the reputation weights of pages pointing to it on the same topic. That is, links emanating from pages with high reputations are weighted more heavily. This allows the extraction of only important parts of Web pages for delivery to mobile devices. This method provides savings in the wireless traffic and downloading time while providing a satisfactory reading experience on the mobile device.
- *Page and form summarization*: This method (Buyukkokten, Kaljuvee, Garcia-Molina, Paepcke, & Winograd 2002) breaks each Web page into text units that each can be hidden, partially displayed, made fully visible, or summarized. A variety of methods are introduced that summarize the text units. In addition, HTML forms are also summarized by displaying just the text labels that prompt the use for input.
- *Page reformatted*: This method changes page format, but not its content. Two commercial products use this method:
- *ACCESS*: ACCESS' NetFront browser includes smart-fit rendering technology (n.d.), which intelligently adapts standard Web pages to fit the screen width of any mobile device enabling an intuitive and rapid vertical scrolling process, without degrading the quality or usability of the pages being browsed. Concretely, the following process is performed:
    - Images larger than the screen width are scaled down to *fit the screen width.*
    - *Tables larger than the screen wid*th are split and laid out vertically, as shown in Figure 7.
- *Opera*: Opera's small-screen rendering technology (n.d.) reformats a Web page to fit it

Figure 7. A Web page table split by ACCESS' NetFront Browser

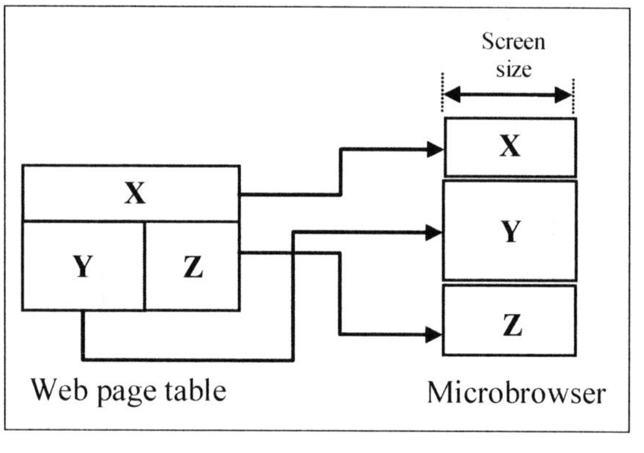

Figure 8. Screen dumps of Opera' small-screen rendering™: (a) before rendering and (b) after rendering

(a) (b)

inside the screen width and eliminates the need for horizontal scrolling. All the content and functionality is still available, it is only the layout of the page that is changed. Figure 8 shows an example of Opera's method.

- *Page structure*: One example of this method analyzes the structure of an existing Web page and splits it into small and logically related units that fit into the screen of a mobile device. Chen, Ma, and Zhang (2005) organize a Web page into a two level hierarchy with a thumbnail representation at the top level for providing a global view and index to a set of subpages at the bottom level for detail information.
- *Visitor's context*: This method displays the Web content based on the visitor's context such as location, information interests, and device capabilities. For example, nearest gas stations or affordable motels are displayed based on a mobile user's location and preferences. Pashtan, Kollipara, and Pearce (2003) propose a method adapting Web content to a user's dynamic context for Motorola Lab's Web-enabled museum system.

At this moment, it is almost impossible to predict which methods will prevail in the future because device or microbrowser manufacturers tend to use their own technologies.

CONCLUSION

It is widely acknowledged that mobile commerce is a field of enormous potential. However, it is also commonly admitted that the development in this field is constrained. There are still considerable barriers waiting to be overcome. One of the barriers is awkward browsing of large text Web pages by using handheld devices, which include six components: (i) mobile operating systems, (ii) mobile central processing units, (iii) microbrowsers, (iv) input/output devices, (v) memory, and (vi) batteries. Unable to effectively view Web pages, mobile commerce cannot be brought to a higher level. This research applies Web usage mining technologies to adaptive Web viewing for handheld devices. Web usage mining is the application of data mining techniques to the usage logs of large Web data repositories in order to produce results that can be applied to many practical subjects, such as improving Web sites/pages, making additional topic or product recommendations, user/customer behavior studies, and so forth. A Web usage mining system must be able to perform five major functions: (i) data gathering, (ii) data preparation, (iii) navigation pattern discovery, (iv) pattern analysis and visualization, and (v) pattern applications. This approach improves the readability and download speed of mobile Web pages.

REFERENCES

ACCESS. (n.d.). *Small-fit rendering*. Retrieved June 14, 2007, from http://www.access-company.com/products/netfrontmobile/contentviewer/mcv_tips.html#Anchor-Smar-45765

Buyukkokten, O., Kaljuvee, O., Garcia-Molina, H., Paepcke, A., & Winograd, T. (2002, January). Efficient Web browsing on handheld devices using page and form summarization. *ACM Transactions on Information Systems*, *20*(1), 82-115.

Canalys. (2004a). *A world of difference*. Retrieved April 14, 2006, from http://www.canalys.com/pr/2004/r2004061.pdf

Canalys. (2004b). *Global mobile device market shows tremendous growth*. Retrieved March 22, 2006, from http://www.canalys.com/pr/2004/r2004081.pdf

Canalys. (2004c). *Global smart phone shipments treble in Q3*. Retrieved December 3, 2006, from http://www.canalys.com/pr/2004/r2004102.pdf

Canalys. (2005a). *Global smart mobile device sale surge past 10 million in quarter*. Retrieved April 25, 2006, from http://www.canalys.com/pr/2005/r2005041.pdf

Canalys. (2005b). *Smart phones up, handhelds down globally in Q2*. Retrieved January 15, 2006, from http://www.canalys.com/pr/2005/r2005071.pdf

Canalys. (2005c). *Global mobile device shipments hit new peak in Q4 2004*. Retrieved May 02, 2006, from http://www.canalys.com/pr/2005/r2005012.pdf

Canalys. (2005d). *Worldewide smart phone market soars in Q3*. Retrieved December 3, 2006, from http://www.canalys.com/pr/2005/r2005102.pdf

Canalys. (2006). *Smart mobile device market growth remains steady at 55%*. Retrieved December 3, 2006, from http://www.canalys.com/pr/2006/r2006071.pdf

cellular-news. (2006). *Worldwide mobile phone sales up—except in Japan*. Retrieved May 3, 2006, from http://www.cellular-news.com/story/20573.php

Chakrabarti, S., Dom, B. E., Kumar, S. R., Raghavan, P., Rajagopalan, S., Tomkins, A., et al. (1999). Mining the Web's link structure. *IEEE Computer*, *32*(8), 60-67.

Chen, Y., Ma, W.-Y., & Zhang, H.-J. (2005). Adaptive Web pages for small-screen devices. *IEEE Internet Computing*, 9(1), 50-56.

Cooley, R. W. (2000). *Web usage mining: Discovery and application of interesting patterns from Web data*. Unpublished doctoral thesis, University of Minnesota.

Gartner, Inc. (2006). *Gartner says worldwide combined PDA and smartphone shipments market grew 57 percent in the first half of 2006*. Retrieved October 30, 2006, from http://www.gartner.com/it/page.jsp?id=496997

Hu, W.-C., Yang, H.-J., Lee, C.-W., & Yeh, J.-H. (2005a). A technical perspective of World Wide Web usage mining. In J. Wang (Ed.), *Encyclopedia of data warehousing and mining* (pp. 1242-1248). Hershey, PA: Idea Group Publishing.

Hu, W.-C., Yeh, J.-H., Chu, H.-J., & Lee, C.-W. (2005b). Internet-enabled mobile handheld devices for mobile commerce. *Contemporary Management Research, 1*(1), 13-34.

Korfhage, R. R. (1997). *Information storage and retrieval* (pp. 105-144). John Wiley & Sons.

Opera Software ASA. (n.d.). *Opera's small-screen rendering*. Retrieved June 23, 2007, from http://www.opera.com/products/mobile/smallscreen/

Pashtan, A., Kollipara, S., & Pearce, M. (2003). Adapting content for wireless Web services. *IEEE Internet Computing*, *7*(5), 79-85.

Silicon Valley Daily. (2006). *HP regains lead in global PC sales*. Retrieved June 11, 2007, from http://www.svdaily.com/gartner1.html

Symbian Limited. (2006). *Fast facts*. Retrieved December 10, 2006, from http://www.symbian.com/about/fastfacts/fastfacts.html

Xie, X., Miao, G., Song, R., Wen, J.-R., & Ma, W.-Y. (2005, March). Efficient browsing of Web search results on mobile devices based on block importance model. In *Proceedings of the 3rd IEEE International Conference on Pervasive Computing and Communications (PerCom 2005)*, Kauai Island, HI.

Yin, X., & Lee, W. (2004). Using link analysis to improve layout on mobile devices. In *Proceedings of the 13th International conference on World Wide Web*, New York City (pp. 338-344).

Chapter XI
Integration of Public University Web Sites and Learning Management Systems

Bernhard Ostheimer
University of Giessen, Germany

Axel C. Schwickert
University of Giessen, Germany

ABSTRACT

Internet technology has found its way into all areas of business and research. The World Wide Web is also used at universities to achieve different goals. On the one hand, it acts as a means of outer appearance, on the other hand, as an instrument of knowledge transfer and knowledge examination. Of course other purposes in addition to those named above do exist. Often different systems are used to achieve the different goals; usually, Web content management systems (WCMS) are used for the outer appearance and learning management systems (LMS) for transfer and examination of knowledge. Although these systems use the same medium (i.e., the WWW), it can be stated that often there is a heterogeneous landscape of systems. Resultant is the object of investigation of the present chapter. The chapter analyses the challenges concerning the integration of public Web sites and LMS a typical European university has to face.

INTRODUCTION

Internet technology has found its way into all areas of business and research. The World Wide Web is also used at universities to achieve different goals. On the one hand, it acts as a means of outer appearance (target groups are potential, current, and former students, researchers, lecturers, press, the interested publicity, etc.), and on the other hand, as an instrument of know-ledge transfer and knowledge examination (target groups are potential, former, and current students and lecturers). There exist other purposes in addition to those named above.

Often different systems are used to achieve the different goals; usually, Web content management systems (WCMS) are used for the outer appearance and learning management systems (LMS) for transfer and examination of knowledge. Although these systems use the same medium (i.e., the WWW), it can be stated that often there is a heterogeneous landscape of systems. Resultant is the object of investigation of the present chapter. The chapter analyses the challenges concerning the integration of public Web sites and LMS a typical European university has to face.

The research framework used for this investigation thus can be divided into two categories regarding the system types used: WCMS and LMS. In praxis, there is more than one system per system category implemented at a university because of the organisational conditions explained in this chapter.

BACKGROUND

To analyse the object of investigation for the present chapter systematically, it is necessary to consider the typical organisational conditions at a European university. A typical European university consists of a large number of hierarchically ordered or networked organisational units such as, for example, the central administration, library, departments, departments, research institutes, and programmes (Bajec, 2005). Each organisational unit presents itself and provides information specific to the unit on an individual public Web site in the Internet. All these Web sites can be accessed by directly entering the URL or navigating through the university's Web portal. In many cases several hundred independent Web presences represent various levels of hierarchy within the university's organisation. 'Technology follows organisation!' is the principle according to which a large number of public Web presence islands represent the organisational units of the university, which are highly decentralised and often—to some extent—autonomous (as illustrated in Figure 1) (Schwickert, 2004).

Within a university, organisational, management, and task structures are mostly highly decentralised. Consequently, the organisational units can make decisions and act at a high degree of autonomy. Resultant is that layout, design, and navigational and functional concepts of Web sites do greatly vary within a university. There is no doubt that a patchwork of public Web sites cannot project a professional image of the university as a whole to the outside world (Ostheimer, 2007). This results in the problem that a corporate design needs to be integrated into the numerous university public Web sites and aggregated information needs to be efficiently accessible.

The quality of a university is communicated by and reinforced through public Web sites to the general public. The quality, however, originates from the core business of a university: research and teaching, which can be divided into disciplines, subjects, and programmes and then can be managed in independent departments, departments, research institutes, and so forth within a university. Universities have by now realised that quality and efficiency of their research and teaching can benefit from a support programme of online elements. This requires electronic teaching and learning environments, that is, 'learning management systems,' which are available to students and researchers online (Claussen, 2004, Grohmann, 2006).

The fact that the university's core business units can, as described above, largely make decisions and act autonomously has led to a situation where several LMS in the various departments and departments of a university operate separately from each other. Financial, personal, and technical resources are implemented to introduce and operate these LMS 'islands,' to a great extent

Figure 1. Typical structure of a university's Web presence

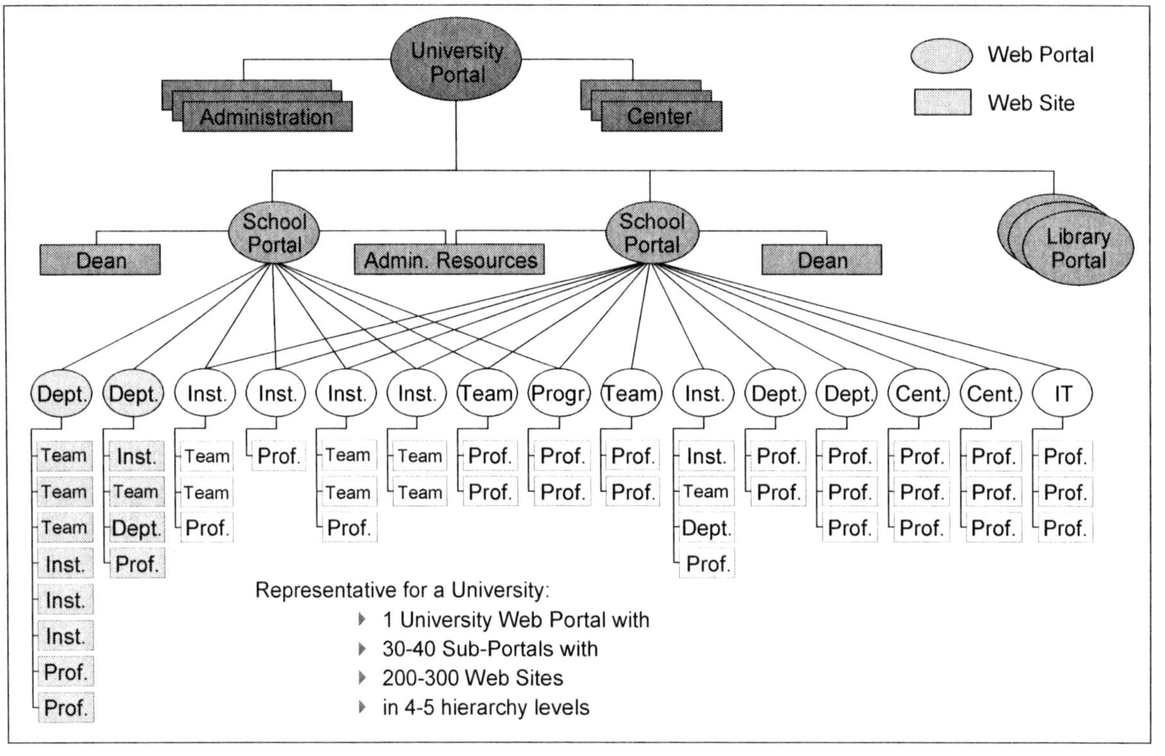

redundantly. Different systems also prevent network effects which would be beneficial for the experience and learning curves of the operators and users when working with the LMS within a university. As a result, the growing number of students attending interdisciplinary programmes in particular, where different subjects and courses are involved, are confronted with several LMS, mainly open source systems such as Moodle, StudIP, Ilias, ATutor, OpenLMS, and others (Bett & Wedekind, 2003; Claussen, 2004; Dittler & Bachmann, 2003; Schulmeister, 2005).

Consequently, in addition to the integration of navigational and functional concepts and a corporate design into university Web sites, there is also the question of how an LMS is established and which LMS suits best the university's teaching areas. The two problems have a great deal in common: the addressee and the medium, that is, the students and the Web.

To sum up, the following challenges can be filtered from the analysis of the object of investigation:

- In a university there are often several hundred Web portals and Web sites, which differ greatly regarding design, navigation, and features, and which represent several levels of hierarchy within the organisation. This does not present an advantageous picture of the university to the outside world.
- In the organisational units of the university's core business of 'research and teaching,' several LMS are operated independently

alongside each other. This is economically inefficient and prevents network effects within the university.

EMBEDDED LMS IN CORPORATE WEB SITE FARMS

The university Web portal lists Web portals of all departments. The function of the university portal is, on the one hand, to present the university as a whole to the outside world and, on the other hand, to enable navigation inside of the university. All information and features which concern the departments internally are presented at the departments' portals and, if necessary, more detailed on subordinate Web sites. The entire Web portal and Web site structure of a university is structured in hierarchies and networks according to the subsidiary principle, that is, each organisational layer and unit is (solely) in charge of the area for which it is competent and responsible. The entire Web portal and Web site structure (for academic and organisational content) should be designed as completely as possible in accordance with the university's corporate design. For this reason, it is necessary that the university provides its organisational units with a design template, which can be used throughout the university to integrate the corporate design. It is also taken into consideration that, within this framework, each organisational unit maintains a certain degree of individuality, for example, by using its own pictures, colour combinations, forms of content, and so forth. Owing to the fact that autonomy is important for the university organisational units' self-esteem, it would not be a good idea to force all the Web sites to become totally uniform. Experience has shown that departments, research institutes, and even work and research groups very much appreciate individual customisable university Web-styleguide templates because producing such a template would require technical know-how that the organisational units as a rule do not have.

Taking the university as a whole into consideration, it emerges as a clearly structured (for academic and organisational content) and homogeneously designed mass of Web portals and Web sites, that is, a 'corporate Web site farm.' The technical production of such a Web site farm requires a WCMS, which enables the allocation of the responsibility for the individual Web sites' content and functions top-down to the organisational units, to whom the Web sites 'belong' (Nakano, 2002; Zschau, Traub, & Zahradka, 2002). Equally, the Web content management system must be able to aggregate the accountability for higher responsibilities bottom-up to superior organisational layers within the university hierarchy. The WCMS should reinforce the organisation's flexibility (Ostheimer, 2007).

Ideally, a student should be able to find all the descriptive information and interactive functions which are relevant to the student's individual program and classes in the department's Web portal (e.g., enrolment, hours, places, lecturers, description of content, tasks required, literature recommendations, etc.) (Conrad & Donaldson, 2004). This information should be made available not only to registered students, but also to people interested in studying at the university, in order to contribute effectively towards the acquisition of more students. Department or department portals group together general information that the student needs for planning studies in course, room, and enrolment directories. Specific information on the courses can be found at the subordinate Web site of the body offering the course (institutes, professors, lecturers etc.), as long as the relevant Web site complies with the standards of the superior Web portal as far as design, navigation, and functionality are concerned.

The students learn in departments of a classical university either by attending courses which require presence, or by working alone as an individual learner, or working cooperatively in groups. The term 'e-learning' aims to support these three organisational forms of learning with the aid of electronic media. At university departments with a high proportion of courses which require presence, e-learning programs are generally combined to offer 'blended learning.' For example, a lecturer makes downloadable digital material available for self-study to accompany the lectures requiring presence. Moreover, the students can discuss the contents of the lecture with each other or with the lecturer in Web forums. The sum of all the parts is more than a traditional lecture; 'electronic' learning therefore becomes 'extended' learning. E-learning is thus primarily effective internally for a university and targets in the first place registered students and lecturers who are active in the core business of 'research and teaching' of a university. For the e-learning program the lecturers require practical technical instruments, teaching materials that have been prepared especially for e-learning, and the relevant didactic qualifications.

The Web already integrates a broad line-up of e-learning instruments into the online area, which are serving as a technical basis. A common characteristic of the majority of LMS is that there is a limit in features on both electronic and blended learning for students and lecturers, that is, internal university addressees. Their e-learning requirements cannot be fulfilled by a public Web presence, which is separated from the e-learning environments for students and lecturers, both technically and in terms of content and design (Schmidt, 2003). The symptoms of this type of disintegration are unfortunately too often evident online: departments/institutes operate Web presence islands with public contents and functions and link up within this public Web presence to an LMS, which is completely separate in terms of content and function and optically different as well.

Teaching at universities, in particular, with its high proportion of courses requiring presence involves a mix of information, communication, and cooperation, which, on the one hand, deliberately contains public elements and, on the other hand, contains specific elements that are deliberately only intended for registered students. Thus, for example, courses with descriptions of the content, references to sources, a large quantity of dates, places, lecturers, university regulations, ECTS information, and so forth can generally be seen at the provider's Web site and in central directories. Discussion forums, course evaluations, examination enrolments and results, digital reading materials, and so forth for the same courses should, however, be deliberately made accessible to (registered and possibly paying) participants of the relevant courses only, in a limited access LMS. The separate operation of an organisational unit's public Web site and its LMS will therefore inevitably lead to discontinuities in media, recording data several times, redundancies, and inconsistencies, due to the fact that data are being kept separately.

Within a university 'corporate Web site farm,' which is clearly structured from an academic and organisational point of view and homogeneously designed, students and lecturers should be able to transfer transparently between optically, navigationally, and functionally adapted e-learning environments. The WCMS and LMS which are used for the technical production must be based on a common database so that the public and protected contents can be used consistently many times and beyond.

The chapter's key objectives are:

- The university requires a homogeneous 'corporate Web site farm.' The technical production of this requires a WCMS, which can allocate the responsibility for content and functions within the university hierarchy top-down and aggregate bottom-up.
- The public corporate Web site farm and the LMS in operation should be integrated so that public and protected contents are presented

together and consistently on a homogeneous interface. The LMS offer the students the exact contents and functions that they require for their individual study situations with a uniform set of operating procedures.

SOLUTION: EXTENDABLE WCMS WITH PLUG-IN LMS

For the e-business presence of an organisationally decentralised university, the presentation of a collection of general information and links to the decentral organisational units is the basic task of a portal, which constitutes the common public starting point for the entire decentralised group of organisations. Beyond its function of branching out, the portal is all the more useful for visitors, if more individual information and functions on the decentralised organisational units can be consistently integrated into the portal. In time-saving 'one-stop vi-sits' to the portal, visitors should be able to inform themselves as comprehensively, concisely, clearly, and reliably as possible on the entire decentralised group of organisations. In decentralised organisational environments, a WCMS must be able to ensure the efficient production and top-down classification of individual single Web sites and, in addition, bottom-up integration for consistent portals.

From an economic point of view, the speciality of these tasks is justified by the fact that the content integration of various individual Web sites should enable the portal to operate as efficiently as possible in an automated manner with only a limited number of manual interventions. From a technical and functional perspective, a WCMS must ensure that the portal with decentralised organisational units is configured for the operator with selectionable information and functions. In a multilayer decentral organisational environment the WCMS must be able to amalgamate a variable number of hierarchy-based or network-based portals in a united portal structure.

A classical university can be seen as a predestined area for the implementation of this kind of 'extendable' WCMS. In the following description, the Justus-Liebig-University (JLU) Giessen, Germany and the WCMS 'Web portal system' (WPS), which is used there, will be described as an example for the objectives described above (WPS, 2006). The WPS has been working successfully since April 2002 for the portal of the entire Department of Business Management and Economics at JLU and its currently 25 individual organisational units. By mid-2005, the WPS was implemented for all the areas of three other departments and two research institutes of JLU. In total, more than 150 portals and Web sites are currently being produced, filled with content and operated by the WPS at JLU.

A Web portal which navigationally integrates all the Web sites of its organisational units and aggregates its public contents can be considered as the central starting point for every department. A department portal, for example, lists all the courses offered by its organisational units in a course directory for the entire department. The same applies for the people, publications, rooms, materials, forums, and a lot of other information. In order to successfully aggregate all the contents, it is necessary that the organisational units that are assigned to the portal autonomously fill their own decentral Web sites with contents using the WPS, thus making the contents available to the central WPS database. A department course directory is, for example, only of use if it really does list all the department's courses. This is where the role of the WPS as a basis becomes clear: each decentral organisational unit keeps its individual contents up-to-date and records them all in the WPS and the contents are then centrally aggregated completely automatically by the WPS for the entire department. The responsibility for the 'quality of the portal' is thus largely handed over to the organisational units (Schwickert & Grund, 2004).

The extendability of a WPS for a university portal structure is scalable from two perspectives. On the one hand, the WPS can gradually be extended to the university's decentral organisational units. Several WPS, for example, for different departments can be technically operated parallel and ensure that content, function, and navigation are aggregated not only within the department, but in the central university portal. Implementation covering all the areas of the university is therefore not necessary. Alongside scalability in the areas covered, the quantity of functions offered is also scalable. The WPS offers an organisational unit a multitude of prefabricated and ready-to-use functions for the production and filling of its public Web site (e.g., newsboards, a download centre, online editions, online forums, marketplace and shop solutions, administration of lists of people, publications, projects and links, administration of courses and special events, online enrolment, online evaluation, as well as site, page, template, and menu editors) (Claussen, 2004; Grohmann, 2006; Rosenberg, 2006). For each organisational unit a certain selection of these functions can either be put into use or closed down. The editors, as authorised administrators, can control whether and which organisational units can go online with individually designed Web sites, or whether all the Web sites of a department, for example, have to use a uniform corporate design template.

Moreover, the quantity of functions can be scaled by the selective activation of integrated LMS functionalities (Ostheimer, 2007) as soon as the public portal/site structure has been established in the university's organisational units. The WPS produces these LMS functionalities by optionally plugging in the students' personal information center (SPIC) and teachers' administration center (TAC).

- SPIC is the personalised environment for a student, enabling the student to occupy and track courses. Each student can register as a SPIC user via a link on the public portal/site and maintain the student's own SPIC environment. SPIC registration and SPIC access are, for example, integrated in the department's portal at the Department of Economics at JLU Giessen.
- The TAC is a personalised environment for a lecturer, enabling the lecturer to manage the course program. The WPS administrator gives the lecturer the relevant TAC authorisation for the courses. The lecturer can also go through the TAC user account via the public portal/site to reach the lecturer's own personal TAC environment.

The SPIC/TAC environment is illustrated in Figure 2. A student occupies the courses the student is attending in the current semester in the student's SPIC environment (by enrolling online either with or without authentication), puts together timetables and appointments, subscribes to the relevant newsboards, downloads, forums, and evaluates the student's courses.

The lecturer is offered extended functions for the lecturer's courses in the TAC environment. With these functions, the lecturer can activate online enrolment for the courses if the lecturer wishes. Either admission for the course is unrestricted, or participants have to give information in several steps, such as password, real name, or even smart card authentication within the framework of the public key infrastructure (PKI) at JLU Giessen (Treber, Berg, & Schwickert, 2004). The lecturer can configure the system according to the lecturer's requirements, either unrestricted public lectures or a controlled group of participants. The lecturer can include general information, newsboards, downloads and uploads, forums, bookmarks, evaluations, and ECTS descriptions for each of the courses. The lecturer can either be permitted to use or denied use of each individual TAC function for each course.

Beyond the functions relevant to the administration of teaching, the WPS offers in addition a broad palette of community functions

Figure 2. SPIC/TAC environment in the Department of Business Management and Economics at JLU

for registered SPIC/TAC users. Each user can participate, using several profiles and names, in various forums, discussion, and study groups via integrated instant messaging, and also run buddy groups and place classifieds.

All the information, content, and functions offered on the courses can either be made available only in personalised environments, or alternatively in public Web sites and portal aggregations for the relevant organisational units, too. The WPS thus seamlessly integrates the WCMS and the embedded LMS functions for public and protected areas of a university in the Web. In this way, the university avoids operating several unlinked systems parallel.

The introduction of SPIC for the students of the Department of Business Management and Economics started in December 2004 and ran very smoothly without any instructive measures. There are currently (June 2007) 2,900 active registered SPIC users and this constitutes approximately 98% of the students enrolled in the department. At the time SPIC was introduced, all the courses (approximately 300 in the 2004/2005 Winter term) and the relevant contents and functions were already available for the students in SPIC. Feedback from the students and lecturers was positive in every respect. Ultimately, the comprehensive coverage of all the courses, the integrated community functions, and the intuitive operability of SPIC contributed decisively towards the extremely quick adoption of the system by students and lecturers.

Since the introduction of the WPS and the SPIC/TAC, new structures have emerged automatically regarding the necessary allocation of the tasks of recording and maintaining content within the university. The simple and intuitive operability of the system enables people in the individual organisational units (i.e., professors, research institutes, work groups, andoffice of the dean) to carry out Web publishing, without having any technical know-how. The number of people responsible for the content has increased to such an extent that the people producing the content are also those that record and maintain it in the system. The processes of Web publishing have also inevitably changed. Whereas previously technically experienced Webmasters were occupied with formatting and releasing contents, leading to bottlenecks, now each author can and should be responsible for the author's own content and put it online.

To sum up, the solution's key messages are:

- The Web portals and Web sites can be quickly and efficiently restructured into Web presences for the general public which are homogeneous in terms of content, function, and design. Each organisational unit is in charge of its own Web site; they are not forced to comply with a central WCMS solution, but the responsibility is allocated step-by-step and optionally.
- Owing to the fact that the WPS automatically aggregates content, Web portals emerge that contain a wealth of information which is up-to-date at all times. Individual Web sites can be added and organised in hierarchy-based or network-based portals or be 'closed down' at all times. Any number of portals within any number of layers of hierarchy can be created.
- The LMS interfaces are transparent and integrated in the public portals and sites with access only for authorised users. LMS-relevant data are recorded in the system only once and multiple uses of this data are consistently and precisely defined.
- All WCMS and LMS functions of the WPS can be individually activated or deactivated for each user group and organisational unit. Each WPS user can therefore be assigned a personalised quantity of functions.

CONCLUSION: TECHNOLOGY ENABLES ORGANISATION!

If each author is personally responsible for putting content online—that is, with no further editorial checks—there is, in principle, the danger that the quality of content suffers. Experience in the departments of JLU Giessen using WPS has shown the contrary. Empirical research into the user behaviour of people working with WPS clearly indicates that people are all the more careful when publishing 'their own content.' The addressees of the contents, chiefly students, confirm that the quality of the content in the departments concerned has improved categorically in comparison to the quality of the content before the implementation of the WPS.

Furthermore, the impact on the quality and a considerable impact on quantity can be observed for the departments' portals and sites. Monitoring the publishing process has proved that the quantity of published news, downloads, info pages, forums, announcements, directories, and so forth has multiplied and continues to grow in all the departments. In the same way, the number of visitors to the public portals and individual Web sites is permanently rising. Here the performance of the Web site is reinforced by success: more and better contents make a visit to the site more interesting; as more visitors are recorded at the site, it becomes more 'effective' and interesting to publish information on the site.

Besides the 'time to Web' has also been significantly reduced due to the use of the WPS. The complete publishing process with the WPS has been automated with the use of user-friendly Web forms. Owing to automation and the fact that there are no more bottlenecks with the 'Webmaster,' who was responsible for producing, updating, and publishing individual pages as well as the whole Web site in the conventional publishing process, the time lapse between the production of content and its publishing has been reduced considerably.

The positive impact on quality, quantity, and time largely leads to a reduction in costs. When organisational units with their own Web site start using the WPS as application-service-providing solution in the department, they no longer need their own system administrator and save costs for the individual maintenance and servicing of redundant system technology within the organisational units. Initial acquisition and customisation costs and a limited running expense for the maintenance of the system software for the WPS are covered by a multitude of organisational units. The simplicity of the publishing process per WPS largely makes it no longer necessary to employ experienced technicians. As a result personnel costs for this—mostly for student jobs—can be reduced considerably.

As regards the portal, costs have also decreased, due to the fact that filling and maintaining the portal is carried out by the WPS with more than 80% completely automatically drawn from the contents of the connected subportals and individual sites. Only a few pages with static information are produced manually in the portal. In comparison to a portal which is produced manually or only partly automated, the portal master has a lot less work, because the content is aggregated by the WPS. In the areas where the WPS is in operation at JLU Giessen, it has been established that the reduction of cost induced by the WPS in the field of Web content management pays off the necessary investments of the department within 2 years and largely overcompensates for the running costs induced by the WPS.

Up to now WPS and SPIC/TAC have made an integrated system environment available, making it possible to produce consistent public portals and sites as well as closed areas for teaching administration. In 2007, an extension of the functions provided by the WPS is on the agenda with learning support for the students in the form of ongoing online examinations. WPS interfaces will be added to the SPIC/TAC, enabling the lecturers to supplement their blended learning program of presence courses and WBT with interactive versatile online exams and tests. Students' learning success will be individually available for each student and teachers get aggregated overviews of their students' learning progress. The objective is to have created an e-university platform with the WPS, which is flexibly scalable in terms of capacity and functions. Via this platform, the university will be able to consistently produce and efficiently operate its entire public Web presence, its closed teaching administration, and blended learning.

REFERENCES

Bajec, M. (2005). Educational portals: A way to get an integrated, user-centric university information system. In A. Tatnall (Ed.), *Web portals: The new gateways to Internet information and services* (pp. 252-269). Hershey, PA: IGI Global, Inc.

Bett, K., & Wedekind, J. (Eds.). (2003). *Lernplattformen in der Praxis*. Muenster, Germany: Waxmann.

Claussen, S. (2004). *Entwurf einer personalisierten virtuellen Lernumgebung*. Aachen, Germany: Shaker.

Conrad, R., & Donaldson, J. (2004). *Engaging the online learner*. San Francisco: Jossey-Bass.

Dittler, M., & Nachmann, G. (2003). Entscheidungsprozesse und Begleitmassnahmen bei der Auswahl und Einfuehrung von Lernplattformen—Ein Praxisbericht aus dem LearnTechNet der Universitaet Basel. In K. Bett & J. Wedekind

(Eds.), *Lernplattformen in der Praxis* (pp. 175-192). Muenster, Germany: Waxmann.

Grohmann, G. (2006). *Learning management*. Koeln, Germany: Josef Eul.

Nakano, R. (2002). *Web content management*. Boston: Addison-Wesley.

Ostheimer, B. (2007). *Verteilende eBusiness-systeme—organisatorische flexibilisierung am beispiel eines verteilenden eUniversity-systems*. Wiesbaden, Germany: Deutscher Universitaets-Verlag.

Rosenberg, M. (2006). *Beyond e-learning*. San Francisco: John Wiley.

Schmidt, I. (2005). *Blended e-learning*. Saarbruecken, Germany: VDM.

Schulmeister, R. (2005). *Lernplattformen fuer das virtuelle Lernen*. Muenchen, Germany: Oldenbourg.

Schwickert, A. C. (2004). Dezentrales Web content management. In S. Geberl, S. Weinmann, & D. F. Wiesner (Eds.), *Impulse aus der Wirtschaftsinformatik* (pp. 247-260). Heidelberg, Germany: Physica.

Schwickert, A. C., & Grund, H. (2004). *Web content management—Grundlagen und Anwendung mit dem Web portal system V. 2.5* (Arbeitspapiere WI, No. 3/2004). Giessen, Germany: Justus-Liebig-University.

Treber, U., Berg, J. H., & Schwickert, A. C. (2004). *Smart-card-anwendungen am fachbereich wirtschaftswissenschaften der justus-liebig-universitaet giessen* (Arbeitspapiere WI, No. 7/2004). Giessen, Germany: Justus-Liebig-University.

WPS. (2007). *Informationen zum WPS*. Retrieved June 1, 2007, from http://www.Web-portal-system.de/

Zschau, O., Traub, D., & Zahradka, R. (2002). *Web content management*. Bonn, Germany: Galileo Press.

Chapter XII
Innovating through the Web: The Banking Industry Case

Chiara Frigerio
Università Cattolica del Sacro Cuore, Italy

ABSTRACT

In recent years, the financial services industry has been witness to considerable consolidation (Berger & Udell, 2006; De Nicolò, Bartholomew, Zaman, & Zephirin, 2004; Figueira, Neills, & Schoenberg, 2007) and organizational progress in order to sustain two main objectives: efficiency and commercial effectiveness (Epsten, 2005; Sherman & Rupert, 2005). In order to sustain customer-oriented and efficiency strategies, banks have started to explore new ways of conducting their business, introducing areas of innovation in their services, practices, and structures to offer the most complete array of services possible (Quinn et al., 2000). On the other hand, new services and products drive retail banks to explore new ways of producing or delivering these novelties. This is true especially for Internet banking services that offer services to customers 24/7, and it becomes clear that adding new services, that is, trading online or bill payments, is easily and quickly geared towards improving commercial effectiveness. The following chapter aims at describing to what extent the Internet has developed new services and businesses, and what are the main figures of the phenomenon in Europe. Moreover, the Internet has introduced new coordination processes within each financial institution. Let us think about Intranet portal, content management tools, and business process management suites, which are now quite spread in banks due mainly to their technological ease-of-use. Thus, Internet is representing an innovation wave extremely relevant for the financial industry as a whole, and the effects on banks' performance is emerging. What do we expect in the near future? In all probability, the usage of Web-based application will be bigger and bigger also in other contexts of the bank processes, even if some risks could occur when clear strategies and change management practices do not direct the innovation.

INTRODUCTION: INNOVATION AND WEB BANKING

The banking system is undergoing remarkable strategic and organizational change processes, which will consolidate and spread even in the future. Such changes are not to be considered as a contingent change in the design and development criteria of the business, but they point out an innovation character which extends far beyond those boundaries. The application fields of innovation criteria are numerous (e.g., technological, organizational, commercial) even though two main critical issues often hinder the creation and spread of innovation: a) the origin of innovation, that is, the characteristics of the innovation source; and b) the possibility to know beforehand the effects of innovation and to keep its sustainability in the long run.

With reference to the subject of the origin of innovation, the most traditional theoretic version counters the organizational process innovation with the innovation of product and service. "Product innovation is given by the introduction of a new product or service to meet the market request or a particular external customer, while process innovation is defined as a new element which is introduced into the organizational processes in order to produce a product or to distribute a service" (Damanpour & Gpalakrishanan, 2001, pp. 47-48). Organizations develop product or service innovation in order to increase their market shares or to improve their strategic positioning, while they innovate in the process in order to achieve economies of scale and increase profitability (Utterback & Abernathy, 1975). The innovation of the first type allows the development of the first phases of the life cycle of a product/service, therefore it involves relevant investments and high risks. On the contrary, organizations that operate process innovations generally develop themselves at the time in which the product/service is in a maturity phase, therefore in the long term. For this reason, process innovation generally requires less investment (Anderson a& Tushman, 1991; Barras, 1990).

Moreover, by overcoming the dichotomy between product/service innovation and process innovation it is possible to consolidate an approach which takes into consideration the combination of both drives within an innovation policy (See Figure 1). Therefore, both product/service and process innovation may be combined and may generate added value within a short period and with minor costs (board 3). Therefore, in some cases, the ability to innovate is measured both

Figure 1. Product/service/process innovation

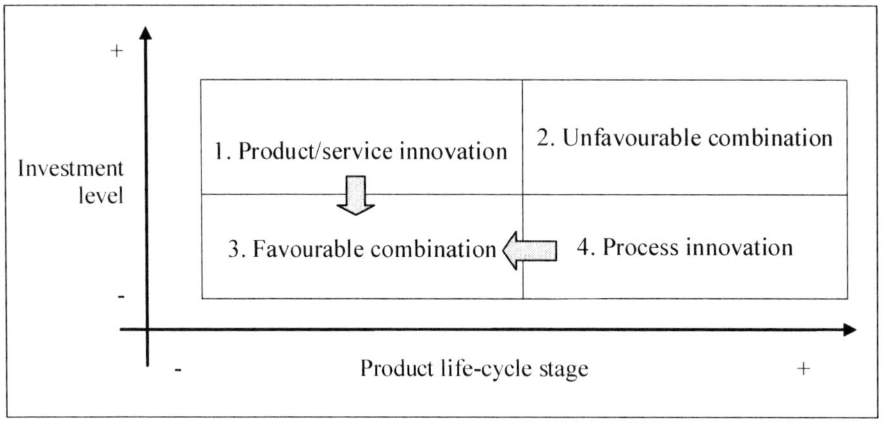

in the product/service part as well as in the process one. The convergence of the innovation strategic approaches is ensured by organizational mechanisms which have the role of coordinating and managing skills, relations, and innovation processes as a whole.

With reference to the second critical issue source, managerial literature associates to innovation the characteristics of "radicality." An innovation is radical when its adoption process over time develops in nonlinear maturity phases, characterized by times of performance growth on discontinuous effects. The main obstacle in the innovation management is given by the difficulty of previously predicting moments in which a performance leap and the relevant extent of the performance growth will take place. The effects of innovation may be even more uncertain if the innovation recipient is a little controllable, being it external (e.g., the customer). As a matter of fact, the lack of control over the external party prevents knowing the cause-effect relationship of the leverages affecting the innovation adoption process. Such leverages may also be generated or limited by solicitations which are outside the bank-to-client relationship, difficult to foresee and to manage.

All the above mentioned introductory considerations may be useful in evaluating one of the main innovation areas developed by the banking industry over the last 20 years, that is, the Web.

As a matter of fact, the introduction of this technology immediately generated a great application interest from financial institutions and was directed mostly towards:

a. Development and management of both relationships with the outside (i.e., stakeholder and clients) and the inside (i.e., cooperators and employees), through the use of the Internet and the Web as innovations in the publication and circulation of contents or services.
b. The development of "core banking" applications and information systems in Web-based environments characterized by integrability and pace of important achievements.

This chapter aims at exploring the first applicability area, by describing how and in which way the Web development in such a context has been innovative. Specifically, it analyzes first the characteristics of *Internet banking* and then those of *Intranet banking*. By the first term, we refer to the development of distribution channels of product/services alternative to the physical presence, based on Web technology and distributed through Internet. In this area, product innovations went and go hand-in-hand with process/channel innovations. As a matter of fact, certain products specifically created for the Internet channel, like c/a, loans, and financial services, are supported by new management processes, aimed at providing procedure efficiency and standardization. The joint product and process innovation enables to support further innovation strategies aimed, on the one side, at creating increasingly customer behaviour-oriented products and, on the other, at improving processes even in nonefficient areas. To support the theory concerning the mutual relationship of product and service innovation, one may consider the evidence substantiated by the "multichannel service strategies," characterized by the presence of cross-channel and cross-customer product/services and processes.

Therefore, though the evolution of the Internet, the banking phenomenon is continuous and steady; another application area of the Web use, less visible but equally active, is the one of Intranet banking. Intranet is an information support available for members of an organization aimed at providing easy and immediate access to organizational and training information, thus reducing distribution costs and the level of information dispersion which potentially increases the job complexity and employee dissatisfaction.

The development of information technology-based innovations, like Web banking, raises the issue of the importance of an assessment of the threats of technology innovation. It is enough, for instance, to look at the problem of trade (products and channels), organization (production processes and innovation delivery), and information (data and information flows) integration, deriving from the implementation of an Internet strategy. The lack of integration represents one of the main defeating threats of the innovation implementation process (Ciborra, 1996). As a matter of fact, the latter is effective only when it is a part of a whole and is not out of context. Another technology innovation challenge is represented by the speed and the direction of the standard implementation process. Specifically in the Web banking industry, the interoperability requirements are essential. Technology innovations which may vary the widespread standards are particularly critical and difficult to implement.

INTERNET BANKING: A HISTORICAL SUMMARY

Internet banking started growing at the beginning of the 90s and saw its maximum development at the end of that same decade.

Its importance in the definition of bank commercial and marketing strategies was immediately clear. The development of applications which would allow customers to remotely operate by means of their home PC was minimally complex and fairly cheap. But the decisional factor of the main business plans on Internet banking was the evaluation of the transaction marginal efficiency. As a matter of fact, since the first analysis it was clear that the cost of the process of a transaction opened on a virtual channel was considerably lower than the physical one (e.g., that of a promoter branch agency). This scenario initially led some banks and then most of them to adopt an Internet strategy.

Especially in the leading banking groups in Europe, the potentials made available by Internet led to the creation of units fully dedicated to the management of virtual customers (both retail and corporate). Such units were often legally autonomous companies, that is, subsidiaries of the parent bank. However, the strategy of direct banks (i.e., those which were exclusively present on the virtual channel) does not have a long life. As a matter of fact, on the one hand, besides the totally dedicated channel, banking groups keep the possibility of operating their accounts via alternative channels (often by using the same virtual bank technology). On the other hand, certain companies born to represent the mere virtual channel of their parent bank start creating, little by little, within a multichannel perspective "light" branches with employees able to support customers. This was, for example, the Italian case of Banca 121. The current situation is characterized by a higher number of universal banks integrating Internet service into their offer. Only a few operators are solely virtual and they are generally specialized in online trading. This last service has long been considered by most operators as the killer-application of Internet banking. At a time when investments in negotiable securities had high returns, online trading developed considerably and, in certain cases, it replaced the figure of the financial promoter or of the financial manager. Subsequently, the development of further services, not only linked to brokerage, allowed also the development of more complete Internet strategies for retail banking (e.g., c/a transactions, circulars, requests for consulting support) and also for private banking, characterized by sophisticated and demanding clients.

The e-commerce phenomenon for banks has not yet given the expected results. Some banks have activated e-commerce initiatives both in the business-to-business (B2B) as well as in the business-to-consumer (B2C) areas. English and Spanish banks proved to be the most active in this situation. Besides payment management, often accompanied by the creation of special credit cards in the case of B2C, banks also offer partnership formal guarantee services (especially in the case of B2B portals) and of special financing. We have progressively witnessed the passage from an activity of incubator and partnership, still present in the less developed areas, to the activity of facilitating or aggregating, though the few existing cases do not allow us to consider e-commerce for banks as a consolidated phenomenon.

THE PRESENCE ONLINE: AN ANALYSIS OF WEB SITES

A recent survey carried out in 2003 allowed the comparison of the Web world experience of the European banking key-players, with particular reference to strategies and behaviours (Carignani & Frigerio, 2003).

The survey targeted seven European countries (i.e., France, Germany, Italy, Scandinavian Countries, Spain, Switzerland, UK), selected among those having a higher level of Internet usage penetration and characterized by important investments through alternative channels. To this purpose, the survey analyzed: first the German market, considered as the widest and most developed market of the European scenario; the British one, whose financial and banking tradition is, in many aspects, similar to the US banking system; the Scandinavian system, important for the extremely widespread Internet use among the population; and the Swiss one, characterized by a peculiar banking tradition in retail and private banking. Other significant experiences closer to the Italian context are the one of France, characterized by a good spread of alternative channels, and the one of Spain which, regardless of the disadvantage compared to its European competitors, is now coming up strongly into the direct banking sector with interesting offers.

With regards to the individual banks selected within each chosen country, in Germany, whose

Figure 2. Internet banking strategies

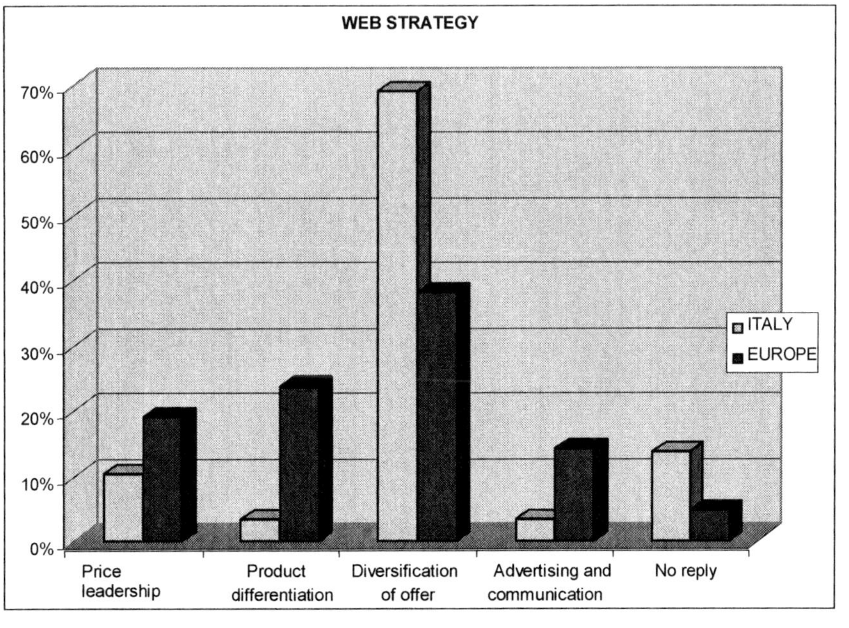

banking system still represents today the reference model of all operators of the Old Continent, and in the UK, the survey investigated traditional banks which chose to invest hugely in new channels by implementing rather aggressive strategies, and the most significant examples of direct banks born as spin-offs of the major banking groups of their respective countries. This way it is possible to monitor and compare the direct banking initiatives of both leading banks and followers. With reference to the French market, the survey focused merely on the main traditional banks in terms of customers, as well as on volumes and on the three most successful initiatives in the e-banking field. In the Scandinavian banking industry, which, as previously mentioned, is strongly geared towards the use of innovative channels, the attention was focused on those banking institutions which have adopted the most interesting and innovative initiatives from a technological point of view and which, for this reason, may represent a reference model for banks of other countries like ours, where Internet penetration is still low, though growing, and the population is still little accustomed to use technological channels. Considering the central role given to the Italian situation within our research, a wide sample of banking institutions was selected. In particular, attention was given to those banks which form part of the major groups of the country, as well as to single institutions having a considerable volume of customers.

From the analysis of data provided in the field of the strategic choice made, both Italian and European banks show the trend of looking at electronic channels as a way of diversifying their offer. The search for new services which may be offered online, thus contributing to increasing the added value given by the bank, is pursued by 68% of Italian banks and 38% of European banks. Price leadership is a primary value just for

Figure 3. Value-added services offered by online banks

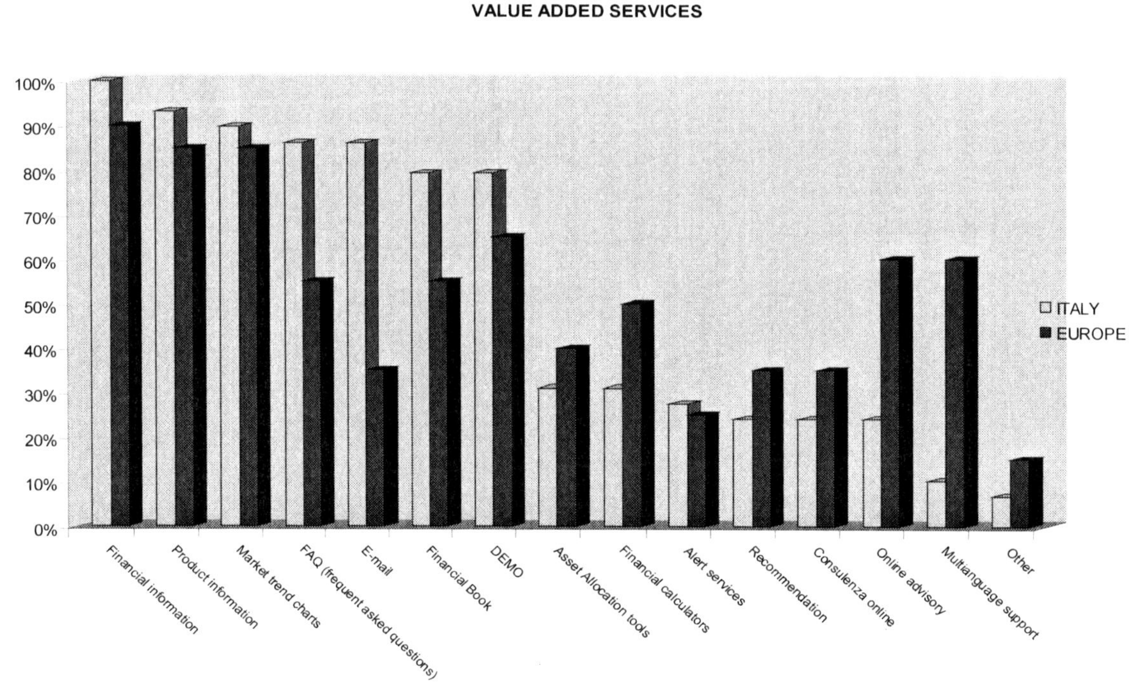

10% of Italian banks (typically those which are merely remote) and of 19% of foreign banks. This value makes even more sense if it is compared to the result given by the same survey carried out in 2000. As a matter of fact, in that occasion, it appeared that price reduction represented the main objective of Italian banks, followed by the advertising channel. Moreover, 24% of European banks are aiming at product differentiation, by introducing new products and services which have a different configuration for the Web.

Going more into details with regards to the implemented strategies, the types of services offered have been analyzed, also comparing the offer of other transfer services by remote banking.

All European and Italian banks which took part in the survey declared they offer online transfer services, information, and value-added services. Call centers basically provide the same offer variety. Actually, the range of operations which may be carried out through the use of the most recent e-banking technologies is not so wide. Among banks which are equipped with Wap technology, only 70% are able to offer transfer services but only 30% more innovative services, which bring a value added to the overall offer of the bank. SMS messages are used mainly as an information tool. Moreover, the few banking institutions which have extended to the Web TV/digital television their interaction means with customers are trying to develop transfer services which are mostly of transactional type.

The online trading expansion has pushed banks to disclose market information. This information area is mostly explored by foreign banking institutions; to this effect, 94% of European sites, compared to 84% of Italian sites, dispose of a dedicated area. The majority of banks offer services for customers while visitors may consult only stock exchange quotations. The service is almost exclusively supplied by a specialized information provider.

Banks also offer many value-added services; simulation tools and communication services (e.g., forum, newsletter, mailing) are among the most available services. Seventy-four percent of Italian Web sites and 53% of foreign Web sites taking part in the sample offer financial simulators which support customers in the most efficient portfolio choices. Here again, one can notice the will of our virtual banks to aim at clients information. In Europe, the percentage is higher in Switzerland and Great Britain where banking institutions have a long tradition in managing savings. Discussion forums are present in some banks, especially in Central Europe, but they are also rapidly developing in southern banking institutions.

Financial information is the most widely value-added service (100%) offered by Italian banks, followed by product information (93%), market trend graphics (90%), and online help like FAQ, e-mail, and demos, which represent 86%, 86%, and 79% of the surveyed population, respectively. On the other end of the rating scale are multilingual supports (10%), trading simulators, and online financial consultancy (24% each). The latter, in particular, shows a certain growth compared to the previous survey, but is still too weak to be considered representative.

The European sample, on the other hand, shows a somehow different trend. The multilingual support is present in all European countries, with the exception of Great Britain, and the same goes for financial stimulators. To the possibility of asking for advice by mail or to finding an answer to frequent doubts, thanks to the FAQs, European financial institutions prefer to offer financial calculators and asset allocation instruments (though still in embryo).

From the institutional analysis of the Web site, it shows that 94% of banks of the sample deem it indispensable to offer visitors and customers information about the bank. A good 92% of Italian sites describe the structure of the banking institution and its history, while just 79% of foreign banks provide this kind of content. Moreover, listed banks offer the possibility to consult and download the latest balances and relating reports, as well as the most significant press articles. Italian banking institutions seem to give this information area a higher value rather than their foreign counterpart, as well as with regards to the array of news available on the site. All the analyzed Web sites contain their product catalogue. In this regard, there is a difference in price listing: while 83% of Italian Web sites make reference to prices, only 21% of European Web sites make this information available to visitors. Generally speaking, however, the information available concerning this issue is poor and incomplete with regards to contents and presentation clarity.

Access speed for the display of the Internet banking service and of online trading has considerably increased (8.75 seconds on average). Italian sites have a higher uploading speed compared to European ones, thanks to the reduced presence of simulators and other value-added services and to the improvement of pages design. In order to optimize the page opening time, HP got equipped with direct service links and improvements were made in the site design. Italian banks got higher scores in the design of Web pages (on average 4.1 against a mean European score of 3.73). The number of images per page has reduced in order to increase browsing speed; legible characters are used, as well as dark tones for the background and light tones for writing. Well aware that the browsing difficulty constitutes a discouraging factor for the average net surfer, banks are paying particular attention to this aspect of their site.

INTRANET BANKING AS AN ORGANIZATIONAL INTEGRATION TOOL

As we were saying, the study on the effects of the use of the Web technologies in the financial and banking industry cannot overlook the intercompany application scope. Banks are traditionally considered as complex, large-size organizations, characterized by geographical widespread location and organizational and product differentiation. Due to the frequent merge and acquisitions transactions (M&As), the organizational complexity has lately grown. As a matter of fact, the search for economies of scale in a minimally differentiated market characterized by end customers being highly sensitive as far as the "price" variable is concerned, pushed the bank top management to increase the volume of products and services offered and to be present in a wide range of markets. The trend is therefore to look for economies through merge and integration projects whose main objective is to increase the company size and to extend the bank geographical boundaries even in international contexts. The search for these situations of strong boost towards volumes has an immediate effect on the competition of the industry, which becomes, on the one hand, increasingly global and, on the other, increasingly based on efficiency and innovation parameters.

Against a rather widespread strategic choice of dimensional growth within the banking industry, the distinguishing factor, on the one hand, and challenging factor, on the other, is represented by the integration pace issue, that is, the time needed to carry out strategic plans. Therefore, the success factor consists of the ability to create the conditions so that organizational systems and mechanisms change and reach a sound configuration in the shortest time possible.

Within this scenario, a role of paramount importance is played by technologies and information systems in triggering changes through the specification of common practices and procedures in the operations and information management. Lorence and Lorsch (1967) had already pointed out that the need for integration originates from organizational differentiation.

Increasing abilities aimed at managing organizational interdependencies are considered one of the most evident effects of the spread of information technology (Agliati, 1996). In this context the importance of an extensive applica-

Figure 4. CM and BPM projects in financial institutions

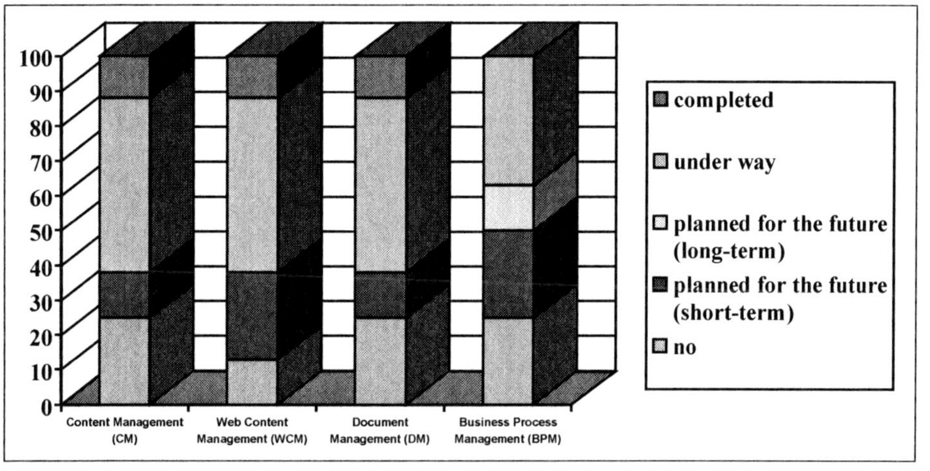

tion of different ICT solutions is stressed, aimed at facing both operational interdependence (i.e., linked to intermediate and final input and output flows which are created as an effect of division of labour) and information interdependence (i.e., referring to the need to exchange information in order to face and solve problems encountered during task performance) (Ferraro, 2000).

A common aspect refers to the progressive increase of internal communication flows channeled through *groupware* applications and, in case of larger organizations, the presence of dedicated Intranet and company portals. The networks (namely *Intranet*) increase the capability of collecting, processing, and redistributing "codified" knowledge. The same infrastructures allow users to exchange messages and different texts, forward requests, consult specialized document files, benefit from remote training and coaching courses, and so forth. The use of such instruments for communication purposes is led by the growing use of solutions and services coming from third parties (i.e., external suppliers or companies referring to the same banking group).

The introduction of the Internet paradigm for the communication and exchange management is not so strong; as an example, the use of *Extranet* in the relationships with external partners is still marginal. Sometimes the relationship with certain *outsourcers* (with the exclusion only of persons in charge of the management information system) uses the support offered by similar tools.

Entrusting the operational phases to outside bodies, with the consequent need to carry out systematic controls on the work performed by the supplier and on the creation of working *teams* of specialists coming from the extra-banking world, may be the reason behind the development, in the future, of computerized applications addressed to external counterparts both at the top and at the bottom of the company organizational limits. Should this trend gain ground, we may witness a further expansion of the limits of information systems, with clear repercussions on the bodies in charge of the operation and management of the said systems.

Numerous banks have realized how *Intranets* may increase the visibility of the organization, since they make clear the existence of knowledge, *routine,* and interconnections between all parties contributing to the same process. In certain fields the analysis perspective is widened and this ends up by making the same organization function *transparently* towards the different players.

INTRANET APPLICATIONS: THE CASE OF CONTENT MANAGEMENT AND OF BUSINESS PROCESS MANAGEMENT

Information systems, in their totality, are essential tools aimed at achieving coordination. Proof of this are the investments that financial institutions make in order to adopt innovative technologies alongside the most traditional corporate information systems.

A survey carried out by CeTIF[1] on a sample of 9 banking groups and 2 service centers, ended in March 2005, showed the existence of significant innovative projects in the fields of content management (CM) and business process management (BPM). By the first term we refer to a set of solutions that automate, partially or fully, the process of creation and publication of contents (*CMt*) and of documents (*document management*) for users inside the bank (Intranet, internal portals) and outside the bank (Extranet, Web etc.). By the second term we refer to the set of *workflow* technologies which manage and monitor the business process, by integrating it with traditional applications.

Both such technologies may be considered innovative business integration tools, though their role may be considered of coordination support and of automation (or, to a certain extent, of decision support), respectively.

However, their choice must be explained in the light of the organizational change that major credit institutions have started in the past few years. Many of the cases considered show that the implementation of CM or BPM solutions was followed up by information system integrations or migrations (48%), merge or acquisitions (23%), and, in the case of BPM, after process revision projects (25%). Though the implementation of both projects derived from common organizational needs, the data clearly show a different trend of the first ones compared to the second ones (Figure 2).

Sixty-fiver percent of interviewed banks declared that they are presently carrying out, or they will soon carry out, CM projects, while only 54% of the sample declared that they are presently carrying out or they are planning to carry out in the short term BPM projects. About 10% of banks have planned investments in BPM in the long term. Therefore it becomes evident that CM projects are more widespread than BPMs among financial institutions, which means that CM support technology is more widespread than BPM support technology.

When considering the volume of the projects, we realize that if, on the one hand, CM may have reached a good circulation because it is applied quite frequently to performance information and product/market information processes (pic. 3), on the other hand, BPM is spread only in the field of credits and information processes (38%), finance and treasury (25%), and auditing and risks (23%). In certain cases (i.e., purchase invoice cycle, information processes, collectionsm and payments), the BPM extends also outside the boundaries of the credit institution, involving also customers or, often, suppliers.

This Difference May be Due to Different Factors

First of all, the factors may be the causes and the relating expected results which boosted the innovation introduction in both technological applications. The survey shows that both CM and BPM projects are mostly induced by exogenous market forces and by the consequent need to reduce the so called "*time to market*" (71%). If, on the one side, the market pressure and the introduction of the customer-oriented culture get, in theory, banks to look for ways to improve internal processes and communication integration,

Figure 5. The technological innovation stages

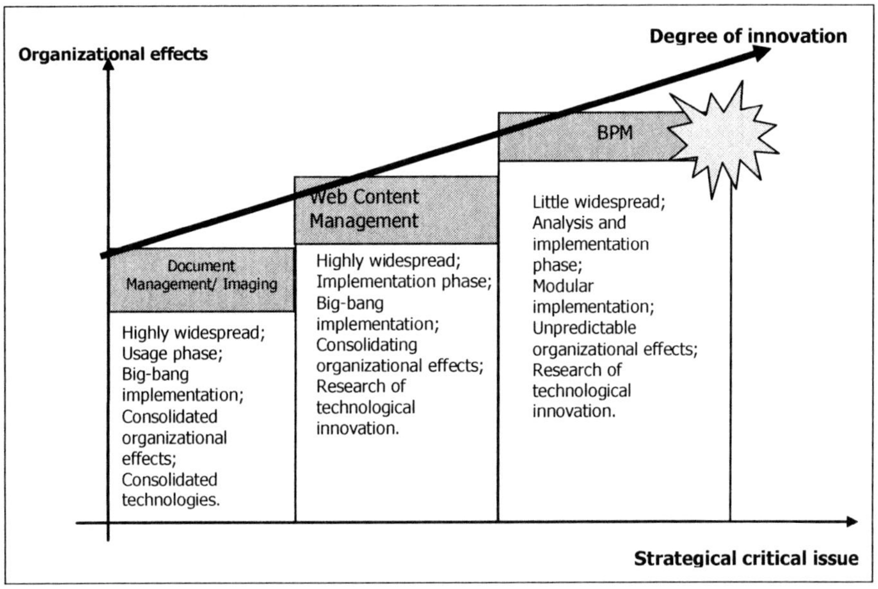

on the other side, reality shows that the equipment with tools bringing immediate and recognizable advantages, mostly outside the bank itself, was made easier at first. Therefore, CM projects, aiming at facilitating information publication and maintenance to the outside public and, only in a second phase, inside the financial institution, have anticipated the BPM projects. The latter are seen as management excellence solutions within processes (68% of interviewed institutions) and as tools implemented to overcome difficulties due to the need for technical modernization of the traditional information systems (44%). The results expected from the two solutions seem to be very different, with consequent differences in the implementation technical difficulties, in costs, and in the identification of organizational effects. If, on the one hand, CM solutions are applied by the bank alongside the existing ones, without the need of organizational changes and important technological integrations, on the other hand, BPM solutions are based on rather more complex technologies, characterized by organizational uncertainty.

Even the analysis of the obstacles faced by CM and BPM projects may be useful to understand the origin of the different spread of such projects. Aggregated data show a common sensitivity towards the cost factor (about 24% of the sample indicated costs as the main obstacle to such initiatives) and the relating difficulty in defining the return on investment (ROI). Another strong obstacle is due to the cultural and organizational reluctance (78%) to the change brought about by these technological innovations. On the other hand, difficulties relating to technologies (18%) do not seem to be a fundamental inhibitor factor. The empiric evidence show, however, a difference between the obstacles to the CM and BPM projects, with particular reference to the cultural and organizational aspects. The reasons are to be found, first of all, in the characteristics of managers and users of these technological innovations and, secondly, in their implementation modalities.

In the case of CM, 44% of respondents state that the people in charge of these projects are active in the lines of business (LOB), while 32% are active in the human resource management (HR) or in the organization. At present, only 18% of financial institutions have "created" a new professional position (i.e., *knowledge officer*) which centralizes content management in one organizational area; the new trend seems to confirm the growth of this organization solution. BPM projects, instead, are managed by the LOB in 24% of the cases, and by the IT area in 32% of the cases. When looking at the characteristics of users of the above mentioned technologies, important differences arise. CM solutions are more and more spread among customers rather than being centrally used. This entail the support of the training structure whose task is to create professional people able to manage the content publication process, as well as, in certain cases (about 50% of the sample), the relevant monitoring process. Even BPM are spread amongst users that manage the processes which make the object of this study. In particular, with regards to processes, a need of coordination and rationalization of business procedures in the most interdependent areas arose. Bank information systems are made up by parallel and integrated applications, often based on different and complex information structures. However, there is no need to create *ad hoc* professional figures which manage the BPM process and project. As a matter of fact, BPM tools often go hand in hand with or replace transactional information systems which play the main role by completing their integration and automation capacity.

Moreover, CM and BPM projects differentiate for their application modality. While CM is characterized by a *big-bang* implementation logic, which introduces technologies in a sole phase, BPM is often characterized by a modular implementation system.

AN INNOVATION EVOLUTION PERSPECTIVE

It is estimated that the difference in the present status of CM and BPM will fade more and more due to the growing awareness that internal processes are the very same business apparatus; this is due even in the light of the recent regulations on operational risks and also because the two technologies may eventually complement each other.

In particular, the survey shows that the differences between the two technologies may be due, on the one hand, to a lesser knowledge of the BPM technological tool compared to the CM one, and on the other, to the lack, limitedly to the BPM projects, of the domino (or *isomorphism)* effect which is common during the technological innovation application phases and which leads to a resemblance of technological choices by financial institutions.

We hereby would like to hypothesize a hidden link between the two technologies and the presence of an "incremental" effect between them. In order to verify this hypothesis, one must look first of all at the technological and organizational innovation which CM on the one side and BPM on the other bring about.

CM technological innovation may be summarized with the introduction of documental integration tools for archive management and in distributed environments. From a functional point of view, CM allows the automated management of the lifecycle of a document, as well as the need to trace the documents and information contained therein. For this reason, such tools are considered as tools enabling the spread of the internal knowledge of the credit institution.

From a technological point of view, BPM introduces innovation integration between the information systems. Moreover, from a functional point of view, they lead to the identification of process interdependencies and to their management harmonization. Moreover, it is a preparatory tool aimed at the process mapping, which can be used for a range of activities (e.g., risk management, auditing system, definition of internal regulations, etc.).

According to this short synthesis, an evolution perspective for the two technologies seems to take form. The common characteristics may be outlined as follows:

- **Process concept:** Both technologies imply a strong idea of organization through processes. In the case of CM, this merely refers to the document creation and publication processes, while in the case of BPM, this concept is also extended to other business processes.
- **Culture of change:** It shows that the organizational change following the introduction of such technologies is difficult to manage, though it is characterized by different factors. The bank should be able to change and, for this purpose, the management *sponsorship* is important.
- **Organizational and technological monitoring:** The technology innovation seems to be linked to the need of monitoring its implementation from both the technological and organizational point of view.
- **Technological flexibility:** Technological innovation goes hand in hand with the flexibility concept, seen as a capability to "adapt" to the company structures and systems.

We therefore believe that BPM is an evolution and extension of the CM concept (Figure 5). CM projects are therefore destined to evolve towards BPM projects extended to other processes, whenever the financial institution is able to recognize the CM as a technological innovation tool.

SOME USEFUL INSIGHTS

The considerations made on Web-based technologies, both Internet and Intranet, enable one to draw interesting conclusions already outlined at the beginning of this chapter and concerning the innovation characteristics.

Empirical evidence leads us to focus on how the theory may explain the following:

1. Which are the technological innovation signs? Can they always be measured? Does technological innovation progresses through stages or is it an on-going process?
2. Is technological innovation unique? Does it make sense to speak about innovation of each technology sector (automation, decision-making support, coordination)?
3. What is the relationship between technological innovation and organizational efficiency/effectiveness?

Innovation is a synonym of "change or improving evolution of a situation" and surely not a novelty. Some writers consider technological innovation as technology flexibility or as an improvement of performance. Here we would rather exclude such definitions, since the definitions of flexibility and performance may be ambiguous and "contaminated" by other concurrent factors (Maggi, 2003). For this reason, it is therefore believed that technological innovation is not measurable per se. However, the innovation process may be seen as an *on-going process*, more or less fast, characterized by a constant change in technology, players, and structures. This concept leads us to consider technological innovation as a natural process, only little conscious of the fact that it is stimulated by the implementation of certain technologies which may affect the pace (by increasing it or stopping it) and the process awareness.

Hence, we conclude that the answer to the second question is negative. The division in classes, given at the beginning of this chapter and recognized as valid by the organizational literature, seems not to be relevant in the case of innovation. We could at least "defend" such classification by construing it as typology, that is, as a set of types (rather than classes) which do not have precise demarcation limits and which constitute some archetypes. The concept of technological innovation, however, remains an extensive concept, not referable to any typology.

The third questions leads to wider considerations. Though the concept of organizational efficiency and technological innovation has been highlighted by many interlocutors, the gathered data, however, do not provide empirical evidence. The reason shall be found in the lack of a longitudinal analysis over a longer temporal span, for the determination of effects produced by information technology. However, a qualitative consideration can be drawn: The two technologies showed different efficiency values against the same technology use. This proves, as a first estimate, that information technology does not have a certain and determining role for reaching organizational efficiency, though it may influence both the information need, as well as the information processing capability. Therefore, its effects seem mediated by organizational and contextual variables, leading to different values with regards to effectiveness.

Therefore, the objective of this chapter is to define to what extent information technology may affect the organizational efficiency, that is, the capability to reduce the operating costs of the organizational structure. On this point, the interviewed persons stressed the high importance given to the role of information technology. IT enables the processing of great volumes of data, a deeper control, the cost cutting in terms of communication and transactions. Hence a new question arises: Does any information technology have an impact on transaction costs and on operating costs? The answer seems to be yes, though the impact magnitude on efficiency is different. While automation and communication technologies highly influence the economic management of the institution, decision-making support technologies have an indirect impact, more difficult to asses beforehand.

CONCLUSION

This study dealt with technology innovation, with particular reference to Web technologies. The different empirical evidence shows that technology may be a tool to improve the service/product offered, as well as an efficiency tool. As a matter of fact, the implementation of Internet technologies brought about, on the one hand, the development of a considerable number of Internet strategies for products and services of Italian banks. On the other hand, Internet technologies supported the coordination as well as the organizational integration phases of the same banks.

Process and content innovation mainly implies a change in the cultural attitude and, for the time being, it is still determined mostly by endogenous variables.

The analysis of the collected data leads us to draw some last conclusions:

- Information systems are affected by information technology; in order to determine the information system, one cannot do without knowing the development and implementation status of the technology.
- Information systems are affected by the task uncertainty, by coordination and control mechanisms, and by communication tools; the organizational (micro-) variables affect the definition of the information system.
- The information system is an organizational planning factor; it is able to affect the organizational efficiency and therefore the organizational structure.

- Besides information systems, the organizational structure is affected by other variables; this observation leads us to consider the absence of technologic determinism.
- Information technology has a mediated role in defining the organizational effectiveness and a direct role in the organizational efficiency management.

However, the role played by the said technologies affects the organizational structures in a different way.

REFERENCES

Agliati, M. (Ed.). (1996). *Tecnologia dell'informazione e sistema amministrativo.* Milan: Egea.

Anderson, P., & Tushman, M. L. (1991). Managing through cycles of technological change. *Research Technology Management, 34*(3), 26-31.

Barras, R. (1990). Interactive innovation in financial and business services: The vanguard of the service revolution. *Research Policy, 19,* 215-237.

Berger, A. N., & Udell, G. F. (2006). A more complete conceptual framework for SME finance. *Journal of Banking and Finance, 30*(11), 2945-2966.

Buckingham, R., Hirschheim, R. A., Land, F. F., & Tully, C.J. (1987). Information systems curriculum. A basis for course design. In Buckingham et al. (Ed.), *Information systems education: Recommendations and implementation* (pp. 72-113). Cambridge: CUP.

Carignani, A., & Frigerio, C. (2003). *European Web banking strategies.* Università Cattolica, Milan, CeTIF Research Centre.

Ciborra, C. U. (1996). Improvisation and information technology in organizations. In *Proceedings of the 17th International Conference on Information Systems*, Cleveland, OH (pp. 369-380).

Damanpour, F., & Gopalakrishnan, S. (2001). The dynamics of the adoption of product and process innovation in organizations. *Journal of Management Studies, 38*(1), 45-65.

De Nicolò, G., Bartholomew, P., Zaman, J., & Zephirin, M. (2004). Bank consolidation, internationalization, and conglomeration: Trends and implications for financial risk. *Financial Markets, Institutions, & Instruments, 13*(4), 173-217.

Epsten, M. J. (2005). The determinants and evaluation of merger success. *Business Horizons, Elsevier, 48*, 37-46.

Ferraro, F. (2000). L'analisi organizzativa: l'individuo. In R. Mercurio & F. Testa (Eds.), *Organizzazione, assetto e relazioni nel sistema di business* (pp. 52-76). Torino: Giappichelli.

Figueira, C., Neills, J., & Schoenberg, R. (2007). Travel abroad or stay at home? Investigating the patterns of bank industry M&As in the EU. *European Business Review, 19*(1), 23-39.

Frigerio, C. (2005). *Process IT Innovation in banking.* Università Cattolica, Milan, CeTIF Research Centre.

Lawrence, P. R., & Lorsch, J. W. (1967). *Organization and environment, graduate school of business administration.* Boston: Harvard University.

Maggi, B. (2003). *De l'Agir Organisationnel.* Toulose: Editions Octares.

Quinn, J. B., Anderson, P., & Finkelstein, S. (2005). Leveraging intellect. *Academy of Management Executive, 19*(4), 78-94.

Sherman, H. D., & Rupert, T. J. (2006). Do bank mergers have hidden or foregone value? Realized and unrealized operating synergies in one bank merger. *European Journal of Operational Resources, 168*(1), 253-268.

Utterback, J., & Abernathy, W. (1975). A dynamic model of product and process innovation. *Omega, 3*(6), 639-656.

ENDNOTE

1 The research mentioned was carried out by the author between March 2004 and October 2004 through a quali-quantitative analysis, in the form of semi-structured questionnaires and interviews (Frigerio, 2005).

Section III
Applications of Web Strategy

The six chapters of the third section present the results of research work and empirical evidence on the application of Web strategy's principles and methodologies in various operational contexts.

Chapter XIII
An Action Research Case Study of the Facilitators and Inhibitors of E-Commerce Adoption

Orla Kirwan
National University of Ireland, Ireland

Kieran Conboy
National University of Ireland, Ireland

ABSTRACT

This research has studied an established Irish retail business as it takes its first tentative steps into the e-commerce arena. Although the adoption of e-commerce is widely studied in the academic world, only a small percentage of these studies focus on the small to medium size enterprise (SME) retail sector. SMEs account for 97% of Irish companies and employ up to 800,000 people (Chamber of Commerce Ireland, 2006). While examining the SME's adoption of e-commerce, the factors that affected the adoption process were specifically identified and understood. This was achieved by conducting an action research case study. Action research merges research and practice, thus producing exceedingly relevant research findings. This chapter demonstrates how the research was undertaken, and also discusses the justification, benefits, and limitations of using action research. The research concluded that the adoption of e-commerce within the SME sector tends to be slow and fragmented, the presence of a 'Web champion' is paramount to the success of the project, and Internet adoption is faster with the recognition of a business need. It also supported the evidence that a SME is more likely to adopt e-commerce when the SME owner has a positive attitude towards IT.

INTRODUCTION

Zwass (2003) defines e-commerce as 'the sharing of business information, maintaining business information, and conducting business transactions by means of Internet-based technology.' Figures released from Forrester Research (2006) reveal that e-commerce is surging. In the U.S., online sales for the second quarter of 2005 reached nearly $39 billion, a 25% year-over-year increase and a slight bump up from first quarter sales of $38 billion. Total online sales in Europe will top $100 billion this year; this represents a 21% increase over 2005 (Forrester Research, 2006). eBay has already seen its European sales volume soar 85% in the past year.

Internet adoption by small to medium size enterprises (SMEs) is important to the generation of critical mass for e-commerce. While research has shown that SMEs are adopting the Internet and e-commerce (Coltman, Devinney, Latukefu, & Midgley, 2001; Levy, Powell, & Yetton, 2001; Scupola, 2002), there is little systematic research into *how* such companies are adopting e-commerce. Egan, Clancy, and O Toole (2003) recognised the need for such research: 'More work with SME owners might usefully address a method for overcoming e-commerce adoption barriers; this research would be very useful if done in an action research mode.' In answer to Egan's call, and in recognition of the important role that SMEs play in the Irish economy, the overall objectives of the study are to identify the factors that facilitate or inhibit e-commerce adoption in an Irish SME, and to understand how these factors affect the e-commerce adoption process.

Literature Review

SMEs are an important sector of any economy as they contribute to economic growth, social cohesion, and employment as well as regional and local development (Scupola, 2002). The European Union (EU) has 18 million SMEs, which are responsible for 67% of overall employment and 59 % of GDP (Ritchie & Brindley, 2005).

A Forfás study in 1999 found that 99.4% of enterprises in Ireland are SMEs, and that they account for just under half of total enterprise employment in Ireland. Therefore, significant opportunities exist for the Irish SME sector to create efficiencies in communications, shift the trading power balance in its favour, and create new markets (Egan et al., 2003). There has also been an increasing interest in the role and contribution of SMEs for the development and sustainability of a knowledge-based economy in Ireland.

E-Commerce and SMEs

Traditionally, SMEs are lacking in resources, such as time and money (Levy et al., 2001). For this reason, they are relatively late adopters of new technology, and tend to adopt a 'wait and see' attitude to e-commerce adoption (Forfás, 2002).

Forfás (2002), an Irish research agency, found that in Ireland 81% of Irish SMEs have Internet access and 46% of the companies surveyed had a Web site. E-commerce is widely argued to have the potential to transform the marketplace, and to provide SMEs with a wider variety of opportunities to engage in business activities (Levy et al., 2001).

For the last number of years the Internet and e-commerce have been offering exciting new competitive opportunities for SMEs to extend their customer base into the global marketplace, and broaden their involvement into new markets (Lewis & Cockrill, 2002).

Internet adoption by small businesses is important to the generation of critical mass for Internet commerce (Poon & Swatman, 1999). E-commerce is likely to have the most fundamental impact on SMEs by offering tremendous opportunities to high growth SMEs, but successive research studies have shown that significant business benefits are not being realised (Poon & Swatman, 1999, Levy & Powell, 2000). The literature states that

Table 1. Factors affecting the decision to adopt e-commerce

Factor	Reference	Facilitator/ Inhibitor
Perceived Benefits	Poon and Swatman (1999) found that perceived benefits were a key reason why participants adopted and continued to use the Internet.	Facilitator
Business Need	There is little evidence of business strategy driving Internet adoption among SMEs. However, Internet adoption is faster when SMEs recognise a business need (Levy & Powell, 2003).	Facilitator
Owner/Manager Influence	The literature on e-commerce adoption has emphasised the importance of a 'corporate champion' often identified as the owner/manager of an SME (Cragg & King, 1993; Fink, 1998; Poon & Swatman, 1999; Scupola, 2002). An SME that is likely to adopt e-commerce will most often have an owner who has a positive attitude towards IT (Levy & Powell, 2003; Mehrtens, Cragg, & Mills, 2001).	Either
Web Champion	The Web champion or 'corporate champion' may or may not be the SME owner. In some cases, it is an internal member of the staff.	Facilitator
Organisational Structure	For most SMEs, their failure to plan the introduction and exploitation of new technology is due to management limitations (Levy et al., 2001). Min and Galle (2001) found that organisation size is believed to influence the e-commerce adoption practices of firms.	Either
Organisational readiness	This is the extent to which an organisation feels ready to adopt e-commerce (Fink, 1998; Scupola, 2002). Among the factors that determine this are skills and knowledge of the technology, internal IT support, and support from external parties such as IT vendors (Chapman, James-Moore, Szczygiel, & Thompson, 2000; Mehrtens et al., 2001; Scupola, 2002).	Either
Organisational Skills	E-commerce imposes new skills requirements on the retail sector (Lewis & Cockrill, 2002). The successful uptake of e-commerce requires IT expertise, coupled with strong business applications skills and, therefore, a flexible, multiskilled work force is required (Lewis & Cockrill, 2002).	Either
Lack of time and resources	SMEs lack the willingness to dedicate the time and resources necessary to resolve their lack of understanding and skills (Chapman et al., 2000; Lewis & Cockrill, 2002). These resources involved are human, financial, and time. SMEs are regarded as 'poor' in human, financial, and material resources (Levy et al., 2001).	Inhibitor
Lack of Understanding	The lack of understanding of the need to adopt innovations, such as ICT and e-commerce, prevents SMEs using them to overcome existing performance gaps or exploit new opportunities (Chapman et al., 2000).	Inhibitor
Cost Factors	Strategically, information systems are used in order to lower the costs of production, coordination, and transactions or to add value to the product, process, or service (Levy et al., 2001). As the cost of IS falls, SMEs are starting to reap the benefits of these systems, as they are more accessible (Levy et al., 2001). Yet, the tendency in SMEs is still to view IS investment as a cost (Levy & Powell, 2000; Wymer & Regan, 2006).	Either
Government Initiatives	The Irish Government has enacted a number of critical Acts to underpin e-business development including the Electronic Commerce Act 2000, the Copyright and Related Rights Act 2000, and the Communications Regulation Act 2002. Ireland has made major investments in national information infrastructures, including international telecommunications connectivity, Internet data centres, and other support services	Facilitator

(continued on following page)

Table 1. continued

Security Concerns	Daniel, Wilson, and Myers (2002) report that the concern about security was inhibiting adoption, as were lack of customer satisfaction and use. In a 2001 report by the Chamber of Commerce Ireland, it was reported that 34% of businesses felt that concerns about security were an obstacle in their path to e-business (Scott, Golden, & Hughes, 2003).	Inhibitor
External Pressure/External Environment	This pressure comes from existing Internet users, particularly customers and competitors, but also suppliers and potential employees (Cragg & King, 1993; Daniel et al., 2002; Mehrtens et al., 2001).	Either

there is little evidence that SMEs do more than develop Web sites and adopt e-mail (Levy & Powell, 2000).

Factors Affecting the Decision to Adopt E-Commerce

A decision to adopt e-commerce is not taken lightly, especially when the organisation is an SME. There are a variety of factors that facilitate *or* inhibit the adoption of e-commerce, and also several factors that could be classified as *either* a facilitator or an inhibitor, depending on the situation of the SME at that time. These factors need to be addressed *prior* to the decision to adopt e-commerce being made. These factors are outlined in Table 1.

RESEARCH METHODOLOGY

This study is concerned with the factors that inhibit or facilitate Internet adoption within an Irish SME context, and also to understand how these factors affected the e-commerce adoption process. This was achieved by conducting an action research study on River Deep Mountain High (RDMH), Galway. RDMH is a retail SME based in Galway. The shop has specialised in outdoor clothing and equipment since 1991. RDMH expanded in 1995, opening a second branch in Limerick. Including both shops, there are a total of 20 employees. The researcher worked in the SME throughout the research process, and had been employed there for the previous 5 years. At the time of the research, there was no Web site and hence, no e-commerce or e-business capabilities. The SME was in the process of adopting e-commerce, making it a highly suitable SME in which to conduct the proposed research. In order to increase awareness of the shop and hopefully expand the business, the company was moving towards having an online presence, with a view to online shopping facilities in 3 years time. The research took place from August 2003 until August 2004, and it involved three distinct action research cycles, as outlined in Figure 1.

Benefits of using Action Research

The following paragraph outlines the benefits of using action research as a research methodology. The high involvement of the researcher with the research subjects allows for access to rich and in-depth research data. Researchers can overcome the problem of trying to understand the ill-structured, fuzzy world of complex organisations by applying action research, as it addresses real-life problems and the immediate concerns of practitioners (Avison, Lau, Nielsen, & Myers, 1999). Since the topic of the research is partly selected by the

Figure 1. Timeline of RDMH's research framework

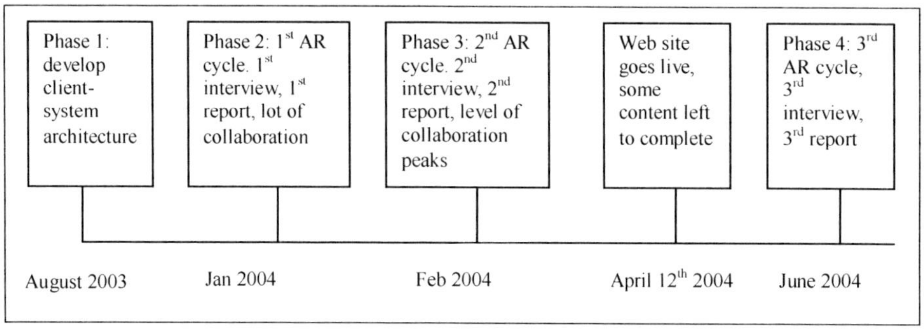

client (e.g., a company in a specific industry), its findings are likely to be of high relevance to at least a section of the practitioner community (e.g., the immediate research client and other companies in the same industry). The real world orientation of the approach offers a singular opportunity to recruit students who hold positions in organisations facing a problem whose solution can lead to relevant research findings. The problem-solving orientation of the research increases chances of obtaining research funding. Action research is usually participative. This implies a partnership between the researcher and the client. This approach may be deemed more satisfying and also more occupationally relevant (McNiff, 2000). Westbrook (1995) presents action research as an approach that could overcome three deficiencies associated with 'traditional research methods':

- It has broad relevance to practitioners.
- It is applicable to unstructured or integrative issues.
- It can contribute to theory.

Action Research and Information Systems (IS) Research

The discipline of IS seems to be a very appropriate field for the use of action research methods (Myers, 1997). IS is a highly applied field, almost vocational in nature. Action research methods are particularly clinical in nature, and place IS researchers in a 'helping-role' within the organisations that are being studied. It should not be surprising that action research is the 'touchstone of most good organisational development practice' and remains the primary methodology for the practice of organisational development.

Action research is one of several qualitative research methods used in the field of IS (Galliers, 1992). Such qualitative research is important for studying complex, multivariate, real-world phenomena that cannot be reduced for study with more positivist approaches (Baskerville & Pries-Heje, 1999). Action research is especially important in situations where participation and organisational change processes are necessary (Baskerville & Wood-Harper, 1996).

Given the frequent calls for IS research to be more relevant to practice (McTaggert, 1998; Rademacher, 2001), it is believed that action research has the potential to contribute to making IS research relevant. Action research merges research and practice thus producing exceedingly relevant research findings. Such relevance is an important measure of the significance of IS research.

Justification of using Action Research in this Study

The researcher decided that action research was the preferred method for this research based on the factors listed in Table 2.

Limitations of using Action Research

Like most qualitative research methodologies, action research is difficult to do well and easier to do atrociously. Action research is much harder to report and the researcher also has to justify the overall approach. This has to be done well enough so that even if examiners do not agree with the approach, they have to acknowledge that the researcher has provided an adequate rationale. Action research involves heavy involvement in the research situation, with the opportunity for good learning, but at the potential cost of objectivity.

Table 3 outlines how this research identified and addressed these possible limitations. This was undertaken to increase the credibility of the research findings.

Table 2. Factors influencing research methodology

Factor influencing research methodology	Reason for Influence
There is a unique opportunity available to the researcher to conduct the research in a dynamic SME environment.	The SME involved, RDMH, was at that time initiating the process of e-commerce adoption. The researcher was employed there, and had been for 5 years, and also had full access to the SME.
In this case, the researcher is solely responsible for the e-commerce implementation process.	The RDMH management had very limited participation. Action research is a research method that solves immediate practical problems while expanding scientific knowledge (Avison et al., 1999).
High suitability to the research topic and to how the research will be carried out.	Egan et al. (2003) recognise the need for such research: 'More work with SME owners might usefully address a method for overcoming e-commerce adoption barriers; this research would be very useful if done in an action research mode.'
As the research area is concerned with rich, subjective, qualitative data, not quantitative data, the research philosophy is interpretivist in nature.	Action research tends to be both participative and qualitative. The virtue of action research is in its *responsiveness to the research situation* (Greenwood & Levin, 1998). One of the key principles of action research is: let the data decide. At each step of the process, the information collected so far will determine the next step of the process.
The research approach will be inductive.	This research was concerned with generating theories not hypothesis testing.
The research situation demands responsiveness during the research project as the research occurs in a changing environment in real-time.	Action research is appropriate when the research question relates to describing an unfolding series of actions over time in a given group, community or organisation; understanding as a member of a group how and why their action can change or improve the working of some aspects of a system; and understanding the process of change or improvement in order to learn from it (Coghlan & Brannick, 2001). The proposed research in RDMH, where the strategic focus of the SME is short-term and is therefore in a constant state of flux, makes action research suitable for this research.
Participation of researcher in the research environment.	One of the reasons for the emergence of action research and its subsequent use in the IS field is the recognition, largely motivated by the early work of ethnographers, that a research environment can be more deeply understood if the researcher becomes part of that environment (McNiff, 2000).
The researcher serves two masters.	A key characteristic of action research sets it apart from other research approaches. In action research, investigators try to fulfill the needs of their study subjects and, at the same time, generate new knowledge. As such, IS action researchers have to serve two masters: their immediate research clients, who directly benefit from the research while it is being conducted, and the IS academic community in general.

FINDINGS

Factors Affecting RDMH's E-Commerce Adoption Process

From the research, the following factors were identified as having affected RDMH's adoption process:

Perceived benefits: It was reported in the Literature Review that the recognition of perceived benefits by an SME will facilitate the e-commerce adoption process. These findings are consistent with the literature review as both the SME owner and the researcher felt that there were perceived

Table 3. Limitations of using action research and how the researcher overcame these

Possible limitations of AR	How to address the limitation	How this was achieved in the study of RDMH
Achieving rigour – way of assuring the quality of the data collected	Use of action research cycles to provide adequate iterations	There were three definite action research cycles during the collection of the results. Susman and Evered (1978), two leading authors in this field, advise the use of numerous action research cycles within a given piece of research.
↓	Planning of methodical data collection methods	Interviews were used as the data collection method. The researcher was careful not to influence the owner regarding the answers, as this would increase researcher bias.
↓	Careful nurturing of collaboration with subjects	The collaboration process with the owner began in August 2003, with the establishment of the client-system infrastructure, this collaboration continued throughout the research process.
↓	Interpretations will be developed as part of the data collections	The researcher and owner developed interpretations on the findings. This occurred when meeting to come to a consensus on the results.
↓	The relevant literature will be accessed as part of the interpretation, to widen the dialectic	For the generation of each report for the SME owner, the literature was accessed
↓	Multiple data sources will be accessed to provide a dialectic	For the findings, the primary sources of data were the researcher and the SME owner. This was supplemented by documentation and literature from the organisation, and relevant academic literature.
↓	Establish an ethical client-system infrastructure and research environment	This was established in August 2003. It was agreed that the SME would be used as the proposed research environment, and it was also agreed that the researcher would have unlimited access to the information and documentation contained within the SME that was related to the research project.
Validity – Are the findings really about what they appear to be about?	Conscious and deliberate enactment of the action research cycles	Each stage of each action research cycle was followed with deliberate purpose, and each stage was adhered to correctly according to the academic research that was carried out on the action research cycles.
Researcher bias/Lack of impartiality	Researcher needs to consider the extent to which the story is a valid presentation of what has taken place	The researcher was very careful not to prompt the SME owner during the interviews, with respect to the researcher's answers, as the researcher had already independently answered the questions **prior** to interviewing the owner.
Generalisability – the extent to which the findings are equally applicable in other settings	It will not be a problem if the researcher does not claim that the results, conclusions, or theory can be generalised.	As SMEs tend to be industry specific, there is a lot of diversity across the sectors. This study was concerned with the retail sector.

benefits available from adopting e-commerce.

Web champion: The literature on e-commerce adoption has emphasised the importance of a 'corporate champion' often identified as the owner/manager of an SME (Cragg & King, 1993; Fink, 1998; Poon & Swatman, 1999; Scupola, 2002). The 'Web champion' in the case of RDMH was the researcher, not the SME owner, who initiated and completed the e-commerce site. These findings are consistent with the literature, as the literature states that this role may or may not be assumed by the owner, and also the process of e-commerce adoption will be facilitated by the presence of one.

Organisational skills: The level of knowledge among IT and non-IT professionals is important in the e-commerce adoption process (Fink, 1998; Mehrtens et al., 2001; Scupola, 2002). Organisations with more IT experience or greater IT already in use are more likely to adopt IT (Fink, 1998). The findings from RDMH are consistent with this, as the researcher was employed by the organisation and completing a master's degree in e-commerce at the same time. The SME owner had also computerised all of the stock in both branches of RDMH, and so is knowledgeable about IT and has a positive attitude to IS/IT.

Lack of time and resources: SMEs lack the willingness to dedicate the time and resources necessary to resolve their lack of understanding and skills (Chapman et al, 2000; Lewis & Cockrill, 2002). The researcher experienced a significant lack of time during the project. This was due to the fact that the researcher continued to work on the shop floor, as well as developing the Web site. This is consistent with the literature as RDMH is a typical SME.

Lack of understanding: The lack of understanding of the need to adopt innovations, such as ICT and e-commerce, prevents SMEs using them to overcome existing performance gaps or exploit new opportunities (Chapman et al., 2000). Again, this is similar to above and consistent with the literature. The researcher also lacked some understanding of the new technologies being used in the adoption process.

External pressures: There is little evidence of business strategy driving Internet adoption among SMEs. However, Internet adoption is faster when SMEs recognise a business need (Levy & Powell, 2003). The SME owner did not feel that it was a business need, therefore there was nothing done to adopt or develop a Web site for the shop until the researcher approached with the research proposal. This is consistent with the literature findings.

How the Factors Actually Affected RDMH's Adoption Process?

As is evident, not all of the factors listed in Table 1 were deemed relevant in the context of RDMH. Why were only 6 out of the 13 possible factors listed as relevant? It was found that some were not applicable due to the nature of the Web site (i.e., front-end, information only). Others were not applicable due to the researcher embracing the role of Web champion, and not the SME owner. Finally, others were not relevant due to the present state of the SME involved (e.g., financial situation, level of IT already adopted and future strategic expectations of having an online presence). See Table 4 for an explanation of how the factors identified were actually relevant.

CONCLUSION

This research has investigated the process that RDMH went through in order to add a Web site to the organisation. After analysing an SME that has moved from off-line to online, a number of similarities and differences from the literature have been identified. The following conclusions have been drawn from the findings of this research.

Conclusions Regarding the E-Commerce Adoption Process

This research supports Poon and Swatman's (1999) and Levy and Powell's (2000) evidence that the adoption of IS/IT tends to be slow, fragmented, and occurs in stages. In addition, this research confirms Levy et al's concept that most SMEs will pursue a business strategy in which a company combines online e-commerce with a traditional retail outlet.

This research also supports Mehrtens et al.'s (2001) and Levy and Powell's (2003) research evidence that an SME that is likely to adopt e-commerce will most often have an owner who has a positive attitude towards IT adoption, who is innovative, and who is knowledgeable about IT. The literature also emphasised the importance of a 'corporate champion' often identified as the owner/manager of an SME (Cragg & King, 1993; Fink, 1998; Poon & Swatman, 1999; Scupola, 2002).

Table 4. How the factors identified actually affected the adoption process

Factor affecting the decision to adopt	Facilitated/ Inhibited	How was it relevant?
Perceived benefits	Facilitated	These helped the SME owner to see some future benefit from the project. It also helped to increase the owner's enthusiasm for the Web site.
Web champion	Facilitated	All of the incentive for this Web site had come from the researcher. This had started in August 2003 when the researcher approached the SME owner and proposed the Web site development.
Organisational skills	Facilitated	The researcher was employed by the organisation, so the skills that the researcher had as a result of the first year of the MBS programme benefited the SME.
Lack of time and resources	Inhibited	The researcher worked in the shop and developed the Web site, and as a result, experienced a serious time shortage during the day trying to balance both activities.
Lack of understanding	Inhibited	The researcher spent extra time on the development of the site, as it was the first time using the software involved, Dreamweaver MX.
External pressures	Inhibited	The researcher saw that customers were ready for the Web site, having been asked at different times over the previous 3 years if the company had one or not. The SME owner did not agree in this instance, as a lack of a business was stated as the reason why the organisation had not previously gone online.
Role of the SME owner ➢ This factor did affect the adoption process, although there was conflict between the researcher and the SME owner when agreeing on the role that it played.	Facilitated	The research has shown that the owner had a very positive attitude to IS/IT adoption, as when the researcher approached the owner with the Web site proposal, the owner agreed to the suggestion immediately. The owner's positive attitude facilitated the adoption of e-commerce, even though it was stated that the owner did not see the Web site as a business need.
Cost factors ➢ Again, this factor did affect the adoption process, although there was conflict between the researcher and the SME owner when agreeing on the role that it played	Inhibited	Initially, the owner did not feel that cost was an issue as the researcher was completing it in-house. In reality, cost was the deciding factor of the type of Web site that would be adopted and developed. The owner perceived the development of the Web site as free until the first report was studied. It was then decided to adopt a front-end, information-only site with no online sale capabilities, as the researcher could develop it in-house at no extra cost. This is consistent with the literature findings as the tendency in SMEs is still to view IS investment as a cost (Levy & Powell, 2000; Wymer & Regan, 2006).

Most SMEs will find that Web technologies will not be one of their core competences (Riemenschneider & McKinney, 1999). This is not consistent with the findings, as the researcher was employed by the SME while concurrently pursuing a master's degree in e-commerce.

Conclusions Regarding the Use of Action Research as a Methodology

From the research methodology perspective, the researcher encountered a number of difficulties that are well documented limitations of using action research. This included a significantly longer case study chapter, due to the need to present and analyse the three action research cycles in detail. The findings chapter was also quite complex as each issue needed to consider the researchers view, the SME owners view, and the final amalgamation of the two. This was overcome by eventually splitting the findings chapter into two separate components.

The degree of documentation was also high as the researcher maintained a journal throughout the adoption process, and each meeting was recorded in detail. As a result, the researcher had a significant amount of data to analyse. The researcher also had to ensure that when reflecting on the data, perceptions and interpretations were nonbiased and accurate.

The benefits from having undertaken the research using this methodology included an increased depth of knowledge about the research process. Also, the extra depth, documentation, and analysis were rewarded through what the researcher believes were more credible research findings. The researcher learned some valuable lessons regarding the research process. Action research incorporated a significant amount of extra work for the researcher, a point that was raised by the supervisor at the start of the study. However, the researcher feels that the extra work was rewarded as it resulted in richer and more credible findings.

REFERENCES

Avison, D., Lau, F., Nielsen, P. A., & Myers, M., (1999). Action research. *Communications of ACM, 42*(1), 94-97.

Baskerville, R., & Pries-Heje, J. (1999). Grounded action research: A method for understanding IT in practice. *Accounting, Management and Information Technologies, 9*, 1-23.

Baskerville, R., & Wood-Harper, A. T. (1996). A critical perspective on action research as a method for information systems research. *Journal of Information Technology, 11*(4), 235-246.

Baskerville, R., & Wood-Harper, A. T. (1998). Diversity in information systems action research methods. *European Journal of Information Systems, 7*(2), 90-107.

Chambers of Commerce. (2006). Retrieved December 13, 2006, from www.chambers.ie

Chapman, P., James-Moore, J., Szczygiel, M., & Thompson, D., (2000). Building Internet capabilities in SME's. *Logistics Information Management, 13*(6), 353-361.

Coghlan, D., & Brannick, T. (2001). *Doing action research in your own organization*. Sage Publications.

Coltman, T., Devinney, T., Latukefu, A., & Midgley, D. (2001). E-business: Revolution, evolution or hype? *California Management Review, 44*(1), 57-85.

Cragg, P., & King, M. (1993). Small-firm computing: Motivators and inhibitors. *MIS Quarterly, 17*(1), 47-61.

Daniel, E., & Grimshaw, D. (2002). An exploratory comparison of electronic commerce adoption in large and small enterprises. *Journal of Information Technology, 17*, 133-147.

Daniel, E., Wilson, H., & Myers, A. (2002). Adoption of e-commerce by SME's in the UK.

International Small Business Journal, 20(3), 253-270.

Egan, T., Clancy, S., & O' Toole, T. (2003). The integration of e-commerce tools into the business processes of SMEs. *Irish Journal of Management, 24*(1), 139-154.

Fink, D. (1998). Guidelines for the successful adoption of information technology in small and medium enterprises. *International Journal of Information Management, 18*(4), 243-253.

Forfás. (2002). *Ebusiness, where we are and where do we go from here?*

Forrester Research. (2006, June 29). *Europe's eCommerce forecast: 2006-2010.* Author.

Galliers, R. (1992). *Information systems research issues, methods and practical guidelines*. Blackwell Scientific Publications.

Greenwood, D., & Levin, M. (1998). *Introduction To action research.* Sage Publications.

Levy, M., & Powell, P. (2000). Information systems strategy for small and medium sized enterprises: An organizational perspective. *Journal of Strategic Information Systems, 9*, 63-84.

Levy, M., & Powell, P. (2003). Exploring SME Internet adoption: Towards a transporter model. *Electronic Markets, 13*(2), 171-181.

Levy, M., Powell, P., & Yetton, P. (2001). SME's: Aligning IS and the strategic context. *Journal of Information Technology, 16*, 133-144.

Lewis, R., & Cockrill, A. (2002). Going global-remaining local: The impact of e-commerce on small retail firms in Wales. *International Journal of Information Management, 22*(3), 195-209.

McNiff, J. (2000). *Action research in organizations*. London: Routledge.

McTaggart, R. (1998). Is validity really an issue for participatory action research? *Studies in Cultures, Organizations & Societies, 4*(2), 211-237.

Mehretens, J., Cragg, P., & Mills, A. (2001). A model of Internet adoption by SME's. *Information & Management, 39*, 165-176.

Min, H., & Galle, W. (2001, April). Electronic commerce-based purchasing: A survey on the perpetual differences between large and small organizations. *International Journal of Logistics, 4*(1), 79-96.

Myers, M. (1997). Qualitative research in information systems. *MIS Quarterly, 21*(2), 241-243.

Poon, S., & Swatman, P. (1999, January). An exploratory study of small business Internet issues. *Information & Management, 35*(1), 9-18.

Rademacher, R. (2001). The changing profile of information systems research: 1995-2000. *Journal of Computer Information Systems, 42*(1), 13-17.

Riemenschneider, C., & McKinney, V. (1999). Assessing the adoption of Web-based e-commerce for businesses. *Electronic Markets, 9*(1/2), 9-13.

Ritchie, B., & Brindley, C. (2005). ICT adoption by SMEs: Implications for relationships and management. *Work & Employment, 20*(3), 205-217.

Scott, M., Golden, W., & Hughes, M. (2003). E-tailing in Ireland: A review of Ireland's top 25 retailers. *Irish Marketing Review, 16*(1), 15-24.

Scupola, A. (2002). Adoption issues of B2B Internet commerce in European SMEs. In *Proceedings of the 35th Hawaii International Conference on Systems Science, 2002.*

Susman, G., & Evered, R. (1978). An assessment of the scientific merits of action research. *Administrative Science Quarterly, 23*(4), 582-603.

Westbrook, R. (1995). Action research, a new paradigm for research on production and operations management. *International Journal of Operations and Production Management, 15*(12), 6-20.

Wymer, S., & Regan, E. (2006). Factors influencing e-commerce adoption and use by small and medium businesses. *Electronic Markets, 15*(4), 438-453.

Zwass, V. (2003). Electronic commerce and organizational innovation: Aspects and opportunities. *International Journal of Electronic Commerce, 7*(3), 7-37.

Chapter XIV
Acceptance of the Mobile Internet as a Distribution Channel for Paid Content in Germany

Svenja Hagenhoff
University of Goettingen, Germany

Christian Kaspar
University of Goettingen, Germany

Lutz Seidenfaden
University of Goettingen, Germany

Björn Ortelbach
University of Goettingen, Germany

ABSTRACT

This chapter is about a survey based on 7,178 valid responses which analyses the mobile content usage in Germany. Key findings are that paid mobile contents will not be a mass market in the medium term. Nevertheless, we found that respondents that are familiar with mobile radio and handset technology and read specialized printed media on a regular basis showed the highest acceptance of mobile paid contents. The mobile Internet is perceived as a chance for the media industry to generate additional revenues from paid contents. Successful business models for the mobile Internet will only be possible if mobile content formats generate added consumer value. In this context, media companies planning to establish mobile services for content distribution are facing the problem that acceptance of mobile services has not yet been researched thoroughly.

INTRODUCTION

The publishing industry is characterized by oligopolistic structures, differentiated products, and fierce competition. The growing range of contents available in digital distribution channels like the Internet has led to decreasing circulations, revenues, and earnings in the publishing industry. Despite concentrated efforts to generate direct revenues from online recipients, the traditional freebie-mentality of Internet users seems to be overwhelming compared to the small online revenues. This development may lead to a strategic threat of the publishers' core business if not to an enduring threat to their existence.

To find a way out of this situation, publishers can pursue two strategies in order to generate additional revenues (Seidenfaden, Kahnwald, Kaspar, & Ortelbach, 2005): they can enforce the diversification into new business areas such as providing Internet service or embrace measures to increase customers' willingness to pay for online content (Gregg, 2001). The willingness to pay does not only depend on the quality of the content but also on its usefulness and the quality of its format (Fink, 2002). Examples for such value added services are mobile Internet services which provide location-independent access to contents (Zobel, 2001). In order to generate revenues from the utilization of content in the mobile Internet it is necessary that the content provided addresses existing customer requirements.

In practice, it is difficult for media businesses to identify relevant areas for mobile services, because their acceptance is only poorly understood. Therefore, the decision problem of a publishing house in the mobile services area can be divided into three parts:

- First, the customers who are likely to use the service have to be identified.
- Second, the content which is supposed to satisfy the requirements of the potential customers needs to be determined.
- Third, the service formats which promise to be successful given the preferences of the potential customers need to be identified.

The chapter on hand answers the three questions based on a representative survey of German Internet users from the viewpoint of acceptance research. The survey was done in May 2005. The following gives an overview of the related work and the status of acceptance research related to the mobile Internet. Section 3 describes the underlying methodology of the survey in detail. In section 4 the results of the survey are described. Section 5 analyses the results of the survey and points out the major findings. In section 6 findings from another current survey are presented shortly in order to complement the authors' findings for drawing conclusions. The second to last addresses the future trends concerning mobile services. In the last section, the conclusion is given.

BACKGROUND

In this survey, mobile Internet means having access to the Internet via mobile devices like cell phones or personal digital assistants. Such devices are cordless, small enough to carry around, and it is not necessary to boot them. Therefore an ubiquitous access to the internet is possible.

Acceptance represents a form of consenting behavior or consenting attitude towards an innovation (Anstadt, 1994). Acceptance research in business administration can be understood as an approach that tries to identify reasons on the customer side that lead to either acceptance or rejection of an innovation (Govindarajulu, Reithel, & Sethi, 2000; Venkateshb & Davis, 2000). Popular approaches in acceptance research concerning innovations are given by Davis (1989), Nielsen (1993), and Rogers (1995). Acceptance of an innovation depends on three factors: the (cognitive) knowledge of potential users, interest and a positive (affective) attitude, and a specific (conative) decision to buy (Müller-Böling & Müller, 1986).

Two basic groups of studies which contribute to research of acceptance of the mobile Internet can be identified: expert groups (e.g., Davison, Walsh, & Brown, 2000; Müller-Verse, 1999; TIMElabs, 2000) and customer surveys (e.g., Kearney, 2003; Nokia, 1999; Sirkinh & Dean, 2000; Wohlfahrt, 2004). In the studies mentioned above no shared argumentation can be found. Therefore, they do not enable the reliable deduction of promising mobile services. Instead, it appears that the acceptance of mobile services needs to be examined empirically and experimentally for each individual case. Furthermore, the present status of acceptance research in the mobile Internet concerning mobile service provision by companies from different market segments, for example, publishers or financial service providers, is rather unfocussed or has only little significance due to small samples.

Therefore, a lack of research could be identified in this field. The chapter on hand is a first step to address this issue from the viewpoint of the mass media industry.

CONCEPT OF THE RESEARCH APPROACH

The main goal of the following acceptance analysis is to determine a decision map which can be used by media companies to identify promising distribution forms for their content in the mobile Internet. In order to achieve this, three central questions are examined:

- With regards to the diversity of distribution formats in the mobile Internet (i.e., MMS, SMS, WAP, etc.), it seems necessary to identify those with the highest user acceptance.
- The question of which content is used by which type of customer and to which extent in a mobile setting is at the core of the examination.
- Furthermore, the necessity to examine affective acceptance criteria suggests finding out which motives lead to the usage of mobile services and which do not.

These three questions address problem areas which require validation through field research. Since the underlying central questions of the analysis characterize quantifiable product/market combinations, the following survey was conducted as a standardized written survey. The primary goal is the detection and description of the verifiable usage of mobile services by the costumers of magazine publishers. With this, the survey mainly addresses cognitive and also affective aspects of acceptance.

The survey was done together with the Association of German Magazine Publishers (Verband Deutscher Zeitschriftenverleger, VDZ). We created a standardized online questionnaire which was advertised on the homepages of 16 popular German consumer magazines in order to achieve a high degree of representation. The survey was divided into four sections. First, participants were asked about their usage of 13 common types of printed media (i.e., newspapers, news magazines, (tv)program magazines, women's magazines, family magazines, music and youth magazines, living magazines, adult magazines, lifestyle magazines, motor magazines, sports magazines, science/travel/culture magazines, and IT magazines). The answers were used to create a typology of users of mobile content. The central question of the second section was the degree to which the acceptance of mobile radio technology influences the usage of mobile Internet applications. Questions included the mobile devices owned by the interviewee and the preferred mode of Internet access. The third section, which was the main focus of the survey, was centered around the use of contents on the mobile Internet. We examined potential combinations of content and different formats, including SMS, MMS, and WAP. Besides, mobile music and video services as well as specialized contents such as guidebooks, classifieds, and location-based services (LBS) were surveyed. The questionnaire was concluded by questions related to the motivation leading to the use or nonuse of the formats SMS, MMS, WAP, and LBS.

The reason for the suggestion of actions for publishers who want to provide mobile value-added services to their customers is an assumed logical connection between media specific customer segmentation (e.g., using format and content type as segmentation criteria) and the usage of mobile services. The verification of this hypothesis is not a matter of this survey but a matter of our further research.

DESCRIPTION OF THE RESEARCH RESULTS

Out of the 13,402 viewers of the questionnaire, 7,178 completed the survey. Approximately 75% of the participants were male. On the whole, they were evenly distributed among the age and income brackets (for age and income of the sample see following tables).

In the following parts the survey results will be presented concerning to usage frequency of mobile service formats per reader segment, usage frequencies of mobile content services per reader segment, and motivation for the usage of the mobile Internet.

Usage Frequencies of Mobile Service Formats per Reader Segment

In order to segment the respondents, we tested the regular use of 13 different types of printed media. Daily newspapers were used most frequently followed by news, program, and IT magazines. These results correspond to present statistics of the German mass media market, which state that the circulation and the revenues of the newspaper segment significantly exceed those of the magazine market (Figure 1).

Concerning to the diffusion of devices the people can use to receive content online, the following results were found: stationary PCs are most widely distributed (87% of the participants), followed by mobile phones (79%) and notebook PCs (54%). Mobile phones are most widely distributed among the women's, youth, music, and lifestyle segments. Only 4% of the interviewees use "Smartphones," which combine the features of mobile phones and PDAs. Users of these devices are considered to have a high acceptance of mobile services. Smartphones have an above average diffusion among readers of motor and adult press.

Concerning paid mobile Internet services, it turned out that WAP-pages supplied by the mobile service provider's portals are the most widely used services. Contents supplied by SMS, MMS, and LBS were popular as well. Mobile music and

Table 1. Age and income of the sample

Age of the sample	Percentage	Absolute
14-19 years	3.90%	279
20-29 years	23.55%	1,687
30-39 years	30.06%	2,153
40-49 years	22.66%	1,623
50-59 years	13.86%	993
60+ years	5.98%	428
Sum	**100.00%**	**7,163**

Income of the sample	Percentage	Absolute
> EUR 5,000	12.80%	752
EUR 3,601-5,000	18.21%	1,070
EUR 2,601-3,600	21.46%	1,261
EUR 1,601-2,600	22.24%	1,307
EUR 900-1,600	14.58%	857
< EUR 900	10.70%	629
Sum	**100.00%**	**5,876**

Figure 1. Distribution of print titles relative to the survey population (n = 7,178)

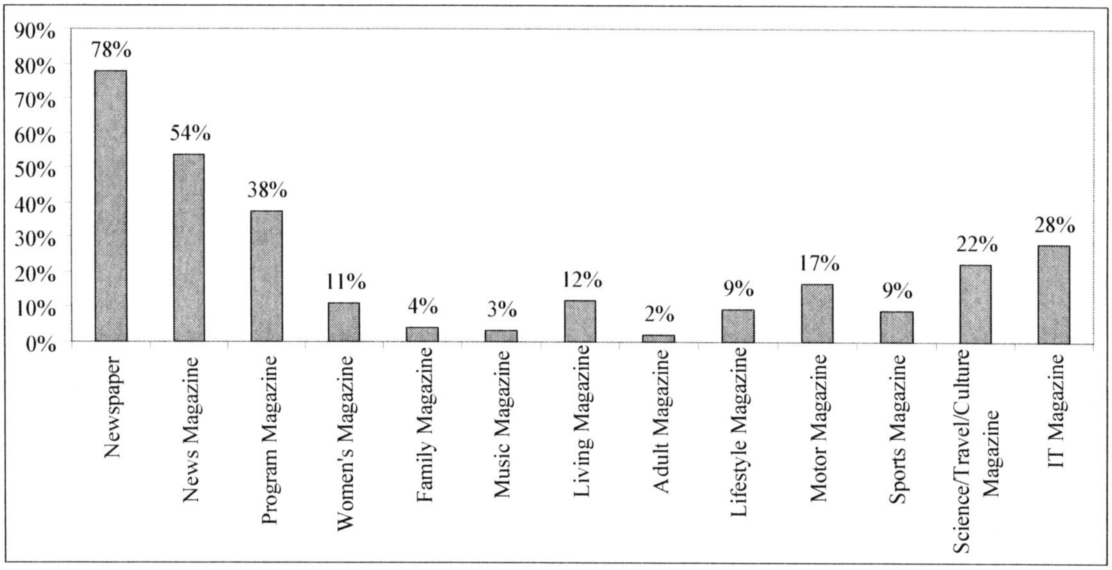

Figure 2. Distribution of mobile services relative to survey population (n = 7,178)

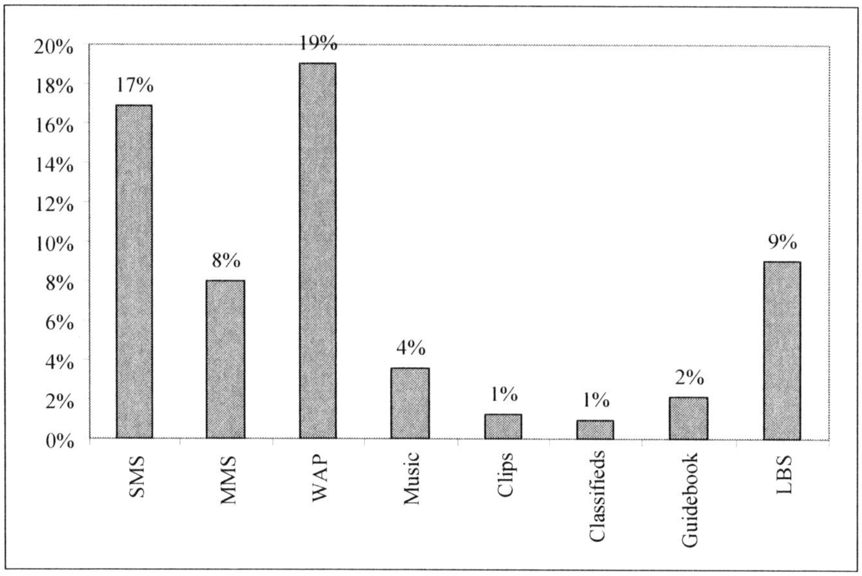

video services, guidebooks, and classifieds are of little significance (Figure 2).

The mass market segments such as daily newspapers, news, and program magazines show above average usage rates of all service categories compared to the population. The evaluation of service usage related to the size of the segment gives a different impression. Figure 3 shows the usage rates of the four most widely spread mobile service formats in relation to the size of each reader segment, and each mean frequency of response (the value in brackets indicates the number of respondents of each reader segment).

Particularly, readers of niche media with lower circulation, such as music, adult, lifestyle, motor, and travel magazines, display an above average willingness to use paid mobile content. For example, 28% of the readers of sports press use paid SMS services at least occasionally and 35% of the readers of adult press use paid WAP services. Music services display above average popularity among readers of youth and music press (11.9%) and adult press (13.8%).

Usage Frequencies of Mobile Content Services per Reader Segment

In the following, we will examine the usage frequencies of different services offered via SMS, MMS, WAP, and LBS in detail in order to achieve a deeper understanding of their acceptance among different user groups. Video services, classifieds, and guidebooks will be neglected due to their sparse usage, as well as music services which were not examined in detail.

Figure 3. Mobile service usage by reader segments (in %)

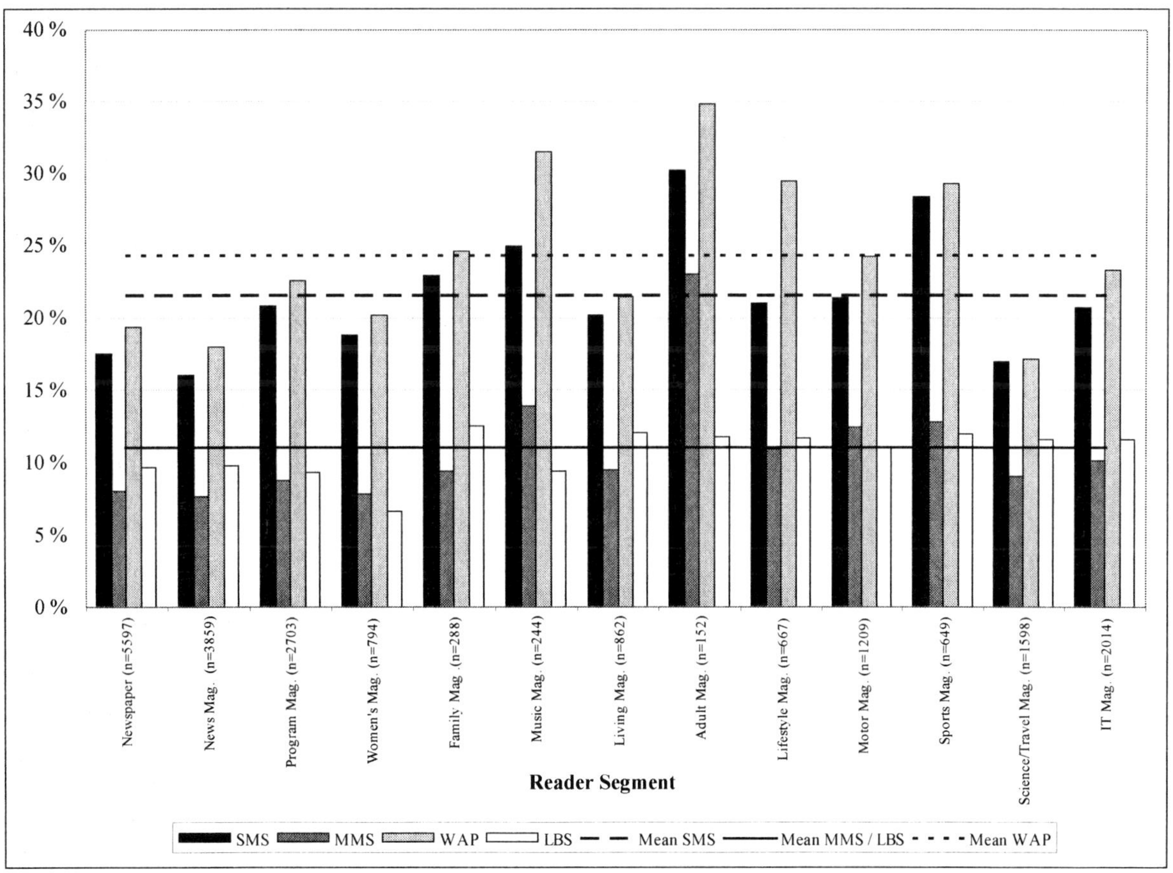

SMS (Figure 4)

On average, SMS services are used by 17% of the interviewees. Readers of adult magazines display the most noticeable usage (30%). Readers of sports magazines (28.3%) as well as music titles (25%) are also frequent users. Seven specific versions of paid SMS services were examined, including both entertainment and information. On average, the most popular services are sports news (8.6%) and general news (6.4%). Financial news and horoscopes are of minor importance (2.5% and 2.4%). Aside from the significant use of SMS sports news by readers of sports press (21%), readers of music and youth magazines display a noticeable willingness to use entertainment content (12%). Readers of adult and lifestyle magazines are frequent users of sports news (14% and 8.5%).

Figure 4. Usage frequencies of mobile content services (SMS) per reader segment (in %)

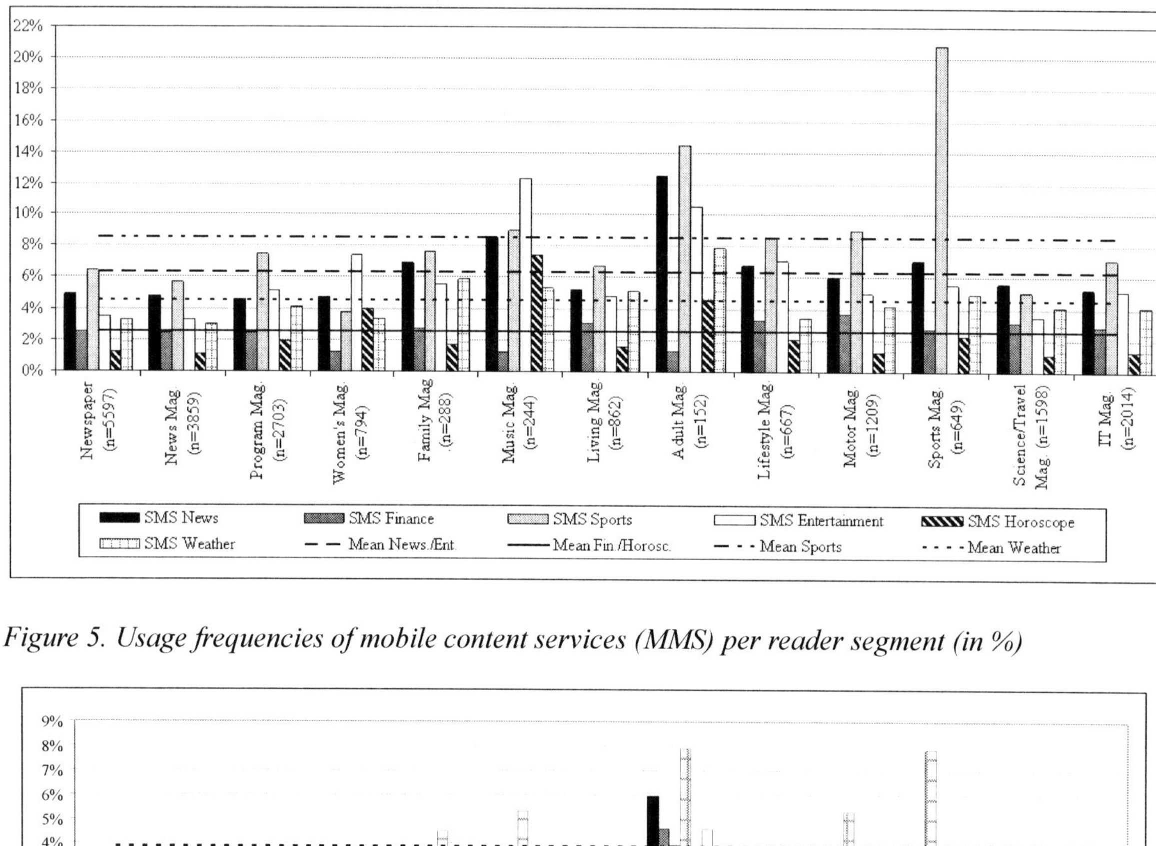

Figure 5. Usage frequencies of mobile content services (MMS) per reader segment (in %)

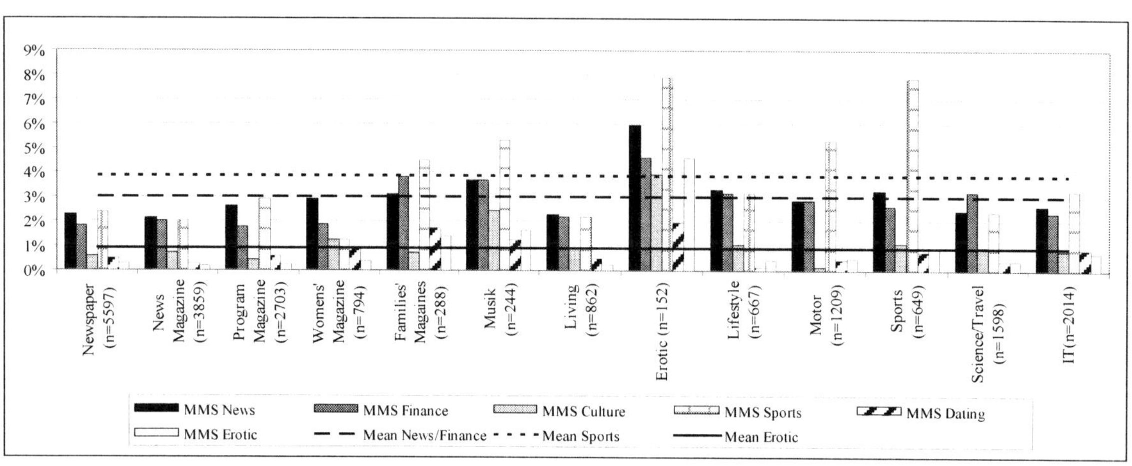

MMS (Figure 5)

The usage of paid MMS services is similar to the case of SMS, albeit on a lower average level (8%) of the respondents. Again, sports news and general news have the highest average usage rates. Readers of adult magazines (23%), music and youth magazines (14%), and motor/sports magazines display an above average acceptance of MMS services in general. MMS-based sports news is consumed by readers of sports and adult magazines (8% and 7.8%). Readers of adult magazines (6%), music and youth magazines (3.6%), and sports press (3.1%) frequently consume general news.

WAP (Figure 6)

As stated above, WAP-based portal services see the strongest demand among the target group. Twelve typical WAP offerings were surveyed. Aside from media-specific information and entertainment offerings such as general news (9.1%), sports news (8.9%), leisure (3.1%), and travel information (3.2%), device-specific products such as ring tones (9.4%), e-mail (7.1%) and games (6.7%) were purchased most frequently by the population. Entertainment contents such as ring tones and games are used above average by the readers of youth and music magazines (8.2%). Again, information content such as general news and sports are purchased most frequently by readers of adult (15.8% and 14.5%), lifestyle (12.6% and 10.3%), and sports press (8.5% and 20.2%).

LBS (Figure 7)

Respondents were able to choose among four kinds of LBS which have already been realized by the mobile service providers: location-based directories and yellow pages, so called "community services," navigation services, and location-based mobile payment. The usage level was generally low, with directories and navigation being the most popular services (3.6% and 3.4%). Community and payment services are of minor importance (1.6% and 0.4%). There is a noticeably high usage frequency of these services among readers of family magazines (e.g., 6.3% using directory services) and living magazines (e.g., 5% using navigation). Further segments displaying above average demand for LBS are music and youth and adult publications.

Figure 6. Usage frequencies of mobile content Services (WAP) per reader segment (in %)

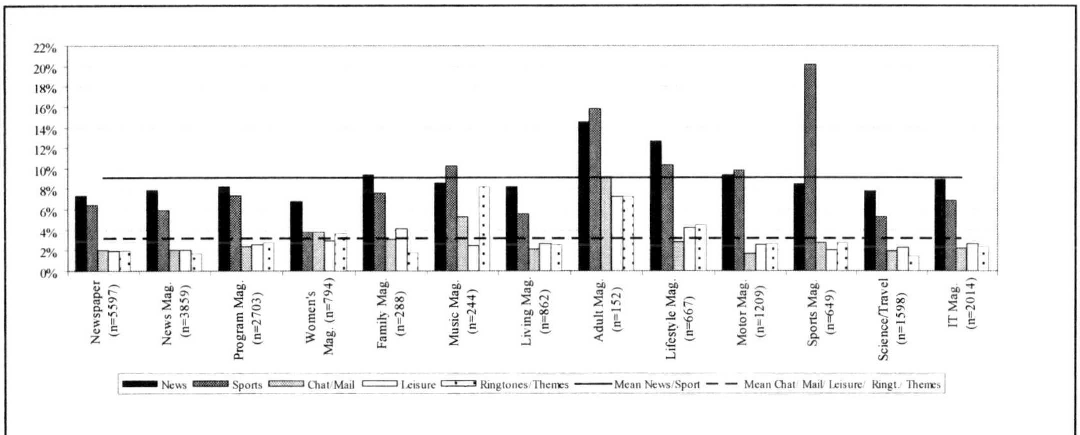

Figure 7. Usage frequencies of mobile content services (LBS) per reader segment (in %)

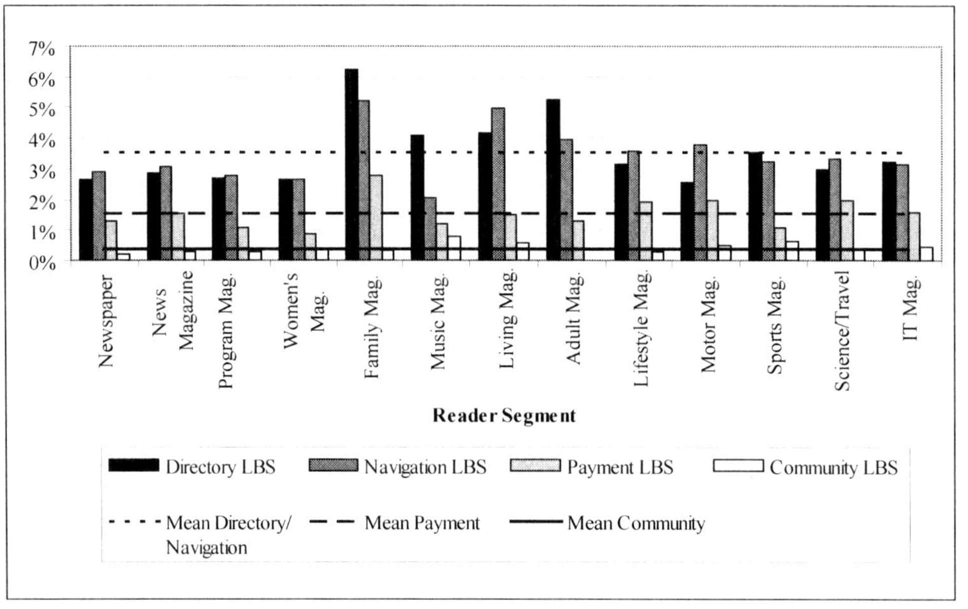

Motivation for the Usage of the Mobile Internet

If the specific variants of mobile services were indicated to be not used, respondents were asked on which conditions they would consider using the services in the future. The survey contained items testing the expected utility (price and added value), the perceived quality of the content, and aspects of technical usability (e.g., speed, display quality, and device usability). There were no significant differences between the reasons stated by different segments and concerning different services. As displayed in Table 2, price and added value were dominant.

For each service that a respondent indicated to use at least casually criterions were available that allow conclusions about the respondent's motive of using the certain service. In detail, eight criterions have been made available. Thereof, four criterions indicate the respondent's purpose-neutral interest (e.g., practical value of the service, curiosity about the service, fascination from mobile technology in general, and habitual usage) and four criterions indicate the respondent's earmarked interest (e.g., mobility, information, entertainment, and interactivity).

Figure 8 shows that the attraction of the mobile Internet is mainly determined by the respondents' curiosity and general fascination of technology.

Table 2. Conditions for future usage of the mobile Internet

	Mean	Deviation
Lower prize	41.60%	13.40
Added value	18.50%	6.40
Content quality	18.10%	7.00
Higher bitrate	10.80%	6.50
Higher display quality	8.20%	4.80
Better device usability	7.00%	3.40

Figure 8. Purpose-neutral interests (above) and content-oriented gratifications (below) for the usage of the mobile Internet (in %)

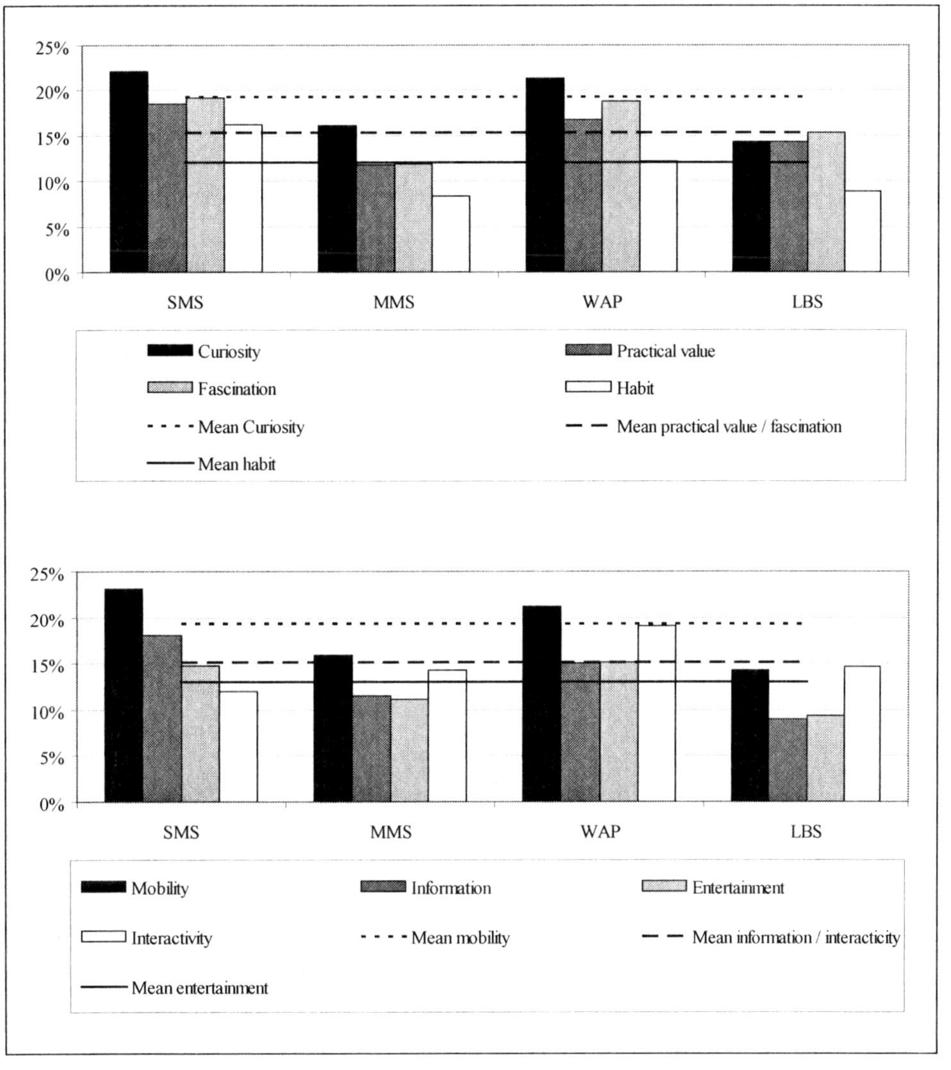

Thereby, services based on WAP and SMS show highest practical value leading to higher habitual usage of these service formats. Furthermore, the figure shows that mobile services serve needs that are connected to the reason of mobility. While SMS-based services, as well as surprisingly music services, serve information needs best, MMS, WAP, and LBS are meant to interact with other users.

FINDINGS FROM THE SURVEY

Against the background of the research questions listed in section 2 we will analyze the survey's results in order to identify significant coherences of indicated behavior and to conclude from this behavior upon the respondents' acceptance of contents in the mobile Internet. According to the first question we will examine the survey's results for systematic dependencies between the consumption of printed contents and the usage of mobile content formats. Afterwards, we will study the motives indicated for the usage of mobile contents to identify differences of consumers' needs.

Analysis of the Dependencies between Consumption of Printed Contents and Usage of Mobile Content Formats

Table 3 leads to the assumption that respondents reading different print titles will show different usage patterns of mobile Internet content beyond the differences explained by their choice of mobile devices. In particular, readers of music, adult, lifestyle, motor, and sport magazines show an above average willingness to use mobile services. Table 3 shows confidence coefficients resulting from (Chi-Square) tests of statistical independence between the reading habits and the willingness to use mobile contents. The test results prove that there is a significant dependency. However, the interrelation is very low for all reader groups, as indicated by the coefficients which are very small.

In Figure 9, we compare the values of the contingency coefficients of the five reader groups with the highest significance and the three most popular mobile service formats (i.e., SMS, MMS,

Table 3. Confidence coefficients for magazine readership and mobile services use

	Newspaper	News Magazine	Program Magazine	Womens' Magazine	Families' Magaines	Musik	Living	Erotic	Lifestyle	Motor	Sports	Science/ Travel	IT
SMS	0.028	0.037	0.080	0.051	0.030	0.053	0.036	0.068	0.046	0.056	0.100	0.027	0.068
MMS	0.018	0.040	0.030	0.028	0.034	0.044	0.027	0.082	0.045	0.072	0.072	0.033	0.063
WAP	0.031	0.033	0.073	0.015	0.030	0.070	0.029	0.065	0.086	0.061	0.091	0.028	0.071
LBS	0.040	0.041	0,026	0.033	0.024	0.020	0.055	0.018	0.052	0.035	0.043	0.053	0.051

significant on 95%-level

significant on 99%-level

Figure 9. Coefficients for magazine readership and mobile services use

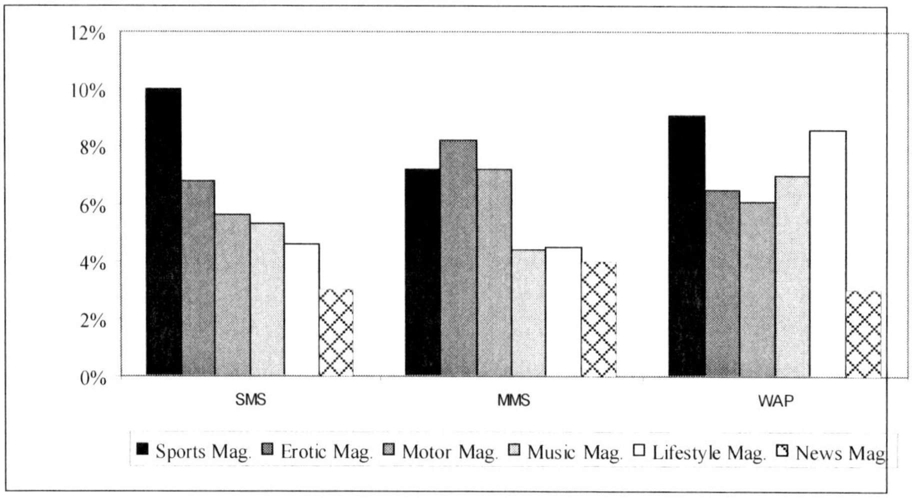

Table 4. Confidence coefficients for magazine readership and mobile content use

		Music Magazines	Adult Magazines	Lifestyle Magazines	Sports Magazines
SMS	Sports	0.026	0.051	0.039	0.205
	News	0.038	0.053	0.034	0.040
	Entertainment	0.085	0.057	0.058	0.029
MMS	Sports	0.038	0.061	0.020	0.124
	News	0.018	0.037	0.031	0.026
	Entertainment	0.042	0.059	0.025	0.026
WAP	Sports	0.033	0.061	0.059	0.183
	News	0.012	0.040	0.071	0.016
	Entertainment	0.087	0.059	0.058	0.017

significant on 95%-level

significant on 99%-level

Figure 10. Relation of information and entertainment (above) and curiosity and habitual usage (below)

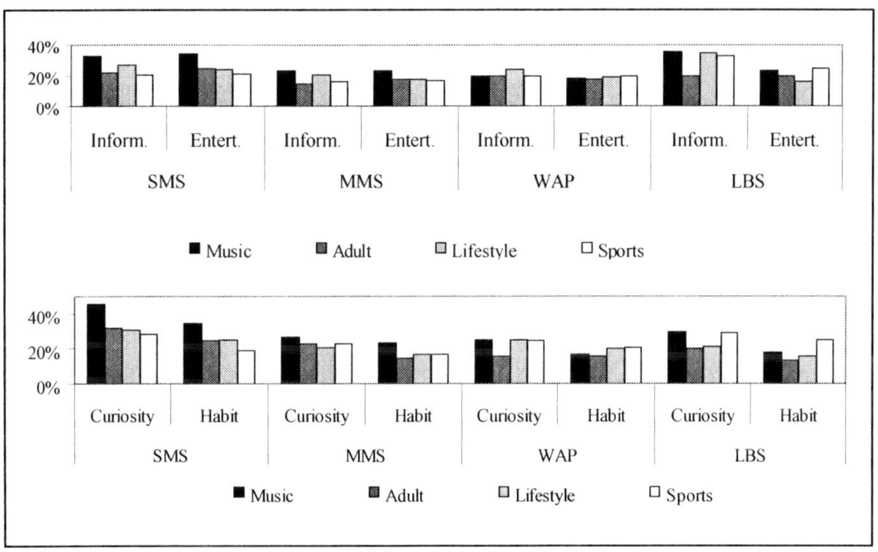

and WAP). Readers of sports magazines have the strongest interrelation with SMS- and WAP-based services while readers of adult magazines prefer MMS-based services.

To answer the second research question, the interrelations identified on the level of service formats need to be verified on the level of different contents distributed in these formats. As we have detailed before, news, sports, and entertainment contents are consumed most frequently using SMS-, MMS-, and WAP-based services. Therefore, the results of independence tests for these contents in the four reader groups with highest willingness to use mobile services are shown in Table 4. It reveals that reading music, adult, and sports magazines is significantly interrelated with using mobile contents. However, besides the interrelations between usages of mobile sports contents by readers of sports magazines, the values of these interrelations are rather small.

Analysis of the Motives Indicated for the Usage of Mobile Contents

In section 4 we described that the attractiveness of the mobile Internet is determined by curiosity and fascination of technology as well as it is determined by the need for ubiquitous information. Examining the respondents' motives in relation to their reading behavior reveals only marginal differences between different reader groups. For example, SMS- and MMS-based contents tend to be perceived as entertainment by readers of music and adult magazines whereas readers of lifestyle magazines use them in order to satisfy information needs (Figure 10). Only readers of adult magazines show a balance between habitual mobile content reception and satisfaction of spontaneous information needs. We did not find any significant differences either between the motives of mobility and ubiquity, dominating interactivity within all reader groups, and between technology fascination and practical value.

Classifying and Summarizing the Findings

Concerning the above mentioned kinds of acceptance (affective and cognitive) we can summarize the findings. With the help of Rogers' (1995) attributes of an innovation and the acceptance criteria of Nielsen (1993),the key findings can be classified as shown in the following matrix.

The general findings of the survey can be summarized as follows:

Mobile contents are used mainly in order to satisfy curiosity and location-independent information needs. In contrast to, for example, Ziegler and Adam (2001), who assume that special interest contents such as guidebooks will dominate users' demand of mobile services, we found that only contents of general interest such as news and sports are being used frequently.

Respondents reading special interest magazines such as adult or sports press show a higher willingness to use mobile contents. In contrast, consumers of mass market segments such as newspapers and news magazines did not show significant demand for mobile contents. This result is disillusioning for three reasons. First, readers of music, adult, lifestyle, or sports magazines are

Table 5. Classification of general findings

	Acceptance criteria			
Dimension of acceptance	**Relative advantage/ Observability**	**Costs**	**Compatibility/Possibility to try**	**Complexity/Reliability**
Affective (interest)	Usage because of curiosity/fascination; Added value as precondition for future usage	Low prices as precondition for future usage	Conditionality of usage of services through traditional media usage and device know how	Less experienced but curiosity driven behavior; usability and format quality as precondition for future usage
Cognitive (usage)	In general little willingness to use; Usage is focused on current and informative content	-	Usage focuses on simple services (SMS, WAP)	

naturally limited in number; this will result in small market sizes. Second, there is only a small degree of interrelation within these groups; this leads to the conclusion that even those readers are not willing to use mobile services to a large degree. Both aspects support the thesis that it is unlikely that mobile content will become a mass market in medium term. Third, our examination of motives revealed that using mobile contents is mainly driven by curiosity and fascination. This leads to the assumption that content providers cannot expect a stable demand in the near future.

FINDINGS FROM FURTHER CURRENT SURVEYS

In order to draw conclusions, the results of this survey shall be complemented by findings from other current surveys covering mobile content. For methodological reasons scientific surveys addressing the German (potential) users of mobile content are considered. Only one survey can be found which was undertaken after 2005. In summer 2006, social scientist Hanekop (2007) started an interesting experiment concerning mobile TV (MI FRIENDS: **M**obile **I**nteractive **TV** (German: **F**ernsehen), **R**adio, **I**nformation, **E**ntertainment, and **N**ew **D**igital **S**ervices). Within 8 weeks, 190 people tested a German mobile TV offer. Concerning the age and income of the sample, the experiment is comparable to our survey.

The most important finding from the mobile TV experiment is the following: Mobile content reception takes place in special situations. Mobile TV is mainly used en route (in the bus, underground, stations) or while waiting for the next date or event. Generally speaking, mobile TV is used for passing the time. This means that people do not have lot of time "in one piece" for consuming content. Therefore they are not able to concentrate a long time on a certain topic, or they do not have that much time to find the content they want to consume. For this reason adequate content needs to be available in "short pieces." Objects like news are much better than continuing or complex stories. Furthermore, the navigation within the content offer needs to be done quickly. Therefore, the navigation has to be simple. Today, cell phones usually are much too complex to handle. Also, individualized offers could be useful. If people only get content corresponding to their individual interests they do not need that much time for searching for the right content.

Hanekop draws the conclusion that mobile TV will have new habits of usage in comparison to the established media.

FUTURE TRENDS

Currently, in general, developing services for mobile usage is still very expensive. There are three reasons for this situation (similar Diekmann, Kaspar, Seidenfaden, & Hagenhoff, 2006). First, there are too many different carrier technologies like universal mobile telecommunications system (UMTS), worldwide interoperability for microwave access (WiMax), or Bluetooth. If the provider of any mobile service will meet critical customer masses, the service has to be compatible to all of these different technologies. This is not practicable. Second, the same situation has to be realized for devices like cell phones. The critical mass aspect will result in designing mobile services for different hardware platforms. Again, this is not practicable. Third, some functionality, like payment, localizing the user, or individualizing the services are not specific for a certain service, but in general needed for a lot of different services from different providers. Developing that functionality again and again means wasting scarce resources.

Giving people mobile access to the Internet for using services like TV, radio, news, or location-based services (e.g., restaurant finder), architectures are needed which are designed in a more intelligent and therefore cheaper way than it is done nowadays. Like for other software applications, a middleware for mobile services is needed. Often needed functionality and technology interfaces are developed once and reused by different service providers. This will result in mobile services with lower costs than nowadays.

CONCLUSION

Both surveys presented in this chapter examined the usage of content accessed via mobile Internet. The first survey, presented in detail, shows that there assumedly is no large market potential (concerning numbers of customers) for mobile content. The second survey comes to the realization that mobile content is used in special situations, namely for passing time. Therefore, content objects are needed which can be consumed within very view minutes (small content pieces). Those small content pieces do not fetch large revenues per piece. From both aspects, managers of publishing houses or TV stations can learn that large revenues neither per piece nor from a bulk business will arise. Providing content for mobile access therefore only is worthwhile in case of very low variable costs, and in case of low fixed costs. At the moment it still has to be assumed that mobile Internet as a distribution channel for content is useful only

1. In case of using intelligent infrastructure like well-established content management systems and middleware which allow to support this channel without much extra costs (otherwise this channel cannot be profitable).
2. In case of using it as an instrument of customer relationship management for supporting the established media business.

REFERENCES

Anstadt, U. (1994). *Determinanten der individuellen Akzeptanz bei Einführung neuer Technologien*. Frankfurt a.M., Germany: Europäischer Verlag der Wissenschaften.

Davis, F. (1989). Perceived usefulness, perceived ease of use, and user acceptance of information technology. *MIS Quarterly*, *2*(13), 319-340.

Davison, J., Walsh, A., & Brown, D. (2000). *Mobile e-commerce: Market strategies*. London: Ovum Ltd.

Diekmann, T., Kaspar, C., Seidenfaden, L., & Hagenhoff, S. (2006). Kontextbewusste Informationsdienste auf Grundlage des Information Beacon Frameworks. In F. Lehner (Ed.), *Multikonferenz wirtschaftsinformatik*. Retrieved August 30, 2007, from http://www.gi-mms.de/mms2006/kurzbeitraege/diekmann.pdf

Fink, L. (2002). *Economies of aggregation*. Unpublished doctoral dissertation, University of Cologne, Germany.

Govindarajulu, C., Reithel, B., & Sethi, V. (2000). A model of end user attitudes and intentions toward alternative sources of support. *Information & Management, 3*(37), 77-86.

Gregg, S. (2001). Die Konvergenz: Telekommunikationsanbieter und Medienunternehmen—Wettbewerber oder partner? In A. Vizjak & M. Ringlstetter (Eds.), *Medienmanagement: Content gewinnbringend nutzen* (pp. 37-44). Wiesbaden, Germany: Gabler.

Hanekop, H. (2007). *DMB-Projekt MI FRIENDS—Ergebnisse der Begleitforschung München*. BLM Schriftenreihe Band 86. München, Germany: Verlag Reinhard Fischer.

Kearney, A. T. (2003). *Mobinet 5: The new mobile mindset.* Retrieved June 11, 2007, from http://www.atkearney.com/shared_res/pdf/Mobinet_Monograph_S.pdf

Müller-Böling, D., & Müller, M. (1986). *Akzeptanzfaktoren der bürokommunikation.* München, Germany: Oldenbourg.

Müller-Verse, F. (1999). Mobile commerce report *(Durlacher research report).* Retrieved December 16, 2006, from http://www.durlacher.com/downloads/mcomreport.pdf

Nielsen, J. (1993). *Usability engineering.* San Diego: Academic Press.

Nokia. (1999). *The demand for mobile value-added services.* Retrieved June 11, 2007, from http://www.telecomsportal.com/Assets_papers/Wireless/Nokia_mobile_vas.pdf

Rogers, E. (1995). *Diffusion of innovations.* New York: Free Press.

Seidenfaden, L., Kahnwald, N., Kaspar, C., & Ortelbach, B. (2005). *Mediennutzung im digitalen Leben: Active content interfaces, paid content und integrierte geschäftsmodelle*, Goettingen, Germany: Business Village.

Sirkinh, H., & Dean, D. (2000). *Mobile commerce: Winning the on-air consumer.* Retrieved June 11, 2007, from http://www.bcg.com/publications/files/M-Commerce_Nov_2000_Summary.pdf

TIMElabs. (2000). *Winning in Mobile eMarkets.* Retrieved February 13, 2007, from http://www.timelabs.de/pdf/TIMElabs_C_mB2B_d.pdf

Venkatesh, V., & Davis, F. (2000). A theoretical extension of the technology acceptance model. Four longitudinal field studies. *Management Science*, *2*(46), 186-203.

Wohlfahrt, J. (2004). *Akzeptanz und wirkung von mobile-business-anwendungen.* Hamburg, Germany: Dr. Kovač.

Ziegler, M., & Adam, B. (2001). Mastering mobile media markets. Eschborn, Germany: Diebold Consulting.

Zobel, J. (2001). *Mobile business and m-commerce.* Munich: Hanser.

Chapter XV
Information Quality Satisfaction of Communication Portals: A Study of Central Cyber Government Office (CCGO) of the Hong Kong Government

Kevin K.W. Ho
The Hong Kong University of Science and Technology, Hong Kong

ABSTRACT

Information quality is critical for a communication portal because there are a myriad of information types, including textual, audio, video, and other complex information types, which an organization has to manage. In this study, we examine whether information generated from an in-house developed communication portal of the Hong Kong government would have higher quality than those sister portals developed by individual government departments using commercial packages. We conducted a survey-based study to understand how users evaluate the information quality of these communication portals. This portal case is interesting because (1) the Hong Kong government has invested millions of U.S. dollars in its implementation and (2) the number of potential users is huge (over 140,000).

INTRODUCTION

To keep governments operating smoothly, swift but careful fine-tuning of public policies and strategies are required. This situation creates a demand for establishing a seamless information flow among government agencies. To facilitate such information exchange, there is a global trend for governments to take advantage of information and Internet technologies for providing their services online. This can improve both efficiency and service quality of business processes within governments. Therefore, in previous years, various governments have heavily invested in information technology infrastructures and software applications. Taking Hong Kong as an example, this far Eastern metropolitan has invested over $100M per year in e-government projects since 2000. For the period between April 1999 and March 2006, Hong Kong has invested $1,094.4M (Ho, 2007). This huge investment helps Hong Kong to advance to one of the top countries/cities in the maturity of e-government development (Accenture, 2004; Melitski, Holzer, Kim, Kim, & Rho, 2005).

Among these $1,094.4M invested by the Hong Kong Government, 16% was invested in projects related to the development of IT infrastructure and software applications, which aimed to support **government-to-government (G2G) and government-to-employee (G2E)** transactions (Ho, 2007). In this chapter, we shall study the information quality satisfaction of one of the key G2G and G2E applications of Hong Kong, that is, the central cyber government office (CCGO).

COMMUNICATION PORTAL IN HONG KONG GOVERNMENT: CENTRAL CYBER GOVERNMENT OFFICE (CCGO)

In 2000, the Hong Kong Government launched its Intranet communication portal, the CCGO, for its employees (HKG, 2007). This portal is designed for handling G2G and G2E transactions. Figure 1 shows the IT infrastructure of the Hong Kong government and Table 1 reports the description the information and applications provided in the main screen of CCGO.

CCGO is an **Intranet communication portal**, which acts as a hub for linking up services provided to internal customers by different government agencies. Its function is similar to GovHK (http://www.gov.hk), the e-government one-stop portal for members of the public. Users can access CCGO via their network terminals, which are connected to the government Intranet. CCGO provides five major services to government officers. First, it serves as an intragovernment information-sharing center. Government officers can obtain the latest version of internal regulations and guidelines issued through CCGO. Second, it provides an instant access to the latest service-wide information, such as circular memoranda and notices. Third, it provides a video-on-demand function for broadcasting important government events, such as press conferences. Forth, it acts as a training portal for government officers who can participate in those self-learning programs organized by the government in-house training agency. Last but not least, CCGO also serves as an intergovernment business center for dissemi-

Figure 1. Schematic diagram of communication networks in the Hong Kong government

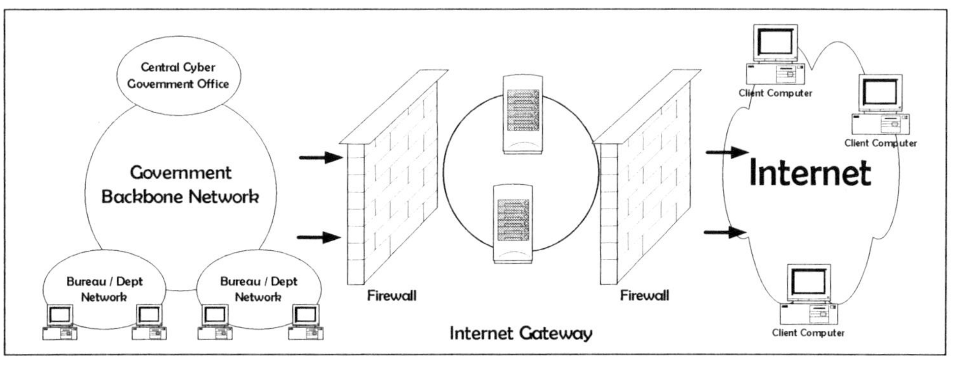

Table 1. Applications and information provided in CCGO

Applications		
✧ Search Engine ✧ Download Sites ✧ G2E applications: e-leave	✧ Bulletin Board ✧ Discussion Forum ✧ Video Channel	✧ Government Telephone Directory ✧ Weather Report
Quick Link to Information		
✧ Regulations ✧ Information Technology ✧ Finance and Procurement ✧ The Legislative Council ✧ Forms ✧ Departmental Intranet	✧ Circulars and Circular Memorandum ✧ Training and Development ✧ Laws and Legal Information ✧ Useful Statistics ✧ Green Management ✧ Transportation	✧ Communication ✧ Civil Service ✧ Public Sector Reform ✧ Other Useful Information ✧ Event Calendar

nating and collecting returns among government agencies (HKG, 2007).

Apart from providing these major services, CCGO also provides other add-on services, ranging from telephone directory and glossaries to bulletin boards and discussion forums. Plus, CCGO acts as a platform for G2E and G2G applications. New applications, such as the electronic leave application system, electronic payroll system, and departmental portals, have been launched in recent years (HKG, 2007).

RESEARCH QUESTIONS

In this chapter, we examine the information quality satisfaction of CCGO. We choose CCGO as the focus of our study based on the following reasons. First, the Hong Kong government is one of the pioneers in implementing e-government in the world. It ranked seventh in the worldwide "Overall Maturity in e-Government" (Accenture, 2004). Thus, we conjuncture CCGO should be well designed. Second, the number of users of CCGO is huge. When we conducted our study in 2003, the number of users of CCGO was around 50,000 and was equal to one-third of the total number of government officers. In the past 4 years, the user population of CCGO has been increased to 140,000 and CCGO is now accessible by nearly all government officers. With more knowledge on how information quality factors affect user satisfaction on CCGO, business firms and governments can obtain insights on how to develop their communication portals. Therefore, this chapter aims at addressing the following two research questions:

1. Does an in-house developed information portal produce higher information quality than portals hosted on commercial packages?
2. Which features of information quality affect **user satisfaction**?

This chapter is organized as follows. The next section presents prior research studies on information quality and the development of our research model. Then, we report our research methodology, data collection, and data analysis. We shall conclude our essay by discussing the future trend of the development of **communication portals**.

BACKGROUND

Research on Information Quality

Information systems (IS) researchers have studied the relationship between information quality and user satisfaction for over 30 years (Melone, 1990). For example, Bailey and Pearson (1983) study user satisfaction and developed 39 factors for measuring computer user satisfaction. These factors include information quality constructs. Srinivasan (1985) examines the implementation of computerized modeling systems of 29 organizations. He notices that system effectiveness can be measured by **information quality constructs**. Plus, DeLone and McLean (1992) note that information quality is one of the six dimensions which have a significant impact on the success of information systems.

IS researchers also develop theoretical models or methodologies to explain the impact of information quality on user satisfaction (Lee, Strong, Kahn, & Wang, 2002), systems development and implementation (Mahmood, 1987; Shim & Min, 2002), and data integrity (Lee, Pipino, Strong, & Wang, 2004). New methods, such as structural equation modeling (SEM) (Bharati, 2003; Khalil & Elkordy, 2005), elaboration-likelihood model (ELM) (Bhattacherjee & Sanford, 2006), and experiments (Aladwani, 2003) are employed to analyze such impact.

With the rapid development of e-commerce in recent years, IS researchers begin to investigate the impact of information quality on the user acceptance and satisfaction on e-commerce Web sites. Negash, Ryan, and Igbaria (2003) and Shih (2004) study the impact of information quality on the effectiveness of Web-based customer support system and the user acceptance of e-shopping, respectively. Sullivan and Walstrom (2001) study the impact of information quality on the consumer perception on the e-commerce Web site. Plus, Park and Kim (2006) study the impact of information quality on information satisfaction on e-commerce in Korea.

Similarly, with the rapid growth of e-government in these years, IS researchers also investigate the impact of information quality on e-government. For example, Cykana, Paul, and Stern (1996) discuss the data quality management guidelines used by the Department of Defense and report that **accuracy**, completeness, consistency, **timeliness**, uniqueness, and validity are the core factors of data quality requirements. Also, Aladwani (2002) studies the user satisfaction on the information systems in public organizations in Kuwait.

Research Model

In our study, we include the four most commonly used information quality constructs in our research model. These four constructs are **information accuracy, presentation format, information timeliness,** and **content relevance**. Table 2 reports the related prior studies.

Table 2. Information quality constructs used in the study

Constructs	Prior Studies
Information Accuracy (A)	Bailey and Pearson (1983) Cykana et al. (1996) Doll and Torkzadeh (1998) Mahmood (1987) Miller and Doyle (1987) Srinivasan (1985)
Presentation Format (P)	Doll and Torkzadeh (1998) Cykana et al. (1986)
Information Timeliness (T)	Bailey & Pearson (1983) Cykana et al. (1996) Doll and Torkzadeh (1998) Mahmood (1987) Miller and Doyle (1987) Srinivasan (1985)
Content Relevance (C)	Bailey and Pearson (1983) Doll and Torkzadeh (1998) Miller and Doyle (1987) Srinivasan (1985)

Based on the result of prior research, we conjuncture the level of user satisfaction is positively related to the level of information quality. Plus, as the Hong Kong Government has invested millions of dollars to develop CCGO, we conjuncture the level of information quality of CCGO is higher compared with the information quality of other Intranet portals developed using commercial packages. Figure 2 shows our research model and our hypotheses developed based on our research models are as follow:

Measures on User Satisfaction

H1a: User satisfaction is positively associated with the level of information accuracy of communication portals.

H1b: User satisfaction is positively associated with the level of presentation format of communication portals.

H1c: User satisfaction is positively associated with the level of information timeliness of communication portals.

H1d: User satisfaction is positively associated with the level of information content relevance of communication portals.

Comparison of Information Quality of between CCGO and other Communication Portals

H2a: The level of information accuracy of CCGO is higher compared with other government portals developed using commercial packages.

H2b: The level of presentation format of CCGO is higher compared with other government portals developed using commercial packages.

H2c: The level of information timeliness of CCGO is higher compared with other government portals developed using commercial packages.

H2d: The level of content relevance of CCGO is higher compared with other government portals developed using commercial packages.

Figure 2. Research model of the study

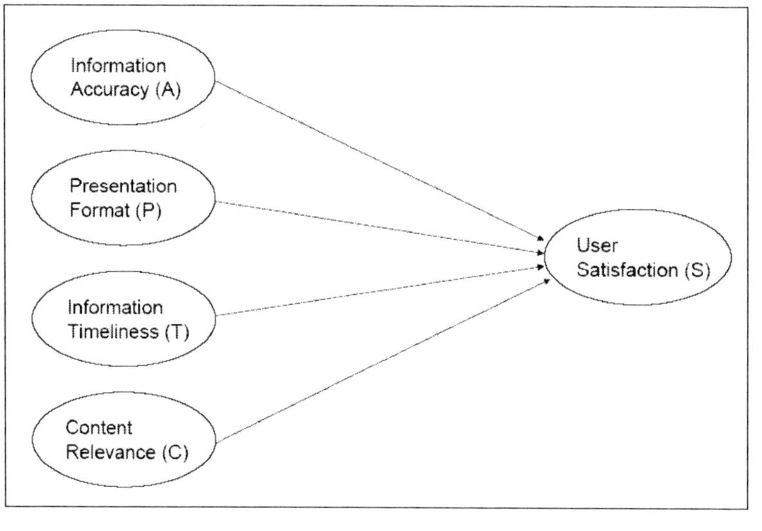

METHODOLOGY AND DATA ANALYSIS

Methodology

The survey was administered in late 2003 and lasted for 12 weeks. The questionnaire is adapted from the scale developed by Doll and Torkzadeh (1988) and Lee et al. (2002), which measures information quality from the systems. The instrument taps overall satisfaction and covers four dimensions: information accuracy, presentation format, information timeliness, and content relevancy.

A pilot test was conducted with a panel of 12 doctoral students. Two of the panelists were also government officers (i.e., part-time doctoral students), and hence, both the views of IS researchers and government officers were considered when we refined the questionnaire. The respondents took approximately 10 minutes to complete the questionnaire. The finalized version was made up of 12 questions, which were measured on a 5-point Likert scale (see Appendix).

We invited various government agencies to participate in this study and agencies from education sector, disciplined force, and post office were participants in this study. Respondents' job nature ranged from operation to administration. We conducted the survey on site and distributed paper questionnaires. Each survey form contained a cover letter explaining the purpose of the study. It also stated that the individual responses of this survey would be kept confidential and that the report would only depict data summaries. Each survey session was conducted on company time. A member of our research team explained the purpose of the survey and helped answer any questions raised by respondents on site.

We received 211 questionnaires, of which 204 (or 96.7%) of them were complete. One hundred and forty-one (or 69%) of the respondents were males, and the average age of the whole sample was 37 years. The summary of their office positions is summarized at Table 3. Twenty subjects (9.8%) used CCGO daily and the others used commercial packages (e.g., Lotus Notes or other Microsoft

Table 3. Background of the respondents of the study

Office Position	Annual Salary in 2003-04	No. of Respondents
Frontline staff, including: ✧ Clericals ✧ Secretaries ✧ Police constables, and so forth.	< US$29,100	29
Frontline managers, including: ✧ Senior clericals (registry heads) ✧ Junior executives (Deputy section heads) ✧ Teachers ✧ Police sergeants, and so forth.	US$29,100 – US$67,600	137
Middle management, such as: ✧ Senior executives (section heads) ✧ Subject heads of school, ✧ Police inspectors, and so forth.	US$67,601 – US$107,100	33
Top management, such as: ✧ Chief executives (branch head) ✧ School principals ✧ Chief police inspectors and superintendents, and so forth	> US$107,100	6

products) for message exchange services and meeting scheduling.

Data Analysis

We used SPSS version 10.0 to conduct factor analysis. Principal component analysis was used for factor extraction with standardized independent variables, and factor rotation was performed with Varimax with Kaiser normalization. The model converged in five iterations. A four-factor solution was obtained with all component eigenvalues greater than one. The independent factors were information accuracy (A), presentation format (P), information timeliness (T), and content relevance (C). These factors explained 78.6% of the total variance in the survey. Items load highly (> 0.7) on their associated factors, as shown in Table 4.

These factors are evaluated for reliability, convergent validity, and discriminant validity. In the reliability test, the internal consistency for each factor is assessed by computing Cronbach's alpha (see Table 5). According to Nunnally (1978), a reliability value of 0.7 or above is acceptable. Hence, all factors are deemed reliable.

Convergent validity is achieved if there are higher correlations between items in the same factor (Campbell & Fiske, 1959). To access this validity, Pearson correlation coefficients are computed with the two-tailed t-test. Table 6 provides the correlation matrix. All correlations between pairs of items within the same factor are statistically different from zero at p-value < 0.01. Thus, convergent validity is demonstrated.

Discriminate validity is demonstrated if an item correlates more highly with items within the same factor than with items in a different factor.

Table 4. Rotated component matrix

Items	Content Relevance (C)	Information Timeliness (T)	Information Accuracy (A)	Presentation Format (P)
A1			0.903	
A2			0.885	
P1				0.865
P2				0.818
T1		0.838		
T2		0.863		
C1	0.876			
C2	0.834			
C3	0.823			
C4	0.735			

Table 5. Reliability scale of the constructs

Constructs	Number of Items	Reliability
Information Accuracy (A)	2	0.916
Presentation Format (P)	2	0.853
Information Timeliness (T)	2	0.876
Content Relevance (C)	4	0.887

Table 6. Correlation matrix

	A1	A2	P1	P2	T1	T2	C1	C2	C3	C4
A1	1.00									
A2	0.85	1.00								
P1	0.33	0.37	1.00							
P2	0.43	0.41	0.74	1.00						
T1	0.36	0.40	0.49	0.50	1.00					
T2	0.38	0.35	0.34	0.42	0.78	1.00				
C1	0.39	0.41	0.42	0.32	0.40	0.33	1.00			
C2	0.47	0.50	0.37	0.34	0.47	0.41	0.76	1.00		
C3	0.36	0.40	0.44	0.32	0.25	0.24	0.72	0.65	1.00	
C4	0.40	0.45	0.25	0.32	0.42	0.51	0.66	0.77	0.50	1.00

Table 7. Regression model with user satisfaction as dependent variable

Constructs	Coefficient	*t*-value
Information Accuracy (A) [H1a]	0.143	2.088 *
Presentation Format (P) [H1b]	0.314	6.216 ***
Information Timeliness (T) [H1c]	-1.331	-3.098 ***
Content Relevance (C) [H1d]	0.083	0.530
Constant	1.701	2.549 ***
R^2(adj)	0.315	
Note: *** $p < 0.001$, ** $p < 0.01$, * $p < 0.05$		

Validity is determined by counting the number of times an item has a higher correlation with an item from another factor than with items in its own factor. Campbell and Fiske (1959) suggest that a count of less than one-half is acceptable. We made comparisons by examining the correlation matrix of items (see Table 6). There is no correlation value between items from different factors higher than those between items within the same factor. Based on this assessment, items in each factor are discriminant from items in another factor.

To have a more thoughtful analysis of the relationship between these information quality constructs and user satisfaction, we conducted regression analysis with satisfaction as the dependent variable (see Table 7). The adjusted R-square of the regression equation is 0.315, which suggests that there exists a high correlation between these information quality constructs and user satisfaction.

As shown in the regression result, content relevance (*t*-value = 0.53, or $p = 0.60$) is not a significant factor affecting user satisfaction and, hence, H1d is not supported. Information accuracy (*t*-value = 2.088, or $p = 0.04$) and presentation format (*t*-value = 6.216 or $p < 0.01$) are significant factors affecting user satisfaction and, hence, our H1a and H1b are supported. Even though information timeliness (*t*-value = -3.098 or $p < 0.01$) is a significant factor, its sign is negative and, thus, the result is opposite to our hypotheses. Therefore, H1c is also not supported.

To examine whether the level of information quality is higher for CCGO compared with other communication portals developed by government agencies, we compare the scores of the four latent factors of both CCGO and these portals. First, we compared the score of information accuracy. Results demonstrate that both CCGO (mean = 4.87) and other portals (mean = 4.20) had high output accuracy compared to the mean of our scale (i.e., 3). In particular, the score for CCGO is statistically higher than that of the other portals ($p = 0.02$). It is generally agreed that new systems often take time and experience to ensure all things are operating at peak accuracy. As CCGO was launched for around 4 years when we conducted our study, most bugs and software problems should have already been fixed. Also, as it is the portal hosted by the central government, users would believe that its information accuracy would be better than those developed by individual departments would. Thus, H2a is supported.

Second, both CCGO (mean = 3.05) and other portals (mean = 3.36) had satisfactory presentation formats. Unlike the case of information accuracy, the score of other communication portal in presentation format is significantly higher than the one of CCGO ($p = 0.04$). This observation may be because most employees had used the commercial software, such as Lotus Notes, for many years and, hence, they were more familiar with its interfaces and presentation formats. Hence, H2b is not supported.

Third, users were dissatisfied with the information timeliness for both CCGO (mean = 2.66) and other portals (mean = 2.45). The score for CCGO is statistically higher than that of the other portals ($p = 0.08$). We noticed that during our study the information disseminated in CCGO usually has a time lag of around 2 working days. As the government was using e-mail and fax to send out information before they were uploaded to CCGO, users might have an impression that timely information could not be provided by CCGO. For those users who used their own portal, they faced the same problem. Therefore, we have a lower score for information timeliness. However, H2c is supported.

Fourth, both CCGO (mean = 2.26) and other portals (mean = 2.59) had the score of content relevance below the midpoint of the scale. The score difference between CCGO and other portals is statistically insignificant ($p = 0.12$). This observation is not surprising because similar to portals on the Internet, CCGO and its sister portals had an information overload problem. The information overload arises from the disbelief of content providers that they should provide as much "useful" information to potential users as possible. Hence, for a general user, the user will find that most of the information provided is irrelevant.

Table 8. Summary of results

Hypothesis	Result
H1a: User satisfaction is positively associated with information accuracy.	Supported
H1b: User satisfaction is positively associated with presentation format.	Supported
H1c: User satisfaction is positively associated with information timeliness	Not Supported
H1d: User satisfaction is positively associated with content relevance	Not Supported
H2a: The level of information accuracy of CCGO is higher than other portals	Supported
H2b: The level of presentation format of CCGO is higher than other portals.	Not Supported
H2c: The level of information timeliness of CCGO is higher than other portals.	Supported
H2d: The level of content relevance of CCGO is higher than other portals.	Not Supported

This will push down the score of information content relevance. Therefore, H2d is also not supported. Table 8 below summarizes the result of our hypotheses.

DISCUSSION AND FUTURE TRENDS

With plenty of in-house IT and development resources, the Hong Kong government was one of the first to implement an advanced communication portal. The benefits of CCGO include more secure creation of and access to information, versioning, information audit trails, approval paths, secured distribution, and workflow and collaboration automation. The major drawback to it is the high software and implementation costs, which costs millions of U.S. dollars.

Our study shows three practical implications. First, in-house developed communication portals have higher information quality in output accuracy. It is because in-house developed systems have higher degree of connectedness with other existing in-house developed components. The output from one subsystem could be easily fed into another subsystem. Therefore, if an organization intends to develop a system generating high-quality information, it should take the in-house development approach if other existing subsystems are in-house developed as well.

It is interesting to know that the presentation format of CCGO was found to be worse than that of commercial systems. There can be two possible reasons. The first reason was that the system development team neglected the importance of interface design. The second reason was that users were more familiar with commercial software packages and, thus, they perceived these portals developed using other software packages to have a better format.

Government officers were not satisfied with the content relevance and information timeliness. There can also be two possible reasons. First, CCGO is an Intranet portal targeted for government officers from more than 90 departments. This makes the information provided by CCGO too general. When we conducted this study in 2003, departmental information was not available at CCGO. With the implementation of departmental portals in CCGO from 2004 onwards, we expect this problem should have been resolved. It is because individual government agencies can create their own portals to disseminate the tailor-made information to their own users using the CCGO platform.

It is also interesting to observe that the coefficient for timeliness in the regression model is negative, which is different from our original prediction. We suggested that the negative coefficient for timeliness is due to adaptation problem. During our interviews with 10 (or 4.9%) random samples of our respondents immediately after the survey, we noticed that when we conducted the study, the Hong Kong government had just started to migrate from paper-based office to paperless office. As a result, government officers were required to adjust their pace to face the fast turnaround of information flow and to collect information actively through Web-surfing. This brought a paradigm shift to government officers. As a result, they were of the view that information timeliness was a burden instead of an advantage to them.

Based on the findings of this study, we suggest that management of both public and private sectors should be reminded of the impact of information quality to user satisfaction of their communication portals. Usually, the information provided in an Intranet portals is decided by the management based on their understanding of the operation of the business. Thus, the management would try to provide the most useful information to their staff through their Intranet portal and perceive that the information quality to user should be relatively higher. However, from what we have observed in our study on CCGO, we notice that it may not be the case. This reflects that management should take a more active role to consolidate the need of their staff, who are the actual users of the portal, so that the management can design a portal with high information quality to users. As shown in our study, a centralized portal, which originally can reinforce the corporate image within the organization, may have negative impacts on user satisfaction on the communication portal due to some unexpected reasons. It is demonstrated by the negative effect of information timeliness to user satisfaction in CCGO. Hence, management should consider delegating the editing right to the frontline managers. This idea is similar to the development of departmental portals in CCGO, which allow the individual agencies to provide their tailor-made Web page for their employees.

CONCLUSION

Our study conducted a survey in a number of government agencies to examine the information quality of an in-house developed communication portal by the Hong Kong government. Information quality is critical for a communication portal because there are a myriad of information types, including textual, audio, video and other complex information types, which an organization has to manage. Results show that information accuracy, presentation formats, and information timeliness are important information quality factors. We suggest that public and business organizations should take into account the impact of information quality on the user satisfaction of their portals when they design their Intranet portals.

REFERENCES

Accenture. (2004). *E-Government leadership: Engaging the customer*. Retrieved July 15, 2007, from https://www.accenture.com/Global/Services/By_Industry/Government/R_and_I/CitizensView.htm

Aladwani, A. M. (2002). Organizational actions, computer attitudes, and end-user satisfaction in public organizations: An empirical study. *Journal of End User Computing, 14*(1), 42-49.

Aladwani, A. M. (2003). A deeper look at the attitude-behavior consistency assumption in information systems satisfaction research. *Journal of Computer Information Systems, 44*(1), 57-63.

Bailey, J. E., & Pearson, S. W. (1983). Development of a tool for measuring and analyzing computer user satisfaction. *Management Science, 29*(5), 530-545.

Bharati, P. (2003). People and information matter: Task support satisfaction from the other side. *Journal of Computer Information Systems, 43*(2), 93-102.

Bhattacherjee, A., & Sanford, C. (2006). Influence process for information technology acceptance: An elaboration likelihood model. *MIS Quarterly, 30*(4), 805-825.

Campbell, D., & Fiske, D. W. (1959). Convergent and discriminant validation by the multitrait-multimethod matrix. *Psychological Bulletin, 56*(2), 81-105.

Cykana, P., Paul, A., & Stern, M. (1996). DoD guidelines on data quality management. In *Proceedings of the Conference on Information Quality*, Cambridge, MA (pp. 154-171).

DeLone, W. H., & McLean, E. R. (1992). Information systems success: The quest for the dependent variable. *Information Systems Research, 3*(1), 60-95.

Doll, W. J., & Torkzadeh, G. (1988). The measurement of end-user computing satisfaction. *MIS Quarterly, 12*(2), 259-274.

Ho, K. K. W. (2007). The e-Government development, IT strategies and portals of Hong Kong SAR Government. *International Journal of Cases on Electronic Commerce, 3*(2), 71-89.

Hong Kong Government (HKG). (2007). *IT infrastructure and standards – infrastructure for e-government – central cyber government office*. Retrieved July 15, 2007, from http://www.ogcio.gov.hk/eng/infra/eccgo.htm

Khalil, O. E. M., & Elkordy, M. M. (2005). EIS information: Use and quality determinants. *Information Resources Management Journal, 18*(2), 68-93.

Lee, Y. W., Pipino, L., Strong, D. M., & Wang, R.Y. (2004). Process embedded data integrity. *Journal of Database Management, 15*(1), 87-103.

Lee, Y. W., Strong, D. M., Kahn, B. K., & Wang, R. Y. (2002). AIMQ: A methodology for information quality assessment. *Information and Management, 40*(2), 133-146.

Mahmood, A. (1987). Systems development methods: A comparative investigation. *MIS Quarterly, 11*(3), 293-311.

Melitski, J., Holzer, M., Kim, S.-T., Kim C. G., & Rho, S. Y. (2005). An e-government assessment of municipal Web sites. *International Journal of Electronic Government Research, 1*(1), 1-19.

Melone, N. P. (1990). A theoretical assessment of the user-satisfaction construct in information systems research. *Management Science, 36*(1), 76-91.

Miller, J., & Doyle, B. A. (1987). Measuring effectiveness of computer-based information systems in the financial services sector. *MIS Quarterly, 11*(1), 107-124.

Negash, S., Ryan, T., & Igbaria, M. (2003). Quality and effectiveness on Web-based customer support systems. *Information and Management, 40*(8), 757-768.

Nunnally, J. C. (1978). *Psychometric theory* (2nd ed.). New York: McGraw-Hill.

Park, C. H., & Kim, Y. G. (2006). The effect of information satisfaction and relational benefit on consumers' online shopping site commitments. *Journal of Electronic Commerce in Organizations, 4*(1), 70-90.

Shih, H. P. (2004). An empirical study on predicting user acceptance of e-shopping on the Web. *Information and Management, 41*(3), 351-368.

Shim, S. J., & Min, B. K. (2002). Organizational factors associated with expert systems implementation. *Journal of Computer Information Systems, 42*(4), 71-76.

Srinivasan, A. (1985). Alternative measures of system effectiveness: Associations and implications. *MIS Quarterly, 9*(3), 243-253.

Sullivan, J. R., & Walstrom, K. A. (2001). Consumer perspectives on service quality of electronic commerce Web sites. *Journal of Computer Information Systems, 41*(3), 8-14.

APPENDIX: SAMPLE QUESTIONNAIRE

User Satisfaction

S1. The system is successful.
S2. I am satisfied with the system.

Information Accuracy

A1. The system is accurate.
A2. I am satisfied with the accuracy of the system.

Presentation Format

P1. The output is presented in a useful format.
P2. The information is clear.

Information Timeliness

T1. I get the information you need in time.
T2. The system provides up-to-date information.

Content Relevance

C1. The system provides the precise information you need.
C2. The information content meets your needs.
C3. The system provides reports that seem to be just about exactly what I need.
C4. The system provides sufficient information.

(1 = Disagree, 5 = Agree)

Chapter XVI
The Evaluation of IT Investments through Real Options

Maria Alice Frontini
University of São Paulo, Brazil

Fernando José Barbin Laurindo
University of São Paulo, Brazil

ABSTRACT

The decisions about IT investments are increasingly more complex, due to technical uncertainties and to the dynamics of organizational and strategic issues. One promising alternative for solving this problem would be the use of real options. Thus, this chapter intends to apply a relatively new methodology, called real options, used in corporate strategy for evaluating and deciding about new investments in IT. In order to do so, an analytic review of the literature is presented. The real options methodology is particularly recommended in two situations: in the case of the existence of a significant level of uncertainty about the benefits to be achieved by IT investments, or when IT benefits do not impact directly the current business but create a platform for future investments, capable of producing future new business impacts.

INTRODUCTION

As information technology (IT) evolves, investment decisions upon it become increasingly more complex. Chief information officers (CIO) face prejudice from top management, concerned about the real benefits that IT can bring up to business. Technological uncertainties make it very difficult to estimate the value that can be extracted from the usage of new IT solutions, and the IT benefit is even more integrated to the organizational, process, and marketing dimensions of a company.

This scenario can bring two consequences to the business environment: on the one hand, the CIO is confronted with greater skepticism about the approval of new projects; on the other hand, this conservatism can be taking the entrepreneurs to a lack of investments in IT infrastructure projects that can represent the foundation for the outcome of other projects, which can be a basis for future opportunities for products and services innovation offered by the company to the market.

This chapter intends to, through an analytic review of the literature, apply a relatively new methodology, called real options, used in corporate strategy for evaluating and deciding about new investments in IT. Particularly, this methodology is highly recommended when there is either a considerable level of uncertainty about the benefits to be achieved by the investment, or when IT benefits do not directly impact the business but create a platform for future investments, hence future business impacts.

PERSPECTIVES ON IT ASSESSMENT IN ORGANIZATIONS

Lately, many aspects have been discussed on the role and governance of IT, either in academic reviews or in managerial environments. On the one hand, there are doubts about results yielded by IT investments. On other hand, IT applications cause enchantment, especially those related to global economy and the Web (Porter, 2001; Tapscott, 2001). Farrell (2003) emphasizes that IT brings significant gains when related to organizational processes, and even further when integrating the industry value chain (Porter & Millar, 1985). In this item, two aspects are highlighted: the strategic role of IT in the organizations and the evaluation of IT effectiveness.

The Strategic Impact of IT Effectiveness

One of the first initiatives to understand IT role in organizations established a model to assess IT organization in alignment to business, according to phases in business growth and/or in business maturity.

Nolan (1979) is the most widely known and disseminated model. Taking into account the introduction of automation and information technology into enterprises, Nolan proposed a framework which establishes four phases in the IT evolving pattern. Afterwards, the author enhanced his model into six phases: initiation, contagion, control, integration, data administration, and maturity.

- Initiation: When first computers are purchased for labor reduction and for reducing paper handling.
- Contagion: When IT is expanded to other functions such as invoice, inventories, and checks emission. There is not, however, the concept of information integration.
- Control: The growth in the use of information systems in the organization becomes explosive; the IT department is professionally administered.
- Integration: The restructuring demanded by the previous stage is completed, in response to the pressure for better administration; information systems are guided to support needs of different managerial levels. Information has better quality, as a direct result of centralization of the IT department under a single administrative structure and as a consequence of the use of databases management systems.
- Data administration: Is dominated by database technology. The IT department recognizes that information is a very valu-

able resource that should be accessible by all users in the organization.

- Maturity: Information is considered an asset of the organization. Users are active and responsible and IT growth is planned and organized. Applications portfolio is complete and its structure reflects the organization and its information flow.

Nolan's model allows for visualizing of the evolution of some characteristics (maturity process) along each of the six phases. For instance, according to the improvement in each phase, there is a growing involvement of users, as well as better planning and control of IT. At lower levels of maturity, IT budget controls are not so rigid in order to facilitate its adoption; at higher levels, there is a strategic planning of IT resources and assets, such as data and information. Finally, an organization may present differences in the maturity level, according to each business unit or function.

In order to understand the IT role, it is critical to get competitive advantages in the value chain (Porter & Millar, 1985). A value chain is the set of technological and economically distinct activities each organization applies to run a business. The value system encompasses the value chain of each participant of an industry, from vendors to end consumers. Links among participants connect organizational activities within the value system. IT is embedded in value chains, changing the way of executing value activities and the nature of links between them. Given that, IT can change the competitive scenario in three ways:

- Changing a sector structure by shaping the five competitive forces (Porter, 1979).
- Creating new competitive advantages, reducing costs, improving differentiation, and changing competitive scope.
- Originating new business models based on new value propositions or new ways of doing things.

According to the information needs of business process (value chain) and products, the potential of IT changes so as to change a system value. In organizations in which process and products have a high information density, information systems play a significant role (Porter & Millar, 1985).

However, according to Duhan et al. (2001), knowledge-based organizations (such as consultancy companies) are harmed by the value chain analysis, due to the fact that is difficult to identify the value added in each activity. Under this circumstance, these authors support Prahalad's core competences approach; Hamel's (1990) is more suitable to plan IT as a strategic resource.

McFarlan's (1984) strategic grid allows for identifying how IT is positioned according to a business strategy and its operations. This model analyzes the impact of current and future applications into business. There are four grids, each representing an organization status: support, factory, transition, and strategic.

- Support: IT has a small influence in current and future corporate strategy.
- Factory: Current IT applications definitely contribute to an organization success; however, there are no future applications which may impact the strategy.
- Turnaround: IT changes from an irrelevant position (support grid) to a significant role in corporate strategy due to new applications with strategic impact.
- Strategic: IT has a significant role in an organization overall strategy. Either current applications or future ones are strategic and directly impact business performance.

In order to evaluate the strategic impact of IT, McFarlan proposed to analyze IT effects into the five competitive forces (Porter, 1979); if IT impacts at least one of the forces, IT applications will be considered strategic. Hence, IT can have a more or less significant strategic position, according to a business industry or process or product.

A broader perspective about the interaction between IT and business was elaborated by Henderson and Venkatraman (1993), which presented the strategic alignment model, which emphasizes and analyzes the strategic relevance of IT in organizations. This model is based not only on internal factors (enterprise driven) but also on external ones (market driven). The authors offer not only an analysis based on IT impact on the business (traditional model) but also the IT potential to change business trends. Furthermore, it requires the monitoring of the new technologies delivered in the marketplace (Burn & Szeto, 2000).

Thus, according to this model to plan IT, four relevant aspects should be considered: business strategy, IT strategy, organizational infrastructure and processes, and IT infrastructure and processes.

The strategic alignment model takes into account the assumption that an IT effective management requires continuous balance about the decisions made among the four factors. The novelty of this model is based on the fact that IT can impact business strategy and plans; in general, the strategic planning is the other way round, that is, business strategy is the starting point for IT planning.

According to Henderson and Venkatraman (1993), there are four perspectives for the strategic alignment, each taking into account a different starting point, either business strategy or IT strategy. The most relevant aspect is that there is an alignment between business and IT, whatever it is. According to the stream of alignment, the perspective is different, that is:

- Strategy Execution: *business strategy ⇒ business structure ⇒ IT structure*
- Technological Transformation: *business strategy ⇒ IT strategy ⇒ IT structure*
- Competitive Potential: *IT strategy ⇒ business strategy ⇒ business structure*
- Service Level: *IT strategy⇒ IT structure ⇒ business structure*

The traditional IT strategy perspective is represented in the strategy execution alignment, in which a business strategy defines an organizational structure which demands IT infrastructure requirements. Service level perspective considers IT as a business within another business, which might become a spin-off.

New business models, which IT has been enabling recently, might be explained based on the alignment patterns of competitive potential and technological transformation. In competitive potential perspective, an organization, which builds IT competences and knowledge as a result of its IT strategy, may search for new business strategies or models, for example, Amazon.com. In the technological transformation perspective, an organization defines a business strategy which requires a new IT strategy, due to new requirements in the IT infrastructure. Companies which seek to integrate and automate the relationship between the supply chain and vendors and/or customers through Internet are examples of this perspective.

Approaches and Models for IT Evaluation

In the last two decades, different models emerged to assess IT, according to their different specialization, that is, data processing, information systems, and information and communication technology.

According to Walrad and Moss (1993) and Wetherbe (1979), there are two different approaches to evaluate IT: oriented toward efficiency, which is strongly focused on in-house system development activities; and toward effectiveness, which seeks IT organizational impact.

Another classification of IT evaluation is conceived by Smithson and Hirscheim (1998). Besides establishing some categories, they put into perspective many different ideas, techniques, and models to evaluate IT. The authors establish three zones to categorize the different methods to evaluate IT: efficiency, effectiveness, and understanding zones.

Hence, the effectiveness concept is very important to asses IT, considering the alignment between corporate and IT strategies. Maggiolini (1981) defines efficiency as the measure of a conversion process of inputs into outputs and effectiveness as the measure of adherence of outputs to requirements. For Walrad and Moss (1993), efficiency is related to "doing things right," emphasizing cost and quality factors of the information systems; effectiveness is related to "doing the right things," that is, focusing on the enhancement of business performance as a result of applying information systems.

This work primarily focuses on IT assessment based on effectiveness. In general, assignments related to effectiveness evaluation present some characteristics differentiated among themselves, which can therefore be classified into three groups (Laurindo, 2002):

- IT assessment based on financial metrics: These models correlate IT investments with financial performance indicators of an organization.
- IT assessment in alignment with corporate strategy: Models of this group link business strategy with IT strategy or with IT applications portfolio.
- IT assessment processes and categories: This group contains assignments which discuss the IT assessment process itself and the categories/dimensions to assess IT.

Analyzing the theory base, Laurindo (2002) emphasizes common aspects which appear frequently and in the center of attention of each proposition:

- Strategic alignment, indicated by Benko and McFarlan (2003), Willcocks and Lester (1997), Earl, Rockart, and Ross (1996), Luftman (1996), and Henderson and Venkatraman (1993).
- Effectiveness beyond efficiency, focused on by Porter (2001), Willcocks and Lester (1997), Farbey, Land, and Targett (1995), and Rockart (1979).
- Intense and close relationship between business and IT management, suggested by Li and Ye (1999), Willcocks and Lester (1997), Prairie (1996), Earl et al. (1996), and McFarlan (1990).
- Different IT evaluation criteria according to the application, revealed by Smithson and Hirscheim (1998), Willcocks and Lester (1997), and Farbey et al. (1995).
- Productivity indicators related to business productivity, specified by Willcocks and Lester (1997), Broadbent and Weil (1997), Byrd and Marshal (1997), Mahmood (1993), Kaplan and Norton (1992) (not specifically for IT, but for integrated systems as a productivity effort), and Rockart (1979).
- Dynamic (as a continuous process) and flexible IT governance as well as its assessment, presented by Smithson and Hirscheim (1998), Willcocks and Lester (1997), Farbey et al. (1995), Henderson and Venkatraman (1993).
- Productivity paradox, revealed by Stratopoulos and Dehning (2000), Brynjolfsson and Hitt (1998), Willcocks and Lester (1997), and Henderson and Venkatraman (1993).

When assessing IT, its application characteristics should be considered, as it is proposed into the ladder for assessing benefits by Farbey et al. (1995). In this model, there is a categorization of eight types of IT applications, based on the benefits steps. On each benefit step there is an indication of the factors that should be considered to evaluate IT. These factors can be more objective, that is, more focused on efficiency (lower steps), or more subjective, that is, more focused on effectiveness (higher steps).

According to Earl (1989), and Willcocks and Lester (1997), IT assessment must be treated within a broaden perspective, involving different aspects. Under this consideration, integrative assessment models might be applied (Laurindo et al, 2001), because they take into consideration and aggregate a variety of elements from different approaches to evaluate IT, composing a broad analytical structure of assessment. Some examples are Laurindo (2002) and Willcocks and Lester (1997).

According to Turban, Rainer, and Potter (2003), IT investments can be justified (and therefore chosen) through the analysis of costs, benefits, and the comparison between them. In these authors' perspective, IT investments can be of two different types: IT infrastructure and IT applications.

IT infrastructure encompasses physical resources, components, services, and management that support all the information systems of an organization (e.g., data-warehouses, a datacenter, networks, and a corporative database). Investments of IT infrastructure are programmed to last a long time and can thus be shared by many applications.

IT applications are computer programs designed for supporting a specific task, a business process, or even another application. IT applications can be used by one department or can be shared by many departments, which is much more difficult to evaluate.

From some authors' perspective, there are some alternatives to perform costs-benefits analysis.

One alternative is the use of financial methods, such as the internal rate of return, the net present value. These methods face many difficulties when applied to IT projects, due to the great proportion of intangible benefits present in this situation. Another problem is the high degree of uncertainties in projects involving new technologies, as the case of e-business applications.

Recently, one method that has been adopted by many organizations is the "business case." This method gained notoriety during the boom of the dot.com companies in the 1990s. The idea of the business case is to produce a document with the justifications of the managers for the implementation of IT applications, considering financial, organizational, and strategic aspects. It is useful, for instance, to detect if the company should implement an IT application that is already in use by competitors, in order to keep its competitiveness.

IT projects can be evaluated under the perspective of benchmarking, that is, based on reports of best practices or on the performance indicators of an industry that consulting firms or sector associations can provide.

The value analysis technique can also be used (Turban, McLean, & Wetherbe, 2002). This method considers two phases: first, the construction of a cheap prototype (for understanding the value of benefits of the system); in the second phase, a complete system is built.

In situations of great uncertainty, the real options approach can be interesting, since it considers the opportunities that a project can provide in the future, even if in the present costs overcome the benefits (Campbell, 2002). This method is particularly attractive in the case of IT projects with great technological uncertainties, as the case of many Internet-based applications. The use of the real options approach will be described in the next section.

BACKGROUND: "REAL OPTIONS"—UNCERTAINTY CREATING OPPORTUNITY

Basic Concepts

According to Amram and Kulatilaka (1999), option is a right, not an obligation, of taking an action in the future. Options possess value when there are uncertainties (for the concepts presented in this work, uncertainty is the random nature of external events on which managers have no influence). For instance, an option contract negotiated in the finance market gives the right of buying a stock at a certain price on a certain date. This option will be exercised (used) only if the stock is superior to the value contracted on the exercise date.

A "real option" is the extension of the concept of financial option for the real assets. The difference is that a financial option is a contract, while the "real option" needs to be identified and specified, so that the internal decisions of investments are made so as to create future opportunities of investment or disinvestment, regarding external events (uncertainties), and addressing project management decisions in order to take advantage of the developed real options. Thus, in a broader sense, " real option" is not just an evaluation model, it is also the bases for the investments process of the organization, since it allows the creation of an aligned vision between the strategy of projects investments and the business strategy of the company, besides integrating project value and risk to the risk analysis of the whole company.

The value of an option basically depends on three factors: the expected value of a variable (result of a phase of a project) for a future decision, the period of time until the decision has to be made, and, finally, the related volatility. The greater the uncertainty or the volatility, the greater the value of the option will be, because it will leverage benefits originated from a satisfactory result and it limits the losses if the result does not cope with the expected value. So, the larger the uncertainty, the greater is the value created by an investment decision that is equivalent to the option of deciding about future investments or disinvestments.

In summary, the evaluation of a real option requests the following information that is applied to the differential equation of Black and Scholes (1973), which granted them the Nobel Prize in 1997:

- The current value of the portfolio of assets that is equivalent to cash incomes/outcomes of the investment.
- The date on which the investment decision will be taken.
- The cost of the investment or the striking price of the investment option (strike price).
- The risk-free rate.
- The volatility of the corresponding assets (the only estimate information).
- Financial returns (e.g., payment of dividends) obtained by the retention of assets.

There are advantages and disadvantages in evaluating based on options. One of the main advantages is the small amount of information to be estimated (just volatility of the assets should be estimated; the other variables are obtained from the market), if compared to the discounted cash flow. The risk is completely reflected in the model through the volatility of the corresponding assets, which indicate the impacts of the variations and uncertainties to the value of the investment in subject. Other relevant points are the restrictive opportunities for subjective evaluations and the fact that the evaluation is mathematically carried out.

On the other hand, a disadvantage of this model is the abstraction degree requested to identify the existing options in the real assets, and to translate the investment in a portfolio of financial assets or group of assets, that that are equivalent to cash incomes/outcomes of the investment of the project and the volatility corresponding to the uncertainties of this investment. It is important to highlight that the traditional models based on projections of cash flows work well for projects where the uncertainty is small or the changes are predictable.

Types of Real Options and a Parallel One with Investments in IT

There are basically five categories of real options that concern the way of evaluating and managing investments.

- *Waiting to invest options.* The value of waiting for an investment is superior to the accomplishment of the investment in the immediate present.
- *Growth options* investments in projects that allow future opportunities of new projects that will affect the growth of the company or of their incomes.
- *Flexibility options.* Investments in projects that allow flexibility in the use of existing assets, regarding external uncertainties.
- *Exit options.* Investments in projects with high uncertainties that tend to present negative results when evaluated through traditional methods; however, if the value of the option of abandoning the project (in case the result is not satisfactory) would be taken into account, the global result of the investment would become attractive.
- *Learning Options.* Typical of phased investments, in which the approval of continuity of the project depends on the result of the previous phase.

The Process for Applying Real Options Model

Amram and Kulatilaka (1999) suggest a four-stage process (Figure 1) for investment decisions through the real options approach. All of the stages are oriented for the finance market, yet the most critical stage is the first, in that the problem is structured and the main aspects of the investment should be considered.

- 1st stage: Problem structuring. Careful description of the decision, the period for

Figure 1. Real options application process (adapted from Amram and Kulatilaka [1999])

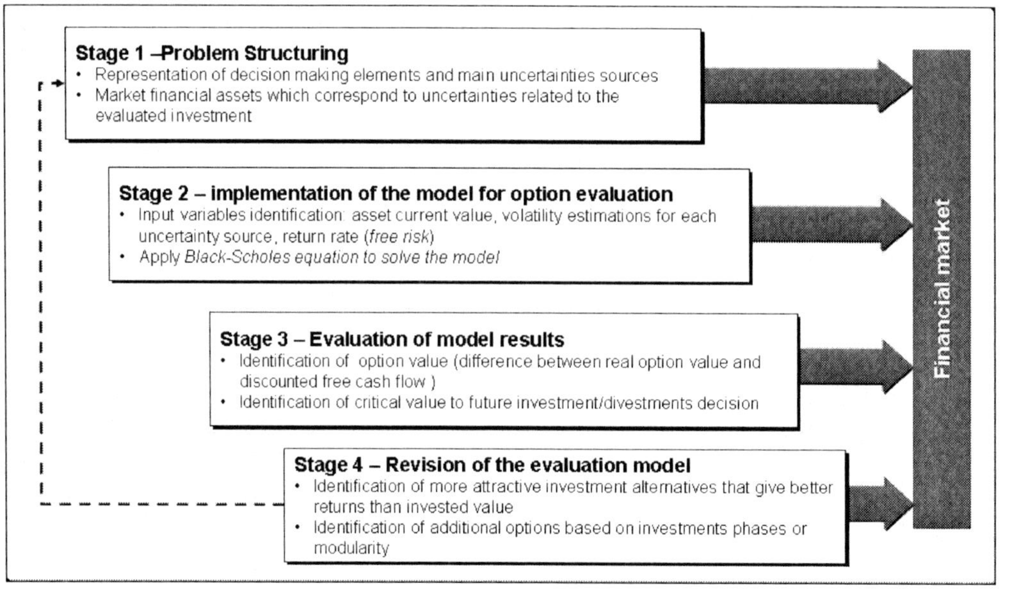

decision, and who will do it. Identification of the sources of uncertainties and also the risks peculiar to the investment, that is, inherent risks to the project and the external uncertainties. Identification, in the finance market, of the group of assets that best responds to the uncertainties of the real option. The model should be comprehensible and simple, reflecting the decision rationale.

- 2nd stage: Implementation of the model for option evaluation. Identification of input variables (i.e., current value of the corresponding assets, volatility of each uncertainty source, cash incomes/outcomes of the investmen,t and risk free rate) and application of the Black-Scholes (1973) equation.
- 3rd stage: Evaluation of model results. Comparison with the result obtained using traditional discounted cash flow, revision of reference values for investment decisions along time, identification of risk profile, and probabilities of giving up or of investing.
- 4th stage. Revision of the evaluation model. Identification if it is possible to expand the alternatives of investments or if the strategy should be redesigned, in order to leverage positive results and to limit negative results.

Real Options and Traditional Evaluation Methods for Project Investments

Craveiro, Severo, and Sbragia (2000) and Cooper, Edgett, and Kleinschmidt (2000) synthesize models and methods to evaluate and to support decision making in project investments, specifically in project portfolio management. Such as:

- Financial Models: These models are usually oriented to value maximization of investments; through the selection of projects the estimated value of which are above a reference value, for instance, net present value (NPV) or economic value added (EVA). These models allow for the evaluating of investments return and its contributions to key performance indicators of a business. The critical points of these models lie on how precisely key variables are estimated and how adherent the hypotheses are toward a market dynamic environment. The positive aspects consist of identifying the most relevant and competitive aspects of an investment in order to ensure its adequate execution and investment returns.
- Strategic Models: These models evaluate the alignment of projects to be invested toward a corporate strategy. The alignment is checked in two dimensions: one evaluates the alignment between project objectives and corporate strategy objectives; another assesses the alignment between investments and organization priorities. The great advantage of this model is its simplicity to ensure that the projects invested are in alignment with a corporate strategy; however, it does not allow comparing of the investments between themselves.
- Scoring Models: They are qualitative models which allow top management to assess and compare different investments. Basically, evaluating criteria are formulated and weights are assigned to typify corporate priorities; finally, each investment is assessed with a score. Definitely, the simplicity and flexibility of these models application are encouraging. Flexibility accounts for choosing different evaluation criteria and simplicity in translating top management knowledge and experiences into a score. Limitations are in the subjectivity aspects of the models, in terms of investments objectives and expected results.
- Balance Models: These models intend to evaluate the whole project portfolio selected to be invested, in order to cross key variables

that support decision making, such as: innovation type vs. project values, risk levels vs. expected benefits vs. success rate, and so forth. The advantage of these models is to simultaneously evaluate strategic alignment, financial assessment, and top management qualitative criteria. Disadvantages rely on the amount of possible analyses that might not lead to a conclusion if not well designed.

Tritle et al. (2000) highlight that models to assess portfolio of projects do not incorporate uncertainties. In addition, the authors suggest that additional analysis should be incorporated to address critical factors related to uncertainties and risks. Cooper, Edgett, and Kleinschmidt (1998) conclude that the most effective organizations in management project portfolio take into account multiple hybrid models, which get commitment among top management.

Basically, Amram and Kulatilaka (1999) criticize financial models that are based on expected cash flow:

- Future estimates of cash flow tend to be extremely optimistic or create an illusion about expected results, hence may become biased and disconnected from reality.
- In general, project management and decision making stick to the original model. Although projects evolve and uncertainties of external variables decrease, investment decisions do not change, hence they broaden the gap between plans and reality. Finally, plans tend to be abandoned and decision making relies upon intuition.

However, it is relevant to remark that traditional financial models work well to assess investments in a project which either presents low level of uncertainties or contains predictable changes. Financial models based on discounted free cash flow help to identify key drivers of revenues, expenses, and investments, as well as allow prioritization and selection of projects and their investments.

Real options link uncertainties and expected results with a timeframe (Figure 2). With the passing of time, uncertainties decrease and ex-

Figure 2. Real options vs. traditional evaluation methods (adapted from Amram and Kulatilaka [1999])

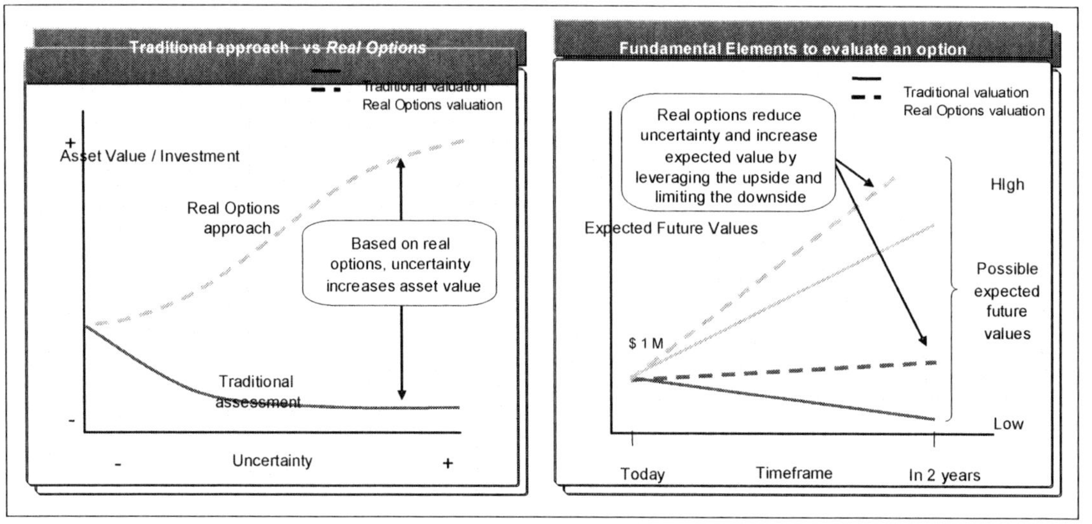

pected values change, hence expected returns on investments change. Therefore, there is a tight link between a model and its reality, which is reflected as the model and the investment project evolve.

Amram and Kulatilaka (1999) state that in industries and projects whose investments are related to future growth, the best model to assess them are the ones that do not fear uncertainties, but leverage the upside and limits the downside, that is, "real options." Some examples are:

- Industries such as pharmaceuticals, high tech (e.g., telecom manufactures, hardware and software), and energy (e.g., gas and oil) invest in technological platforms or in basic research which establish the grounds for future developments; and
- Investments in R&D projects, in IT projects, or in installed capacity, which allow the launching of new products and services, or the offering of new markets.

CONTROVERSIES IN THE EVALUATION OF IT

Looking at the global economy and the value IT has delivered, there is some questioning concerning clear evidences that IT bring significant productivity.

The productivity paradox, widely discussed in the literature, argues about the impact of IT investments on the productivity of the companies. On the one hand, some criticisms, such as Landauer's (1995), advocate that benefits of IT investments were minimal; on the other hand, other authors, such as Brynjolfsson (1993) and Willcoks (1994) highlight that the lack of evidence of benefits from IT investments is related to deficiencies in the evaluation methodologies and metrics of measurement of these impacts.

Brynjolfsson (1993) indicates four elements that led to a misunderstanding of the benefits generated by IT to the businesses: mismeasurement of outputs and inputs (poor evaluation methods), lags due to learning and adjustment (the time necessary to IT benefits can be noted), redistribution and dissipation of profits (increase in the market share of a company, not in the productivity of the industry), and finally, mismanagement of information and technology (poor IT management practices).

In this context, real options can be a method of investment evaluation and project management that can contribute to a better identification and quantification of the results obtained from IT investments.

As Smithson and Hirschheim (1998) state, the evaluation of IT has turned more and more complex and difficult of being accomplished, basically in function of the wide and fundamental role that IT can play in businesses:

- The scope of IS is increasingly wider and more inclusive, involving the whole organization, becoming more complex and extremely integrated, hindering the evaluation of a system separately.
- IT became a basic infrastructure for information and communication of the whole corporation, producing few immediate direct results/benefits to the business (growth option).
- IT is increasingly engendered in companies services and products, making it difficult to evaluate the resulting benefits of IT separately of the business (growth option).
- IT is more and more faced as an element of competitive advantage of the company, where there is a high uncertainty component for the final result, hindering any projection of results or turning them meaningless (exit option).
- IT is one of the components of a wide transformation of businesses that involves products and services lines, processes, systems,

and organizational structure, which hinders the results evaluation process (learning option).

- Traditionally, IT investments were discussed under the approach of systems life cycle, just considering feasibility of the project and post-implementation results, partly because they automated manual activities. Nowadays, most IT investments implicate replacement or evolution of the existing IT platform that requests a more careful evaluation, that is, other aspects should be analyzed, such as costs/benefits of the continuity or replacement of legacy systems, or even the creation of an intermediate solution (e.g., learning option, waiting to invest option).

These elements considerably hinder the application of traditional techniques of economical evaluation for IT investments. However, this work intends to show that, in this situation, real options is a suitable technique that allows the appropriate approach for this problem. Specifically, real options should be applied for selection and management of projects in which it is difficult to quantify benefits obtained with the IT investments; that is, projects that present high uncertainty as for implementation success, or that represent a technological infrastructure/platform, or in projects that result in business transformation or in which elements of competitive advantage of a company are considered.

APPLYING REAL OPTIONS IN THE GOVERNANCE OF IT

Due to the characteristics of real options method, it can be very useful in IT management, not for all types of companies and projects of investments, but for those investments in which a future option to invest or disinvest is created.

Figure 3. Real options applicability according to IT management and system life cycle (elaborated by authors)

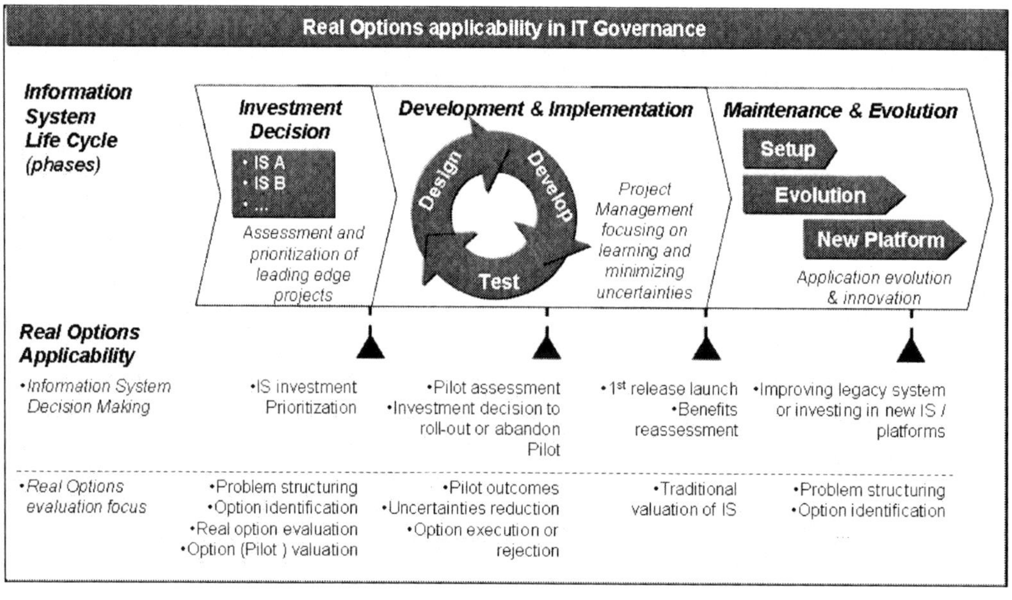

How to Insert Real Options in IT Governance?

Willcoks and Lester (1997) suggest that the deficiencies in IT evaluation derived not only from problems in the metrics used, but also from the absence of an integrated evaluation process. Thus, the real options method cannot just be inserted as one more metric for economical-financial evaluation of investments on IT, but should also be integrated in the decision process and management of IT project investments.

Along the life cycle of an information system, real options can be incorporated in several phases (Figure 3). In the phase of *Investment Decision*, real options can be applied when there are high uncertainties about benefits to the business and a pilot can be implemented to learn with it and reduce uncertainties. Initially, the problem should be structured, future options identified, and their value calculated; according to this, the systems can be prioritized.

In the *Development & Implementation* phase, the real options approach can be applied throughout an IS implementation phased approach, which installs a pilot in order to learn with it and minimize uncertainties. In this case, the strike price of an option is a reference about the amount to be invested in the pilot to learn with it before deciding about the IS roll-out. The strike price represents the amount the management is willing to invest in an IS pilot in order to have the option to decide, in the future, about investing in the IS depending on the pilot outcomes.

Finally, in the *Maintenance & Evolution* phase, it is relevant that the IS portfolio is frequently reevaluated. In this phase of IS life cycle, real options can appropriately support evaluation of the new investments in order to decide about evolving current IS or innovating IS/platforms to support future business strategic objectives.

One example of how to apply real options in IT is illustrated by a LATAM telecom operator, the telco operator intended to improve and also to innovate its Web presence by enabling an *avatar* that would interact with customers in a personalized way in order to support self-caring and buying activities. The direct impact of this *avatar* in sales goals and in migrating customers to self-caring was not clear. A traditional IT or financial evaluation method would not appropriately assess the potential of this Web technology. Instead of declining the idea upfront, the top management decided to create a real option and launched a pilot, which would apply the *avatar* in few areas of the Web site to support some specific activities. The decision of investing in a pilot allowed the telco company to learn and minimize uncertainties in terms of customers adoption for avatar support and its business impact, in order to decide later to or not to invest in the solution, without compromising the budget.

It is relevant to identify characteristics of IT governance in this telco operator for this particular project. IT almost always plays a strategic role regarding present and future impacts in telecom business. In this example, a business strategy required both a technological vision and a new Web architecture to be implemented. Finally, in this particular case, business and IT areas handled a common responsibility in this project, though with very loose budgeting controls.

Under these circumstances, the establishment of the real options method in IT governance can contribute for effectiveness and efficiency of IT. Considering effectiveness, it will allow the achievement of IT objectives, that is, it will support decisions about IT investments in terms of "the right thing to be done." Considering efficiency, it will integrate investment decisions to IS management, guiding the implementation of IS with the rationale identified during the option identification, in order to learn during the IS life cycle, minimize uncertainties, and "do things right."

Under which Conditions to Apply Real Options in IT Governance?

Besides being a complex technique for evaluating investments, real options should not always be used to base the decision process for IT investments. Considering the levels of evaluation of IT proposed by Laurindo and Shimizu (2000), it is possible to analyze the conditions in which the use of real options is adapted to the evaluation of these IT investments.

IT Relevance to a Company Strategic Competitiveness

Considering the strategic grid of McFarlan (1984) that classifies companies regarding present and future impacts of IT in the businesses, it can be inferred that real options is quite appropriate to the companies classified in the *strategic* and *turnaround* grids. In companies positioned in the *strategic grid*, on which IT has great influence on present and future business results, investments tend to generate opportunities of future incomes (growth option), and many projects are related to leading edge technologies or innovative solutions that naturally bring uncertainties for implementing and/or operating them properly (learning option). In companies positioned in the *turnaround grid*, in which IT can create a foundation for future competitive transformations of the company, real options can be applied in projects that establish a platform for future business developments/innovations, or in projects that support a company to face significant changes in industry competitiveness. In companies positioned in the *factory grid*, IT investments neither represent a future option of business nor future improvement of its efficiency, so IT investments should be evaluated through traditional techniques (e.g., NPV, payback, etc.).

Figure 4. Application of real options according to IT strategic positioning (adapted from McFarlan [1984])

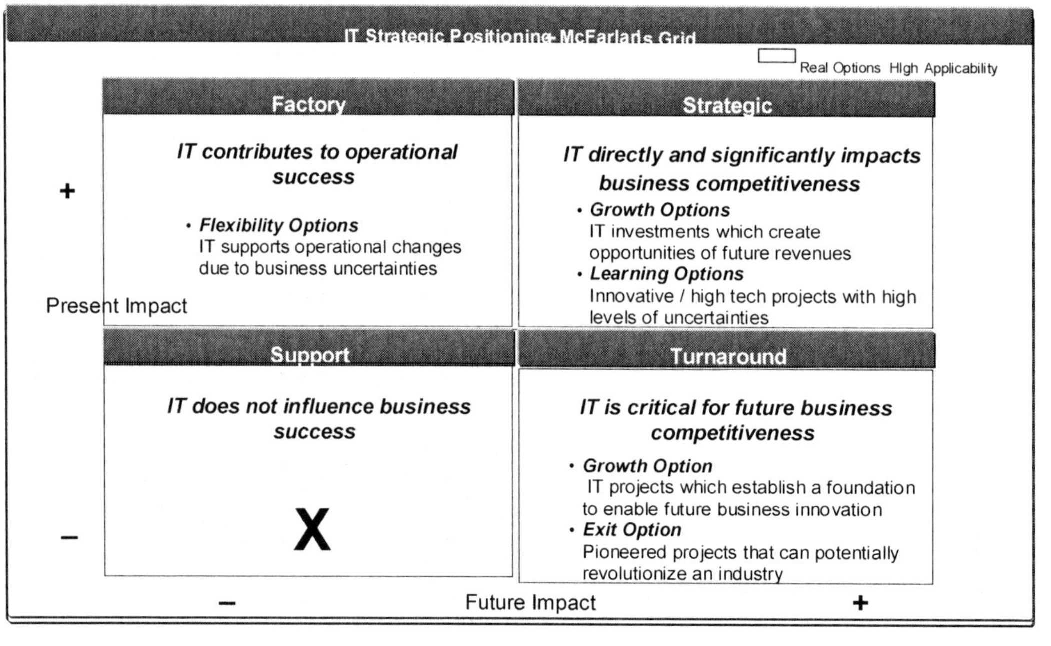

Finally, in companies positioned in the *support grid*, IT impact in businesses does not involve uncertainties or future opportunities; thus real options should not be applied (Figure 4).

IT Strategic Alignment within a Company

From the point of view of the strategic alignment model proposed by Henderson and Venkatraman (1993), there should be an alignment between strategy and structure of businesses and of IT. It is possible to use the real options approach in cases in which IT alignment is related to *technological transformation* or *competitive potential* (Figure 5). In the first case, *technological transformation,* regarding the innovation of a businesses strategy, IT has to envision a new architecture in terms of applications and/or infrastructure in order to support a business innovation. In the second case, *competitive potential,* the application of leading edge technologies may lead to the development of IT competences and knowledge, which, combined with a visionary attitude from top management, may create new business opportunities. Consequently, in both cases, there are high levels of uncertainties; therefore, decision process and project management may request an approach to deal with those uncertainties appropriately, leveraging possible opportunities instead of avoiding to invest in them.

IT Organizational Evolving Phase

Taking into account the IT organizational framework proposed by Nolan (1979) and the contribution of real options in assessing IT investments (Figure 6), real options can be an appropriate evaluation method when a company has a formal IT planning and control process (phase 3 - *control*). It would even be more useful when businesses and IT areas jointly accept the responsibilities about IT investments and consider IT as part of the strategic planning (phase 6 - *Maturity*). It does not

Figure 5. Real options applicability according to IT Strategic Alignment (adapted from Henderson and Venkatraman [1993])

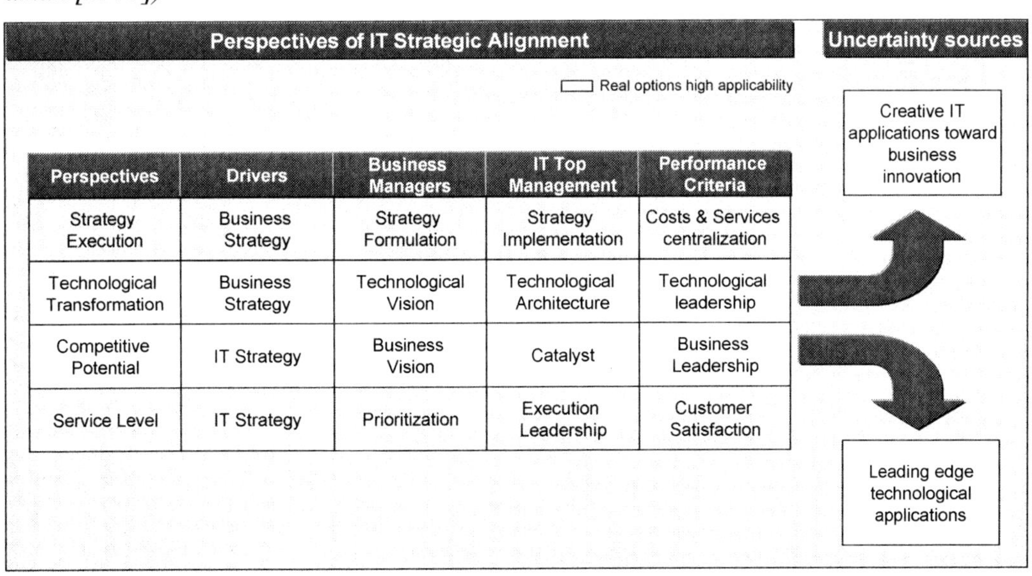

Figure 6. Real options applicability according to IT organizational evolving phase (adapted from Nolan [1979])

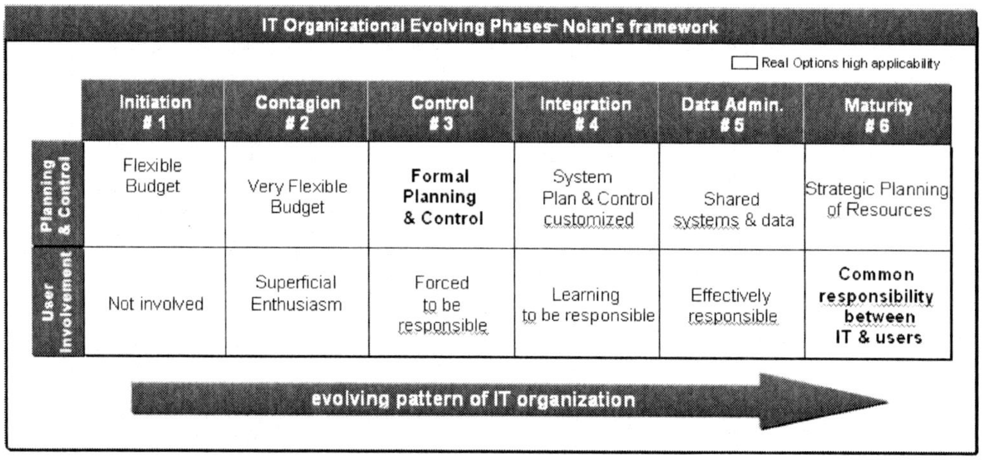

Figure 7. Decision tree for real options application (elaborated by authors)

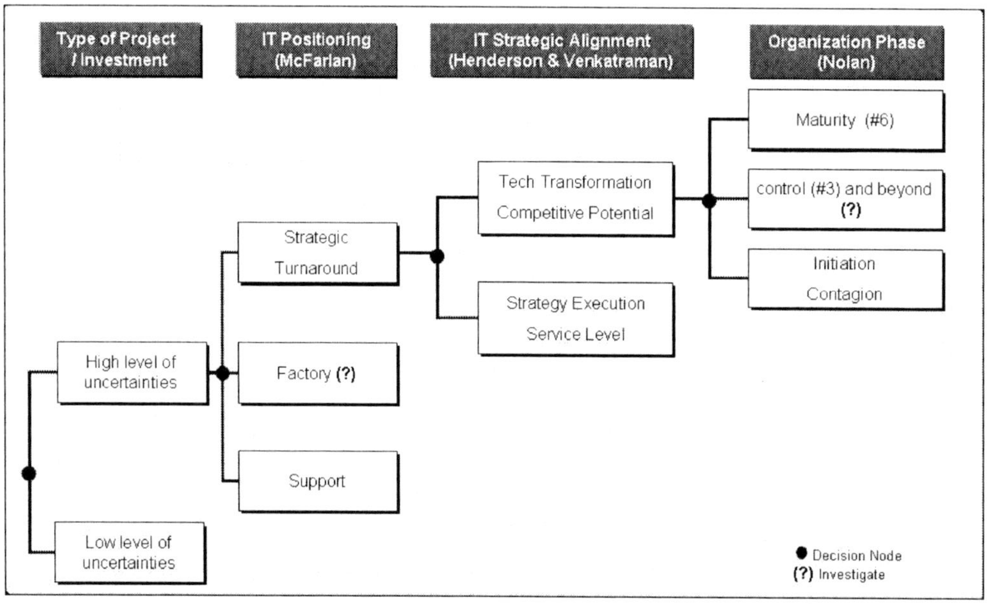

mean that real options is not relevant in *initiation* and *contagion* phases; however, it is less likely that investors identify benefits of IT application, since budget practices do not concern IT results. In addition, it does not mean that real options is always applicable as an evaluation technique for companies organizations located in Phases 3 to 6, because investments under evaluation should represent a future option for other sources of incomes or embed uncertainties that can be reduced with knowledge and know-how acquisition during a project execution.

Summing up, according to some characteristics of IT governance there is an indication of the likelihood of applying real options to support IT investment decision making, according to a decision tree presented in Figure 7.

One possible example of an IT project that would benefit from a real options approach was the case of the Web portal of Companhia Vale do Rio Doce (CVRD), a huge Brazilian mining company that operates round the world (Guimarães, 2001; Paduan, 2001).

At a first moment, CVRD decided to launch a portal for selling minerals in a joint action with 13 large international mining companies that represent 60% of the global mining companies. Operating in the Internet was a new situation for CVRD, thus, it was critical to share the risks with other firms.

At a second moment, this portal allowed CVRD to sell logistics services, which started to play a pivotal role in its business portfolio.

A conventional approach for evaluation could not capture the possibilities that the portal would bring to CVRD business strategy. The real option method would be a way of translating the new business scope brought by the technology in a structured decision process.

Regarding CVRD and the case of its Web portal, one can identify the conditions of which applying real option makes sense. Around 2000, Web portals benefits were not clear, there were uncertainties in terms of how to set up a Web presence, create and capture value in mining companies; IT tends to be positioned in the turnaround grid or even factory grid. Around 2000, Web technologies, competences, and knowledge were enabling *technological transformations* and CVRD, as an open capital company, already had formal IT planning and control. Therefore, it sounds reasonable to evaluate the mining Web portal for CVRD within a real option approach; the crucial aspects are to structure the evaluation and to identify and to assess the real options embedded under this investment.

FUTURE TRENDS

In the last years, it has been possible to notice some changes regarding the main uncertainties and sources of innovation in Internet applications and business Web strategies. At a first moment, the adoption of the new technology was innovative, new businesses models which combined bricks and clicks were emerging, and results were quite uncertain.

Nowadays the adoption of Internet is quite ubiquitous and, therefore, is not an innovation in itself. Thus, new aspects become both sources of innovation and of uncertainties: the media for using Internet, new business models that seek to create and extract value from communities and from user generated content, and new forms of interaction among people and companies.

In the business-to-consumer (B2C) arena, companies are consciously creating real options to learn or decide later on to or not to invest in virtual worlds and communities. Some significant examples in virtual worlds are the rush of companies to build a presence in Second Life, a place where visitors create virtual personalities willing to interact in a social and/or commercial sphere. In this case, the investment may be structured as a *waiting to invest* option, in which without compromising a big amount of investment a company ensures its stake in the environment. However, there are companies that are immediately launching their own virtual environment without exactly knowing how to exploit it and how to extract value from it, which characterizes a growth option that is an investment that will allow future revenues opportunities.

Other examples of communities building are related to marketers and advertisers engaging in exploring the popularity of social-networking Web sites such as MySpace, Facebook, and Orkut. In addition, some companies are pushing for creating communities in order to promote and commercialize user-generated content such as video, music, advertising, and so forth. In both cases, a real option approach could limit the downside of an investment or guide the management of these new endeavors in order to reduce uncertainties and decide upon further investments.

In addition, the convergence among telecom, media, and computing industries enabled by the IP protocol brings up a new stream of innovation in terms of products and services. WiFi and broadband usage are changing the reach and richness of Internet applications, as well as the competitive landscape. Meanwhile, companies are adopting new structures and organization, and deciding about new investments in order to develop new business models and sustain profitability within a digital convergent environment.

In the business-to-business (B2B) and supply chain management (SCM) arena, IT applications and new business models have enhanced relationships among companies and customers. In fact, nowadays, big companies competing in the global market extensively use SCM and B2B software in their daily operations. Actually, they depend on these tools to perform their activities. Therefore, the use of these Internet applications is a requirement or a qualifying criterion for these companies. Consequently, many of the risks that companies faced in the past during their decision-making processes regarding the adoption of Internet-based business models are nowadays greatly reduced. However, these companies should seek other sources of innovation based on the use of IT applications in order to obtain competitive advantages.

These aspects should be studied in new researches to be developed in the future.

CONCLUSION

The approach of real option can significantly contribute to IT management, inserting a new point of view to investments decisions and management of projects implementation, in order to effectively contribute with IT function, particularly in investment in infrastructure projects, new technologies, or innovative solutions in the use of IT to the businesses.

Considering companies effectiveness, application of real options collaborates in the sense that it avoids the immediate rejection of investment in projects with high uncertainty, cases that would be discarded through the application of traditional techniques for evaluation investments. Additionally, this approach can change IT managers' mentality, forcing them to identify options in future investments, that is, conceive projects with the inclusion of mechanisms that minimize losses and leverage results in function of the high inherent uncertainty to the investment.

In relation to the "productivity paradox," the proposal in this chapter can also be very useful. First, it emphasizes new metrics in the evaluation of IT impacts on businesses, metrics that monitor the learning of the organization in relation to the inherent uncertainties to these projects. Second, the methodology of real option is perfectly adapted to evaluating benefits of projects that constitute an option for future investments that will bring direct benefits to the business.

However, the present work has some limitations. For instance, it should apply this proposal to some actual cases of IT investments, testing the decision tree for the application of real option, pointing out the differences of the proposed method to the traditional methods of economic-financial evaluation, and, finally, indicating the implications to IT project management. Other points to be analyzed are related to the applicability of the methodology to IT area vs. IT managers' maturity, or even to the degree of development of the specific stock market of the company (element that bases the uncertainty evaluation in the financial and real option) and of the corporate management.

In spite of the criticisms above, this work illustrates a way of achieving IT effectiveness through organizational dimension and incorporating real option to the investment decision process.

REFERENCES

Amram, M., & Kulatilaka, N. (1999). *Real option, managing strategic investment in an uncertain world.* Cambridge: H.B.S Press.

Benko, C., & Mcfarlan, F. W. (2003). *Connecting the dots* (p. 246). Boston: Harvard Business School Press.

Black, F., & Scholes, M. (1973). The pricing of options and corporate liabilities. *Journal of Political Economy, 81*, 637-659.

Broadbent, M., & Weil, P. (1997, Spring). Management by Maxim: How business and IT managers can create IT infrastructures. *Sloan Management Review, 38*(3), 77-92.

Brynjolfsson, E. (1993). The productivity paradox of information technology. *Communications of the ACM, 36*, 67-77.

Brynjolfsson, E., & Hitt, L. M. (1998, August). Beyond the productivity paradox: Computers are the catalyst for bigger changes. *Communications of the ACM.*

Burn, J. M., & Szeto, C. (1999, September). A comparison of the views of business and IT management on success factors for strategic alignment. *Information & Management, 37*, 197-216.

Byrd, T. A., & Marshall, T. T. (1997). Relating information technology investment to organizational performance: A causal model analysis. *Omega, International Journal of Management Science, 25*(1), 43-56.

Campbell, J. A. (2002). Real options analysis of the timing of IS investment decisions. *Information & Management, 39*(5), 337-344.

Cooper, R. G., Edgett, S. J., & Kleinschmidt, E. J. (1998, July-August). Best practice for managing R&D portfolio. *Research Technology Management.*

Cooper, R.G., Edgett, S. J., & Kleinschmidt, E. J. (2000, March-April) New problems, new solutions: Making portfolio management more effective. *Research Technology Management.*

Craveiro, A. M., Severo, A. C. R., & Sbragia, R. A. (2000, November). Priorização de projetos e avaliação de portfolio. In *Proceedings of XXI Simpósio de Gestão da Inovação Tecnológica,* São Paulo.

Earl, M. J., Rockart, J. F., & Ross, J. W. (1996, Fall). Eight imperatives for the new IT organization. *Sloan Management Review, 38*(1), 43-55.

Farbey, B., Land, F. F., & Targett, D. A. (1995). Taxonomy of information systems applications: The benefits evaluation ladder. *European Journal of Information Systems, 4*(1), 41-50.

Farrell, D. (2003, October). The real new economy. *Harvard Business Review, 81*(10), 104-112.

Guimarães, C. (2001, February). Vale tudo. *Revista Exame, 733.*

Henderson, J. C., & Venkatraman, N. (1993). Strategic alignment: Leveraging information technology for transforming organizations. *IBM System Journal, 32*(1), 4-16.

Kaplan, R. S., & Norton, D. P. (1992, January/February). The balanced scorecard: Measures that drive performance. *Harvard Business Review, 70*(1), 71-79.

Landauer, T. (1995). *The trouble with computers: Usefulness, usability and productivity.* Cambridge: MIT Press.

Laurindo, F. J. B. (2002). *Tecnologia da informação: Eficácia nas organizações* (p. 248). São Paulo.

Laurindo, F. J. B., & Shimizu, T. (2000). *Evaluating strategies in information technology.* Paper presented at the Performance Measurement 2000

Conference - Past, Present and Future (pp. 323-330). Cambridge: Inglaterra.

Li, M., & Ye, L. R. (1999, January). Information technology and firm performance: Linking with environmental, strategic and managerial contexts. *Information & Management, 35*(1), 43-51.

Luftman, J. N. (1996). Applying the strategic alignment model. In J. N. Luftman (Ed.), *Competing in the information age: Strategic alignment in practice* (pp. 43-69). New York: Oxford University Press.

Maggiolini, P. (1981). *Costi e benefici di un sistema informativo.* Itália: Etas Libri.

Mahmood, M. A. (1993). Associating organizational strategic performance with information technology investment: An exploratory research. *European Journal of Information Systems, 2*(3), 185-200.

McFarlan, W. E. (1984). Information technology changes the way you compete. *Harvard Business Review, 62*(3), 98-103.

McFarlan, W. F. (1990, Summer). The 1990s: The information decade. *Business Quarterly, 55*(1), 73-79.

Nolan, R. L. (1979). Managing the crises in data processing. *Harvard Business Review, 57*(2), 115-126.

Paduan, R. (2001, June). Como a Internet está transformando (de verdade) a vida nas empresas. *Revista Exame, 742.*

Porter, M. E. (1979, March-April). How competitive forces shape strategy. *Harvard Business Review, 57*(2), 81-92.

Porter, M. E. (2001, March). Strategy and the Internet. *Harvard Business Review, 79*(3), 63-78.

Porter, M. E., & Millar, V. (1985, July-August). How information gives you competitive advantage: The information revolution is transforming the nature of competition. *Harvard Business Review, 63*(4), 149-160.

Prahalad, C. K., & Hamel, G. (1990, May-June). The core competencies of the corporation. *Harvard Business Review, 68*(3), 79-90.

Renkema, T. J. W., & Berghout, E. W. (1997). Methodologies for information system investment evaluation at the proposal stage: A comparative view, *Information and Software Technology, 39*(1), 1-13.

Rockart, J. F. (1979, March-April). Chief executives define their own data needs. *Harvard Business Review, 57*(2), 81-92.

Smithson, S., & Hirschheim, R. (1998). Analyzing information systems evaluation: Another look at an old problem. *European Journal of Information System, 7,* 158-174.

Stratopoulos, T., & Dehning, B. (2000). Does successful investment in information technology solve the productivity paradox? *Information & Management, 38*(2), 103-117.

Tapscott, D. (2001). Rethinking strategy in a networked world. *Strategy + Business, 24,* 8.

Turban, E., McLean, E., & Wetherbe, J. (2002). *Information technology for management: Transforming business in the digital economy* (3rd ed.). New York: John Wiley and Sons Inc.

Turban, E., Rainer, R., & Potter, R. (2003). *Introduction to information technology* (3rd ed.). New York: John Wiley & Sons, Inc.

Walrad, C., & Moss, E. (1993) Measurement: The key to application development quality. *IBM Systems Journal, 32*(3), 445-460.

Wetherbe, J. C. (1979). *Systems analysis for computer-based information systems.* St. Paul: West Publishing CO.

Willcoks, L. (1994). *Information management: Evaluation of information systems investments.* London: Chapman and Hall.

Willcoks, L. P., & Lester, S. (1997). In search of information technology productivity: Assessment issues. *Journal of the Operational Society, 48,* 1082-1094.

Chapter XVII
Strategic Use of the Internet and Organizational Structure for E-Business:
"Celta" Case at GM Brazil

Sílvia Novaes Zilber
UNINOVE, Brazil

ABSTRACT

The Internet provides a global network infrastructure that is shifting business models, strategies, and processes. Many authors reflect on the importance of incorporating e-business into the firm's global strategy. This chapter deals with these issues in discussing the introduction of e-business activities by General Motors Brazil, specifically in connection with the launch of the Celta car, an entry-level car designed to be sold on the Internet. A historical examination of e-business strategy shows that many organizations have formulated excellent conceptual strategies for e-business but failed to deliver sound execution. A key to successful Internet strategies is the leadership shown by senior management. Technological demands may also conflict with the successful implementation of e-business initiatives, requiring greater interaction between the CEO and CIO. The organizational structure implemented for the launching and sales of Celta car warranted the integration between employees on the business side and in IT in the context of GM Brazil's strategic objective of growing the market share for lower-priced cars.

INTRODUCTION

According to a study by McKinsey (2003), in the next 10 years, the automotive industry will be shattered by a third "revolution," following the invention of the assembly-line production by Henry Ford and the lean production of Toyota. Customers expect "more car" for the same money, which means continuous cost pressure and innovation for OEMs. This fact leads to a range of transformations in the structure of the automotive supply chain. For example, in order to improve customer satisfaction and increase revenue growth and shareholder value, large OEMs and their suppliers are forced to build large automotive networks.

Cars are developed and manufactured by OEMs and their supplier networks, which produce as much as 70% of the value of a vehicle. Consequently, the cost and quality of a vehicle are a function of the productivity of a network of firms working in collaboration. As far as the tendencies in the **automotive industry** are concerned, Benko and McFarlan (2003) explain that the automotive industry is transforming itself, from the design studios of Germany to the modular assembly of cars in Brazil to the new relationships with suppliers in Detroit to the factory floors of Japan. Three trends stand out: (1) the manufacturer's changing relationship with its customers; (2) new partnerships with suppliers; and (3) the reinvention of the factory floor. They represent a convergence of progressive thinking from major players around the globe.

Perhaps the biggest change in the auto industry is its **relationship with its customers**. Increasingly, customers are being given the opportunity to customize their purchases, including everything from the exterior shape to the interior features. In addition, as manufacturers and dealers build stronger relationships with their customers and better understand their needs, they can personalize their interactions with them. For both the manufacturer and the dealer, this opportunity is substantial.

The **Internet** has been a driver of change in business relationships. **Electronic business** enables consumers to interact directly with corporate information systems through the public infrastructure of the Web. The interconnectivity and interactive nature of the Internet make it a unique medium in a strategic context that differs from previous applications of information technology to business communications, such as electronic data interchange (EDI), which in its traditional form is based on rigid standards of information exchange over private networks between preexisting business partners.

According to Dutta and Biren (2001), the manufacturing sector, dominated by the **automotive industry**, has embraced the Internet as a new medium for growth and customer expansion. Customers purchasing a new vehicle spend a large amount of time researching the various models and making price comparisons among similar companies. Typically, approaching a car dealer to inquire about car features, or worse to negotiate price, is viewed as an unfavorable, often dreaded, experience. By providing consumers with the ability to shop and even finance a new purchase online (excluding actual signing of the final papers), the companies have managed to turn purchasing a car into a fun and interesting experience.

In this context, the Internet is able to provide an enormous amount of information about the client to the industry: persuasive evidence has described the strategic use of information resources in organizations; information systems are strategic to the extent that they support a firm business strategy.

Strategy can be defined as a "quest to match a firm's resources and capabilities to the opportunities and risks created by its external environment" (Grant, 1991). During the 1980s the dominant Porterian *competitive forces* approach (Porter, 1980) emphasized the relation of a company to its external environment, while more recently the resource-based view (RBV) highlighted the need to consider the firm's internal resources and capabilities. Barney (1991) adds that for resources to create sustained competitive advantage they must be valuable, rare, imperfectly imitable, and not strategically substitutable.

Strategic use of IS and related information technologies, such as Internet, can impact organizational-level variables such as entry barriers, suppliers and customers, industry rivalry, search and switching costs, and intra- and inter-organizational efficiency (Bakos &Treacy, 1986; Barua, Konana, Whinston, & Yin, 2004; Mahmood & Soon, 1991; Porter & Miller, 1985).

In this context, Internet and **e-business** emerge as a source of competitive advantage creation. Its emergence created opportunities for two types of firms: so-called "Internet pure plays" or start-ups, that is, firms that were born on the Internet and that had never had facilities of any kind in the physical world and they operate online; and conventional organizations, many of which saw a chance of increasing market share by expanding into e-business.

This chapter focuses on the latter category, discussing the specific case of an automotive manufacturer **(GM Brazil)** that used the Internet to achieve a strategic goal, that is, the growing of its entry-level car market share.

In this research, we address the results of **e-business** adoption within the domain of Internet enabled business initiatives.

When **General Motors Brazil** decided to increase its market share for entry-level cars, it opted for e-business as a means of making this possible; selling cars directly to consumers via the Internet in what is commonly called **B2C**, or business-to-consumer marketing.

This chapter has the objective of evaluating the applicability of a model of e-business operations in a company that has long operated in the physical world ("brick-and-mortar" company) and that has implemented business activities using the Internet.

The questions arose from the use of this e-business model are explained below:

*How has **electronic business** addressed a market need? Which **organizational structure** was used to implement the **e-business** activities?*

The case chosen to investigate these questions was the launching of **Celta** car by **GM Brazil.** This company was able to opt for direct marketing of its new entry-level car model, the Celta, without dealers or other intermediaries for two reasons. First, it had a flexible manufacturing facility with assembly lines configured to allow small modifications in accordance with a certain amount of customization and configurations selected online by the consumer. The second key facilitator of **direct sales** was the Internet.

The key management-related factors were the commitment of senior executives to the idea (which actually came from the marketing director of GM Brazil), and the existence of a clearly defined business model on which it was based the strategy, coupled with an organizational structure tailored to the pursuit of this objective.

LITERATURE REVIEW

E-Business as a Strategic Tool

Contemporary organizations are aligning IS and business strategies to improve organizational performance (Kearns & Lederer, 2000).

Charles Schwab and Amazon.com have leveraged existing processes with online access and employed a first-mover strategy to realize record online revenues.

Dell Computer, in a first–mover strategy, has achieved success by aligning its core competency in supply-chain expertise with Internet technology.

In this context, the Internet and e-business emerge as a source of competitive advantage.

According to Turban (2000), e-business is the delivery of information, products and services, or payments via telephone lines, computer networks, or other electronic means. Cunningham (2001) defines e-business as commercial transactions conducted over public or private networks, including public and private transactions that use the Internet as the means of implementation. These transactions include funds transfers, online exchanges, auctions, product and service distribution, supply chain activities, and integrated corporate networks.

This new way of doing business produces instruments that enable many things to be done, from developing Internet portals for the buying and selling of products and services to supply chain management, customer relationship management (CRM), and even integration of different business units within the same organization via Intranets (i.e., using Internet technology in enterprise mode so as to restrict access to members of the organization via a firewall).

According to Barua et al. (2004) any customer with access to the Internet is able to gather information interactively regardless of time and location, (possibly) customize and order products/services, change orders dynamically, check order status, and seek online advice.

In theory, according to a study by Freeland and Stirton (2000) for The Boston Consulting Group, many established companies with solid experience in the "real world" or traditional "brick-and-mortar" companies are well positioned to succeed in e-commerce. They have critical assets, such as strong brands, customer relationships, and logistics systems that can give them an edge over startup competitors. In practice, however, traditional bricks-and-mortar firms will not be able to exploit these assets unless they are effectively organized for e-business. Indeed, for large firms the most difficult challenges of e-business are not so much strategic as organizational.

During the process of migration to the Internet by traditional companies, they will face a number of challenges such as corporate culture and leadership, channel conflicts (bricks-and-mortar vs. online), cost savings, the need to find suitable professionals to work in e-business, and the need to invest in IT infrastructure. In tackling these challenges, firms have to consider and experiment with a "portfolio" of potential e-business initiatives. The literature presents several business models designed to assist them in this migration process and evaluate the results.

According to Applegate (2001), a business model describes succinctly how the business is structured, what kind of people is needed for that business, and what roles they perform. Thus the description of a firm's e-business activities in its business model facilitates the analysis of this business structure and of the roles people play in it.

The model proposed by Applegate (2001) was used for this study, a model that consists of three components: business concept, capabilities, and value.

Chesbrough and Rosenbloom (2001) stresses that the creation of a business model differs from the conventional notion of developing a strategy, since a business model is more than an attempt to hypothesize an exploratory initiative within a given market, rather, it is a fully worked out, well-defined plan of action.

Most large corporations have not one but many e-business initiatives. In such cases a business model can be useful to coordinate e-business activities, set out their overall goals, and serve as a framework for evaluating their results, particularly measuring whether the results match the goals set in the model.

As for aspects relating to organizational structure, according to Oliveira (2001), it describes the arrangement and grouping of activities and resources to achieve the established goals and results.

When a firm decides to implement e-business as an additional activity it will have an impact on existing processes and operations. According to Freeland and Stirton (2000) in their analysis of a study by The Boston Consulting Group on organizing for e-business, the most difficult challenges are organizational rather than strategic. The success of a company's e-business strategy depends on its ability to organize appropriately, these authors conclude. In many situations a sound strategy founders on organizational problems. More and more large firms are focusing on getting their online organizations up and running but many of them neglect to make the changes and linkages to their offline businesses that are necessary for success in both.

One of the factors that make e-business different from traditional business is the need for convergence, that is, convergence in technical platforms and convergence in business proper. The possible organizational arrangements that can be adopted when implementing e-business depend on broader issues, such as what the company aims to do with its online business. In other words, how can e-business help the firm achieve its overall objectives, as established in its general business strategy? Thus it is important to bear in mind that the company's overall strategy and global objectives are the foundation on which its initiatives are based on every direction, including the approach adopted when implementing e-business.

Several authors (e.g., Kalakota & Robinson, 1999; Lientz & Rea, 2001; Plant, 2000; Robert & Racine, 2001; Tapscott, Lowy, & Ticoll, 1998; Turban et al., 2000) stress the importance of integrating the firm's objectives and global strategy with the e-business design adopted. Success in e-business requires that investment in technological infrastructure for e-business be linked to the organization's plans, strategies, and tactics. The organization must be clear about all the requirements for success in accomplishing its goals in both online and offline operations.

The key strategic issues that should be raised by top management include: How can we transform the old firm into a new one? How can we build a bridge to fill the capability gap between the physical needs of today and the digital needs of tomorrow?

According to Plant (2000), successful implementation of e-business by a firm derives from a business model that entails every content-providing area of the organization contributing to the global business model. In terms of organizational structure, any firm engaged in e-business contains various content owners, that is, members of the organization who deliver the information posted on the Web site or whose functions are affected by the site and its use (e.g., logistics, purchasing, services).

Traditional firms with a predominance of horizontal command and control structures are not suitable for online operations, which require a far more adaptive structure where the functional head of e-business is central. The author suggests a Web-centered organizational structure, in which the head of e-business or e-commerce is well-served by having close and agile relationships with other management groups.

In this structure there is a Vice President for E-Commerce with several subordinates who provide content relating to their respective areas. These are the content owners mentioned above. The head of IT for e-commerce does not need to have a technical background. This area often needs to be run by a creative person rather than a technician.

Another solution is to outsource e-business activities, keeping minimal in-house staff and naming e-commerce groups in the contractor as needed. The profile of the people involved in e-commerce should combine technical aspects and aspects related to business management.

As for the question of governance, Freeland and Stirton (2000) suggest the creation of a small e-commerce center with decision-making authority within the organizational structure. In the early phases of e-commerce, many companies had several business or functional units pursuing e-commerce strategies with relatively little coordination, they note. The approach was to "let a thousand flowers bloom" on the theory that this was the best way to encourage experimentation, learning, and fast responses to new competitive threats. Now, however, many companies have discovered that this decentralization leads to waste and duplication and may also compromise the effectiveness of their e-commerce efforts. Thus, as the authors found in their study of hundreds of large firms in several sectors, there is a tendency to create a small but powerful central unit to coordinate e-commerce, led by a senior executive who reports directly to the CEO.

Barua et al. (2004) propose a model of business value for Internet-enabled business transformation that suggests that Internet-enabled business performance is ultimately judged by traditional financial performance measures such as revenue per employee, gross profit margin, return on assets, and return on invested capital. Further, they posit that the improved financial performance is a result of day-to-day excellence in interactions with customers, suppliers, and other business partners.

Operational excellence measures (e.g., % online business, % online procurement, % customers service provided online, etc.) are "intermediate" level performance measures that are conceptually similar to critical success factors. The strength of the relationship between operational excellence and financial performance may depend on factors such as channel conflict, competitive response, and economic conditions that are beyond the scope of their study.

RESEARCH METHDOLOGY

The present study can be considered as having an exploratory nature due to the contemporary character of the phenomenon studied and because of the limited amount of academic knowledge accumulated on the topic to date. Therefore, resulting from the genre of the research undertaken, there is no concern here in establishing relations between dependent and independent variables in order to prove or disprove predetermined hypotheses.

Exploratory studies are conducted with the objective of providing overviews of an approximate nature regarding a given phenomenon. The present research was based on a case study method that focused on the launching of the **Celta**, a small and low cost car produced by GM Brazil and studied in depth by Zilber and Vasconcellos (2005). According to Yin (1990) the case study is the preferred strategy when the questions are presented in the form of "How?" or "Why?," which is the present case, where the questions are: "How is the e-business structured?" and "Why was a given model adopted for the e-business operations?"

It was decided to adopt the business model proposed by Applegate (2001) for studying the issues proposed in this study. That author sums up the importance of business models with the following chart showing the building blocks of a business model and the relationships among them.

The categories of analysis for each component of the model are shown below:

An organization's ***business concept*** defines its market opportunities, products and services

Figure 1. Components of a business model

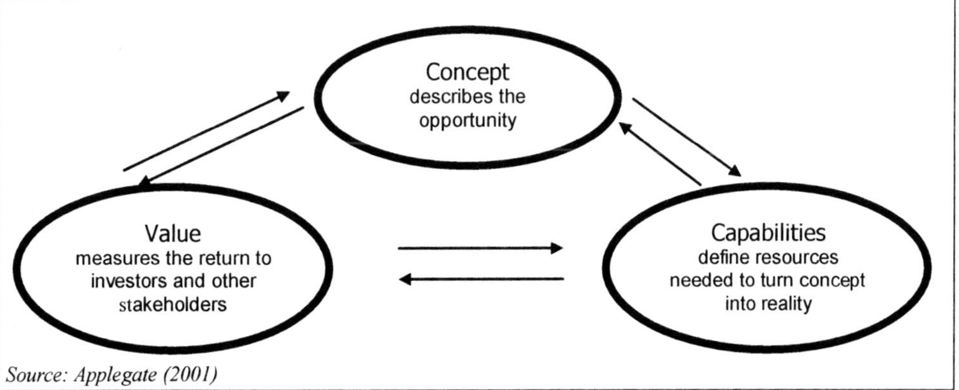

Source: Applegate (2001)

offered, competitive dynamics, strategy for capturing a dominant position, and strategic options for evolving the business.

An organization's ***capabilities*** are built and delivered through its people and partners, organization and culture, operating model, marketing/sales model, management model, business development model, and infrastructure model.

Value is measured by benefits returned to all stakeholders, benefits returned to the firm, market share and performance, brand and reputation, and financial performance.

The factors selected from this model of study were:

- Within the ***"business concept"*** component: *market opportunities, competitive dynamics, and strategic options to lever business*
- Within the ***"capabilities"*** component: ***organizational structure*** *and marketing/sales model*
- Within the ***"value"*** component*: financial result*

The case chosen to be studied in depth was that of General Motors Brazil and, more specifically, its launching of the Celta, a lower-priced automobile. This choice was made, in the first place, because this car was sold entirely over the Internet, which constitutes a case of the use of the Internet in a strategic way for carrying out the company's business.

Managers and directors of GM were interviewed whose activities had been affected by e-business operations. They were persons responsible for IT areas in the company, as well as of the areas of marketing, purchasing, sales, and the e-business area, defined as such. These interviews were based on a semistructured guideline, drawn up with the purpose of obtaining replies to the questions being studied.

CASE PRESENTATION: THE LAUNCHING OF THE CELTA BY GM BRAZIL

The Role of the Company's Global Strategy in Defining the E-Business Model

There are two major blocks of e-business operations at GM Brazil: a) B2C operations, which involve direct sale of cars on the Internet, including the case described in this chapter, the launching of the Celta, a lower-priced car that was released with sales only on the Internet; and b) business-to-business (B2B) operations, which involve the company's purchasing area and its relationships with suppliers.

There is no business coordinator for all e-business operations. Therefore, B2C activities are coordinated by an area subordinated to the marketing department, while B2B operations are subordinated to the purchasing area, with no specific coordination. It is interesting to note that, right from the conception phase, the sale of the Celta on the Internet was an idea generated by the company's senior marketing board, with backing from upper management. An organizational structure dedicated to this project was created, which, even after having been concluded, continued to operate by providing maintenance to the routine e-business operations related to B2C. This new area was also responsible for generating new ideas.

But the B2B operations were not centralized in a single dedicated organizational structure. They were inserted into the existing organizational structure. Perhaps for this reason, they failed to attain the same financial results as the B2C. Therefore, there is an e-business management totally dedicated to B2C operations, subordinated to the director of marketing. Under this e-business management are staff and technical areas, including a total of approximately 10 persons.

On the IT side, there is an e-commerce management that is also fully dedicated to these activities. Other e-business initiatives are exercised in the area of purchasing (B2B) but do not have the volume of business generated by the B2C area.

The business model used was a *value creation model* based on business-to-consumer (B2C) retailing and using metrics for results and infrastructure suited to e-business. B2C means direct sales via the Internet. Results are to be measured using a specific set of metrics for e-commerce. The model also calls for investment in e-business infrastructure in the form of close links between the company's IT department and the business units involved in direct marketing of entry-level cars via the Internet.

The project triggered changes in the company's organizational structure, including hiring of new personnel and the creation of new units. It is worth noting the clear evidence of a link between company strategy and the formulation of the e-business project. The idea of selling cars via the Internet arose from the strategic goal of increasing market share in the entry-level segment. GM Brazil invested more than $800 million in the Celta, which therefore embodied significant confidence in the Brazilian market on the part of this manufacturer over several years. The aim was to improve GM's positioning against its competitors and in particular to challenge its main competitor, the market leader in the entry-level segment.

The success of the new car would also mean the difference between the success and failure of a strategy established in 1992. At that time GM Brazil decided to expand its offering to include products for all segments from entry-level cars to heavy-duty trucks. The strategy proved successful. GM's market share in Brazil rose 21% to 25%. The problem was that the strategy could be sustained without a significant presence in the market for small or entry-level cars, which accounts for 70% of all car sales in Brazil. The aim of the Celta project was to increase GM's competitiveness in this segment. The price would be higher than that of its competitor, but GM expected superior design and technology to offset the higher price.

In addition to superior technology, GM focused on using a low-cost production process rather than an inexpensive design. This was achieved by implementing a modular plant in consortium with suppliers and flexible production methods to turn out 120,000 units per year. GM's modular consortium has 17 suppliers strategically located in the same plant and connected in real time. This enables suppliers, for example, to know just when their products are needed on the assembly line.

The purpose of this short introduction before our detailed presentation of the B2C model developed by GM Brazil is to highlight the connection between the carmaker's e-business initiative and its very clear strategy of growing market share using technology that enables a product to be sold directly to the consumer.

When the project began at GM around 2000, there was a worldwide movement in favor of giving priority to e-business and the parent company created a new structure called "e-GM." Thus the initiative of direct sales via the Internet matched the parent company's expectations. A directive clearly came from headquarters, indicating that subsidiaries should "invest in e-business projects." At the same time, the Brazilian subsidiary adopted a strategy of becoming a market leader in entry-level cars.

The marketing director of GM Brazil then had the idea of combining the two policy directives—investment in e-business worldwide and growing market share in small cars in Brazil—with the existence of a new plant in Brazil that made flexible manufacturing possible. The initial idea was to use the "e-shop" system already put in place by the parent company in the U.S. This system assisted the buyer in configuring the product, calculated the price, and told the buyer where the product was located. However, it did not enable direct sales via the Internet.

The marketing department in Brazil was bolder. It created a Web site that completed the process by adding direct sales to configuration and pricing. Thus GM Brazil created a structure dedicated to developing this project of direct marketing via the Internet.

The next section describes the process for direct sales of small cars to consumers via the Internet developed by GM Brazil marketing department in collaboration with IT area.

GM Brazil's Process for Selling the Car "Celta" Online

The business model for the B2C project to sell the Celta online had four pillars:

- Consumers were to be billed directly and not through dealerships.
- Pricing was to be unified for the entire country (prices may now differ depending on the region).
- Rapid delivery.
- Given the need to develop IT systems for this strategy, it was decided to use the Internet. The expected volume of customer interactions was large, justifying an automated and integrated process. A concrete benefit would be direct knowledge of consumer wants and needs.

Dealerships were involved only in final delivery of the car to the purchaser.

The selling process was direct to the final consumer but delivery remained in the hands of dealers in order to avoid channel conflict. The system adopted by GM Brazil includes rather than excludes dealers. The consumer buys a car and can select the dealership that will deliver it. Dealers receive a delivery fee, which is smaller than the commission on traditional sales but may be commercially more attractive since the dealer is not required to hold inventory. However, dealers need to have a good costing system to realize that the new procedure is more profitable.

The system presents a number of advantages for GM:

1. GM saves 5% of the commission normally paid to dealerships, transforming this saving into a discount on the price to the consumer;
2. Producing cars for delivery by dealers means holding inventory, and inventory equals cost. Producing for direct sale to consumer reduces inventory and hence cost; 3) Direct contact with consumers enables GM to obtain customer information, which can be used to improve CRM. The type of information relates to color, model, and accessories. This goes into a customer database and becomes business intelligence for use in future transactions; and 4) Direct marketing leads to better knowledge of customers wants and needs, reducing the number of models and options required, and facilitating production, which can be tailored to demand.

Advantages for consumers: The consumer pays less (thanks to pass-through of the manufacturer cost saving) and gets faster delivery.

Dealership's Involvement in Online Sales

- Customers with Internet access make choices online, save the configuration, and go to the dealership to pick up the car.
- Customers without Internet access go to a dealership to use the sales site, configuring the car they want to buy. The benefit to the dealer is selling cars without needing to have them physically on the premises.

Possible Problems that could Arise from this Sales System

- Resistance on the part of the consumer public, used to "trying out" the product on dealers premises, where they can get into the car and check its equipment, take test drives, and have other types of direct personal contact

- Resistance on the part of the dealers, who would receive a lower commission for delivering cars sold on the Internet than if they made the sales themselves.
- Since consumers can configure certain items according to their needs and preferences, the factory could run into problems if it was not flexible enough, and this could cause delays in delivery.

The results obtained with the launching of the Celta went far beyond expectations: sales exceeded initial estimations and the company sold more products directly to consumers in this way, some months later. The site has to be updated every time a new product is included. There are also logistics issues that have to be analyzed carefully before new models are put on the market. Thus, this e-business model is intrinsically related to the company's strategy in the sense that it is directly linked to its core-business.

In 2002, 30% of GM Brazil's total sales were carried out online. In 2001, sales of the Celta via the Internet accounted for almost half the revenue generated by B2C overall in Brazil.

Organizational Structure for Online Car Sales

One of the first difficulties encountered in the implementation of the project was the fact that the company organizational structure at the time had no area that could be responsible for the project of setting up B2C operations in the company, more specifically, the sale of cars directly over the Internet.

For this reason, the e-business structure has evolved as follows, in terms of B2C:

- 1998 – Internet area set up inside strategic marketing
- 1999 – two departments set up exclusively for e-business

Figure 2 shows the organizational structure in 1999:

The dotted box around Sales Support recalls that in late 2001, the director of e-business was promoted to a job at the parent company and this department was incorporated into Sales Support. Two new departments were set up in the same year: a department of e-business reporting to the CIO, and a department of e-business reporting to the director of marketing and sales.

The e-business director in the marketing area was already a GM employee. He was formerly a marketing manager and was reallocated to this position. The e-business director in the IT area was hired from outside with a specific remit to implement the online selling project. In 2001 the e-business director acquired control over sales support in general, not just Internet sales support.

The new area was set up to increase the hierarchical status of e-business and the focus on online selling of the Celta.

This entire new structure originated with the idea of selling the new small model (Celta car) directly to consumers via the Internet. The marketing director of GM Brazil had the idea, which was approved by the parent company executive committee, together with a global budget, resulting in a world pilot of e-commerce and direct online marketing.

The project organizational configuration was as follows. The "business owners," that is, those responsible, were the CIO and the director of marketing and sales. They reported to hierarchical superiors (regional and global) if any problems arose. The Brazil project manager, another business owner, took operating decisions on Web site design and the entire selling process involving

Figure 2. E-business structure of GM Brazil in 1999

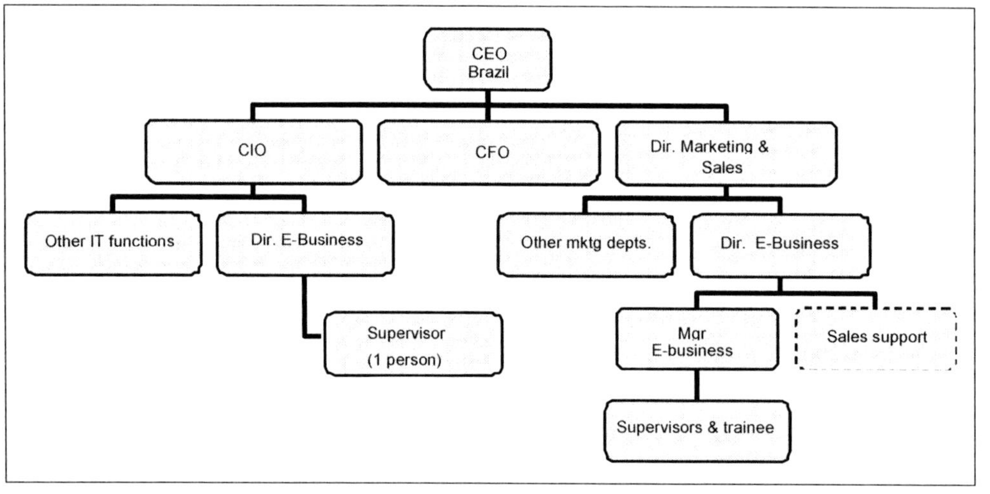

both the company and end users (customers). The next level down was the program management office (PMO), comprising project managers whose function was to integrate the various activities. Each unit responsible for a given functionality included a person responsible for IT and another for marketing (business). Most IT personnel worked for contractors (outsourcing).

This e-business project structure was put in place at the end of 1999. The B2C Web site was launched in the second half of 2000. The project was up and running in less than a year. Once it went live the e-business area was conserved as originally created with the addition of a new unit for IT maintenance (the IT e-business director plus one or two other people).

GM partnered with UOL, one of Brazil's largest Internet service providers, in communication and hosting of the institutional site for other cars. At the end of 2001, the director of e-business in Brazil was promoted to a position with the parent company and the e-business structure was merged into sales support. There are synergies between these areas because of Internet sales.

The e-business area needs information from other parts of the company, such as brand or product management, pricing, vehicle distribution (which cars can be delivered where, deciding how many cars go to distribution centers), billing and accounts receivable, and sales. E-business personnel are constantly working to familiarize themselves with the day-to-day operations of other areas and make or collect suggestions on new projects.

Figure 3 shows the structure of the e-business group (B2C) at GM Brazil (2002).

The e-business manager has a certain amount of decision making authority, but in certain cases issues are resolved by the head of department, director, board and so on, depending on the type of decision required.

The e-business director in the marketing area was already a GM employee. He was formerly a marketing manager and was reallocated to this position. The e-business director in the IT area was hired from outside with a specific mandate to implement the online selling project. In 2001, the e-business director acquired control over sales support in general, not just Internet sales support.

The new area was set up to increase the hierarchical status of e-business and the focus

Figure 3. Organizational structure of e-business group (B2C) - GM Brazil - 2002

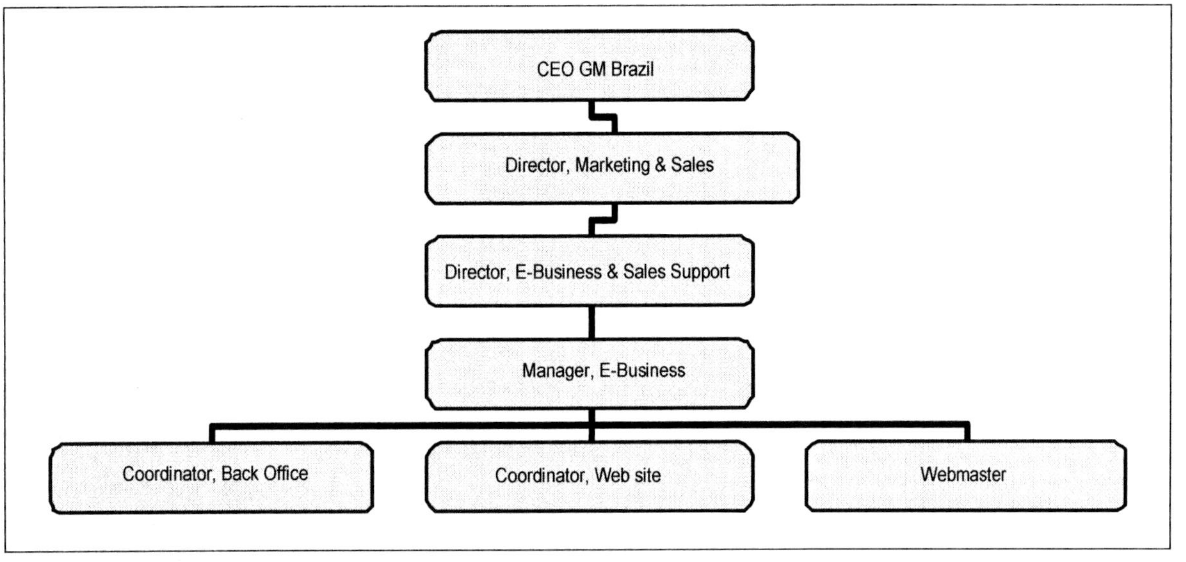

Figure 4. Organizational structure of GM Brazil and e-business areas — early 2002

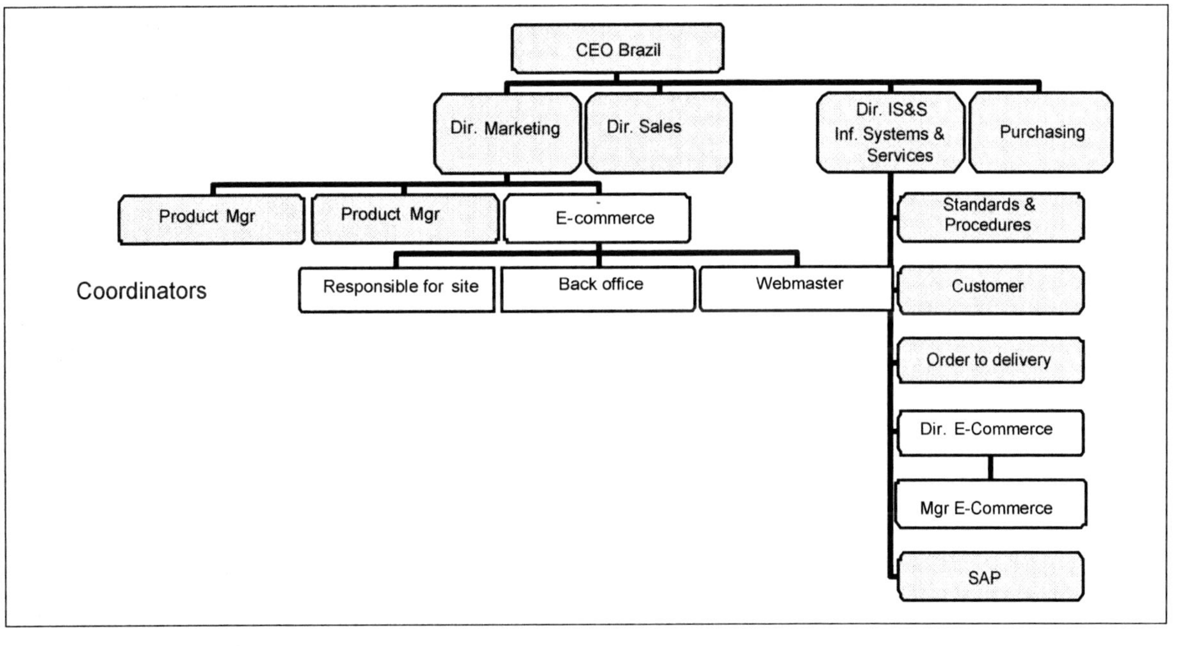

on online selling of the Celta. This entire new structure originated with the idea of selling the new entry-level model directly to consumers via the Internet. The marketing director of GM Brazil had the idea, which was approved by the parent company's executive committee, together with a global budget, resulting in a world pilot of e-commerce and direct online marketing. The world's first direct sales e-commerce site for entry-level cars went live in September 2000, selling the Celta.

The fact that there was a centralized structure for B2C operations allow the information coming in from the content providers to be analyzed and made use of, keeping information from becoming obsolete and allowing periodic input into this information, updating the sites of sales, and communication to consumers.

Figure 4 shows the organization chart of GM Brazil and the e-business areas in early 2002.

ANALYSIS OF GM-CELTA CASE

This section presents the conclusions of our analysis of GM Brazil's Celta project from the standpoint of the e-business model implemented.

The business model adopted in this work was that conceived by Applegate (2001), where a business model describes succinctly how the business is structured, what kind of people are needed for that business, and what roles they perform. This model consists of three components: business concept, capabilities, and value. We will analyze some of these components in the Celta case, as following.

Considering the *business model concept*, the basic concept in the Celta case was direct marketing of small cars using the Internet and delivery of value to the customer via B2C, with clearly defined metrics to measure results and the use of an appropriate IT infrastructure to ensure connectivity with consumers, besides guarantee the updating of information by the content providers. B2B activities do not use the same model and are not centrally coordinated.

The market opportunity was identified to increase the market share in the small car segment.

The small car segment in Brazil is extremely competitive, with many competitors, and the price factor is essential for success.

Strategic options for levering business are shown with the use of the Internet for direct sales of cars to final consumers; a lower final sale price is possible because of the use of electronic channels to reach the final consumer. The possibility for final consumers to configure the car according to their preferences is a factor that gives the model a competitive advantage in relation to the competition.

Considering the *business model capabilities,* capabilities were sought in house and from outside, including reallocation of executives to e-business and hiring of an outside IT specialist to head IT for e-business. IT infrastructure is outsourced.

In terms of *organizational structure used*, it could be seen that to develop the project for selling the Celta on the Internet, a matrix structure was set up with persons from the head office and the branches involved, and the decisions were taken jointly by one person responsible for the IT area and another for the business area. The operation of sales activities over the Internet uses a centralized structure dedicated to e-business, with a central coordinator of e-business operations who responds directly to the company senior management of marketing.

There is coordination of the **IT** operations related to sales on the Internet by a director of IT dedicated to e-business.

There are other factor that could be seen as advantage for GM. There is a *flexible factory* using the concept of modular consortium for the production of the Celta. The Celta car was conceived to be produced in this "modular" factory, which was a new plant with a new concept; in this way, there were no problems of integration of new IT applications with the legacy ones and new business processes with existing ones once the planning of the car production was conceived since the beginning to be made at this flexible plant.

In terms of *value delivered by the model,* it was seen that revenue generation was via direct sales and the financial results went beyond expectations from the first month of sales. Other value aggregated was more and better customer relationships, faster processes, and cost savings to GM.

The dynamics involving the launching of the Celta shows an identification of a market opportunity, specifically, the possibility of increasing the market share of small cars which, in Brazil, is one of the automotive industry most important segments. It is a segment where, until then, GM Brazil had not been competitive and held no significant slice in it.

The director of marketing at the time had the strategic objective of increasing the company's share in this segment and it was this guideline that oriented the company's efforts. The use of the Internet to make direct sales available to final consumers proved to be a very innovative possibility at the time, because the maximum that was done abroad was for the final consumer to configure the model, without carrying out the entire sale.

This operation allowed a reduction in costs, since one tax on the transactions was excluded (i.e., the IPI Tax on industrialized products), thus reducing the cost of the car by about 5%, and this reduction was passed on to the final consumer, generating a competitive advantage for GM, since the popular car segment is very price sensitive.

The company built into this model a concern for metrics and for the integration of its IT infrastructure, providing added value to the customer. It thus showed a concern for clear results; the initiative of e-business must be related to results. In this regard, the concept of value chain, developed by Porter (2001), was used, with the need to connect all operations in a way that would deliver distinctive value to the customer.

According to Amit and Zott (2001), value is created by e-business through the way in which transactions are carried out. The authors drew up a model to evaluate the creation of value in e-business where there are four conductors of value: efficiency (reducing asymmetry of information between sellers and buyers, speed of transmission of information, etc.); complementarity (when a set of factors provides more value than each of them individually); retention (value in e-business is increased to the extent that consumers are motivated to buy again); and novelties or innovations (such as setting up value by connecting parts that were not connected before the innovation).

Analyzing the model used by GM from the point of view of these authors, the company was seen to be efficient in using the Internet as a way to facilitate transmission of information among manufacturer, customer, and dealer (for delivery of the car); a high level of retention was attained, since buyers were able to customize details of their cars by configuring them on the Internet, besides the innovation of having the customer connected directly to the manufacturer.

In terms of the abilities of the model, GM has a new factory in southern Brazil that uses more flexible processes, a fact that allowed sales over the Internet. Customers can order customized details thanks to this flexibility of the factory and to the use of modular consortium (suppliers inside the factory) that guarantee faster manufacturing.

In this aspect, Brazil was the pioneer in the model of direct sales to consumers, since, in other countries, customers can only configure the car over the Internet, even though the respective sites are known as direct sales channels.

According to Timmers (1998), the definition of innovative operation on the Internet is where the use of the Internet allows measures to be taken that would be impossible without it. Direct sales to consumers would be very difficult for an assembly plant were it not for the facilities afforded by the Internet.

Another aspect to be observed was the fact that the use of executives in the business area coming from the company itself and the employment of partnerships with consolidated companies in the IT area, outsourcing of technical aspects of IT. But to coordinate these partners there was an IT director fully dedicated to the e-business project inside the company. The construction of this model took into consideration the establishment of strategic alliances with successful partners on the Internet (such as AOL, edmunds.com, etc.). This made it possible to:

- Facilitate the generation of traffic.
- Promote GM products in a favorable way, that is, construct a presence of the company that became widespread.
- Establish a partnership with "first class" companies in the area of the Internet.

The objectives of GM Brazil for e-business can be described as to increase and broaden relationships with consumers, see growth in revenue, respond to consumers faster, and show leadership. These objectives were attained with the sale of the Celta on the Internet. To demonstrate the importance of the Internet for GM, the company's worldwide CEO reports that 52% of the consumers go shopping for a car first, and go to a dealer later. The expectation at the time was that this number would rise to 75% in two years, representing approximately 11 million vehicles whose purchase processes began with the Internet.

In 2002, 30% of GM Brazil's total sales were carried out over the Internet.

The respondent did not say, however, if there is an evaluation as to whether sales over the Internet are less expensive than conventional sales.

This case of success showed that some of the critical factors for successful sales over the Internet are: 1) compatibility between the core competences of the company and the use of e-business operations; 2) level of commitment of upper management regarding the implementation of e-business; and 3) and adequate organizational structure that guaranteed that the decisions taken could be effectively concluded for the efficient functioning of the e-business operations.

This enterprise–direct sales over the Internet–surpassed expectations in terms of results. With a deduction of 5.5% in sales of this popular model of car over the Internet, GM Brazil registered a total 2,050 units sold (50% of the total number produced) in the first 14 days operating on the Web. The results of the initiative had worldwide repercussions and are used by the CEO at the company's headquarters as an example of a successful undertaking in the area of e-business.

The success of this initiative prompted a "race" by competitors to also carry out their sales over the Internet. The sales of this car on the Internet were equivalent to almost half of the revenue generated by B2C business in 2001 in Brazil.

The importance of the involvement of upper management in the project of selling cars on the Internet was commented above. The idea started from the senior management of marketing and was a tool to put a strategic objective into operation; it increased the company's market slice, and more specifically, increased the sales of entry-level cars in Brazil. As a result, as Porter (2001) has commented, the Internet is nothing more than a powerful tool that serves to attain the company's strategic objectives.

GM Brazil's upper management perceived the usefulness of this instrument for attaining effective competitive advantage ahead of its competitors. The direct sale of cars on the Internet produced a reduction in the price due to lower costs (no middleman), in the first stage, besides responding to the objective of providing buyers with customizing services by which they can use the Internet to configure their car the way they want. The company's upper management had the responsibility of guaranteeing the resources needed for this project (sale of cars directly on the Internet) in order to make it successful, both in the sense of supporting an organizational structure dedicated to e-business, in causing changes in the existing structure in a way that would provide the status needed for e-business operations, and in providing financial and human resources (hiring of persons and the allocation of funds where necessary), as well as technological resources, with the engaging of partners in the IT area.

The organizational structure created—with an e-business manager subordinated to the marketing division—guaranteed that the decisions taken could be effectively concluded for the adequate functioning of the e-business operations.

GM Brazil made use of a matrix-type structure for its e-business project to sell the Celta directly via the Internet, but with a combination of e-business structure, IT, and business resources, plus an area called "customer experience" to provide the required customer focus. Resources were allocated by the parent company in the U.S., including capital and specialized personnel. IT was also given priority, with the structure clearly showing the necessary integration between the business and the IT areas.

The e-business project to sell cars directly online involved approximately 250 people at GM Brazil and GM United States, besides partners responsible for developing the Web technology. The project organization was highly sophisticated, with a project manager known as a Process Information Officer (PIO) in charge of overall coordination, and several "business owners."

Since the project was successful, with success being measured through the results obtained with sales, a small team was maintained, with a central coordinator for e-business operations, in order to sustain, update, and continue to prospect new operations in e-business. It is interesting to note that only the activities related to direct sales on the Internet have this coordinator. Other e-business operations, such as B2B (relationship with suppliers, and procurement) and others of B2C (such as post-sales relationships) remain decentralized throughout the company, responding to the functional managers of areas that already existed before the advent of the e-business operations.

The results obtained with these activities are lower than those for the direct sale operations in terms of financial gain, showing that a small structure of coordination of these operations can bring about positive results.

CONCLUSION

The main objective of this chapter was to analyze the implementation of e-business operations in a company that has long operated in the physical world, by making use of an e-business model that especially took into account the following dimensions:

- In regard to the business concept component: business opportunities and competitive dynamics
- In regard to the capabilities: the organizational structure for e-business
- In regard to the value component: financial results

The segment studied was the automotive industry, which, in terms of its dynamics, is highly competitive, with its players all looking for innovations and opportunities that will give them competitive advantages over their competitors. The segment of small and entry-level cars is responsible for 70% of car sales in Brazil, and GM Brazil had no significant share in this segment, before the Celta car. There was, therefore, an attractive opportunity for growth and the company went looking for strategic options that would lever its sales in this segment.

GM Brazil was able to opt for direct marketing of its new entry-level model, the Celta, without dealers or other intermediaries, for two reasons. First, it had a flexible manufacturing facility with assembly lines configured to allow small modifications in accordance with a certain amount of customization, with configurations selected online by the consumer. The second key facilitator of direct sales was the Internet.

It was an innovation in the clear definition of "core competence" of the assembly plant. To put this direct sales initiative into practice, one of the company strategic options was to present a clear link between e-business initiatives and corporate strategy. The aim of the B2C project was to increase sales of small cars, and this strategic alignment brings competitive advantage to the company according to many authors (e.g., Chan & Huff, 1993; Henderson & Venkatraman, 1993; Kearns & Lederer, 2000).

The *competitive dynamic* of this segment shows that price is a fundamental factor in consumers' final decisions. Therefore, the use of the Internet as a tool enabled a reduction in the price of the Celta; through reduction in costs (reduction in the IPI Tax because of direct sale, and reduction in the fees paid to dealers) proved to be a form of obtaining competitive advantage.

Besides the price factor, the possibility of the final consumer of configuring the car before delivery proved to be an extra factor that adds value. It is important to emphasize that this type of configuration was only possible due to the existence of an innovative manufacturing model at the plant, which employs a flexible production process.

Top management involvement was total, with the original idea coming from the marketing director of GM Brazil. The project was successful in financial and marketing terms, exceeding expectations. This success was apparently due to the combination of an appropriate business model and a highly committed top management.

Partnering with dealerships played a fundamental role by avoiding channel conflict. The role allotted to dealers was to deliver cars purchased on the Web. As noted by Weil and Vitale (2001), one way of avoiding channel conflict is to realign the distributor functions.

Porter (2001) comments on the complementary role that should be played by the Internet show that this tool reinforces the competitive advantage of the "real" part of the business. When this is the case, results are satisfactory. This conclusion undoubtedly applies to the case of GM Brazil, where the e-business initiative was in line with a clear strategic objective, that is, to increase market share in the small car segment. This was the first project in the world to sell cars directly online, stimulating global acclaim and recognition by the parent company.

Also in terms of *strategic options used to lever business*, the company made an innovative use of the Internet. In fact, linking consumer to manufacturer for direct marketing of passenger cars would have been practically impossible without the Internet. In addition, the Internet suits the strategies that the automotive car automobile manufacturers are currently pursuing, according to a study by Santos and Medeiros (2001) on differentiation (e.g., the Internet facilitates differentiation in services), associations and alliances (facilitated by online interconnectivity), and, above all, geographic expansion (the Internet is a global network).

In their constant pursuit of lower cost and higher profit, the automakers have begun using a new global division of labor in which emerging-country subsidiaries have considerable importance. This favors Brazil, enabling subsidiaries here to offer personalized models for direct sale over the Internet, as discussed in this chapter.

In regard to the capabilities, the e-business initiative made a significant difference, as reflected in the organizational structure utilized. The company created (and has kept in place), especially for this project, an IT unit totally dedicated to supporting direct sales of this small car model via the Internet, thus guaranteeing adequate IT infrastructure. On the business side, a totally dedicated structure was also created to guarantee the necessary functionalities and meet the needs of customers, dealers, and manufacturing.

As for the use of structures by project, GM Brazil provides an excellent example of joint coordination of the e-business project by the IT and business areas. The project implementation methodology was completely formalized and created entirely in house on the basis of project management concepts covering every stage, such as scope, planning, design, construction, implementation, support, and maintenance. A structure was used where the integration of IT operations with business operations created more competitive business.

In terms of delivery of *value*, of special importance is the establishment of precise metrics to evaluate the results in each phase of the project, which made it possible to identify problems in a precise way and help solve them, delivering an efficient final result.

CONTRIBUTIONS

The following aspects of the Celta case can be tallied up as contributions to the understanding of the mechanism of implementation of e-business operations in a company that until then had only operated in the physical world:

- The fundamental role of the company's global strategy in defining the model to be adopted for e-business. This strategic objective (increased sales of small cars) oriented the implementation of the B2C operations, that is, direct sales via the Internet.
- Involvement of top management in providing the needed resources, including financial, technological, and human.
- The e-business project to sell cars directly online involved approximately 250 people at GM Brazil and GM United States, besides partners responsible for developing the Web technology. The project organization was highly sophisticated, with a project manager known as a PIO in charge of overall coordination, and several "business owners."
- Use of an adequate organizational structure in conception and implementation stages of the matrix structure project, with central coordination of the e-business operations by one person responsible for IT and another for the business area.
- Resources were allocated by the parent company in the U.S., including capital and specialized personnel. IT was also given priority, with the structure clearly showing the necessary integration between the business and the IT areas.
- After the project of direct sales online finished, a small team was maintained, with a central coordinator for e-business operations, responding directly to the marketing director, in order to sustain, update, and continue to prospect new operations in e-business. This structure guarantees the necessary functionalities and meets the needs of customers, dealers, and manufacturing.
- Excellent communication between the area responsible for e-business and the content providers. This fact allowed for updating of information and searching for new opportunities in e-business. This communication was facilitated by the centralized organizational structure with a centralized coordinator of e-business operations.
- Existence of an assembly plant that has a concept of modular consortium and flexible production which allows changes in the production line, and let customers configure their cars over the Internet according to their needs.
- The company was efficient in using the Internet as a means of facilitating the traffic of information between manufacturer, customer, and dealer (for car delivery). A high level of retention was attained by allowing customers to customize the details of their cars over the Internet, besides having offered innovations by putting customers into direct contact with the manufacturer.
- The innovative use of the Internet, which, until then, had been used worldwide only to let customers configure the model of the car but not for complete processing of the sale over the Internet, as was the innovative case of the Celta car in Brazil.
- Conflicts in channels were avoided, with the involvement of the dealers in the process of configuration, test drives, and delivery of the cars to the customers; there was payment of a fee to the dealers for each delivery handled.
- GM Brazil showcased its own use of the Internet as a marketing tool.

- In 2002, 30% of GM Brazil's total sales were carried out over the Internet, showing the success of this new activity (sales online)

The following can be considered the main limits involved:

- Centralizing of all e-business operations under a single coordinator. For example, B2B operations have a decentralized coordination, and post-sale operations do not use the same structure. Results obtained through these operations were not as successful as those attained through sales over the Internet.
- Measurement of other e-business operations: Direct sale over the Internet generate revenue, whereas other operations that consist purely of relationships are harder to measure.
- Competitors have since launched their own direct sales sites. Therefore, GM is no longer the sole reference for this innovation

REFERENCES

Applegate, L. M. (2001). *Emerging e-business models: Lessons from the field.* Harvard Business School. Retrieved June 30, 2008, from http://monkey.icu.ac.kr/sslab/course/ICE720/data/subject-Presentation/EmergingE-BusinessModel.pdf

Bakos, J. Y., & Treacy, M. E. (1986). Information technology and corporate strategy: A research perspective. *MIS Quarterly, 10*(2), 107-119.

Barney, J. (1991). Firm resources and sustained competitive advantage. *Journal of Management, 17*(1), 99-120.

Barua, A., Konana, P., Whinston, A. B., & Yin, F. (2004). Assessing Internet enabled business value: An exploratory investigation. *MIS Quarterly.* Retrieved June 30, 2008, from http://crec.mccombs.utexas.edu/works/articles/barua_konana_whinston_yin_01311.pdf

Benko, C., & McFarlan, W. (2003). *Metamorphosis in the auto industry.* Retrieved June 30, 2008, from http://www.connectingthedotsbook.com/articles/CTD_Metamorphosis_in_the_Auto_Industry.pdf

Chesbrough, H., & Rosenbloom, R. S. (2001). *The role of the business model in capturing value from innovation: Evidence from Xerox Corporation's technology spinoff companies.* Retrieved June 30, 2008, from www.hbs.edu/dor/papers2/0001/01-002.pdf

Cunningham, M. J. (2001). *B2B business-to-business: Como implementar estratégias de e-commerce entre empresas.* Rio de Janeiro: Ed. Campus.

Dutta, S., & Biren, B. (2001). *Business transformation on the Internet: Results from the 2000 study.* Retrieved June 30, 2008, from http://mslab.hau.ac.kr/mgyoon/ec02/paper/9.pdf

Freeland, G. D., & Stirton, S. (2000). *Organizing for e-commerce* (discussion paper). *The Boston Consulting Group Inc.* Retrieved June 30, 2008, from www.bcg.com/publications/files/organizing%20ECommerce%20Apr%2000.pdf

Grant, R. M. (1991, Spring). The resource-based theory of competitive advantage: Implications for strategy formulation. *California Management Review,* 114-135.

Henderson, J. C., & Venkatraman, N. (1993). Strategic alignment: Leveraging information technology for transforming organizations. *IBM Systems Journal, 32*(1) 4-16.

Kalakota, R., & Robinson, M. (1999). *E-business-roadmap for success.* MA: Addison Wesley Longman, Inc.

Kearns, G. S., & Lederer, A. L. (2000). The effect of strategic alignment on the use of IS-based resources for competitive advantage. *Journal of Strategic Information Systems.* Retrieved June 30,

2008, from http://web.staffs.ac.uk/~bstpjw/ebusiness/downloads/stman04.pdf

Kraemer, K. L., & Dedrick, J. (2002). Strategic use of the Internet and e-commerce: Cisco systems. *Journal of Strategic Information Systems*, 5-29. Retrieved June 30, 2008, from http://www.doai.uom.gr/mai/Course_Material/E_Commerce/folinas/7.pdf.

Lientz, B. P., & Rea, K. P. (2001). *Transform your business into E*. CA: Academic Press.

Mahmood, M. A., & Soon, S. K. (1991). A comprehensive model for measuring the potential impact of information technology on organizational strategic variables. *Decision Sciences, 22*(4), 869-897.

McKinsey. (2003). *Study HAWK 2015 – Wissensbasierte Veränderung derautomobilen Wertschöpfungskette, VDA*. Retrieved June 30, 2008, from http://www.wk-tim.de/fachtagungen/2004/unterlagen/10_MuellerFreiling.pdf

Oliveira, Djalma de Pinho Rebouças de. (2001). *Sistemas, organização e métodos: uma abordagem gerencial*. São Paulo: Atlas.

Plant, R. (2000). *E-commerce formulation of strategy*. NJ: Prentice Hall.

Porter, M. E. (1980). *Competitive strategy: Techniques for analyzing industries and competitors*. New York: The Free Press.

Porter, M. E. (2001, March). Strategy and the Internet. *Harvard Business Review*, 63-78.

Porter, M. E., & Millar, V. E. (1985). How information gives you competitive advantage. *Harvard Business Review*, 149-160.

Robert, M., & Racine, B. (2001). *E-strategy pure & simple*. New York: McGraw-Hill.

Santos, A. M., & Medeiros, M. (2001). *Reestruturação da indústria automobilística na América do Sul*. BNDES Setorial, Rio de Janeiro, n. 14, September, www.bndes.gov.br/conhecimento/bnset/set1403.pdf

Slack, N., & Lewis, M. A. (2001). *Operations strategy*. FT Prentice Hall.

Tapscott, D., Lowy, A., & Ticoll, D. (2000). *Plano de Ação para uma economia digital*. Makron Books do Brasil Ltda.

Turban, E. et al. (2000). *Electronic commerce: A managerial perspective*. NJ: Prentice-Hall Inc.

Weil, P., & Vitale, M. R. (2001). *Place to space: Migrating to e-business models*. Harvard Business School Publishing Corporation.

ZIilber, S. N., & Vasconcellos, E. (2005). The adoption of e-business from a strategic perspective: The case of the Celta at General Motors Brazil. *International Journal of Automotive Technology and Management (IJATM), 5*(4), 411-429.

Chapter XVIII
On the Use of Soft Computing Techniques for Web Personalization

G. Castellano
University of Bari, Italy

A. M. Fanelli
University of Bari, Italy

M. A. Torsello
University of Bari, Italy

ABSTRACT

Due to the growing variety and quantity of information available on the Web, there is urgent need for developing Web-based applications capable of adapting their services to the needs of the users. This is the main rationale behind the flourishing area of Web personalization that finds in soft computing (SC) techniques a valid tool to handle uncertainty in Web usage data and develop Web-based applications tailored to user preferences. The main reason for this success seems to be the synergy resulting from SC paradigms, such as fuzzy logic, neural networks, and genetic algorithms. Each of these computing paradigms provides complementary reasoning and searching methods that allow the use of domain knowledge and empirical data to solve complex problems. In this chapter, we emphasize the suitability of hybrid schemes combining different SC techniques for the development of effective Web personalization systems. In particular, we present a neuro-fuzzy approach for Web personalization that combines techniques from the fuzzy and the neural paradigms to derive knowledge from Web usage data and represent the knowledge in the comprehensible form of fuzzy rules. The derived knowledge is ultimately used to dynamically suggest interesting links to the user of a Web site.

INTRODUCTION

The growing explosion in the amount of information and applications available on the World Wide Web has made more severe the need for effective methods of personalization for the Web information space. The abundance of information combined with the heterogeneous nature of the Web makes Web site exploration difficult for ordinary users, who often obtain erroneous or ambiguous replies to their requests. This has led to a considerable interest in Web personalization which has become an essential tool for most Web-based applications. Broadly speaking, Web personalization is defined as any action that adapts the information or services provided by a Web site to the needs of a particular user or a set of users, taking advantage of the knowledge gained from the users' navigational behavior and individual interests, in combination with the content and the structure of the Web site. In other words, the aim of a Web personalization system is to provide users with the information they want or need, without expecting them to ask for it explicitly (Nasraoui, 2005; Mulvenna, Anand, & Buchner, 2000).

The personalization process covers a fundamental role in an increasing number of application domains such as e-commerce, e-business, adaptive Web systems, information retrieval, and so forth. Depending on the application context, the nature of personalization may change. In e-commerce applications, for example, personalization is realized through recommendation systems which suggest products to clients or provide useful information in order to decide which products to purchase (Adomavicius & Thuzilin, 2005; Baraglia & Silvestri, 2004; Cho & Kim, 2004; Mobasher, 2007b; Schafer, Konstan, & Riedl, 2001). In e-business, Web personalization additionally provides mechanisms to learn more about customer needs, identify future trends, and eventually increase customer loyalty to the provided service (Abraham, 2003). In adaptive Web sites, personalization is intended to improve the organization and presentation of the Web site by tailoring information and services so as to match the unique and specific needs of users (Callan, Smeaton, Beaulieu, Borlund, Brusilovsky, Chalmers et al., 2001; Frias-Martinez, Magoulas, Chen, & Macredie, 2005). In practice, adaptive sites can make popular pages more accessible, highlight interesting links, connect related pages, and cluster similar documents together (Perkowitz & Etzioni, 1997). Finally, in information retrieval, personalization is regarded as a way to reflect the user preferences in the search process so that users can find out more appropriate results to their queries (Kim & Lee, 2001; Enembreck, Barthès, & Ávila, 2004).

The development of Web personalization systems gives rise to two main challenging problems: how to discover useful knowledge about the user's preferences from the uncertain Web data and how to make intelligent recommendations to Web users. A natural candidate to cope with such problems is **soft computing** (SC), a consortium of computing paradigms that work synergistically to exploit the tolerance for imprecision, uncertainty, approximate reasoning, and partial truth in order to provide flexible information processing capabilities and obtain low-cost solutions and close resemblance to human-like decision making. Recently, the potentiality of SC techniques (i.e., neural networks, fuzzy systems, genetic algorithms, and combinations of these) in the realm of Web personalization has been explored by researchers (e.g., Jespersen, Thorhauge, & Pedersen, 2002; Pal, Talwar, & Mitra, 2002; Sankar, Varun, & Pabitra, 2002; Yao, 2005).

This chapter is intended to provide a brief survey of the stat-of-art SC approaches in the wide domain of Web personalization, with special focus on the use of hybrid techniques. As an example, we present a neuro-fuzzy Web personalization framework. In such a framework, a hybrid approach based on the combination of techniques taken from the fuzzy and the neural paradigms is employed in order to identify **user profiles** from Web usage data and to provide dynamical predictions about Web pages to be suggested to the current user, according to the user profiles previously identified.

The content of chapter is organized as follows. In Section 2 we deal in depth with the topic of Web personalization, focusing on the use of Web usage mining techniques for the development of Web applications endowed with personalization functions. Section 3 motivates the use of soft computing techniques for the development of Web personalization systems and overviews existing systems for Web personalization based on SC methods. In Section 4 we describe a neuro-fuzzy Web personalization framework and show its application to a Web site taken as case study. Section 5 closes the chapter by drawing conclusive remarks.

WEB PERSONALIZATION

Web personalization is intended as the process of adapting the content and/or the structure of a Web site in order to provide users with the information they are interested in (Eirinaki & Vazirgiannis 2003; Mulvenna et al., 2000; Nasraoui 2005). The personalization of services that a Web site may offer is an important step towards the solution of some problems inherent in Web information space, such as alleviating information overload and making the Web a friendlier environment for its individual user, and, hence, creating trustworthy relationships between the Web site and the visitor-customer. Mobasher, Cooley, and Srivastava (1999) simply define Web personalization as the task of making Web-based information systems adaptive to the needs and interests of individual users. Typically, a personalized Web site recognizes its users, collects information about their preferences, and adapts its services in order to match the users' needs. Web personalization improves the Web experience of a visitor by presenting the information that the visitor wants to see in the appropriate manner and at the appropriate time.

In literature, many different approaches have been proposed for the design and the development of systems endowed with personalization functionality (Kraft, Chen, Martin-Bautista, & Vila, 2002; Linden, Smith, & York, 2003; Mobasher, Dai, Luo, & Nakagawa, 2001;). In the majority of the existing commercial personalization systems, the personalization process involves substantial manual work and, most of the time, significant effort for the user. A better way to expand the personalization of the Web is to automate the adaptation of Web-based services to their users. Machine learning methods have a successful record of applications to similar tasks, that is, automating the construction and adaptation of information systems (Langley, 1999; Pohl, 1996; Webb, Pazzani, & Billsus, 2001). Furthermore, the integration of machine learning techniques in larger process models, such as that of **knowledge discovery** in data (KDD or data mining), can provide a complete solution to the adaptation task. Data mining has been used to analyze data collected on the Web and extract useful knowledge leading to the so-called Web mining (Eirinaki & Vazirgiannis, 2003; Etzioni, 1996; Kosala & Blockeel, 2000; Mobasher, 2007a; Pal et al., 2002). Web mining refers to a special case of data mining which deals with the extraction of interesting and useful knowledge from Web data. Three important subareas can be distinguished in Web mining:

- *Web content mining*: Extraction of knowledge from the content of Web pages (e.g., textual data included in a Web page such as words or also tags, pictures, downloadable files, etc.).
- *Web structure mining*: Extraction of knowledge from the structural information present into Web pages (e.g., links to other pages).
- *Web usage mining*: Extraction of knowledge from usage data generated by the visits of

the users to a Web site. Generally, usage data are collected into Web **log files** stored by the server whenever a user visits a Web site.

In this chapter, we focus mainly on the field of Web usage mining (WUM) that represents today a valuable source of ideas and solutions for the development of Web personalization systems. Overviews about the advances of research in this field are provided by several other authors (e.g., Abraham, 2003; Araya et al., 2004; Cho & Kim, 2004; Cooley, 2000; Facca and Lanzi, 2005; Mobasher, 2006, 2005; Mobasher, Nasraoui, Liu, & Masand, 2006; Pierrakos, Paliouras, Papatheodorou, & Spyropoulos, 2003). In general, regardless the application context, three main steps are performed during a WUM personalization process (Mobasher, Cooley, & Srivastava, 2000):

- **Preprocessing:** Web usage data are collected and preprocessed in order to identify **user sessions** representing the navigational activities of each user visiting a Web site.
- **Knowledge discovery**: The session data representing the users' navigational behaviour are analysed in order to discover useful knowledge about user preferences in the form of user categories of user profiles.
- **Recommendation**: The extracted knowledge is employed to customize the Web information space to the necessities of users, that is, to provide tailored recommendations to the users depending on their preferences.

While preprocessing and knowledge discovery are performed in an off-line mode, the employment of knowledge for recommendation is carried out in real time to mediate between the user and the Web site the user is visiting. In the following subsections, each step of the personalization process is more deeply examined.

Preprocessing

Access log files represent the most common source of Web usage data. All the information concerning the accesses made by the users to a Web site are stored in log files in chronological order. According to the common log format (www.w3.org/Daemon/User/Config/Loggin.htm#common-logfile-format) each log entry refers to a page request and includes information such as the user's IP address, the request's date and time, the request method, the URL of the accessed page, the data transmission protocol, the return code indicating the status of the request, and the size of the visited page in terms of number of bytes transmitted. By exploiting such information, models of typical user navigational behavior can be derived and used as input to the next step of knowledge discovery. The derivation of navigational patterns from log data is achieved through a preprocessing activity that filters out redundant and irrelevant data, and selects only log entries related to explicit requests made by users. Cooley (2000) extensively discusses the methods adopted to execute data preparation and preprocessing activity. Typically Web data preprocessing includes two main tasks, namely, data cleaning and user session identification.

The aim of data cleaning is to remove from log files all records that do not represent the effective browser activity of the connected user, such as those corresponding to requests for multimedia objects embedded in the Web page accessed by the user. Elimination of these items can be reasonably accomplished by checking the suffix of the URL name (all log entries with filename suffixes such as gif, jpeg, GIF, JPEG, jpg, JPG and map are removed). Also, records corresponding to failed user requests and accesses generated by Web robots are identified and eliminated from log data. Web robots (also known as Web crawlers or Web spiders) are programs which traverse the Web in a methodical and automated manner, downloading complete Web sites in order to update the index of a search engine. This task is performed by maintaining a list of known spiders and through heuristic identification of Web robots. Tan and Kumar (2002) propose a robust technique which is able to detect, with a high accuracy, Web robots by using a set of relevant features extracted from access logs (e.g., percentage of media files requested, percentage of requests made by HTTP methods, average time between requests, etc.).

The next task of Web log preprocessing is the identification of user sessions. Based on the definitions found in different works of scientific literature, a user session can be defined as a finite set of URLs corresponding to the pages visited by a user from the moment the user enters a Web site to the moment the same user leaves it (Suryavanshi, Shiri, & Mudur, 2005). The process of segmenting the activity of each user into sessions, called sessionization, relies on heuristic methods. Spiliopoulou (1999) divides the sessionization heuristics into two basic categories: time-oriented and structure-oriented. Time-oriented heuristics establish a timeout to distinguish between consecutive sessions. The usual solution is to set a minimum timeout and assume that consecutive accesses within it belong to the same session, or set a maximum timeout, where two consecutive accesses that exceed it belong to different sessions. On the other hand, structure-oriented heuristics consider the static site structure or they refer to the definition of conceptual units of work to identify the different user sessions. More recently, Spiliopoulou, Mobasher, Berendt, and Nakagawa (2003) have proposed a framework to measure the effectiveness of such heuristics and the impact of different heuristics on various Web usage mining tasks.

Knowledge Discovery

After preprocessing, the next step of a Web personalization process consists in discovering knowledge from data in the form of user models or profiles embedding the navigational behavior by expressing the common interests of Web visitors. Statistical and data mining techniques have been widely applied to derive models of user navigational behavior starting from Web usage data (Facca & Lanzi 2005; Mobasher, 2005; Pierrakos et al., 2003). In particular, analysis techniques of Web usage data can be grouped into three main paradigms: association rules, sequential patterns, and **clustering** (Han and Kamber (2001) detail an exhaustive review).

Association rules are used to capture relationships among Web pages which frequently appear in user sessions, without considering their access ordering. Typically, an association rule is expressed in the form:

"A.html, B.html $\Rightarrow$ C.html"

which states that if a user has visited page A.html and page B.html, it is very likely that in the same session the same user also visits page C.html. This kind of approach has been used in Joshi, Joshi, and Yesha (2003), and Nanopoulus, Katsaros, and Manolopoulos (2002), while some measures of interest to evaluate association rules mined from Web usage data have been proposed by Huang, Cercone, and An (2002a), and Huang, Ng, Ching, Ng, and Cheung (200a). Fuzzy association rules, obtained by the combination of association rules and fuzzy logic, have been extracted by Wong and Pal (2001).

Sequential patterns in Web usage data detect the set of Web pages that are frequently accessed by users in their visits, considering the order that they are visited. To extract sequential patterns, two main classes of algorithms are employed: methods based on association rule mining and methods based on the use of tree structures and Markov chains. Some well-known algorithms for mining association rules have been modified to obtain sequential patterns. For example, the Apriori algorithm has been properly extended to derive two new algorithms: the AprioriAll and GSP proposed by Huang et al. (2002a) and Mortazavi-Asl (2001). An alternative algorithm based on the use of a tree structure has been presented by Pei, Han, Mortazavi-asl, and Zhu (2000). Tree structures have been also used by Menasalvas, Millan, Pena, Hadjimichael, and Marban (2002).

Clustering is the most widely employed technique to discover knowledge in Web usage data. An exhaustive overview of Web data clustering methods is provided by Vakali, Pokorný, and Dalamagas (2004). Two forms of clustering can be performed on usage data: user-based clustering and item-based clustering.

User-based clustering groups similar users on the basis of their ratings for items (Banerjee & Ghosh, 2001; Heer & Chi, 2002; Huang et al., 2001). Each cluster center is an n-dimensional vector (being n the number of items) where the *i*-th component is the average rating expressed by users in that cluster for the *i*-th item. The recommendation engine computes the similarity of an active user session with each of the discovered user categories represented by cluster centroids to produce a set of recommended items.

Item-based clustering identifies groups of items (e.g., pages, documents, products) on the basis of similarity of ratings by all users (O'Connor & Herlocker, 1999). In this case a cluster center is represented by a m-dimensional vector (being m the number of users) where the *j*-th component is the average rating given by the *j*-th user for items within the clusters. Recommendations for users are computed by finding items that are similar to other items the user has liked.

Various clustering algorithms have been used for user- and item-based clustering, such as K-means (Ungar & Foster, 1998) and divisive hierarchical clustering (Kohrs & Merialdo, 1999). User-based and item-based clustering are typically used as alternative approaches in Web personalization. Nevertheless, they can also be integrated and used in combination, as demonstrated by Mobasher, Dai, Nakagawa, and Luo (2002).

In the context of Web personalization, an important constraint to be considered in the choice of a clustering method is the possibility to derive overlapping clusters. The same user may have different goals and interests at different times. It is inappropriate to capture such overlapping interests of the users in crisp clusters. This makes fuzzy clustering algorithms more suitable for usage mining. In fuzzy clustering, objects which are similar to each other are identified by having high memberships in the same cluster. "Hard" clustering algorithms assign each object to a single cluster that is using the two distinct membership values of 0 and 1. In Web usage profiling, this "all or none" or "black or white" membership restriction is not realistic. Very often there may not be sharp boundaries between clusters and many objects may have characteristics of different classes with varying degrees. Furthermore, a desired clustering technique should be immune to noise, which is inherently present in Web usage data. The browsing behavior of users on the Web is highly uncertain and fuzzy in nature. Each time the user accesses the site, the use may have different browsing goals. The main advantage of fuzzy clustering over hard clustering is that it can capture the inherent vagueness, imprecision, and uncertainty in Web usage data. Fuzzy clustering has been largely used in the context of user profiling for Web personalization (Joshi & Joshi, 2000; Suryavanshi et al., 2005). Castellano, Mesto, Minunno, and Torsello (2007e) prove the applicability of the well-known fuzzy C-means algorithm to extract user profiles. Nasraoui, Krishnapuram, and Joshi (1999) propose a relational fuzzy clustering algorithm named relational fuzzy clustering–maximal density estimator (RFC-MDE). Nasraoui and Frigui (2000) propose a competitive agglomeration relational data (CARD) algorithm to cluster user sessions. A hierarchical fuzzy clustering algorithm has been proposed by Dong and Zhuang (2004) to discover the user access patterns in an effective manner.

Recommendation

Once user preferences are understood by analyzing the derived user profiles, personalized services can be provided to each user, such as sending targeted advertisement to the connected users, adapting the content/structure of the Web site to the user needs, providing a guide to the user navigation, and so forth. Personalization functions can be accomplished in a manual or in an automatic and transparent manner for the user. In the first case, the discovered knowledge has to be expressed in a comprehensible manner for humans, so that knowledge can be analyzed to support human experts in making decisions. To accomplish this task, different approaches have been introduced in order to provide useful information for personalization. An effective method for presenting comprehensive information to humans is the use of visualization tools such as WebViz (Pitkow & Bharat, 1994) that represents navigational patterns as graphs. Reports are also a good method to synthesize and to visualize useful statistical information previously generated. Personalization systems as WUM (Spiliopoulou & Faulstich, 1998) and WebMiner (Cooley, Tan, & Srivastava, 1999) use SQL-like query mechanisms for the extraction of rules from navigation patterns.

Nevertheless, decisions made by the user may create delay and loss of information. A more interesting approach consists of the employment of Web usage mining for personalization. In particular, the knowledge extracted from Web data is automatically exploited to adapt the Web-based system by means of one or more of the personalization functions.

Various approaches can be used for generating a personalized experience for users. These are commonly distinguished in rule-based filtering, content-based filtering, and collaborative or social filtering (Mobasher et al., 2000). In rule-based filtering, static user models are generated through the registration procedure of the users. To generate personalized recommendations, a set of rules is specified, related to the content which is provided to the users with different models. Among the several products which adopt the rule-based filtering approach, Yahoo (Manber, Patel, & Obison, 2000) and Websphere Personalization (IBM) constitute two valid examples. Content-based filtering systems generate recommendations on the basis of the items previously rated by a user. The user profile is obtained by considering the content description of the items and it is exploited to predict a rating for previously unseen items. Examples of systems which adopt this personalization approach are represented by Personal WebWatcher (Mladenic, 1996), NewsWeeder (Lang, 1994), and Letizia (Liebermann & Letizia, 1995). Collaborative filtering systems are based on the assumption that users preferring similar items have the same interests. Personalization is obtained by searching for common features in the preferences of different users which are usually expressed explicitly in the form of item ratings or also in a dynamical manner through the navigational patterns extracted from usage data. Currently, collaborative filtering is the most employed approach of personalization. Amazon.com (Linden et al., 2003) and Recommendation Engine represent two major examples of collaborative filtering systems.

SOFT COMPUTING TECHNIQUES FOR WEB PERSONALIZATION

The term soft computing (SC) indicates a collection of methodologies that work synergistically to find approximate solutions for real-world problems which contain various kinds of inaccuracies and uncertainties. The guiding principle is to devise methods of computation that lead to an acceptable solution at low cost by seeking for an approximate solution to an imprecisely/precisely formulated problem. Computing paradigms underlying SC are:

- Neural computing that supplies the machinery for learning and modeling complex functions;
- Fuzzy logic computing that gives mechanisms for dealing with imprecision and uncertainty underlying real-life problems; and
- Evolutionary computing that provides algorithms for optimization and searching.

Systems based on such paradigms are neural networks (NN), fuzzy systems (FS), and genetic/evolutionary algorithms (GA/EA). Rather than a collection of different paradigms, SC is better regarded as a partnership in which each of the partners provides a methodology for addressing problems in a different manner. From this perspective, the key-points and the shortcomings of SC paradigms appear to be complementary rather than competitive. Therefore, it is a natural practice to build up integrated strategies combining the concepts of different SC paradigms to overcome limitations and exploit advantages of each single paradigm (Hildebrand, 2005; Tsakonas, Dounias, Vlahavas, & Spyropoulos 2002). This relationship enables the creation of hybrid computing schemes which use neural networks, fuzzy systems, and evolutionary algorithms in combination. An inspection of the multitude of hybridization strategies proposed in literature which involve NN, FS, and GA/EA would be somewhat impractical. It is however straightforward to indicate **neuro-fuzzy** (NF) **systems** as the most prominent representatives of hybridizations in terms of the number of practical implementations in several application areas (Lin & Lee, 1996; Nauck, Klawonn, & Kruse, 1997). NF systems use NN to learn and fine tune rules and/or membership functions from input-output data to be used in a FS (Mitra & Pal, 1995). With this approach, the main drawbacks of NN and FS are the black box behavior of NN and the lack of learning mechanism in FS are avoided. NF systems automate the process of transferring expert or domain knowledge into fuzzy rules, hence, they are basically FS with an automatic learning process provided by NN, or NN provided with explicit form of knowledge representation.

In the last few years, the relevance of SC methodologies to Web personalization tasks has drawn the attention of researchers, as indicated in a recent review (Frias-Martinez et al., 2005). Indeed, SC can improve the behavior of Web-based applications, as both imprecision and uncertainty are inherently present in the Web activity. Web data, being unlabeled, imprecise/incomplete, heterogeneous, and dynamic, appear to be good candidates to be mined in the SC framework. Besides, SC seems to be the most appropriate paradigm in Web usage mining where, being human interaction its key component, issues such as approximate queries, deduction, personalization, and learning have to be faced. SC methodologies, being complementary rather than competitive, can be successfully employed in combination to develop intelligent Web personalization systems. In this context, NN with self organization abilities are typically used for pattern discovery and rule generation. FS are used for handling issues related to incomplete/imprecise Web data mining, understandability of patterns, and explicit representation of Web recommendation rules. EA are mainly used for efficient search and retrieval. Finally, various examples of combination between SC techniques can be found in the literature concerning Web personalization, ranging from very simple combination schemas to more complicated ones. An example of simple combination is by Lampinen and Koivisto (2002), where user profiles are derived by a clustering process that combines a fuzzy clustering (the fuzzy C-means clustering) and a neural clustering (using a self-organising map). A Kuo and Chen (2004) discuss a more complex form of hybridization using all the three SC paradigms together, and also design a recommendation system for electronic commerce using fuzzy rules obtained by a combination of fuzzy neural networks and genetic algorithms.

Figure 1. The scheme of the proposed Web personalization system

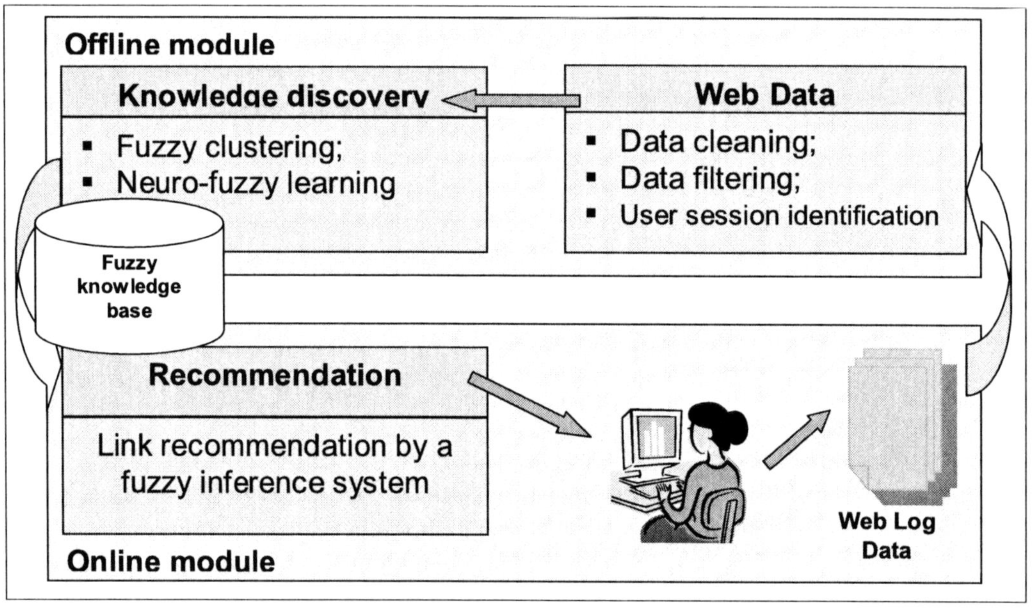

Here, fuzzy logic has also been used to provide a soft filtering process based on the degree of concordance between user preferences and the elements being filtered.

NF techniques are especially suited for Web personalization tasks where knowledge interpretability is desired. One of these tasks is the extraction of association rules for recommendation. Gyenesei (2000) explores how fuzzy association rules understandable to humans are learnt from a database containing both quantitative and categorical attributes by using a neuro-fuzzy approach like the one proposed by Nauck (1999). Lee (2001) uses a NF system for recommendation in an e-commerce site. Stathacopoulou, Grigoriadou, and Magoulas (2003) and Magoulas, Papanikolau, and Grigoriadou (2001) use a NF system to implement a classification/recommendation system with the purpose of adapting the contents of a Web course according to the model of the student. Recently Castellano, Fanelli, and Torsello (2007d) have proposed a Web personalization approach that uses fuzzy clustering to derive user profiles and a neural-fuzzy system to learn fuzzy rules for dynamic link recommendation. The next section is devoted to outlining the main features of our approach, in order to give an example of how different SC techniques can be used synergistically to perform Web personalization.

A NEURO-FUZZY WEB PERSONALIZATION SYSTEM

In this section, we describe a WUM personalization system for dynamic link suggestion based on a neuro-fuzzy approach. A fuzzy clustering algorithm is applied to determine user profiles by grouping preprocessed Web usage data into session categories. Then, a hybrid approach based on the combination of the fuzzy reasoning with a neural network is employed in order to derive fuzzy rules useful to provide dynamical predictions about Web pages to be suggested to the active user, according to user profiles previously identified.

According to the general scheme of a WUM personalization process described in section 3, three different phases can be distinguished in our approach:

- **Preprocessing** of Web log files in order to extract useful data about URLs visited during user sessions.
- **Knowledge discovery** in order to derive user profiles and to discover associations between user profiles and URLs to be recommended.
- **Recommendation** in order to exploit the knowledge extracted through the previous phases to dynamically recommend interesting URLs to the active user.

As illustrated in Figure 1, two major modules can be distinguished in the system: an off-line module that performs log data preprocessing and knowledge discovery, and an online module that recommends interesting Web pages to the current user on the basis of the discovered knowledge. In particular, during the preprocessing task, user sessions are extracted from the log files which are stored by the Web server. Each user session is represented by one record which registers the accesses exhibited by the user in that session. Next, a fuzzy clustering algorithm is executed on these records to group similar sessions into session categories representing user profiles. Finally, starting from the extracted user profiles and the available data about user sessions, a knowledge base expressed in the form of fuzzy rules is extracted via a neuro-fuzzy learning strategy. Such a knowledge base is exploited during the recommendation phase (performed by the online module) to dynamically suggest links to Web pages judged interesting for the current user. Specifically, when a user requests a new page, the online module matches the user's current partial session with the session categories identified by the off-line module and derives the degrees of relevance for URLs by means of a fuzzy inference process. In the following, we describe in more detail all the tasks involved in the Web personalization process.

Preprocessing

The aim of the preprocessing step is to identify user sessions starting from the information contained in a Web log file. Preprocessing of access log files is performed by means of log data preprocessor (LODAP) (Castellano, Fanelli, & Torsello, 2007a), a software tool that analyzes usage data stored in log files to produce statistics about the browsing behavior of the users visiting a Web site and to create user sessions by identifying the sequence of pages accessed by each visitor. LODAP preprocesses log data into three steps: data cleaning, data structuration, and data filtering. During data cleaning, Web log data are cleaned from the useless information in order to retain only records corresponding to the explicit requests of the users (i.e. requests with an access method different from "GET," failed and corrupt requests, requests for multimedia objects, and visits made by Web robots are removed). Next, significant log entries are structured into user sessions. In LODAP, a user session is defined as the finite set of URLs accessed by a user within a predefined time period (in our work, 25 minutes). Since the information about the user login is not available, user sessions are identified by grouping the requests originating from the same IP address during the established time period. The set of all users (IP) is defined by $U = \{u_1, u_2, ..., u_{n_U}\}$ and a user session is defined as the set of accesses originating from the same user (IP) within a predefined time period. Formally, a user session is represented as a triple $\mathbf{s}_i = \langle u_i, t_i, \mathbf{p}_i \rangle$ where $u_i \in U$ represents the user identifier, t_i is the total time access of the i-th session, and $\mathbf{p}_i$ is the set of all pages requested during the i-th session. More in detail,

$$\mathbf{p}_i = \left\langle (p_{i1}, t_{i1}, N_{i1}), (p_{i2}, t_{i2}, N_{i2}), ..., (p_{in_i}, t_{in_i}, N_{in_i}) \right\rangle$$

where p_{ik} is the k-th URL visited during the i-th session, t_{ik} is the total access time to page p_{ik}, and N_{ik} represents the number of accesses to page p_{ik} during the i-th session. Summarizing, after data structuration, a collection $S = \langle \mathbf{s}_1, \mathbf{s}_2, ..., \mathbf{s}_{n_s} \rangle$ of n_s sessions is identified from the log data. Finally, LODAP applies a data filtering process in order to remove requests for very low support URLs, that is, requests to pages which do not appear in a sufficient number of sessions, and requests for very high support URLs, that is, requests to pages which appear in nearly all sessions. Also, all sessions that include a very low number of visited URLs are removed. Hence, after data filtering, only m page requests (with $m \leq n_p$) and only n sessions (with $n \leq n_s$) are retained.

Once user sessions have been identified by LODAP, we create a visitor behavior model by defining a measure expressing the interest degree of the users for each visited page during a session. In our approach, we measure the interest degree for a page as the average access time on that page. Precisely, the interest degree for the j-th page in the i-th user session is defined as $ID_{ij} = t_{ij} / N_{ij}$ where t_{ij} is the overall time spent by the user on the j-th page and N_{ij} is the number of accesses to that page during the i-th session. Hence, we model the visitor behavior of each user through a pattern of interest degrees for all pages visited by that user. Since the number of pages visited by different users may vary, visitor behavior patterns may have different dimensions. To obtain a homogeneous behavior model for all users, we translate behavior patterns into vectors having the same dimension equal to the number m of pages retained by LODAP after page filtering. In particular, the behavior of the i-th user $i = 1,...,n$) is modeled as a vector $\mathbf{b}_i = (b_{i1}, b_{i2}, ..., b_{im})$ where

$$b_{ij} = \begin{cases} ID_{ij} & \text{if page } p_j \text{ is visited during session } \mathbf{s}_i \\ 0 & \text{otherwise} \end{cases}$$

Summarizing, we model the visitor behaviors by a $n \times m$ matrix $\mathbf{B} = \left[b_{ij} \right]$ where each entry represents the interest degree of the i-th user for the j-th page. Based on this matrix, visitors with similar preferences can be successively clustered together to create user profiles, as described in the following subsection.

Knowledge Discovery

In our approach, the knowledge discovery phase involves the creation of user profiles and the derivation of recommendation rules. This is performed by rule extraction for Web recommendation (REXWERE) (Castellano, Fanelli, & Torsello, 2007b), a software tool designed to extract knowledge from user sessions identified by LODAP. REXWERE employs a hybrid approach based on the combination of fuzzy reasoning and neural learning to extract knowledge in two successive phases: user profiling and fuzzy rule extraction. In user profiling, similar user sessions are grouped into clusters (user profiles) by means of a fuzzy clustering algorithm. Then, a neuro-fuzzy approach is applied to learn fuzzy rules which capture the association between user profiles and Web pages to be recommended. These recommendation rules are intended to be exploited by the online component of a WR system that dy-

Figure 2. The start-up panel of REXWERE

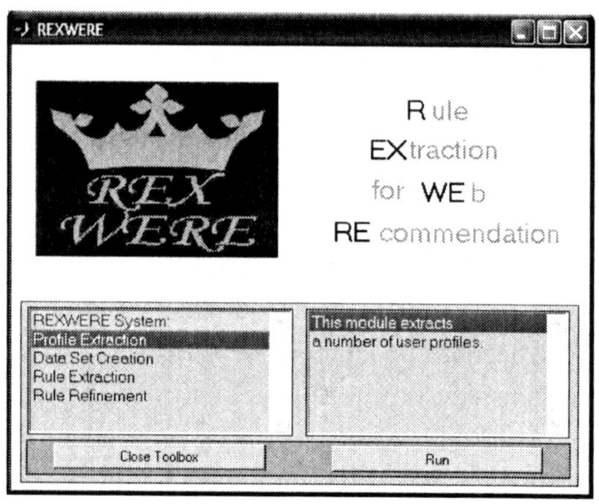

namically suggests links to interesting pages for a visitor of a Web site, according to the profiles the user belongs to. A key feature of REXWERE is the wizard-based interface that guides the execution of the different steps involved in the extraction of knowledge for recommendation. Figure 2 shows the start-up panel of REXWERE.

Starting from the behavior data derived from user sessions, REXWERE extracts recommendation rules in two main phases:

1. User profiling, that is, the extraction of user profiles through clustering of behavior data.
2. Fuzzy rule extraction, that is, the derivation of a set of rules that capture the association between the extracted user profiles and Web pages to be recommended. This task is carried out through three modules:
   - The *dataset creation* module which creates the training set and the test set needed for the learning of fuzzy rules;
   - The *rule extraction* module that derives an initial fuzzy rule base by means of an unsupervised learning; and
   - The *rule refinement* module that improves the accuracy of the fuzzy rule base by means of a supervised learning.

As result, REXWERE provides in output a set of fuzzy recommendation rules to be used as knowledge base in an online activity of dynamic link suggestion.

Discovery of User Profiles

The first task of REXWERE is the extraction of user profiles that categorize user sessions on the basis of similar navigational behaviors. This is accomplished by means of the *profile extraction* module that is based on a clustering approach. Clustering algorithms are widely used in the context of user profiling since they have the capacity to examine large quantity of data in a fairly reasonable amount of time. In particular, fuzzy clustering techniques seem to be particularly suited in this context because they can partition data into overlapping clusters (user profiles). Due to this peculiar characteristic, a user may belong to more than one profile with a certain membership degree. Two fuzzy clustering algorithms are implemented in REXWERE to extract user profiles:

- The well-known fuzzy C-means (FCM) algorithm (Castellano et al., 2007d), that belongs to the category of clustering algorithms working on object data expressed in the form of feature vectors.
- The CARD+ algorithm (Castellano, Fanelli, & Torsello, 2007c), a modified version of the competitive agglomeration relational data algorithm (Nasraoui & Frigui, 2000), which works on relational data representing the pairwise similarities (dissimilarities) between objects to be clustered.

These two algorithms differ in some features. While the FCM directly works on the behavior matrix **B** containing the interest degrees of each user for each page, CARD+ works on a relation matrix containing the dissimilarity values between all pairs of behavior vectors (rows of matrix **B**). Moreover, one key feature of CARD+ is the ability to automatically determine the final number of clusters starting from an initial random number. On the contrary, the FCM requires the number of clusters to be fixed in advance. In this case, the proper number of profiles is established by calculating the Xie-Beni index (Halkidi, Batistakis, & Vazirgiannis, 2002) for different partitions corresponding to different number of clusters; the partition with the smallest value of the Xie-Beni index corresponds to the optimal number of clusters for the available input data.

Both the FCM and the CARD+ provide the following results:

- C cluster centers (user profiles) represented as vectors $\mathbf{v}_c = (v_{c1}, v_{c2}, ..., v_{cm})$ with $c = 1,...,C$
- A fuzzy partition matrix $\mathbf{U} = [u_{ic}]_{i=1,...,n}^{c=1,...,C}$ where each component u_{ic} represents the membership degree of the i-th user to the c-th profile.

These results are used in the successive knowledge discovery task performed by REXWERE.

Discovery of Recommendation Rules

Once profiles have been extracted, REXWERE enters in the second knowledge extraction phase, that is, the extraction of fuzzy rules for recommendation. Such rules represent the knowledge base to be used in the ultimate online process of link recommendation. Each recommendation rule expresses a fuzzy relation between a behavior vector $\mathbf{b} = (b_1, b_2, ..., b_m)$ and relevance of URLs in the following form:

IF (b_l **is** A_{lk}) AND ... AND (b_m is A_{mk}) THEN (relevance of URL_1 is y_{lk}) AND ... AND (relevance of URL_m is y_{lk})

for $k = 1,..,K$ where K is the number of rules, A_{jk} (j=1,..., m) are fuzzy sets with Gaussian membership functions defined over the input variables b_j, and y_{jk} are fuzzy singletons expressing the relevance degree of the jth URL.

The main advantage of using a fuzzy knowledge base for recommendation is the readability of the extracted knowledge. Actually, fuzzy rules can be easily understood by human users since they can be expressed in a linguistic fashion by labelling fuzzy sets with linguistic terms such as LOW, MEDIUM, and HIGH. Hence, a fuzzy rule for recommendation can assume the following linguistic form:

IF (the degree of interest for URL_1 is LOW) AND ... AND (the degree of interest for URL_m is HIGH) THEN (recommend URL_1 with relevance 0.3) AND ... AND (recommend URL_m with relevance 0.8)

Such fuzzy rules are derived through a hybrid strategy based on the combination of fuzzy reasoning with a specific neural network that encodes in its structure the discovered knowledge in form of fuzzy rules. The network is trained on a set of input-output samples describing the association between user sessions and preferred URLs. Precisely, the training set is a collection of n input-output vectors: $\mathbf{T} = \langle(\mathbf{b}_i, \mathbf{r}_i)\rangle_{i=1,...,n}$ where the input vector $\mathbf{b}_i$ represents the behavior vector of the i-th user and the desired output vector $\mathbf{r}_i$ expresses the relevance degrees associated to the m URLs for the i-th visitor. To compute such relevance degrees, we exploit information embedded in the profiles extracted through fuzzy clustering. Precisely, for each behavior vector $\mathbf{b}_i$ we consider its membership values $\{u_{ic}\}_{c=1,...,C}$ in the fuzzy partition matrix $\mathbf{U}$. Then, we identify the two top matching profiles $c_1, c_2 \in \{1,..,C\}$ as those with the highest membership values. The relevance degrees in the output vector $\mathbf{r}_i = (r_i^1, r_i^2, ..., r_i^m)$ are hence calculated as follows: $r_i^j = u_{ic} v_{ic_1} + u_{ic} v_{ic_2}$ for $j = 1,...,m$ and $i = 1,...,n$. Once the training set has been constructed, the neural network can enter the learning phase to extract the knowledge embedded into training set and represent it as a collection of fuzzy rules. The learning is articulated in two steps. The first step is based on an unsupervised learning, based on a rival penalized mechanism, which provides a clustering of the behavior vectors and the definition of an initial fuzzy rule base. In this step, the structure and the parameters of fuzzy rules are identified. Successively, the obtained knowledge base is refined by a supervised learning process. Here, fuzzy rule parameters are tuned via supervised learning to improve the accuracy of the derived knowledge. Major details on the algorithms underlying the learning strategy can be retrieved in the work of Castellano, Castiello, Fanelli, and Mencar (2005).

Recommendation

The ultimate task of our Web personalization approach is the online recommendation of links to Web pages judged interesting for the current user of the Web site. Specifically, when a new user accesses the Web site, an online module matches the user's current partial session against the fuzzy rules currently available in the knowledge base and derives a vector of relevance degrees by means of a fuzzy inference process.

Formally, when a new user has access to the Web site, an active user's current session is created in the form of a vector $\mathbf{b}^0$. Each time the user requests a new page, the vector is updated. To maintain the active session, a sliding window is used to capture the most recent user's behavior. Thus, the partial active session of the current user is represented as a vector $\mathbf{b}^0 = \left(b_1^0, \ldots, b_m^0\right)$ where some values are equal to zero, corresponding to unexplored pages.

Based on the set of K rules generated through the neural learning described above, the recommendation module provides URL relevance degrees by means of the following fuzzy reasoning procedure:

(1) Calculate the matching degree of current behavior vector $\mathbf{b}^0$ to the k-th rule, for $k = 1,.., K$ by means of product operator:

$$\mu_k\left(\mathbf{b}^0\right) = \prod_{j=1}^{n} \mu_{jk}\left(b_j^0\right)$$

(2) Calculate the relevance degree r_j^0 for the j-th URL as:

$$r_j^0 = \frac{\sum_{k=1}^{K} r_{jk}\mu_k\left(\mathbf{b}^0\right)}{\sum_{k=1}^{K} \mu_k\left(\mathbf{b}^0\right)}, \; j = 1 \ldots m$$

This inference process provides the relevance degree for all the considered m pages, independently on the actual navigation of the current user. In order to perform dynamic link suggestion, the recommendation module first identifies URLs that have been not visited by the current user, that is, all pages such that $b_j^0 = 0$. Then, among unexplored pages, only those having a relevance degree r_j^0 greater than a properly defined threshold α are recommended to the user. In practice, a list of links is dynamically included in the page currently visited by the user.

A Case Study

The proposed Web personalization approach was applied on a Web site targeted to young users (average age 12 years old), that is, the Italian Web site of the Japanese movie Dragon Ball (www.dragonballgt.it). This site was chosen because of its high daily number of accesses (thousands of visits each day).

The LODAP system was used to identify user sessions from the log data collected during a period of 24 hours. After data cleaning, the number of requests was reduced from 43,250 to 37,740 that were structured into 14,788 sessions. The total number of distinct URLs accessed in these sessions was 2,268. Support-based data filtering was used to eliminate requests for URLs having a number of accesses less than 10% of the maximum number of accesses, leading to only 76 distinct URLs and 8,040 sessions. Also, URLs appearing in more than 80% of sessions (including the site entry page) were filtered out, leaving 70 final URLs and 6,600 sessions. In a further filtering step, LODAP eliminated short sessions, leaving only sessions with at least three distinct requests. We obtained a final number of 2,422 sessions. The 70 pages in the Web site were labeled with a number (see Table 1) to facilitate the analysis of results. Once user sessions were identified and visitor behavior models were derived by calculating the interest

Table 1. Description of the pages in the Web site

Pages	Content
1	Home page
2	Comments by users
3,…,12	Pictures related to the movie
13,…,18	Pictures of characters
19, 26, 27	Matches
20, 21, 36, 47, 48	Services (registration, login,…)
22, 23, 25, 28, ..., 31, 50, 51	General information about the movie
32, ..., 35, 55	Entertainment (games, videos,...)
37, ..., 46, 49, 52, ..., 54, 56	Description of characters
57, ..., 70	Galleries

Table 2. Clusters of visitor behaviour

Cluster	Cardinality	Visited pages	Interest degree
1	906	(28, 12, 15, 43, 17, 22, 13, 50)	(11.1, 7.3, 6.9, 6.6, 6.59, 5.14, 4.5, 4.4)
2	599	(28, 26, 22, 55, 12, 33, 49, 32)	(80.8, 43.3, 30.1, 25.9, 24.5, 19.2, 18.1, 14.3)
3	917	(28, 12, 26, 13, 22, 9, 19, 16)	(5.3, 4.0, 3.4, 3.1, 2.6, 2.60, 2.3, 2.2)

degrees of each user for each page, leading to a 2422x70 behavior matrix.

Next, the two fuzzy clustering algorithms implemented in REXWERE were applied to the behavior matrix in order to obtain clusters of users with similar navigational behavior. Several runs of FCM were carried out with different number of clusters (C=30, 20, 15, 10). For each trial, we analyzed the obtained cluster center vectors and we observed that many of them were identical. Hence, an actual number of three clusters were found in each run. Also, a single run of the CARD+ was carried out by setting a maximum number of clusters equal to C=15. As a result, this clustering algorithm provided three clusters, confirming the results obtained by the FCM algorithm. This demonstrated that three clusters were enough to model the behavior of all the considered users. Table 2 summarizes the three clusters obtained by CARD+ that are very similar to those obtained after different trials of FCM. For each cluster, the cardinality and the first eight (most interesting) pages are displayed. It can be noted that some pages (e.g., Pages 12, 22, and 28) appear in more than one cluster, thus showing the importance of producing overlapping clusters. In particular, Page 28 (i.e., the page that lists the episodes of the movie) appears in all the three clusters with the highest degree of interest.

An interpretation of the three clusters revealed the following profiles:

- Profile 1. Visitors in this profile are mainly interested in pictures and descriptions of characters.
- Profile 2. These visitors prefer pages that link to entertainment objects (games and video)
- Profile 3. These visitors are mostly interested in matches among characters.

A qualitative analysis of these profiles made by designer of the considered Web site confirmed that they correspond to real user categories reflecting the interests of the typical site users.

The next step was the creation of recommendation rules starting from the extracted user profiles. A neural network with 70 inputs (corresponding to the components of the behavior vector) and 70 outputs (corresponding to the relevance values of the Web pages) was considered. The network was trained on a training set of 1,400 input-output samples derived from the available 2,000 behavior patterns and from the three user profiles, as described in Section 5.2.2. The remaining 600 samples were used for testing. The training of the network was stopped when the error on the training set dropped below 0.01, corresponding to a testing error of 0.03.

The derived fuzzy rule base was integrated into the online recommendation module to infer the relevance degree of each URL for the active user. These relevance degrees were ultimately used to suggest a list of links to unexplored pages retained interesting to the current user. To perform link recommendation, the navigational behavior of the active user was observed during a temporal window of 3 minutes in order to derive the behavior pattern corresponding to the user's partial visit. Such behavior pattern was used as input to the fuzzy rule inference process that computes the relevance degrees for all the considered 10 pages. Then, among the unexplored pages, only those having a relevance degree greater than a = 0.7 were included in the list of links to be suggested. As an example, Figure 3 shows a page of the considered Web site after the

Figure 3. A personalized page of the Web site. The recommended links are displayed in the up-right corner (inside the red circle)© 2008 Fabrizio Mesto. Used with permission.

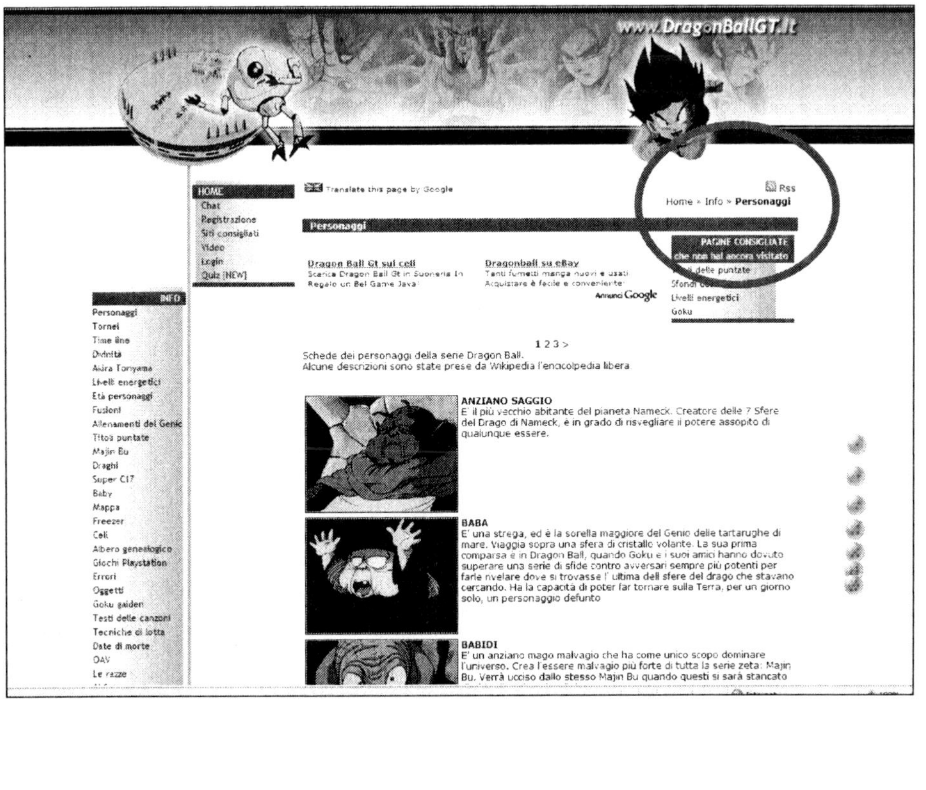

online recommendation module has dynamically included the list of suggested links.

CONCLUSION AND FUTURE TRENDS

The rapid development of the World Wide Web as a medium for information dissemination has generated a growing interest in the domain of Web personalization that may offer a variety of functionalities in several context, such as customization, task performance support, personalized guidance, and so forth. Specifically, in personalized guidance, the knowledge acquired from the analysis of users' navigational behavior (usage data) can be conveniently exploited in order to customize the Web information space to the necessities of users. As a consequence, there is growing interest in tools for automatic identification of user profiles by modeling the preferences of different user categories. Once user preferences are understood by analyzing the discovered user profiles, personalized services can be provided to each user.

In the Web personalization context, soft computing techniques emerge as valid tools to handle the ambiguity and uncertainty inherent in Web usage data. A brief survey of recent approaches to Web personalization that employ SC techniques has been presented. The survey emphasizes how most of Web personalization applications developed so far are based on combinations of SC techniques. As an example, a Web personalization system for dynamic link recommendation joining techniques from the neural and the fuzzy paradigms has been described. This neuro-fuzzy personalization system extracts knowledge from Web usage data in a twofold form: a set of fuzzy user profiles that capture preferences of similar users and a collection of fuzzy rules that describe associations between user profiles and links to be recommended.

In addition to those discussed in this chapter, there are some other aspects of Web personalization where SC is likely to play a key role. For example, user profiles generated by Web mining techniques are typically represented in a simplistic manner, by means a vector of ratings. More expressive models based on fuzzy logic could be explored in order to represent the vague and heterogeneous information characterizing user preferences. Another important facet is the ability to identify the continuous changes in interests of users and dynamically adapt user profiles according to these changes. Neural network learning algorithms based on online schemas could be investigated to cope with this issue. On the whole, hybrid approaches that synergistically combine SC methods show great potential for Web personalization, opening new research directions within the area of Web intelligence.

REFERENCES

Abraham, A. (2003). Business intelligence from Web usage mining. *Journal of Information & Knowledge Management*, *2*(4), 375-390.

Adomavicius, G., & Thuzilin, A. (2005). Towards the next generation of recommender systems: A survey of the state of the art and possible extensions. *IEEE Transaction on Knowledge and Data Engineering*, *17*(6), 734-749.

Banerjee, A., & Ghosh, J. (2001). Clickstream clustering using weighted longest common subsequences. In *Proceedings of the Web Mining Workshop at the 1st SIAM Conference on Data Mining*.

Baraglia, R., & Silvestri, F. (2004). *An online recommender system for large Web sites*. Paper presented at the IEEE/WIC/ACM International Conference on Web Intelligence, Beijing, China.

Callan, J., Smeaton, A., Beaulieu, M., Borlund, P., Brusilovsky, P., Chalmers, M., & et al. (2001). Personalization and recommender systems in digital libraries. In *Proceedings of the 2nd DELOS Workshop on Personalization and Recommender Systems in Digital Libraries.*

Castellano, G., Castiello, C., Fanelli, A. M., & Mencar, C. (2005). Knowledge discovering by a neuro-fuzzy modelling framework. *Fuzzy Sets and Systems, 149,* 187-207.

Castellano, G., Fanelli, A. M., & Torsello, M. A. (2007a, February 16-19). LODAP: A log data preprocessor for mining Web browsing patterns. In *Proceedings of the 6th WSEAS International Conference on Artificial Intelligence, Knowledge Engineering and Data Base (AIKED 2007),* Corfu, Greece (pp. 12-17).

Castellano, G., Fanelli, A. M., & Torsello, M. A. (2007b). REXWERE: A tool for fuzzy rule extraction in Web recommendation. In *Proceedings of NAFIPS 2007,* San Diego.

Castellano, G., Fanelli, A. M., & Torsello, A. M. (2007c, July 15-22). Web user profiling using relational fuzzy clustering. In *Proceedings of 12th International Conference on Fuzzy Theory and Technology (FTT 2007),* Salt Lake City.

Castellano, G., Fanelli, A. M., & Torsello, M. A. (2007d) A neuro-fuzzy collaborative filtering approach for Web recommendation. *International Journal of Computational Science, 1*(1), 27-39. ISSN: 1992-6669 (Print)/ISSN: 1992-6677.

Castellano, G., Mesto, F., Minunno, M., & Torsello, M. A. (2007e, July 7-10). Web user profiling using fuzzy clustering. In *Proceedings of 7th International Workshop on Fuzzy Logic and Applications (WILF 2007),* Portofino Vetta (Genova), Italy.

Cho, Y. H., & Kim, J. K. (2004). Application of Web usage mining and product taxonomy to collaborative recommendations in e-commerce. *Expert Systems with Applications, 26,* 233-246.

Cooley, R. (2000). *Web usage mining: Discovery and application of interesting patterns from Web data.* Unpublished doctoral thesis, University of Minnesota.

Cooley, R., Tan, P. N., & Srivastava, J. (1999). *Discovering of interesting usage patterns from Web data* (Tech. Rep. No. 99-022). University of Minnesota.

Dong, Y., & Zhuang, Y. (2004, June 15-19). Fuzzy hierarchical clustering algorithm facing large database. In *Proceedings of the 5th World Congress on Intelligent Control and Automation,* Hangzhou, P.R. China.

Eirinaki, M., & Vazirgiannis, M. (2003). Web mining for Web personalization. *ACM TOIT, 3*(1), 2-27.

Enembreck, F., Barthès, J-P., & Ávila, B. C. (2004). *Personalizing information retrieval with multi-agent systems* (LNCS 3191/2004, pp. 77-91). Springer, Berlin-Heidelberg.

Etzioni, O. (1996). The World Wide Web: Quagmire or gold mine. *Communications of the ACM, 39*(11), 65-68.

Facca, F. M., & Lanzi, P. L. (2005). Mining interesting knowledge from Weblogs: A survey. *Data & Knowledge Engineering, 53,* 225-241.

Frias-Martinez, E., Magoulas, G., Chen, S., & Macredie, R. (2005). Modeling human behavior in user-adaptive systems: recent advances using soft computing techniques. *Expert Systems with Applications, 29*(2), 320-329.

Gyenesei, A. (2000). *A fuzzy approach for mining quantitative association rule* (TUCS Tech. Rep. 336). Univ. Turku, Dept. Comput. Sci., Lemminkisenkatu 14, Finland.

Halkidi, M., Batistakis, Y., & Vazirgiannis, M. (2002). *Cluster validity methods: Part II.* in SIGMOD Record.

Han, J., & Kamber, M. (2001). *Data mining concepts and techniques.* Morgan Kaufmann.

Hildebrand, L. (2005). Hybrid computational intelligence systems for real world applications. *Series studies in fuzziness and soft computing* (Vol. 179, pp. 165-195). Springer Berlin/Heidelberg.

Heer, J., & Chi, E. H. (2002). Mining the structure of user activity using cluster stability. In *Proceedings of the Workshop on Web Analytics/Second SIAM Conference on Data Mining.* ACM Press.

Huang, X., Cercone, N., & An, A. (2002a). Comparison of interestingness functions for learning Web usage patterns. In *Proceedings of the Eleventh International Conference on Information and Knowledge Management* (pp. 617-620). ACM Press.

Huang, J. Z., Ng, M., Ching, W.-K., Ng, J., & Cheung, D. (2001, August 26). A cube model and cluster analysis for Web access sessions. In R. Kohavi, B. Masand, M. Spiliopoulou, & J. Srivastava (Eds.), *WEBKDD 2001—Mining Web Log Data Across All Customers Touch Points, Third International Workshop,* San Francisco (LNCS 2356, pp. 48-67). Springer.

Jespersen, S. E., Thorhauge, J., & Pedersen, T. B. (2002). A hybrid approach to Web usage mining. In *Proceedings of the 4th International Conference on Data Warehousing and Knowledge Discovery* (pp. 73-82). Springer-Verlag.

Joshi, A., & Joshi, K. (2000). *On mining Web access logs.* Paper presented at the ACM SIGMOID Workshop on Research Issues in Data Mining and Knowledge Discovery (Vol. 63, No. 69).

Joshi, K. P., Joshi, A., & Yesha, Y. (2003). On using a warehouse to analyze Web logs. *Distributed and Parallel Databases, 13*(2), 161-180.

Kim, D.-W., & Lee, K. H. (2001). A new fuzzy information retrieval system based on user preference model. *FUZZY-IEEE,* 127-130.

Kohrs, A., & Merialdo, B. (1999). Clustering for collaborative filtering applications. *Computational intelligence for modelling, control and automation.* IOS Press.

Kosala, R., & Blockeel, H. (2000). Web mining research: A survey. *SIGKDD Explorations: Newsletter of the Special Interest Group (SIG) on Knowledge Discovery & Data Mining, 2*(1), 1-15. ACM.

Kraft, D. H., Chen, J., Martin-Bautista, M. J., & Vila, M. A. (2002). Textual information retrieval with user profiles using fuzzy clustering and inferencing. In P. S. Szczepaniak, J. Segovia, J. Kacprzyk, & L. A. Zadeh (Eds.), *Intelligent exploration of the Web.* Heidelberg, Germany: Physica-Verlag.

Kuo, R. J., & Chen, J. A. (2004). A decision support system for order selection in electronic commerce based on fuzzy neural network supported by real-coded genetic algorithm. *Expert Systems with Applications, 26,* 141-154.

Lampinen, T., & Koivisto, H. (2002). Profiling network applications with fuzzy c-means clustering and self-organising map. In *Proceedings of the 1st International Conference on Fuzzy Systems and Knowledge Discovery: Computational Intelligence for the E-Age* (pp. 300-304).

Lang, K. (1995). Newsweeder: Learning to filter netnews. In *Proceedings of the 12th International Conference on Machine Learning.*

Langley, P. (1999). User modeling in adaptive interfaces. In *Proceedings of the Seventh International Conference on User Modeling,* Canada (pp. 357-370).

Lee, R. S. T. (2001). iJADE IWShopper: A new age of intelligent Web shopping system based on fuzzy-neuro agent technology. In N. Zhong, Y. Yao, J. Liu, & S. Ohsuga *Web intelligence: Research and development* (LNAI 198, pp. 403-412).

Lieberman, H., & Letizia (1995). An agent that assists Web browsing. In *Proceedings of the 14th International Joint Conference in Artificial Intelligence* (pp. 924-929).

Lin, C. T., & Lee, C. S. (1996). *Neural fuzzy systems: A neuro-fuzzy synergism to intelligent systems*. Englewood Cliffs, NJ: Prentice-Hall.

Linden, G., Smith, B., & York, J. (2003). Amazon.com recommendations item-to-item collaborative filtering. *IEEE Internet Computing, 7*(1), 76-80.

Magoulas, G. D., Papanikolau, K. A., & Grigoriadou, M. (2001). Neurofuzzy synergism for planning the content in a Web-based course. *Informatica, 25*(1), 39-48.

Manber, U., Patel, A., & Obison, J. (2000). Experience with personalization on Yahoo. *Communications of the ACM, 43*(8), 35-39.

Menasalvas, E., Millan, S., Pena, J., Hadjimichael, M., & Marban, O. (2002). Subsessions: A granular approach to click path analysis. In *Proceedings of the FUZZ-IEEE Fuzzy Sets and Systems Conference, at the World Congress on Computational Intelligence*, Honolulu (pp. 12-17).

Mitra, S., & Pal, S. K. (1995). Fuzzy multi-layer perceptron, inferencing and rule generation. *IEEE Transactions on Neural Networks, 6*, 51-63.

Mladenic, D. (1996). *Personal Web watcher: Implementation and design* (Tech. Rep. IJSDP-7472). Department of Intelligent Systems, J. Stefan Institute, Slovenia.

Mobasher, B. (2005). Web usage mining and personalization. In M. P. Singh (Ed.), *Practical handbook of internet computing*. CRC Press.

Mobasher, B. (2006). Web usage mining. In B. Liu (Ed.), *Web data mining: Exploring hyperlinks, contents and usage data*. Berlin-Heidelberg: Springer.

Mobasher, B. (2007a). Data mining for personalization. In P. Brusilovsky, A. Kobsa, & W. Nejdl (Eds.), *The adaptive Web: Methods and strategies of Web personalization* (LNCS 4321, pp. 90-135). Berlin-Heidelberg: Springer.

Mobasher, B. (2007b). Recommender systems. *Kunstliche Intelligenz, Special Issue on Web Mining*, (3), 41-43. Bremen, Germany: BottcherIT Verlag.

Mobasher, B., Cooley, R., & Srivastava, J. (1999). *Automatic personalization based on Web usage mining* (Tech Rep. 99010). DePaul University, Department of Computer Science.

Mobasher, B., Cooley, R., & Srivastava, J. (2000). Automatic personalization based on Web usage mining. *Communications of the ACM, 43*(8), 142-151.

Mobasher, B., Dai, H., Luo, T., & Nakagawa, M. (2001). *Effective personalization based on association rule discovery from Web usage data.* Paper presented at the ACM Workshop on Web Information and Data Management, Atlanta.

Mobasher, B., Dai, H., Nakagawa, M., & Luo, T. (2002). Discovery and evaluation of aggregate usage profiles for Web personalization. *Data Mining and Knowledge Discovery, 6*, 61-82.

Mobasher, B., Nasraoui, O., Liu, B., & Masand, B. (2006). *Advances in Web mining and Web usage analysis* (2004 WebKDD Volume) (LNAI 3932). Berlin-Heidelberg: Springer.

Mortazavi-Asl, B. (2001). *Discovering and mining user Web-page traversal patterns.* Unpublished master's thesis, Simon Fraser University.

Mulvenna, M., Anand, S., & Buchner, A. (2000). Personalization on the net using Web mining. *CACM, 43*(8), 123-125.

Nanopoulos, A., Katsaros, D., & Manolopoulos, Y. (2001, August 26). Exploiting Web log mining for Web cache enhancement. In R. Kohavi, B. Masand, M. Spiliopoulou, & J. Srivastava (Eds.), *WEBKDD 2001—Mining Web Log Data Across*

All Customers Touch Points, Third International Workshop, San Francisco (LNCS 2356, pp. 68-87). Springer.

Nasraoui, O. (2005). World Wide Web personalization. In J. Wang (Ed.), *Encyclopedia of data mining and data warehousing.* Hershey, PA: IGI Global, Inc.

Nasraoui, O., & Frigui, H. (2000). Extracting Web user profiles using relational competitive fuzzy clustering. *International Journal on Artificial Intelligence Tools, 9*(4), 509-526.

Nasraoui, O., Krishnapuram, R., & Joshi, A. (1999). Relational clustering based on a new robust estimator with applications to Web mining. In *Proceedings of the International Conf. North American Fuzzy Info. Proc. Society (NAFIPS 99),* New York (pp. 705-709).

Nauck, D. (1999). Using symbolic data in neuro-fuzy classification. In *Proceedings of NAFIPS'99,* New York (pp. 536-540).

Nauck, D., Klawonn, F., & Kruse, R. (1997). *Foundations of neuro–fuzzy systems.* Chichester, UK: Wiley.

O'Connor, M., & Herlocker, J. (1999). Clustering items for collaborative filtering. In *Proceedings of ACM SIGIR '99 Workshop on Recommender Systems: Algorithms and Evaluation.*

Pei, J., Han, J., Mortazavi-asl, B., & Zhu, H. (2000). Mining access patterns efficiently from Web logs. Paper presented at the *Pacific-Asia Conference on Knowledge Discovery and Data Mining* (pp. 396-407).

Pal, S., Talwar, V., & Mitra, P. (2002). Web mining in soft computing framework: Relevance, state of the art and future directions. *IEEE Transactions on Neural Networks, 13*(5), 1163-1177.

Perkowitz, M., & Etzioni, O. (1997). *Adaptive sites: Automatically learning from user access patterns* (Tech. Rep. UW-CSE-97-03-01). University of Washington.

Pierrakos, D., Paliouras, G., Papatheodorou, C., & Spyropoulos, C. D. (2003). Web usage mining as a tool for personalization: A survey. *User Modeling and User-Adapted Interaction, 13*(4), 311-372.

Pitkow, J., & Bharat, K. (1994). WEBVIZ: A tool for World-Wide Web access log visualization. In *Proceedings of the 1st International World-Wide Web Conference.* Geneva, Switzerland (pp. 271-277).

Pohl, W. (1996). Learning about the user modeling and machine learning. In V. Moustakis & J. Herrmann (Eds.), *International Conference on Machine Learning Workshop Machine Learning Meets Human-Computer Interaction* (pp. 29-40).

Sankar, K. P., Varun, T., & Pabitra, M. (2002). Web mining in soft computing framework: Relevance, state of the art and future directions. *IEEE Transaction on Neural Networks, 13*(5), 1163-1177.

Schafer, B., Konstan, J. A., & Riedl, J. (2001). E-commerce recommendation applications. *Data Mining and Knowledge Discovery, 5*(12), 115-152. Kluwer Academic Publishers.

Spiliopoulou, M. (1999). *Tutorial: Data mining for the Web.* Paper presented at the PKDD'99. Prague, Czech Republic.

Spiliopoulou, N., & Faulstich, L. C. (1998). *WUM: A Web utilization miner.* Paper presented at the International Workshop on the Web and Databases, Valencia, Spain (LNCS 1590, pp. 109-115). Springer.

Spiliopoulou, M., Mobasher, B., Berendt, B., & Nakagawa, M. (2003). A framework for the evaluation of session reconstruction heuristics in Web usage analysis. *INFORMS Journal of Computing - Special Issue on Mining Web-Based Data for E-Business Applications, 15*(2), 171-190.

Stathacopoulou, R., Grigoriadou, M., & Magoulas, G. D. (2003). A neurofuzzy approach in student modeling. In *Proceedings of the 9th International Conference on User Modeling, UM2003* (LNAI 2702, pp. 337-342).

Suryavanshi, B. S., Shiri, N., & Mudur, S. P. (2005). An efficient technique for mining usage profiles using relational fuzzy subtractive clustering. In *Proceedings of the 2005 International Workshop on Challenges in Web Information Retrieval and Integration (WIRI'05)* (pp. 23-29).

Tan, P. N., & Kumar, V. (2002). Discovery of Web robot sessions based on their navigational patterns. *Data Mining and Knowledge Discovery, 6*(1), 9-35.

Tsakonas, A., Dounias, G., Vlahavas, I. P., & Spyropoulos, C. D. (2002). *Hybrid computational intelligence schemes in complex domains: An extended review* (LNCS 2308, pp. 494-511).

Vakali, A., Pokorný, J., & Dalamagas, T. (2004). *An overview of Web data clustering practices.* Paper presented at the EDBT Workshops (pp. 597-606).

Webb, G. I., Pazzani, M. J., & Billsus, D. (2001). Machine learning for user modeling. *User Modeling and User-Adapted Interaction, 11,* 19-29. Kluwer.

Wong, S. S. C., & Pal, S. (2001). *Mining fuzzy association rules for Web access case adaptation.* Paper presented at the Workshop on Soft Computing in Case-Based Reasoning/International Conference on Case-Based Reasoning (ICCBR_Ol).

Yao, Y. Y. (2005, May 19-21). Web intelligence: New frontiers of exploration. In H. Tarumi, Y. Liand, & T. Yoshida (Eds.), *Proceedings of the 2005 International Conference on Active Media Technology (AMT 2005),* Takamatsu, Kagawa, Japan (pp. 3-8).

Compilation of References

AA. VV. (2002). *Stanford-Makovsky Web credibility study 2002: Investigating what makes Web sites credible today.* Retrieved June 23, 2008, from www.webcredibility.org

Abraham, A. (2003). Business intelligence from Web usage mining. *Journal of Information & Knowledge Management, 2*(4), 375-390.

Accenture. (2004). *E-Government leadership: Engaging the customer.* Retrieved July 15, 2007, from https://www.accenture.com/Global/Services/By_Industry/Government/R_and_I/CitizensView.htm

ACCESS. (n.d.). *Small-fit rendering.* Retrieved June 14, 2007, from http://www.access-company.com/products/netfrontmobile/contentviewer/mcv_tips.html#Anchor-Smar-45765

Adams, J., Koushik, S., Vasudeva, G., & Galambos, G. (2001). *Patterns for e-business: A strategy for reuse.* IBM Press.

Adomavicius, G., & Thuzilin, A. (2005). Towards the next generation of recommender systems: A survey of the state of the art and possible extensions. *IEEE Transaction on Knowledge and Data Engineering, 17*(6), 734-749.

Agliati, M. (Ed.). (1996). *Tecnologia dell'informazione e sistema amministrativo.* Milan: Egea.

Ahluwalia, K. S., & Jain, A. (2006, October 21-23). *High availability design patterns.* Paper presented at the Thirteenth Conference on Pattern Languages of Programs (PLoP 2006), Portland.

Aladwani, A. M. (2002). Organizational actions, computer attitudes, and end-user satisfaction in public organizations: An empirical study. *Journal of End User Computing, 14*(1), 42-49.

Aladwani, A. M. (2003). A deeper look at the attitude-behavior consistency assumption in information systems satisfaction research. *Journal of Computer Information Systems, 44*(1), 57-63.

Albuquerque, A. B., & Belchior, A. D. (2002, May 7-11). *E-commerce Websites: A qualitative evaluation.* Paper presented at the Eleventh International World Wide Web Conference (WWW 2002), Honolulu.

Alemi, F. (2000, Fall). Management matters: Technology succeeds when management innovates. *Frontiers of Health Service Management, 17*(1), 17-30.

Alexander, S. (1999, June 21). Want to give your career its best skills boost? Invest in these honey pots, say top IT execs. *Computerworld,* 78-82.

Alexandrou, M. (2007). *CRM definition.* Retrieved February 1, 2007, from http://www.mariosalexandrou.com/definition/crm.asp

Allman, D. (2004). *Taking CRM to a higher level of interaction.* Retrieved February 18, 2007, from http://www.destinationcrm.com/articles/default.asp?ArticleID=3849

Alonso, G., Casati, F., Kuno, H., & Machiraju, V. (2004). *Web services concepts, architectures and applications.* Springer.

Ammenwerth, E., Iller, C., & Mahler, C. (2006, January 9). IT-adoption and the interaction of task, technology and individuals: A fit framework and a case study. *BMC Med Inform Decis Mak,* 6-3.

Amram, M., & Kulatilaka, N. (1999). *Real option, managing strategic investment in an uncertain world.* Cambridge: H.B.S Press.

Anand, S., DuLaney, W., & Melling, W. (2004). *Smart advice: Consider long-range vision for IT when upgrading ERP systems.* Retrieved February 18, 2007, from http://www.informationweek.com/security/showArticle.jhtml?articleID=47902493&pgno=1&queryText=Definition+ERP

Anderson, C. (2006). *The long tail: Why the future of business is selling less of more.* Hyperion Books.

Anderson, P., & Tushman, M. L. (1991). Managing through cycles of technological change. *Research Technology Management, 34*(3), 26-31.

Angehrn, A. (1997, August). Designing mature Internet business Strategies: The ICDT model. *European Management Journal, 15*(4), 361-369.

Angehrn, A. A., & Meyer, J. F. (1997, Summer). Developing mature internet strategies (cover story). *Information Systems Management, 14*(3), 109-140.

Ansoff, H. I. (1987). *Corporate strategy* (2nd ed.). Harmondsworth: Penguin.

Anstadt, U. (1994). *Determinanten der individuellen Akzeptanz bei Einführung neuer Technologien.* Frankfurt a.M., Germany: Europäischer Verlag der Wissenschaften.

Applegate, L. M. (2001). *Emerging e-business models: Lessons from the field.* Harvard Business School. Retrieved June 30, 2008, from http://monkey.icu.ac.kr/sslab/course/ICE720/data/subjectPresentation/EmergingE-BusinessModel.pdf

Appleton, B. A. (1997). Patterns and software: Essential concepts and terminology. *Object Magazine Online, 3*(5), 20-25.

Argyris, C. (1998, May-June). Empowerment: The emperor's new clothes. *Harvard Business Review, 76,* 98-105.

Athey, S., & Plotnicki, J. (1998). A comparison of information system job opportunities for it professionals. *Journal of Computer Information Systems, 38*(3), 71-88.

Austvold, E., & Carter, K. (2005). *AMR research: Service-oriented architectures: Survey findings on deployment and plans for the future.* Retrieved November 2, 2007, ftp://ftp.software.ibm.com/software/soa/pdf/amr_files.pdf

Avison, D., Lau, F., Nielsen, P. A., & Myers, M., (1999). Action research. *Communications of ACM, 42*(1), 94-97.

Babcock, C. (2005). *InformationWeek.* Retrieved October 31, 2007, from http://www.informationweek.com;jsessionid=35XHAV0I1RVSWQSNDBCSKH0CJUMEKJVN

Bailey, J. E., & Pearson, S. W. (1983). Development of a tool for measuring and analyzing computer user satisfaction. *Management Science, 29*(5), 530-545.

Bajec, M. (2005). Educational portals: A way to get an integrated, user-centric university information system. In A. Tatnall (Ed.), *Web portals: The new gateways to Internet information and services* (pp. 252-269). Hershey, PA: IGI Global, Inc.

Bakos, J. Y., & Treacy, M. E. (1986). Information technology and corporate strategy: A research perspective. *MIS Quarterly, 10*(2), 107-119.

Bambury, P. (1998). A taxonomy of Internet commerce. *First Monday, 3*(10).

Band, W. (2007). mySAP CRM is a leader in the enterprise CRM suites market. *Forrester research.* Retrieved April 1, 2007, from http://www.forrester.com/Research/Document/Excerpt/0,7211,41054,00.html

Band, W. (2007). The Forrester Wave™: enterprise CRM suites, Q1 2007. *Forrester research.* Retrieved April 1, 2007, from http://www.forrester.com/Research/Document/Excerpt/0,7211,39975,00.html

Banerjee, A., & Ghosh, J. (2001). Clickstream clustering using weighted longest common subsequences. In *Proceedings of the Web Mining Workshop at the 1st SIAM Conference on Data Mining.*

Baraglia, R., & Silvestri, F. (2004). *An online recommender system for large Web sites.* Paper presented at the IEEE/WIC/ACM International Conference on Web Intelligence, Beijing, China.

Baraldi, S., & Memmola, M. (2006). How healthcare organisations actually use the internet's virtual space: A field study. *International Journal of Healthcare Technology and Management, 7*(3/4), 187-207.

Barney, J. (1991). Firm resources and sustained competitive advantage. *Journal of Management, 17*(1), 99-120.

Barquin, R. C. (2006, July 18). *Social network analysis and business intelligence*. Retrieved January 8, 2006, from http://www.b-eye-network.com/view/3088

Barras, R. (1990). Interactive innovation in financial and business services: The vanguard of the service revolution. *Research Policy, 19*, 215-237.

Barua, A., Konana, P., Whinston, A. B., & Yin, F. (2004). Assessing Internet enabled business value: An exploratory investigation. *MIS Quarterly.* Retrieved June 30, 2008, from http://crec.mccombs.utexas.edu/works/articles/barua_konana_whinston_yin_01311.pdf

Baskerville, R., & Pries-Heje, J. (1999). Grounded action research: A method for understanding IT in practice. *Accounting, Management and Information Technologies, 9*, 1-23.

Baskerville, R., & Wood-Harper, A. T. (1996). A critical perspective on action research as a method for information systems research. *Journal of Information Technology, 11*(4), 235-246.

Baskerville, R., & Wood-Harper, A. T. (1998). Diversity in information systems action research methods. *European Journal of Information Systems, 7*(2), 90-107.

Bass, A. (2003, March 15). *Integration management: Cigna's self-inflicted wounds.* Retrieved April 12, 2007, from http://cio.com

Beasty, C. (2007). The enlightened enterprise: Customer relationship management. *Medford, 11*(3), 26-40. Retrieved April 8, 2007, from ABI/INFORM Global database.

Beck, K., & Andres, C. (2005). *Extreme programming explained: Embrace change* (2nd ed.). Addison-Wesley.

Becker, C. (2004). A new game of leapfrog? *Modern Healthcare, 34*, 28-38.

Becker, J., Hassan, N., & Naumann, J. D. (2006). Combating the enrolment downturn in IS/IT programs. In *Proceedings of the Twelfth Americas Conference on Information Systems* (pp. 2289-2298). Atlanta: Association for Information systems.

Benko, C., & Mcfarlan, F. W. (2003). *Connecting the dots* (p. 246). Boston: Harvard Business School Press.

Benko, C., & McFarlan, W. (2003). *Metamorphosis in the auto industry.* Retrieved June 30, 2008, from http://www.connectingthedotsbook.com/articles/CTD_Metamorphosis_in_the_Auto_Industry.pdf

Berger, A. N., & Udell, G. F. (2006). A more complete conceptual framework for SME finance. *Journal of Banking and Finance, 30*(11), 2945-2966.

Bett, K., & Wedekind, J. (Eds.). (2003). *Lernplattformen in der Praxis.* Muenster, Germany: Waxmann.

Bharati, P. (2003). People and information matter: Task support satisfaction from the other side. *Journal of Computer Information Systems, 43*(2), 93-102.

Bhattacherjee, A., & Sanford, C. (2006). Influence process for information technology acceptance: An elaboration likelihood model. *MIS Quarterly, 30*(4), 805-825.

Biba, E. (2005). *Shoppers find less satisfaction online.* Retrieved March 3, 2007, from http://www.pcworld.com/article/id,119778-page,1/article.html?RSS=RSS

Black, F., & Scholes, M. (1973). The pricing of options and corporate liabilities. *Journal of Political Economy, 81*, 637-659.

Bloom, H. (2000). *Global brain: The evolution of mass mind from the big bang to the 21st century* (pp. 42-44). Wiley.

Bollier, D. (1996). *The future of electronic commerce.* Queenstown, MD: The Aspen Institute.

Boroni, C. M., Goosey, F. W., Grinder, M. T., & Ross, R. J. (1998). A paradigm shift! The Internet, the Web, browsers, Java and the future of computer science education. *Association for Computing Machinery Special Interest Group Computer Education Bulletin, 30*(1), 145-152.

Boulding, W., Staelin, R., Ehret, M., & Johnston, W. (2005). A customer relationship management roadmap: What is known, potential pitfalls, and where to go? *Journal of Marketing, 69*(4), 155-166. Retrieved January 31, 2007, from Business Source Premier Database.

Brenner, M., Fontana, C., & Godbout, N. (2003). *Change management.* Retrieved February 18, 2007, from http://www.destinationcrm.com/articles/default.asp?ArticleID=3656+&TopicID=9

Brimson, J. A. (1991). *Activity-accounting: An activity-based costing approach.* New York: John Wiley and Sons, Inc.

Broadbent, M., & Weil, P. (1997, Spring). Management by Maxim: How business and IT managers can create IT infrastructures. *Sloan Management Review, 38*(3), 77-92.

Broenen, J. (2006). Evans Data Corporation. *CodeFutures.* Retrieved June 19, 2008, from www.codefutures.com/weblog/corporate/archives/2006/03/java_developer_survey.html

Brown, J. S., & Duguid, P. (2000). *The social life of information.* Boston: Harvard Business School Press.

Brynjolfsson, E. (1993). The productivity paradox of information technology. *Communications of the ACM, 36*, 67-77.

Brynjolfsson, E., & Hitt, L. M. (1998, August). Beyond the productivity paradox: Computers are the catalyst for bigger changes. *Communications of the ACM.*

Buckingham, R., Hirschheim, R. A., Land, F. F., & Tully, C.J. (1987). Information systems curriculum. A basis for course design. In Buckingham et al. (Ed.), *Information systems education: Recommendations and implementation* (pp. 72-113). Cambridge: CUP.

Burn, J. M., & Szeto, C. (1999, September). A comparison of the views of business and IT management on success factors for strategic alignment. *Information & Management, 37*, 197-216.

Burn, J. M., Ng Tye, E. M. W., Ma, L. C. K., & Poon, R. S. K. (1994). Job expectations of IS professionals in Hong Kong. In J. W. Ross (Ed.), *Proceedings of the 1994 Computer Personnel Research Conference on Reinventing IS: Managing Information Technology in Changing Organizations* (pp. 231-241). New York: Association for Computing Machinery.

Burton-Jones, A., Storey, V. C., Sugumaran, V., & Ahluwalia, P. (2005). A semiotic metrics suite for assessing the quality of ontologies. *Data and Knowledge Engineering, 55*(1), 84-102.

Buschmann, F., Henney, K., & Schmidt, D. C. (2007). *Pattern-oriented software architecture, volume 5: On patterns and pattern languages.* John Wiley and Sons.

Buschmann, F., Meunier, R., Rohnert, H., Sommerlad, P., & Stal, M. (1996). *Pattern oriented software architecture, volume 1: A system of patterns.* John Wiley and Sons.

Busse, D. (2002, April 21). *Usable Web design patterns for world-ready e-commerce sites.* Paper presented at the CHI 2002 Workshop on Patterns in Practice: A Workshop for UI Designers, Minneapolis.

Buttignon, F. (2001). *Strategia e valore nella net-economy.* Milan, Italy: Il Sole 24 Ore.

Buyukkokten, O., Kaljuvee, O., Garcia-Molina, H., Paepcke, A., & Winograd, T. (2002, January). Efficient Web browsing on handheld devices using page and form summarization. *ACM Transactions on Information Systems, 20*(1), 82-115.

Byrd, T. A., & Marshall, T. T. (1997). Relating information technology investment to organizational performance:

A causal model analysis. *Omega, International Journal of Management Science, 25*(1), 43-56.

Callan, J., Smeaton, A., Beaulieu, M., Borlund, P., Brusilovsky, P., Chalmers, M., & et al. (2001). Personalization and recommender systems in digital libraries. In *Proceedings of the 2nd DELOS Workshop on Personalization and Recommender Systems in Digital Libraries.*

Campbell, D., & Fiske, D. W. (1959). Convergent and discriminant validation by the multitrait-multimethod matrix. *Psychological Bulletin, 56*(2), 81-105.

Campbell, J. A. (2002). Real options analysis of the timing of IS investment decisions. *Information & Management, 39*(5), 337-344.

Canalys. (2004). *A world of difference.* Retrieved April 14, 2006, from http://www.canalys.com/pr/2004/r2004061.pdf

Canalys. (2004). *Global mobile device market shows tremendous growth.* Retrieved March 22, 2006, from http://www.canalys.com/pr/2004/r2004081.pdf

Canalys. (2004). *Global smart phone shipments treble in Q3.* Retrieved December 3, 2006, from http://www.canalys.com/pr/2004/r2004102.pdf

Canalys. (2005). *Global smart mobile device sale surge past 10 million in quarter.* Retrieved April 25, 2006, from http://www.canalys.com/pr/2005/r2005041.pdf

Canalys. (2005). *Smart phones up, handhelds down globally in Q2.* Retrieved January 15, 2006, from http://www.canalys.com/pr/2005/r2005071.pdf

Canalys. (2005). *Global mobile device shipments hit new peak in Q4 2004.* Retrieved May 02, 2006, from http://www.canalys.com/pr/2005/r2005012.pdf

Canalys. (2005). *Worldewide smart phone market soars in Q3.* Retrieved December 3, 2006, from http://www.canalys.com/pr/2005/r2005102.pdf

Canalys. (2006). *Smart mobile device market growth remains steady at 55%.* Retrieved December 3, 2006, from http://www.canalys.com/pr/2006/r2006071.pdf

Carignani, A., & Frigerio, C. (2003). *European Web banking strategies.* Università Cattolica, Milan, CeTIF Research Centre.

Carlson Marketing. (2007). *No more limits: On demand CRM goes strategic.* Retrieved March 8, 2007, from http://research.pcpro.co.uk/detail/RES/1163438651l_120.html?src=pcpro

Carr, N. G. (2003). IT doesn't matter. *Harvard Business Review, 81*(5).

Carr, N. G. (2005). The end of corporate computing. *MIT Sloan Management Review, Management of Information Systems, 46*(3), 67-73.

Castellano, G., Castiello, C., Fanelli, A. M., & Mencar, C. (2005). Knowledge discovering by a neuro-fuzzy modelling framework. *Fuzzy Sets and Systems, 149,* 187-207.

Castellano, G., Fanelli, A. M., & Torsello, A. M. (2007, July 15-22). Web user profiling using relational fuzzy clustering. In *Proceedings of 12th International Conference on Fuzzy Theory and Technology (FTT 2007),* Salt Lake City.

Castellano, G., Fanelli, A. M., & Torsello, M. A. (2007, February 16-19). LODAP: A log data preprocessor for mining Web browsing patterns. In *Proceedings of the 6th WSEAS International Conference on Artificial Intelligence, Knowledge Engineering and Data Base (AIKED 2007),* Corfu, Greece (pp. 12-17).

Castellano, G., Fanelli, A. M., & Torsello, M. A. (2007). REXWERE: A tool for fuzzy rule extraction in Web recommendation. In *Proceedings of NAFIPS 2007,* San Diego.

Castellano, G., Fanelli, A. M., & Torsello, M. A. (2007) A neuro-fuzzy collaborative filtering approach for Web recommendation. *International Journal of Computational Science, 1*(1), 27-39. ISSN: 1992-6669 (Print)/ISSN: 1992-6677.

Castellano, G., Mesto, F., Minunno, M., & Torsello, M. A. (2007, July 7-10). Web user profiling using fuzzy clustering. In *Proceedings of 7th International Workshop on Fuzzy Logic and Applications (WILF 2007)*, Portofino Vetta (Genova), Italy.

cellular-news. (2006). *Worldwide mobile phone sales up—except in Japan*. Retrieved May 3, 2006, from http://www.cellular-news.com/story/20573.php

Chakrabarti, S., Dom, B. E., Kumar, S. R., Raghavan, P., Rajagopalan, S., Tomkins, A., et al. (1999). Mining the Web's link structure. *IEEE Computer, 32*(8), 60-67.

Chalmeta, R. (2006). Methodology for customer relationship management. *Journal of Systems & Software, 79*(7), 1015-1024. Retrieved January 31, 2007, from Business Source Premier Database.

Chambers of Commerce. (2006). Retrieved December 13, 2006, from www.chambers.ie

Chan, J. (2005). Toward a unified view of customer relationship management. *Journal of American Academy of Business, 6*(1), 32-39. Retrieved January 30, 2007, from ABI/INFORM Global database.

Chan, S. (2001, August 5-8). *Challenges and opportunities in e-commerce education*. Paper presented at he The Seventh Americas Conference in Information Systems (AMCIS 2001), Boston.

Chan, S., & Fang, X. (2001, August 5-8). *Usability issues in mobile commerce*. Paper presented at the Seventh Americas Conference in Information Systems (AMCIS 2001), Boston.

Chandra, J., March, S. T., Mukherjee, S., Pape, W., Ramesh, R., Rao, H. R., & Waddoups, R. O. (2000). Information systems frontiers. *Communications of the Association for Computing Machinery, 43(1)*, 71-79.

Chang, H., Hsiung, C., & Tsai, R. (2006). The design and analyses of customer relationship management using ARIS technique. *International Journal of Management & Enterprise Development, 3*(1/2), 1. Retrieved March 8, 2007 from Business Source Premier database.

Chapman, P., James-Moore, J., Szczygiel, M., & Thompson, D., (2000). Building Internet capabilities in SME's. *Logistics Information Management, 13*(6), 353-361.

Chen, Y., Ma, W.-Y., & Zhang, H.-J. (2005). Adaptive Web pages for small-screen devices. *IEEE Internet Computing*, 9(1), 50-56.

Chesbrough, H., & Rosenbloom, R. S. (2001). *The role of the business model in capturing value from innovation: Evidence from Xerox Corporation's technology spinoff companies*. Retrieved June 30, 2008, from www.hbs.edu/dor/papers2/0001/01-002.pdf

Cho, Y. H., & Kim, J. K. (2004). Application of Web usage mining and product taxonomy to collaborative recommendations in e-commerce. *Expert Systems with Applications, 26*, 233-246.

Choate, M. (2006, April 28). What makes an enterprise wiki? *cmswatch*. Retrieved at April 15, 2007, from http://www.cmswatch.com/Feature/145-Wikis-in-the-enterprise

Ciborra, C. U. (1996). Improvisation and information technology in organizations. In *Proceedings of the 17th International Conference on Information Systems*, Cleveland, OH (pp. 369-380).

Cigna. (2007). *Cigna reports 4th quarter and full year 2006 results* [News Release]. Retrieved April 14, 2007, from http://cigna.com/about_us/investor_relations/disclosures_archived.html

Claussen, S. (2004). *Entwurf einer personalisierten virtuellen Lernumgebung*. Aachen, Germany: Shaker.

Clear, T. (2006). Valuing computer science education research? In A. Berglund & M. Wiggberg (Eds.), *Proceedings of the 6th Baltic Sea conference on Computing Education Research: Koli Calling 2006* (pp. 8-18). Uppsala, Sweden: University of Uppsala.

Clements, D. (2006). The benefits of customer relationship management software. *Real Estate Finance, 23*(1), 15-17. Retrieved January 31, 2007, from Business Source Premier database.

Coghlan, D., & Brannick, T. (2001). *Doing action research in your own organization*. Sage Publications.

Cohen, P. (2004). *Twelve technical and business trends shaping the year ahead*. Retrieved May 13, 2004, from http://www.babsoninsight.com/contentmgr/showdetails.php/id/687

Coltman, T., Devinney, T., Latukefu, A., & Midgley, D. (2001). E-business: Revolution, evolution or hype? *California Management Review, 44*(1), 57-85.

Conrad, R., & Donaldson, J. (2004). *Engaging the online learner*. San Francisco: Jossey-Bass.

Cooley, R. W. (2000). *Web usage mining: Discovery and application of interesting patterns from Web data*. Unpublished doctoral thesis, University of Minnesota.

Cooley, R., Tan, P. N., & Srivastava, J. (1999). *Discovering of interesting usage patterns from Web data* (Tech. Rep. No. 99-022). University of Minnesota.

Cooper, R. (1988, Summer). The rise of ABC (part I): What is an ABC system. *Journal of Cost Management*, 45-54.

Cooper, R. (1988, Fall). The rise of ABC (part II): When do I need an ABC system? *Journal of Cost Management*, 41-48.

Cooper, R. (1989, Winter). The rise of ABC (part III): How many cost drivers do you need, and how do you do select them? *Journal of Cost Management*, 34-46.

Cooper, R. (1989, Spring). The rise of ABC (Part IV): What do ABC systems look like? *Journal of Cost Management*, 38-49.

Cooper, R. (1990, Spring). Implementing an activity-based cost system. *Journal of Management Accounting*, 33-42.

Cooper, R. G., Edgett, S. J., & Kleinschmidt, E. J. (1998, July-August). Best practice for managing R&D portfolio. *Research Technology Management*.

Cooper, R., & Kaplan, R. S. (1988). Measure costs right: Make the right decision. *Harvard Business Review, 66*(5), 96-103.

Cooper, R.G., Edgett, S. J., & Kleinschmidt, E. J. (2000, March-April) New problems, new solutions: Making portfolio management more effective. *Research Technology Management*.

Couger, J. D., Davis, G. B., Dologite, D. G., Feinstein, D. L, Gorgone, J. T., Jenkins, A. M., et al. (1995). IS'95: Guideline for undergraduate IS curriculum. *Management Information Systems Quarterly, 19*(3), 341-359.

Coveyduc, J., Huang, C., Ostdiek, L., & Reif, J. (2006, October). IBM innovation factory. An integrated solution for accelerating innovation. *IBM*. Retrieved July 3, 2008, from http://ibm.com/websphere/developer/zones/hipods

Cragg, P., & King, M. (1993). Small-firm computing: Motivators and inhibitors. *MIS Quarterly, 17*(1), 47-61.

Craveiro, A. M., Severo, A. C. R., & Sbragia, R. A. (2000, November). Priorização de projetos e avaliação de portfolio. In *Proceedings of XXI Simpósio de Gestão da Inovação Tecnológica*, São Paulo.

Cunningham, C., Song, I., & Chen, P. P. (2006). Data warehouse design to support customer relationship management analyses. *Journal of Database Management, 17*(2), 62-85. Retrieved January 30, 2007, from ProQuest ABI/INFORM Global database.

Cunningham, M. J. (2001). *B2B business-to-business: Como implementar estratégias de e-commerce entre empresas*. Rio de Janeiro: Ed. Campus.

Cykana, P., Paul, A., & Stern, M. (1996). DoD guidelines on data quality management. In *Proceedings of the Conference on Information Quality*, Cambridge, MA (pp. 154-171).

Damanpour, F., & Gopalakrishnan, S. (2001). The dynamics of the adoption of product and process innovation in organizations. *Journal of Management Studies, 38*(1), 45-65.

Daniel, E., & Grimshaw, D. (2002). An exploratory comparison of electronic commerce adoption in large and small enterprises. *Journal of Information Technology, 17*, 133-147.

Daniel, E., Wilson, H., & Myers, A. (2002). Adoption of e-commerce by SME's in the UK. *International Small Business Journal, 20*(3), 253-270.

Davenport, T. H. (2005). *Thinking for a living: How to get better performance and results from knowledge workers.* Boston: Harvard Business School Press.

Davenport, T. H., & Harris, J. G. (2007). *Competing on analytics: The new science of winning.* Boston: Harvard Business School Press

Davenport, T. H., & Prusak, L. (1997). *Information ecology.* Boston: Harvard Business School Press.

Davenport, T. H., & Prusak, L. (2000). *Working knowledge: How organizations manage what they know.* Boston: Harvard Business School Press.

Davis, F. (1989). Perceived usefulness, perceived ease of use, and user acceptance of information technology. *MIS Quarterly, 2*(13), 319-340.

Davis, H. E., & Luehlfing, M. S. (2004, November). Radio frequency identification: The wave of the future. *Journal of Accountancy*, 43-49.

Davis, S., Siau, K., & Dhenuvakonda, K. (2003). A fit-gap analysis of e-business curricula vs. industry needs. *Communications of the Association for Computing Machinery, 46*(12), 167-177.

Davison, J., Walsh, A., & Brown, D. (2000). *Mobile e-commerce: Market strategies.* London: Ovum Ltd.

De Nicolò, G., Bartholomew, P., Zaman, J., & Zephirin, M. (2004). Bank consolidation, internationalization, and conglomeration: Trends and implications for financial risk. *Financial Markets, Institutions, & Instruments, 13*(4), 173-217.

DeLone, W. H., & McLean, E. R. (1992). Information systems success: The quest for the dependent variable. *Information Systems Research, 3*(1), 60-95.

Dervin, B. (1983, May). *An overview of sense-making research: Concepts, methods, and results to date.* Paper presented at the annual meeting of the International Communication Association, Dallas.

Desisto, R. P. (2006). *Magic quadrant for sales force automation, 1H06* [Gartner Research Rep. No. G00141438]. Retrieved March 27, 2007, from http://www.gartner.com/mq/index.jsp?qid=141438

Dibb, S., Simkin, L., Pride, W., & Ferrel, O. (1991). *Marketing, concepts and strategies.* Boston: Houghton Mifflin.

Diekmann, T., Kaspar, C., Seidenfaden, L., & Hagenhoff, S. (2006). Kontextbewusste Informationsdienste auf Grundlage des Information Beacon Frameworks. In F. Lehner (Ed.), *Multikonferenz wirtschaftsinformatik.* Retrieved August 30, 2007, from http://www.gi-mms.de/mms2006/kurzbeitraege/diekmann.pdf

Dignan, L. (2005). *Outsourcing overseers needed.* Retrieved January 15, 2005, from http://www.baselinemag.com/article2/0,3959,1526051,00.asp

Dittler, M., & Nachmann, G. (2003). Entscheidungsprozesse und Begleitmassnahmen bei der Auswahl und Einfuehrung von Lernplattformen—Ein Praxisbericht aus dem LearnTechNet der Universitaet Basel. In K. Bett & J. Wedekind (Eds.), *Lernplattformen in der Praxis* (pp. 175-192). Muenster, Germany: Waxmann.

Doll, W. J., & Torkzadeh, G. (1988). The measurement of end-user computing satisfaction. *MIS Quarterly, 12*(2), 259-274.

Dong, Y., & Zhuang, Y. (2004, June 15-19). Fuzzy hierarchical clustering algorithm facing large database. In *Proceedings of the 5th World Congress on Intelligent Control and Automation*, Hangzhou, P.R. China.

Doucek, P., Novotný, O., Pecáková, I., & Voříšek, J. (2007) *Lidské zdroje v ICT—Analýza nabídky a poptávky po IT odbornících v ČR* [Human resources in ICT—analysis of the demand and supply of ITCT professionals in the CR]. Praha: VŠE.

Drive, E. et al. (2004, August 2). *Road map to an enterprise collaboration strategy.* Forrester Research.

Dromey, R. G. (2003). Software quality - prevention versus cure? *Software Quality Journal, 11*(3), 197-210.

Dubie, D. (2004). Vendors make the utility computing grade. *Network World Fusion.* Retrieved March 13, 2004, from http://www.nwfusion.com/news/2004/0322dellsummit.html

Duke, H. C., & Chul, M. K., & Sang-Il, K., & Soung, H. K. (2006, December). Customer loyalty and disloyalty in Internet retail stores: Its antecedents and its effect on customer price sensitivity. *International Journal of Management, 23*(4), 925-942.

Duncan, D. G. (2006). E-Business curriculum: Literature review reveals opportunities and challenges. *Issues in Information Systems, 7*(1), 264-267.

Dutta, S., & Biren, B. (2001). *Business transformation on the Internet: Results from the 2000 study.* Retrieved June 30, 2008, from http://mslab.hau.ac.kr/mgyoon/ec02/paper/9.pdf

Dyche, J. (2002). Managing your CRM project: The CRM development team. *Informit.com*. Retrieved April 19, 2007, from http://www.informit.com/articles/article.asp?p=25181&seqNum=3&rl=1

Earl, M. J., Rockart, J. F., & Ross, J. W. (1996, Fall). Eight imperatives for the new IT organization. *Sloan Management Review, 38*(1), 43-55.

ebizQ. (2006). Retrieved June 19, 2008, from http://www.actional.com/resources/whitepapers/SOA-Governance-Adoption/SOA-Survey.html

Egan, T., Clancy, S., & O' Toole, T. (2003). The integration of e-commerce tools into the business processes of SMEs. *Irish Journal of Management, 24*(1), 139-154.

Eirinaki, M., & Vazirgiannis, M. (2003). Web mining for Web personalization. *ACM TOIT, 3*(1), 2-27.

Eisenfeld, B., Apfel, A., & Smith, M. (2004, June 2). *Choosing metrics for customer relationship management* (Gartner Research Issue Brief No. DF-22-7541). Stamford, CT: Gartner, Inc.

Elango, B. (2000). Do you have an Internet strategy? *Information Strategy: The Executive's Journal, 16*(3), 32-38.

Enembreck, F., Barthès, J-P., & Ávila, B. C. (2004). *Personalizing information retrieval with multi-agent systems* (LNCS 3191/2004, pp. 77-91). Springer, Berlin-Heidelberg.

Epsten, M. J. (2005). The determinants and evaluation of merger success. *Business Horizons, Elsevier, 48*, 37-46.

Etzioni, O. (1996). The World Wide Web: Quagmire or gold mine. *Communications of the ACM, 39*(11), 65-68.

Evens Data Corporation. (2004, Spring). *North American development survey.* Retrieved June 19, 2008, from www.evensdata.com

Facca, F. M., & Lanzi, P. L. (2005). Mining interesting knowledge from Weblogs: A survey. *Data & Knowledge Engineering, 53*, 225-241.

Fallows, D. (2005, January). *Pew Internet & American Life Project. Search engine users.* Retrieved January 1, 2006 from http://www.pewinternet.org/PPF/r/146/report_ display.asp

Fanberg, H. (2004, Fall). The RFID revolution: Healthcare is ready to embrace this new technology. *Marketing Health Services*, 43-44.

Farbey, B., Land, F. F., & Targett, D. A. (1995). Taxonomy of information systems applications: The benefits evaluation ladder. *European Journal of Information Systems, 4*(1), 41-50.

Fariselli, P., Oughton, C., Picory, C., & Sugden, R. (1997). *Electronic commerce and the future for SMEs in a global marketplace: Opportunities and public policy requirements.* Brussels: CEC.

Farrell, D. (2003, October). The real new economy. *Harvard Business Review, 81*(10), 104-112.

Fernandez, E. B., Liu, Y., & Pan, R. (2001, September 11-15). *Patterns for Internet shops.* Paper presented at the Eighth Conference on Pattern Languages of Programs (PLoP 2001), Monticello.

Ferraro, F. (2000). L'analisi organizzativa: l'individuo. In R. Mercurio & F. Testa (Eds.), *Organizzazione, assetto e relazioni nel sistema di business* (pp. 52-76). Torino: Giappichelli.

Feuerlicht, G., & Voříšek, J. (2003). Key success factors for delivering application services. In *Proceedings of the Systems Integration 2003 Conference* (pp. 274-282). Praha: VŠE. ISBN 80-245-0522-3.

Feuerlicht, G., & Voříšek, J. (2004). Is it the right time for the enterprise to adopt software-as-a-service model. *Information Management, 17*(3/4), 18-21. ISSN 1080-286X.

Fielding, R. T. (2000). *Architectural styles and the design of network-based software architectures*. Unpublished doctoral thesis, University of California, Irvine.

Figueira, C., Neills, J., & Schoenberg, R. (2007). Travel abroad or stay at home? Investigating the patterns of bank industry M&As in the EU. *European Business Review, 19*(1), 23-39.

Fincher, S., Clear, T., Petrova, K., Birch, R., Claxton, G., Hoskyn, C., et al. (2004). Cooperative education in information technology. In R. Coll & C. Eames (Eds.), *International handbook for cooperative education: An international perspective of the theory, research and practice of work-integrated learning* (pp. 111- 121). Hamilton, New Zealand: WACE.

Fink, D. (1998). Guidelines for the successful adoption of information technology in small and medium enterprises. *International Journal of Information Management, 18*(4), 243-253.

Fink, L. (2002). *Economies of aggregation*. Unpublished doctoral dissertation, University of Cologne, Germany.

Fitzgerald, M. (2007). CRM made simple new software to supercharge your sales. *Inc. Boston, 29*(1), 46-47. Retrieved March 26, 2007, from Business -ABI/INFORM Global database.

Fleischer, J. (2006). RightNow technologies' two-sided approach to CRM. *Call Center Magazine*. Retrieved April 6, 2007, from http://www.callcentermagazine.com/showArticle.jhtml?articleID=193500843

Fogg, B. J., Soohoo, C., Danielson, D., Marable, L., Stanford, J., & Tauber, E. R. (2002, October 29). How do people evaluate a Web site's credibility? Results from a large study. *Consumer Reports WebWatch Research Report*.

Fontana, J. (2002, May 17). *Network world fusion*.

Foote Partners LLC Research Group. (2007). *Hot IT skills and certifications pay index*. Retrieved June 15, 2007, from http://www.footepartners.com/htscpi_latest.htm

Forfás. (2002). *Ebusiness, where we are and where do we go from here?*

Forrester Research. (2006, June 29). *Europe's eCommerce forecast: 2006-2010*. Author.

Freeland, G. D., & Stirton, S. (2000). *Organizing for e-commerce* (discussion paper). *The Boston Consulting Group Inc.* Retrieved June 30, 2008, from www.bcg.com/publications/files/organizing%20ECommerce%20Apr%2000.pdf

Frias-Martinez, E., Magoulas, G., Chen, S., & Macredie, R. (2005). Modeling human behavior in user-adaptive systems: recent advances using soft computing techniques. *Expert Systems with Applications, 29*(2), 320-329.

Frigerio, C. (2005). *Process IT Innovation in banking*. Università Cattolica, Milan, CeTIF Research Centre.

Galliers, R. (1992). *Information systems research issues, methods and practical guidelines*. Blackwell Scientific Publications.

Gallivan, M. J., Truex, D. P., & Kvasny, L. (2004). Changing patterns in IT skill sets 1988-2003: a content analysis of classified advertising. *The Association for Computing Machinery DATA BASE for Advances in Information Systems, 35*(3), 64-87.

Gallivan, M., Truex, D. P., & Kvasny, L. (2002). An analysis of the changing demand patterns for information technology professionals. In *Proceedings of the 2002 Association for Computing Machinery Specialist Interest Group Computer Personnel Research Conference on Computer personnel Research* (pp. 11-24). New York: Association for Computing Machinery.

Gangemi, A. (2005, November 6-10). *Ontology design patterns for Semantic Web content*. Paper presented at the Fourth International Semantic Web Conference (ISWC 2005), Galway, Ireland.

Gartner, Inc. (2006). *Gartner says worldwide combined PDA and smartphone shipments market grew 57 percent in the first half of 2006*. Retrieved October 30, 2006, from http://www.gartner.com/it/page.jsp?id=496997

Gartner. (2006). *User survey report: Carriers' service-oriented architecture and Web services*. Author.

Garzotto, F., Paolini, P., Bolchini, D., & Valenti, S. (1999, November 15-18). *"Modeling-by-patterns" of Web applications*. Paper presented at the International Workshop on the World Wide Web and Conceptual Modeling (WWWCM 1999), Paris, France.

George, J. F., Valacich, J. S., & Valor, J. (2005). Does information systems still matter? Lessons for a maturing discipline. *Communications of the Association for Information Systems, 16*, 219-232.

Georgiakakis, P., Psaromiligkos, Y., & Retalis, S. (2006, June 20-23). *The use of design patterns for evaluating personalisable Web-based systems*. Paper presented at the Fifth Workshop on User-Centred Design and Evaluation of Adaptive Systems, Dublin, Ireland.

Geppert, L. (2001, December). *E-Marketplaces: New models*. Paper presented at the First Policy Group Meeting of DEEDS Project, Brussels

Germán, D. M., & Cowan, D. D. (2000, January 4-7). *Towards a unified catalog of hypermedia design patterns*. Paper presented at the Thirty Third Hawaii International Conference on System Sciences (HICSS 2000), Maui.

Gilb, T. (1988). *Principles of software engineering management*. Addison-Wesley.

Giles, J. (2005, December). Internet encyclopaedias go head to head. *Nature, 438*(7070), 900-901.

Gill, G., & Bhattacherjee, A. (2007). The informing sciences at a crossroads: The role of the client. *Informing Sciences: The International Journal of an Emerging Transdiscipline, 10*, 17-39.

Ginige, A., & Murugesan, S. (2001). Web engineering: An introduction. *IEEE Multimedia, 8*(1), 14-18.

Glabman, M. (2004, May). Room for tracking: RFID technology finds the way. *Materials Management in Health Care*, 26-38.

Glaser, J. P. (2002). *The strategic application of information technology in health care* organizations (2nd ed.). San Francisco: Jossey-Bass.

Goff, L. (2000). The e-lusive staff. *Computerworld*, 44-46.

Gómez-Pérez, A., Fernández-López, M., & Corcho, O. (2004). *Ontological engineering*. Springer Verlag.

Govindarajulu, C., Reithel, B., & Sethi, V. (2000). A model of end user attitudes and intentions toward alternative sources of support. *Information & Management, 3*(37), 77-86.

Graham, I. (2003). *A pattern language for Web usability*. Addison-Wesley.

Granger, M. J., Dick, G., Jacobson, C. M., & Van Slyke, C. (2007). Information systems enrollments: Challenges and strategies. *Journal of Information Systems Education, 18* (3), 303-311.

Grant, R. M. (1991, Spring). The resource-based theory of competitive advantage: Implications for strategy formulation. *California Management Review*, 114-135.

Grant, R. M. (2005). *Contemporary strategy analysis 5/e*. Oxford, UK: Blackwell.

Greenwood, D., & Levin, M. (1998). *Introduction To action research*. Sage Publications.

Gregg, S. (2001). Die Konvergenz: Telekommunikationsanbieter und Medienunternehmen—Wettbewerber oder partner? In A. Vizjak & M. Ringlstetter (Eds.), *Medienmanagement: Content gewinnbringend nutzen* (pp. 37-44). Wiesbaden, Germany: Gabler.

Grohmann, G. (2006). *Learning management*. Koeln, Germany: Josef Eul.

Gruber, T. (2005). *Ontology of folksonomy.* Paper presented at the First on-Line Conference on Metadata and Semantics Research (MTSR'05).

Gruber, T. R. (1993). Toward principles for the design of ontologies used for knowledge sharing. *Formal ontology in conceptual analysis and knowledge representation.* Kluwer Academic Publishers.

Guimarães, C. (2001, February). Vale tudo. *Revista Exame, 733.*

Gyenesei, A. (2000). *A fuzzy approach for mining quantitative association rule* (TUCS Tech. Rep. 336). Univ. Turku, Dept. Comput. Sci., Lemminkisenkatu 14, Finland.

Haber, L. (2004). *ASPs still alive and kicking.* Retrieved January 30, 2004, from http://www.aspnews.com /trends/article.php/3306221

Hafiz, M. (2006, October 21-23). *A collection of privacy design patterns.* Paper presented at the Thirteenth Conference on Pattern Languages of Programs (PLoP 2006), Portland.

Halkidi, M., Batistakis, Y., & Vazirgiannis, M. (2002). *Cluster validity methods: Part II.* in SIGMOD Record.

Hamid, R., Nezhad, M., Benatallah, B., Casati, F., & Toumani, F. (2006). Web services

Hamm, S., & Ante, S. E. (2005). Blue world. *Business-Week, 16,* 82

Hammer, M. (1996). *Beyond reengineering.* London: Harper Collins Business.

Hammer, M., & Champy, J. (1993). *Reengineering the corporation: A manifest for business revolution.* London: Nicholas Brealey Publishing.

Hammond, T., Hannay, T., & Lund, B. (2004). The role of RSS in science publishing. *D-Lib Magazine, 10,* 1082-9873.

Han, J., & Kamber, M. (2001). *Data mining concepts and techniques.* Morgan Kaufmann.

Hanekop, H. (2007). *DMB-Projekt MI FRIENDS—Ergebnisse der Begleitforschung München.* BLM Schriftenreihe Band 86. München, Germany: Verlag Reinhard Fischer.

Hart, M. L. (2006). Customer relationship management: Are software applications aligned with business objectives? *South African Journal of Business Management, 37*(2), 17-32. Retrieved January 31, 2007, from Business Source Premier database.

Hasan, L. R., & Abuelrub, E. (2006, June 19-21). *Criteria for evaluating quality of Websites.* Paper presented at the Sixth IBIMA Conference on Managing Information in Digital Economy, Bonn, Germany.

Hayday, G. (2002). *CIOs slowly turning to Web services.* Retrieved June 19, 2008, from http://zdnet.com.com/2100-1106-960985.html

Heer, J., & Chi, E. H. (2002). Mining the structure of user activity using cluster stability. In *Proceedings of the Workshop on Web Analytics/Second SIAM Conference on Data Mining.* ACM Press.

Heigl, R., Glaser, M., & Anja, E. (2006). *Wiki: Web collaboration.* Springer. ISBN 3540229396.

Henderson, J. C., & Venkatraman, N. (1993). Strategic alignment: Leveraging information technology for transforming organizations. *IBM Systems Journal, 32*(1) 4-16.

Hendler, J., Lassila, O., & Berners-Lee, T. (2001). The Semantic Web. *Scientific American, 284*(5), 34-43.

Hildebrand, L. (2005). Hybrid computational intelligence systems for real world applications. *Series studies in fuzziness and soft computing* (Vol. 179, pp. 165-195). Springer Berlin/Heidelberg.

Hinchcliffe, D. (2007). i-Technology viewpoint: Is Web 2.0 the global SOA? *SOA Web Services Journal.* Retrieved February 23, 2007 from http://webservices.sys-con.com/read/164532.htm

Ho, K. K. W. (2007). The e-Government development, IT strategies and portals of Hong Kong SAR Government.

International Journal of Cases on Electronic Commerce, 3(2), 71-89.

Hof, R. D. (2005). Mix, match, and mutate. *Business Week Magazine*. Retrieved February 23, 2007 from http://www.businessweek.com/@@76IH*ocQ34AvyQMA/magazine/content/05_30/b3944108_mz063.htm

Högg, R., Meckel, M., Stanoevska-Slabeva, K., & Martignoni R. (2006). Overview of business models for Web 2.0 communities. In *Proceedings of GeNeMe 2006* (pp. 23-37). Dresden.

Hohpe, G., & Woolf, B. (2004). *Enterprise integration patterns*. New York: Addison-Wesley.

Hong Kong Government (HKG). (2007). *IT infrastructure and standards – infrastructure for e-government – central cyber government office*. Retrieved July 15, 2007, from http://www.ogcio.gov.hk/eng/infra/eccgo.htm

Hoovers. (2007). *CIGNA Corporation*. Retrieved April 14, 2007, from http://hoovers.com/cigna/--ID__10259--/free-co-factsheet.xhtml

How SFIA works. (n.d.). Retrieved December 15, 2007, from http://www.bcs.org/server.php?show=nav.9029

Hu, W.-C., Yang, H.-J., Lee, C.-W., & Yeh, J.-H. (2005). A technical perspective of World Wide Web usage mining. In J. Wang (Ed.), *Encyclopedia of data warehousing and mining* (pp. 1242-1248). Hershey, PA: Idea Group Publishing.

Hu, W.-C., Yeh, J.-H., Chu, H.-J., & Lee, C.-W. (2005). Internet-enabled mobile handheld devices for mobile commerce. *Contemporary Management Research, 1*(1), 13-34.

Huang, J. Z., Ng, M., Ching, W.-K., Ng, J., & Cheung, D. (2001, August 26). A cube model and cluster analysis for Web access sessions. In R. Kohavi, B. Masand, M. Spiliopoulou, & J. Srivastava (Eds.), *WEBKDD 2001—Mining Web Log Data Across All Customers Touch Points, Third International Workshop*, San Francisco (LNCS 2356, pp. 48-67). Springer.

Huang, X., Cercone, N., & An, A. (2002). Comparison of interestingness functions for learning Web usage patterns. In *Proceedings of the Eleventh International Conference on Information and Knowledge Management* (pp. 617-620). ACM Press.

ICFAI. (2007). *CRM implementation failure at Cigna Corporation*. Retrieved April 14, 2007, from http://www.icmr.icfai.org

IDC. (2004). *Western Europe, government sector, Web services and new technologies, levels of adoption and investment plans: An IDC survey*. Retrieved June 19, 2008, from www.idc.com

Ihme, T., & Abrahamsson, P. (2005, July 27-28). *The use of architectural patterns in the agile software development of mobile applications*. Paper presented at the 2005 International Conference on Agility (ICAM 2005), Otaniemi, Finland.

Information Technology Association of America. (2001). *When can you start? Building better information technology skills and careers*. Retrieved August 2, 2005, from http://www.it-careernet.com/betterskills.htm

Intel Suite Two. *The Enterprise 2.0 Suite, powered by Intel*. Retrieved July 3, 2008, from http://www.suitetwo.com.

interoperability specifications. *Computer*. IEEE Computer Society.

Ives, B., Valacich, J., Watson, R., & Zmud, R. (2002). What every business student needs to know about information systems. *Communications of the Association for Information Systems, 9*, 467-477.

Jacobson, I., Booch, G., & Rumbaugh, J. (1999). *The unified software development process*. Addison-Wesley.

Janz, B. D., Pitts, M. G., & Otondo, R. F. (2005). Information systems and health care II: Back to the future with RFID: Lessons learned – some old, some new. *Communications of the Association for the Information Systems, 15*, 132-148.

Jespersen, S. E., Thorhauge, J., & Pedersen, T. B. (2002). A hybrid approach to Web usage mining. In *Proceedings of the 4th International Conference on Data Warehousing and Knowledge Discovery* (pp. 73-82). Springer-Verlag.

Johnson, T. H., & Kaplan, R. S. (1987). *Relevance lost. The rise and fall of management accounting.* Boston: Harvard Business School Press.

Joshi, A., & Joshi, K. (2000). *On mining Web access logs.* Paper presented at the ACM SIGMOID Workshop on Research Issues in Data Mining and Knowledge Discovery (Vol. 63, No. 69).

Joshi, K. P., Joshi, A., & Yesha, Y. (2003). On using a warehouse to analyze Web logs. *Distributed and Parallel Databases, 13*(2), 161-180.

Jukić, N., Jukić, B., Meamber, L., & Nezlek, G. (2002). Employing a multilevel secure approach in CRM systems. *The Journal of Information Technology Theory and Application (JITTA), 4*(2), 17-31. Retrieved March 8, 2007 from Business Source Premier database.

Kalakota, R., & Robinson, M. (1999). *E-business-roadmap for success.* MA: Addison Wesley Longman, Inc.

Kalakota, R., & Whinston, A. (1996). *Frontiers of electronic commerce.* Reading, MA: Addison-Wesley.

Kaluscha, E. A., & Grabner-Kräuter, S. (2003, June 25-29). *Towards a pattern language for consumer trust in electronic commerce.* Paper presented at the Eighth European Conference on Pattern Languages of Programs (EuroPloP 2003), Irsee, Germany.

Kamthan, P. (1999, March 29). E-commerce on the WWW: Prospects and concerns. *Internet related technologies.* Retrieved June 15, 2008, from http://www.IRT.org

Kamthan, P. (2007). Towards a systematic approach for the credibility of human-centric Web applications. *Journal of Web Engineering, 6*(2), 99-120.

Kamthan, P. (2007). Extreme programming for mobile applications. In D. Taniar (Ed.), *Encyclopedia of mobile computing and commerce* (Vol. 1). Hershey, PA: IGI Global, Inc.

Kamthan, P. (2007). Accessibility of mobile applications. In D. Taniar (Ed.), *Encyclopedia of mobile computing and commerce* (Vol. 1). Hershey, PA: IGI Global, Inc.

Kamthan, P. (2008). Patterns for improving the pragmatic quality of Web information systems. In C. Calero, M. Á. Moraga, & M. Piattini (Eds.), *Handbook of research on Web information systems quality.* Hershey, PA: IGI Global, Inc.

Kamthan, P., & Pai, H.-I. (2006). Representation of Web application patterns in OWL. In D. Taniar, & J. W. Rahayu (Eds.), *Web semantics and ontology* (pp. 41-67). Hershey, PA: IGI Global, Inc.

Kamthan, P., & Pai, H.-I. (2006, May 21-24). *Semantic Web-enabled Web engineering: The case of patterns.* Paper presented at the Seventeenth Annual Information Resources Management Association International Conference (IRMA 2006), Washington, D.C.

Kaplan, R. S., & Anderson, S. R. (2004, November). Time driven activity-based costing. *Harvard Business Review*, 131-138.

Kaplan, R. S., & Norton, D. P. (1992, January/February). The balanced scorecard: Measures that drive performance. *Harvard Business Review, 70*(1), 71-79.

Kaplan, R. S., & Norton, D. P. (2004). *Strategy maps.* Boston: Harvard Business School Press.

Kaplan, R. S., & Norton, D. P. (2004). *Strategy maps.* Boston: Harvard Business School Press.

Kappel, G., Pröll, B., Reich, S., & Retschitzegger, W. (2006). *Web engineering.* John Wiley and Sons.

Katz, M. L., & Shapiro, C. (1986). Technology adoption in the presence of network externalities. *Journal of Political Economy, 94*, 822-841.

Kearney, A. T. (2003). *Mobinet 5: The new mobile mindset.* Retrieved June 11, 2007, from http://www.atkearney.com/shared_res/pdf/Mobinet_Monograph_S.pdf

Kearns, G. S., & Lederer, A. L. (2000). The effect of strategic alignment on the use of IS-based resources for competitive advantage. *Journal of Strategic Information*

Systems. Retrieved June 30, 2008, from http://web.staffs.ac.uk/~bstpjw/ebusiness/downloads/stman04.pdf

Kellen, V. (2002). CRM measurement frameworks. *DePaul University.* Retrieved February 18, 2007, from http://www.kellen.net/crmmeas.htm

Kennedy, A. (2006). Electronic customer relationship management (eCRM): Opportunities and challenges in a digital world. *Irish Marketing Review, 18*(1/2), 58-69. Retrieved January 30, 2006, from ProQuest ABI/INFORM Global database.

Khalil, O. E. M., & Elkordy, M. M. (2005). EIS information: Use and quality determinants. *Information Resources Management Journal, 18*(2), 68-93.

Kim, D.-W., & Lee, K. H. (2001). A new fuzzy information retrieval system based on user preference model. *FUZZY-IEEE*, 127-130.

Kline, D., & Burstein, D. (2005). *Blog!: How the newest media revolution is changing politics, business, and culture.* Squibnocket Partners, L.L.C. ISBN 1-59315-141-1.

Kohrs, A., & Merialdo, B. (1999). Clustering for collaborative filtering applications. *Computational intelligence for modelling, control and automation.* IOS Press.

Kolsky, E. (2002). *CRM at Bank One: Becoming customer-centric.* Retrieved February 19, 2007, from Gartner database.

Korfhage, R. R. (1997). *Information storage and retrieval* (pp. 105-144). John Wiley & Sons.

Kosala, R., & Blockeel, H. (2000). Web mining research: A survey. *SIGKDD Explorations: Newsletter of the Special Interest Group (SIG) on Knowledge Discovery & Data Mining, 2*(1), 1-15. ACM.

Kotonya, G., & Sommerville, I. (1998). *Requirements engineering: Processes and techniques.* John Wiley and Sons.

Kraemer, K. L., & Dedrick, J. (2002). Strategic use of the Internet and e-commerce: Cisco systems. *Journal of Strategic Information Systems*, 5-29. Retrieved June 30, 2008, from http://www.doai.uom.gr/mai/Course_Material/E_Commerce/folinas/7.pdf.

Kraft, D. H., Chen, J., Martin-Bautista, M. J., & Vila, M. A. (2002). Textual information retrieval with user profiles using fuzzy clustering and inferencing. In P. S. Szczepaniak, J. Segovia, J. Kacprzyk, & L. A. Zadeh (Eds.), *Intelligent exploration of the Web.* Heidelberg, Germany: Physica-Verlag.

Kruchten, P. (2004). *The rational unified process: An introduction* (3rd ed.). Addison-Wesley.

Kuo, R. J., & Chen, J. A. (2004). A decision support system for order selection in electronic commerce based on fuzzy neural network supported by real-coded genetic algorithm. *Expert Systems with Applications, 26*, 141-154.

Kyte, A. (2001, February). Paper presented at the iEB Europe Gartner Conference 2001, London.

Lampinen, T., & Koivisto, H. (2002). Profiling network applications with fuzzy c-means clustering and self-organising map. In *Proceedings of the 1st International Conference on Fuzzy Systems and Knowledge Discovery: Computational Intelligence for the E-Age* (pp. 300-304).

Landauer, T. (1995). *The trouble with computers: Usefulness, usability and productivity.* Cambridge: MIT Press.

Lang, K. (1995). Newsweeder: Learning to filter netnews. In *Proceedings of the 12th International Conference on Machine Learning.*

Langley, P. (1999). User modeling in adaptive interfaces. In *Proceedings of the Seventh International Conference on User Modeling*, Canada (pp. 357-370).

Laso, I. (2006, October 18). *A Web 2.0 research perspective. DG information society and media.* European Commission Representative at the 6th Internet Next Generation Workshop 2006, Madrid, Spain.

Laurindo, F. J. B. (2002). *Tecnologia da informação: Eficácia nas organizações* (p. 248). São Paulo.

Laurindo, F. J. B., & Shimizu, T. (2000). *Evaluating strategies in information technology.* Paper presented at the Performance Measurement 2000 Conference - Past, Present and Future (pp. 323-330). Cambridge: Inglaterra.

Lawrence, P. R., & Lorsch, J. W. (1967). *Organization and environment, graduate school of business administration.* Boston: Harvard University.

Leavitt, N. (2004, November). Are Web services finally ready to deliver? *IEEE Computer.*

Lee, C. K. (2005). Transferability of skills over the ICT career path. In J. E. Moore & S. E. Yager (Eds.), *Proceedings of the 2005 Association for Computing Machinery Special Interest Group management Information Systems Computer Personnel Research Conference on Computer Personnel Research* (pp. 85-93). New York: Association for Computing Machinery.

Lee, D. M. S., Trauth, E. M., & Farwell, D. (1995). Critical skills and requirements of IS professionals: A joint academic/industry investigation. *Management Information Systems Quarterly, 19*(3), 313-340.

Lee, P. C. B. (2002). Changes in skill requirements of information systems professionals in Singapore. In *Proceedings of the 35th Hawaii International Conference on System Sciences* (pp. 3307-3315).

Lee, R. S. T. (2001). iJADE IWShopper: A new age of intelligent Web shopping system based on fuzzy-neuro agent technology. In N. Zhong, Y. Yao, J. Liu, & S. Ohsuga *Web intelligence: Research and development* (LNAI 198, pp. 403-412).

Lee, S., Koh, S., Yen, D., & Tang, H.-L. (2002). Perception gaps between IS academics and IS practitioners: An exploratory study. *Information and Management, 40*(1), 51-61.

Lee, Y. W., Pipino, L., Strong, D. M., & Wang, R.Y. (2004). Process embedded data integrity. *Journal of Database Management, 15*(1), 87-103.

Lee, Y. W., Strong, D. M., Kahn, B. K., & Wang, R. Y. (2002). AIMQ: A methodology for information quality assessment. *Information and Management, 40*(2), 133-146.

Levitt, J. (2001, October 1). From EDI to XML and UDDI: A brief history of Web services. *Information Week.*

Levy, A. (2004). *What do enterprises want from ASPs?* Retrieved February 13, 2004, from http://www.aspnews.com/analysis/analyst_cols/article.php/2217411

Levy, M., & Powell, P. (2000). Information systems strategy for small and medium sized enterprises: An organizational perspective. *Journal of Strategic Information Systems, 9,* 63-84.

Levy, M., & Powell, P. (2003). Exploring SME Internet adoption: Towards a transporter model. *Electronic Markets, 13*(2), 171-181.

Levy, M., Powell, P., & Yetton, P. (2001). SME's: Aligning IS and the strategic context. *Journal of Information Technology, 16,* 133-144.

Lewis, G., Morris, E., Simanta, S., & Wrage, L. (2007). *SOA common misconceptions.* Retrieved June 19, 2008, from www.iccbss.org/2007/moreinfo/SOA-Common-Misconceptions.pdf

Lewis, R., & Cockrill, A. (2002). Going global-remaining local: The impact of e-commerce on small retail firms in Wales. *International Journal of Information Management, 22*(3), 195-209.

Li, M., & Ye, L. R. (1999, January). Information technology and firm performance: Linking with environmental, strategic and managerial contexts. *Information & Management, 35*(1), 43-51.

Lieberman, H., & Letizia (1995). An agent that assists Web browsing. In *Proceedings of the 14th International Joint Conference in Artificial Intelligence* (pp. 924-929).

Lientz, B. P., & Rea, K. P. (2001). *Transform your business into E.* CA: Academic Press.

Lin, C. T., & Lee, C. S. (1996). *Neural fuzzy systems: A neuro-fuzzy synergism to intelligent systems.* Englewood Cliffs, NJ: Prentice-Hall.

Lin, N., Burt, R. S., & Cook, K. (Eds.). (2001). *Social capital: Theory and research.* New York: Aldine de Gruyter.

Linden, G., Smith, B., & York, J. (2003). Amazon.com recommendations item-to-item collaborative filtering. *IEEE Internet Computing, 7*(1), 76-80.

Litecky, C. R., Arnett, K. P., & Prabhakar, B. (2004). The paradox of soft skills vs. technical skills in IS hiring. *Journal of Computer Information Systems, 45*(1), 69-76.

Litecky, C. R., Prabhakar, B., & Arnett, K. P. (2006). The IT/IS job market: A longitudinal perspective. In K. Kaiser & T. Ryan (Eds.), *Proceedings of the 2006 Association for Computing Machinery Special Interest Group Management Information Systems Conference on Computer Personnel Research* (pp. 50-52). New York: Association for Computing Machinery.

Looney, C. A., & Akbulut, A. Y. (2007). Combating the IS enrollment crisis: The role of effective teachers in introductory IS courses. *Communications of the Association for Information Systems, 19,* 781-805.

Luftman, J. N. (1996). Applying the strategic alignment model. In J. N. Luftman (Ed.), *Competing in the information age: Strategic alignment in practice* (pp. 43-69). New York: Oxford University Press.

Lyardet, F., & Rossi, G. (1998, August 11-14). *Patterns for designing navigable information spaces.* Paper presented at the Fifth Conference on Pattern Languages of Programs (PLoP 1998), Monticello.

Lyardet, F., Rossi, G., & Schwabe, D. (1999, July 8-10). *Patterns for adding search capabilities to Web information systems.* Paper presented at the Fourth European Conference on Pattern Languages of Programming and Computing (EuroPLoP 1999), Irsee, Germany.

Lynch, P. J., & Horton, S. (2002). *Yale CAIM Web style guide* (2nd ed.). Yale University Press.

Maggi, B. (2003). *De l'Agir Organisationnel.* Toulose: Editions Octares.

Maggiolini, P. (1981). *Costi e benefici di un sistema informativo.* Itália: Etas Libri.

Magoulas, G. D., Papanikolau, K. A., & Grigoriadou, M. (2001). Neurofuzzy synergism for planning the content in a Web-based course. *Informatica, 25*(1), 39-48.

Mahmood, A. (1987). Systems development methods: A comparative investigation. *MIS Quarterly, 11*(3), 293-311.

Mahmood, M. A. (1993). Associating organizational strategic performance with information technology investment: An exploratory research. *European Journal of Information Systems, 2*(3), 185-200.

Mahmood, M. A., & Soon, S. K. (1991). A comprehensive model for measuring the potential impact of information technology on organizational strategic variables. *Decision Sciences, 22*(4), 869-897.

Mahmoud, Q. H., & Maamar, Z. (2006). Engineering wireless mobile applications. *International Journal of Information Technology and Web Engineering, 1*(1), 59-75.

Mahoney, F. I., & Barthel, D. (1965). Functional evaluation: The Barthel index. *Maryland State Medical Journal, 14,* 56-61.

Malcolm, C. (2001, February). Making a healthcare Web site a sound investment. *Healthcare Financial Management,* 74-79.

Malec, B., & Friday, A. (2001). *The Internet and the physician productivity: UK & USA perspectives.* Paper presented at the EHMA Congress, Granada.

Malone, T. (2001). The future of e-business. *Sloan Management Review, 43*(1), 104.

Malone, T. W., Jay, J., Legay, L., & Kosorukoff, A. (2006). *Handbook of collective intelligence.* Retrieved at January 15, 2007, from http://www.eu.socialtext.net/mit-cci-hci/index.cgi?handbook_of_collective_intelligence

Manber, U., Patel, A., & Obison, J. (2000). Experience with personalization on Yahoo. *Communications of the ACM, 43*(8), 35-39.

Manolescu, D., & Kunzle, A. (2001, September 11-15). *Several patterns for eBusiness applications.* Paper presented at the Eighth Conference on Pattern Languages of Programs (PLoP 2001), Monticello.

Manolescu, D., Kozaczynski, W., Miller, A., & Hogg, J. (2007). The growing divide in the patterns world. *IEEE Software, 24*(4), 61-67.

Marchand, D. (2006).). Customer relationship management challenging the myth: Focus on people, not the technology. *Perspectives for Managers,* (131), 1-5. Retrieved January 30, 2007, from ProQuest ABI/INFORM Global database.

Markiewicz, M. E., Lucena, C. J. P., Alencar, P. S. C., & Cowan, D. D. (2002). Views and patterns in e-commerce application design. *Annals of Software Engineering, 13*(1-4), 111-140.

Marks, M., & Hong, D. (2006). *Web design patterns collection technical design* (Center for Document Engineering Tech. Rep. CDE2006-TR09). University of California, Berkeley.

Mashing the Web. (2005). *The Economist, 376*(8444), 4.

Mazhelis, O., Markkula, J., & Jakobsson, M. (2005, June 13-15). *Specifying patterns for mobile application domain using general architectural components.* Paper presented at the Sixth International Conference on Product Focused Software Process Improvement (PROFES 2005), Oulu, Finland.

McAfee, A. P. (2005). Will Web services really transform collaboration. *MIT Sloan Management Review, 46*(2).

McAfee, A. P. (2006). Enterprise 2.0: The dawn of emergent collaboration. *MIT Sloan Management Review, 47*(3), 21-28.

McAfee, A. P. (2006). *The impact of information technology (IT) on businesses and their leaders.* Retrieved January 1, 2006, from http://blog.hbs.edu/faculty/amcafee

McAfee, A. P. (2007). Enterprise 2.0 vs. SOA. *Harvard Business School faculty blog.* Retrieved June 8, 2007, from http://blog.hbs.edu/faculty/amcafee/index.php/faculty_amcafee_v3/enterprise_20_vs_soa/

McCabe, L. (2001). *A winning combination: Software-as-services plus business consulting and process services* (Summit Strategies market strategy report). Retrieved October 13, 2001, from http://www.summitstrat.com/store/3ss07detail

McDonald, M. H. B. (1997). *Marketing plans: How to prepare them, how to use them.* Oxford: Butterwoth-Heinemann.

McFarlan, W. E. (1984). Information technology changes the way you compete. *Harvard Business Review, 62*(3), 98-103.

McFarlan, W. F. (1990, Summer). The 1990s: The information decade. *Business Quarterly, 55*(1), 73-79.

McGee, M. (2004). Health-care I.T. has a new face. *Information Week, 988,* 16-26.

McKinsey. (2003). *Study HAWK 2015 – Wissensbasierte Veränderung derautomobilen Wertschöpfungskette, VDA.* Retrieved June 30, 2008, from http://www.wk-tim.de/fachtagungen/2004/unterlagen/10_MuellerFreiling.pdf

McLean, E., & Schneberger, S. (1997). *Georgia ICAPP information technology strategic industry needs assessment.* Unpublished manuscript, special report for the Board of Regents, University System of Georgia.

McNiff, J. (2000). *Action research in organizations.* London: Routledge.

McTaggart, R. (1998). Is validity really an issue for participatory action research? *Studies in Cultures, Organizations & Societies, 4*(2), 211-237.

Medlin, B. D. (2004). Skills crucial to the information technology professionals in the global business environment: An empirical study in the United States. *International Journal of Human Resources Development and Management, 4*(2), 183-199.

Medlin, B. D., & McCarthy, P. (1999). A study of grade distribution differences between gender and use of presentation technology. *Palmetto Review, 2*(1), 35-38.

Mehretens, J., Cragg, P., & Mills, A. (2001). A model of Internet adoption by SME's. *Information & Management, 39*, 165-176.

Melitski, J., Holzer, M., Kim, S.-T., Kim C. G., & Rho, S. Y. (2005). An e-government assessment of municipal Web sites. *International Journal of Electronic Government Research, 1*(1), 1-19.

Melone, N. P. (1990). A theoretical assessment of the user-satisfaction construct in information systems research. *Management Science, 36*(1), 76-91.

Memmola, M. (2007). The development of a Web strategy in a healthcare organization: A case history. In L. Al-Hakim (Ed.), *Web mobile-based applications for healthcare management* (pp. 1-35). Hershey PA: Information Science Publishing.

Menasalvas, E., Millan, S., Pena, J., Hadjimichael, M., & Marban, O. (2002). Subsessions: A granular approach to click path analysis. In *Proceedings of the FUZZ-IEEE Fuzzy Sets and Systems Conference, at the World Congress on Computational Intelligence*, Honolulu (pp. 12-17).

Mendes, E., & Mosley, N. (2006). *Web engineering*. Springer-Verlag.

Messner, W. (2005). Customer relationship management technology. *Business Information Review, 22*(4), 253-262. Retrieved January 31, 2007, from Business Source Premier database.

Meszaros, G., & Doble, J. (1998). A pattern language for pattern writing. In R. C. Martin, D. Riehle, & F. Buschmann (Eds.), *Pattern languages of program design 3* (pp. 529-574). Addison-Wesley.

Microsoft. (2006). Retrieved June 19, 2008, from http://www.microsoft.com/biztalk/solutions/soa/overview.mspx

Miller, J., & Doyle, B. A. (1987). Measuring effectiveness of computer-based information systems in the financial services sector. *MIS Quarterly, 11*(1), 107-124.

Milner, R. (1999). *Communicating and mobile systems: The Pi calculus*. Cambridge: University Press.

Mimoso, M. S. (2004). *More enterprises exposing Web services to Net*. Retrieved June 19, 2008, from www.SearchWebServices.com

Mimoso, M. S. (2004). SOA prominent on 2005 budgets. *Yankee Group of Boston*. Retrieved June 19, 2008, from www.SearchWebServices.com

Min, H., & Galle, W. (2001, April). Electronic commerce-based purchasing: A survey on the perpetual differences between large and small organizations. *International Journal of Logistics, 4*(1), 79-96.

Minard, B. (2002). CIO longevity and IT project leadership. *IT Health Care Strategist, 4*(1), 3-7.

Mindreef, Inc. (2006). Retrieved June 19, 2008, from www.mindreef.com/report

Mithas, S., Krishnan, M. S., & Fornell, C. (2005). Why do customer relationship management applications affect customer satisfaction? *Journal of Marketing, 69*(4), 201-209. Retrieved January 31, 2007, from Business Source Premier database.

Mitra, C. (2002). A strong IT team is essential for CRM success. *Cybermedia India Online Limited (CIOL)*. Retrieved February 18, 2007, from http://www.ciol.com/content/search/showarticle.asp?arid=35658&way=search

Mitra, S., & Pal, S. K. (1995). Fuzzy multi-layer perceptron, inferencing and rule generation. *IEEE Transactions on Neural Networks, 6*, 51-63.

Mladenic, D. (1996). *Personal Webwatcher: Implementation and design* (Tech. Rep. IJSDP-7472). Department of Intelligent Systems, J. Stefan Institute, Slovenia.

Mobasher, B. (2005). Web usage mining and personalization. In M. P. Singh (Ed.), *Practical handbook of internet computing*. CRC Press.

Mobasher, B. (2006). Web usage mining. In B. Liu (Ed.), *Web data mining: Exploring hyperlinks, contents and usage data*. Berlin-Heidelberg: Springer.

Mobasher, B. (2007). Data mining for personalization. In P. Brusilovsky, A. Kobsa, & W. Nejdl (Eds.), *The adaptive Web: Methods and strategies of Web personalization* (LNCS 4321, pp. 90-135). Berlin-Heidelberg: Springer.

Mobasher, B. (2007). Recommender systems. *Kunstliche Intelligenz, Special Issue on Web Mining*, (3), 41-43. Bremen, Germany: BottcherIT Verlag.

Mobasher, B., Cooley, R., & Srivastava, J. (1999). *Automatic personalization based on Web usage mining* (Tech Rep. 99010). DePaul University, Department of Computer Science.

Mobasher, B., Cooley, R., & Srivastava, J. (2000). Automatic personalization based on Web usage mining. *Communications of the ACM, 43*(8), 142-151.

Mobasher, B., Dai, H., Luo, T., & Nakagawa, M. (2001). *Effective personalization based on association rule discovery from Web usage data.* Paper presented at the ACM Workshop on Web Information and Data Management, Atlanta.

Mobasher, B., Dai, H., Nakagawa, M., & Luo, T. (2002). Discovery and evaluation of aggregate usage profiles for Web personalization. *Data Mining and Knowledge Discovery, 6*, 61-82.

Mobasher, B., Nasraoui, O., Liu, B., & Masand, B. (2006). *Advances in Web mining and Web usage analysis* (2004 WebKDD Volume) (LNAI 3932). Berlin-Heidelberg: Springer.

Montero, F., López-Jaquero, V., & Molina, J. P. (2003, September 1-2). *Improving e-shops environments by using usability patterns.* Paper presented at the Second Workshop on Software and Usability Cross-Pollination, Zürich, Switzerland.

Montero, F., Lozano, M., & González, P. (2002, August 5-7). *Designing Web sites by using design patterns.* Paper presented at the Second Latin American Conference on Pattern Languages of Programming (SugarLoafPLoP 2002), Rio de Janeiro, Brazil.

Moore, F. L., & Streib, J. T. (1989). Identifying the gaps between education and training. *Association for Computing Machinery Special Interest Group Computer Education Bulletin,* 52-55. New York: Association for Computing Machinery.

Morris, M., Pohlmann, T., & Oliver, Y. G. (2005, April 8). *How do users feel about technology? Business technographics United States.* Forrester Research.

Mortazavi-Asl, B. (2001). *Discovering and mining user Web-page traversal patterns.* Unpublished master's thesis, Simon Fraser University.

Müller-Böling, D., & Müller, M. (1986). *Akzeptanzfaktoren der bürokommunikation.* München, Germany: Oldenbourg.

Müller-Verse, F. (1999). Mobile commerce report (*Durlacher research report).* Retrieved December 16, 2006, from http://www.durlacher.com/downloads/mcomreport.pdf

Mulvenna, M., Anand, S., & Buchner, A. (2000). Personalization on the net using Web mining. *CACM, 43*(8), 123-125.

Musser, J., & O'Reilly, T. (2006, November). *Web 2.0 principles and best practices.* O'Reilly Radar.

Myers, M. (1997). Qualitative research in information systems. *MIS Quarterly, 21*(2), 241-243.

Nakano, R. (2002). *Web content management.* Boston: Addison-Wesley.

Nanopoulos, A., Katsaros, D., & Manolopoulos, Y. (2001, August 26). Exploiting Web log mining for Web cache enhancement. In R. Kohavi, B. Masand, M. Spiliopoulou, & J. Srivastava (Eds.), *WEBKDD 2001—Mining Web Log Data Across All Customers Touch Points, Third International Workshop,* San Francisco (LNCS 2356, pp. 68-87). Springer.

Nasraoui, O. (2005). World Wide Web personalization. In J. Wang (Ed.), *Encyclopedia of data mining and data warehousing.* Hershey, PA: IGI Global, Inc.

Nasraoui, O., & Frigui, H. (2000). Extracting Web user profiles using relational competitive fuzzy clustering. *International Journal on Artificial Intelligence Tools, 9*(4), 509-526.

Nasraoui, O., Krishnapuram, R., & Joshi, A. (1999). Relational clustering based on a new robust estimator with applications to Web mining. In *Proceedings of the International Conf. North American Fuzzy Info. Proc. Society (NAFIPS 99),* New York (pp. 705-709).

Nauck, D. (1999). Using symbolic data in neuro-fuzzy classification. In *Proceedings of NAFIPS'99*, New York (pp. 536-540).

Nauck, D., Klawonn, F., & Kruse, R. (1997). *Foundations of neuro-fuzzy systems.* Chichester, UK: Wiley.

Negash, S., Ryan, T., & Igbaria, M. (2003). Quality and effectiveness on Web-based customer support systems. *Information and Management, 40*(8), 757-768.

Nevens, M. (2002). The real source of the productivity boom. *Harvard Business Review, 80*(3), 23-24.

New Zealand Department of Labour. (2006). *Information technology professional: Occupational skill shortage assessment.* Retrieved July 22, 2007, from http://www.worklife.govt.nz/PDFs/jvm-prof-ict-2005.pdf

New Zealand Department of Labour. (2007). *Job vacancy monitoring programme reports.* Retrieved July 12, 2007, from http://dol.govt.nz/jvm/job-ad-monthly-report.asp?from=alert)

NextInnovator. (2004). *Top ten technologies for 2005.* Retrieved June 20, 2008, from http://technologyreports.net/nextinnovator/?articleID=3004

Ng Tye, E. M. W., Poon, R. S. K., & Burn, J. M. (1995). Information systems skills: Achieving alignment between the curriculum and the needs of the IS professionals. *Association for Computing Machinery Specialist Interest Group management Information Systems Database, 26*(4), 47-61.

Nguyen, H. Q., Johnson, R., & Hackett, M. (2003). *Testing applications on the Web: Test planning for mobile and Internet-based systems* (2nd ed.). John Wiley and Sons.

Nielsen, J. (1993). *Usability engineering.* San Diego: Academic Press.

Nixon, I., Smith, K., Stafford, R., & Camm, S. (2006). Work-based learning: Illuminating the higher education landscape. *The Higher Education Academy final report.* Retrieved December 5, 2007, from http://www.theksapartnership.co.uk/assets/downloads/KSA_HEA_work_based_learning_final_report.pdf

Nohria, N., & Eccles, R. (1992). *Networks in organizations* (2nd ed.). Boston: Harvard Business Press.

Nokia. (1999). *The demand for mobile value-added services.* Retrieved June 11, 2007, from http://www.telecomsportal.com/Assets_papers/Wireless/Nokia_mobile_vas.pdf

Nolan, R. L. (1979). Managing the crises in data processing. *Harvard Business Review, 57*(2), 115-126.

Nonaka, I., & Takeuchi, H. (1995). *The knowledge-creating company: How Japanese companies create the dynamics of innovation.* Oxford University Press.

Nunnally, J. C. (1978). *Psychometric theory* (2nd ed.). New York: McGraw-Hill.

O'Connor, M., & Herlocker, J. (1999). Clustering items for collaborative filtering. In *Proceedings of ACM SIGIR '99 Workshop on Recommender Systems: Algorithms and Evaluation.*

O'Reilly, D. (2004, 7 May). Web-User satisfaction on the upswing. *PC World.* Retrieved January 1, 2006, from http://www.pcworld.com/news/article/0,aid,116060,00.asp

O'Reilly, T. (2007) *What is Web 2.0.* Retrieved June 9, 2007, from http://www.oreillynet.com/pub/a/oreilly/tim/news/2005/09/30/what-is-web-20.html

OASIS standards. (2005). Retrieved April 20, 2005, from http://www.oasis-open.org

Offutt, J. (2002). Quality attributes of Web software applications. *IEEE Software, 19*(2), 25-32.

OGC. (2007). *ITIL: Service strategy.* London: TSO. ISBN 10: 0113310455.

Oliveira, Djalma de Pinho Rebouças de. (2001). *Sistemas, organização e métodos: uma abordagem gerencial*. São Paulo: Atlas.

Oliveira, L. C. B., & Joia, L. A. (2005, May 26-28). *A model for evaluating B2C e-ocmmerce Websites: Application in the CD e-retailing industry in Brazil*. Paper presented at the Thirteenth European Conference on Information Systems (ECIS 2005), Regensburg, Germany.

Opera Software ASA. (n.d.). *Opera's small-screen rendering*. Retrieved June 23, 2007, from http://www.opera.com/products/mobile/smallscreen/

Oracle financial reports. Retrieved October 13, 2006, from http://www.oracle.com/corporate/investor_relations/analysts/

Ostheimer, B. (2007). *Verteilende eBusiness-systeme—organisatorische flexibilisierung am beispiel eines verteilenden eUniversity-systems*. Wiesbaden, Germany: Deutscher Universitaets-Verlag.

Paduan, R. (2001, June). Como a Internet está transformando (de verdade) a vida nas empresas. *Revista Exame*, *742*.

Pal, S., Talwar, V., & Mitra, P. (2002). Web mining in soft computing framework: Relevance, state of the art and future directions. *IEEE Transactions on Neural Networks, 13*(5), 1163-1177.

Panko, R. (2007, December 6). *Bureau of Labor Statistics 2006-2016 employment projection*. Message posted to ISWORLD Information Systems World Network electronic mailing list, archived at http://lyris.isworld.org/lyris.pl?sub=4931909&id=356489948

Park, C. H., & Kim, Y. G. (2006). The effect of information satisfaction and relational benefit on consumers' online shopping site commitments. *Journal of Electronic Commerce in Organizations*, *4*(1), 70-90.

Parnas, D. L. (1998). Software engineering programmes are not computer science programmes. *Annals of Software Engineering, 6*, 19-37.

Pashtan, A., Kollipara, S., & Pearce, M. (2003). Adapting content for wireless Web services. *IEEE Internet Computing, 7*(5), 79-85.

Paulk, M. C., Weber, C. V., Curtis, B., & Chrissis, M. B. (1995). *The capability maturity model: Guidelines for improving the software process*. Addison-Wesley.

Payne, A., & Frow, P. (2006). Customer relationship management: From strategy to implementation. *Journal of Marketing Management, 22*(1/2), 135-168. Retrieved January 31, 2007, from Business Source Premier:

Pei, J., Han, J., Mortazavi-asl, B., & Zhu, H. (2000). Mining access patterns efficiently from Web logs. Paper presented at the *Pacific-Asia Conference on Knowledge Discovery and Data Mining* (pp. 396-407).

Perkowitz, M., & Etzioni, O. (1997). *Adaptive sites: Automatically learning from user access patterns* (Tech. Rep. UW-CSE-97-03-01). University of Washington.

Perry, M., & Kaminski, H. (2005, July 6-10). *A pattern language of software licensing*. Paper presented at the Tenth European Conference on Pattern Languages of Programs (EuroPloP 2005), Irsee, Germany.

Pertet, S. M., & Narasimhan, P. (2005). *Causes of failure in Web applications* (PDL Tech. Rep. PDL-CMU-05-109). Carnegie Mellon University.

Perzel, K., & Kane, D. (1999, August 15-18). *Usability patterns for applications on the World Wide Web*. Paper presented at the Sixth Conference on Pattern Languages of Programs (PLoP 1999), Monticello.

Petersen, A. W., & Wehmeyer, C. (2005). European qualification strategies in information and communications technologies (ICT)—towards a European (reference) ICT skills and qualification framework. *Studies in Informatics and Control, 14*(1), 37-46.

Petri, C. A. (1962). *Kommunikation mit Automaten*. Unpublished doctoral thesis, Institut f¨ur instrumentelle Mathematik, Bonn.

Petrova, K., & Claxton, G. (2005). Building student skills and capabilities in information technology and eBusiness: A moving target. *Journal of Information Systems Education, 16*(1), 27-43.

Petrova, K., & Medlin, D. (2006). Towards a general framework for the study of ICT skill supply and demand. In M. Khosrow-Pour (Ed.), *Emerging Trends and Challenges in Managing Information Technology: Proceedings of the 2006 Information Resources Management Association Conference* (pp. 255-263). Hershey, PA: International Resource Management Association. .

Pettigrew, A. M. (1985). *The awakening giant: Continuity and change in ICI basil.* Blackwell.

Philipson, G. (2004) *Time to accept the grim reality of IT.* Retrieved March 3, 2004, from http://smh.com.au/articles/2004/03/01/1078117352254.html

Pick, J. B., Mallen, G., Navarrette, C., Guthrie, R., & Valle, L. (1998). Cross cultural comparison of the concordance of IS education/training and IS careers in the United States and Latin America (panel). In *Proceedings of the 1998 Association for Computing machinery Special Interest Group Computer Personnel Research Conference on Computer Personnel Research* (pp. 108-109). New York Association for Computing Machinery.

Pierrakos, D., Paliouras, G., Papatheodorou, C., & Spyropoulos, C. D. (2003). Web usage mining as a tool for personalization: A survey. *User Modeling and User-Adapted Interaction, 13*(4), 311-372.

Pitkow, J., & Bharat, K. (1994). WEBVIZ: A tool for World-Wide Web access log visualization. In *Proceedings of the 1st International World-Wide Web Conference.* Geneva, Switzerland (pp. 271-277).

Plant, R. (2000). *E-commerce formulation of strategy.* NJ: Prentice Hall.

Pohl, W. (1996). Learning about the user modeling and machine learning. In V. Moustakis & J. Herrmann (Eds.), *International Conference on Machine Learning Workshop Machine Learning Meets Human-Computer Interaction* (pp. 29-40).

Poon, S., & Swatman, P. (1999, January). An exploratory study of small business Internet issues. *Information & Management, 35*(1), 9-18.

Porter, M. (1985). *Competitive advantage.* New York: Free Press.

Porter, M. (2001). Strategy and the Internet. *Harvard Business Review.*

Porter, M. E. (1979, March-April). How competitive forces shape strategy. *Harvard Business Review, 57*(2), 81-92.

Porter, M. E. (1980). *Competitive strategy: Techniques for analyzing industries and competitors.* New York: The Free Press.

Porter, M. E. (2001, March). Strategy and the Internet. *Harvard Business Review, 79*(3), 62-79.

Porter, M. E., & Millar, V. (1985, July-August). How information gives you competitive advantage: The information revolution is transforming the nature of competition. *Harvard Business Review, 63*(4), 149-160.

Prahalad, C. K., & Hamel, G. (1990, May-June). The core competencies of the corporation. *Harvard Business Review, 68*(3), 79-90.

Preece, J., Rogers, Y., & Sharp, H. (2002). *Interaction design: Beyond human-computer interaction.* John Wiley and Sons.

Quinn, J. B., Anderson, P., & Finkelstein, S. (2005). Leveraging intellect. *Academy of Management Executive, 19*(4), 78-94.

Rademacher, R. (2001). The changing profile of information systems research: 1995-2000. *Journal of Computer Information Systems, 42*(1), 13-17.

Ragsdale, J. (2007). 2007 service and support technology trends. *SSPA News.* Retrieved April 6, 2007, from http://www.thesspa.com/sspanews/_07Jan/article1.asp

Rainer, E. (2005). New trends & techniques in managing CRM & SAP projects. *ITtoolbox.* Retrieved February 1, 2007, from http://blogs.ittoolbox.com/crm/trends/archives/erp-packages-feature-comparison-4309

Raisinghani, M. S., Tan, E., Untama, J. A., Weiershaus, H., et al. (2005). CRM systems in German hospitals: Illustrations of issues & trends. *Journal of Cases on Information Technology*, *7*(4), 1-26. Retrieved January 30, 2006, from ProQuest ABI/INFORM Global database.

Rand, M. (2004, December 13). Best of the Web: Extreme blogging. *Forbes*. Retrieved January 1, 2006, from http://www.forbes.com/best/2004/1213/bow001.html

Reding, V. (2006, December 3). *The disruptive force of Web 2.0: How the new generation will define the future* (European Commission press release). Hong Kong: Youth Forum, ITU Telecom World.

Reid, A., & Catterall, M. (2005). Invisible data quality issues in a CRM implementation. *Journal of Database Marketing & Customer Strategy Management*, *12*(4), 305-315. Retrieved January 30, 2006, from ProQuest ABI/INFORM Global database.

Reisman, S. (2005, May-June). Quality IT education strikes out. *IT Professional*, 16-17.

Renkema, T. J. W., & Berghout, E. W. (1997). Methodologies for information system investment evaluation at the proposal stage: A comparative view, *Information and Software Technology*, *39*(1), 1-13.

Richards, K. A., & Jones, E. (2006). Customer relationship management: Finding value drivers. *Industrial Marketing Management*. Retrieved February 28, 2007, from ScienceDirect database.

Riemenschneider, C., & McKinney, V. (1999). Assessing the adoption of Web-based e-commerce for businesses. *Electronic Markets*, *9*(1/2), 9-13.

Rigden, C. (1999). The eye of the beholder: Designing for colour-blind users. *British Telecommunications Engineering Journal*, *17*.

Ritchie, B., & Brindley, C. (2005). ICT adoption by SMEs: Implications for relationships and management. *Work & Employment*, *20*(3), 205-217.

Robert, M., & Racine, B. (2001). *E-strategy pure & simple*. New York: McGraw-Hill.

Roberts, M. L., Liu, R. R., & Hazard, K. (2005). Strategy, technology and organizational alignment: Key components of CRM success. *Journal of Database Marketing & Customer Strategy Management*, *12*(4), 315-327. Retrieved January 30, 2006, from ProQuest ABI/INFORM Global database.

Rockart, J. F. (1979, March-April). Chief executives define their own data needs. *Harvard Business Review*, *57*(2), 81-92.

Rogers, E. (1995). *Diffusion of innovations*. New York: Free Press.

Roman, D. et al. (2005). Web service modeling ontology. *Applied Ontology*, *1*(1), 77-106.

Romanosky, S., Acquisti, A., Hong, J., Cranor, L. F., & Friedman, B. (2006, October 21-23). *Privacy patterns for online interactions*. Paper presented at the Thirteenth Conference on Pattern Languages of Programs (PLoP 2006), Portland.

Rosenberg, M. (2006). *Beyond e-learning*. San Francisco: John Wiley.

Rossi, G., & Koch, N. (2002, July 3-7). *Patterns for adaptive Web applications*. Paper presented at the Seventh European Conference on Pattern Languages of Programs (EuroPLoP 2002), Irsee, Germany.

Rossi, G., Lyardet, F., & Schwabe, D. (2000, July 5-9). *Patterns for e-commerce applications*. Paper presented at the Fifth European Conference on Pattern Languages of Programs (EuroPLoP 2000), Irsee, Germany.

Rossi, G., Pastor, O., Schwabe, D., & Olsina, L. (2008). *Web engineering: Modelling and implementing Web applications*. Springer-Verlag.

Rossi, G., Schwabe, D., Danculovic, J., & Miaton, L. (2001, July 4-8). *Patterns for personalized Web applications*. Paper presented at the Sixth European Conference on Pattern Languages of Programs (EuroPLoP 2001), Irsee, Germany.

Salganik, M. J., Dodds, P. S., & Duncan, J. (2006, February). *Experimental study of inequality and unpredict-*

ability in an artificial cultural market. Watts, Science 10, 311(5762), 854-856.

Sankar, K. P., Varun, T., & Pabitra, M. (2002). Web mining in soft computing framework: Relevance, state of the art and future directions. *IEEE Transaction on Neural Networks*, *13*(5), 1163-1177.

Santos, A. M., & Medeiros, M. (2001). *Reestruturação da indústria automobilística na América do Sul*. BNDES Setorial, Rio de Janeiro, n. 14, September, www.bndes.gov.br/conhecimento/bnset/set1403.pdf

Santosus, M. (2001). *Inferiority complex*. Retrieved April 14, 2005, from http://www.cio.com/archive/031504/reality.html

SAP financial reports. Retrieved October 13, 2006, from http://www.sap.com/company/investor/reports/

SAP. (2007). *mySAP CRM: Software for enabling customer relationship management (CRM) without compromise*. Retrieved April 1, 2007, from http://www.sap.com/solutions/business-suite/crm/index.epx

Sartzetaki, M., Psaromiligkos, Y., Retalis, S., & Avgeriou, P. (2003, June 22-27). *Usability evaluation of e-commerce sites based on design patterns and heuristic criteria*. Paper presented at the Tenth International Conference on Human - Computer Interaction, Crete, Greece.

SAS Institute. (1993). *SAS/STAT user guide, version 6* (4th ed.). Cary, NC: Autho.

Schafer, B., Konstan, J. A., & Riedl, J. (2001). E-commerce recommendation applications. *Data Mining and Knowledge Discovery*, *5*(12), 115-152. Kluwer Academic Publishers.

Scheer, A. W. (2000). *ARIS—business process modeling*. Berlin: Springer-Verlag.

Schmidt, D. C., Stal, M., Rohnert, H., & Buschmann, F. (2000). *Pattern-oriented software architecture, volume 2: Patterns for concurrent and networked objects*. John Wiley and Sons.

Schmidt, I. (2005). *Blended e-learning*. Saarbruecken, Germany: VDM.

Schroth, C., & Christ, O. (2007). Brave new Web: Emerging design principles and technologies as enablers of a global SOA. In *Proceedings of the 2007 IEEE International Conference on Services Computing (SCC 2007)*, Salt Lake City.

Schroth, C., & Janner, T. (2007). Web 2.0 and SOA: Converging concepts enabling the Internet of services. *IT Professional*, *9*(3), 36-41.

Schulmeister, R. (2005). *Lernplattformen fuer das virtuelle Lernen*. Muenchen, Germany: Oldenbourg.

Schumacher, M., Fernandez-Buglioni, E., Hybertson, D., Buschmann, F., & Sommerlad, P. (2006). *Security patterns: Integrating security and systems engineering*. John Wiley and Sons.

Schwickert, A. C. (2004). Dezentrales Web content management. In S. Geberl, S. Weinmann, & D. F. Wiesner (Eds.), *Impulse aus der Wirtschaftsinformatik* (pp. 247-260). Heidelberg, Germany: Physica.

Schwickert, A. C., & Grund, H. (2004). *Web content management—Grundlagen und Anwendung mit dem Web portal system V. 2.5* (Arbeitspapiere WI, No. 3/2004). Giessen, Germany: Justus-Liebig-University.

Scott, M., Golden, W., & Hughes, M. (2003). E-tailing in Ireland: A review of Ireland's top 25 retailers. *Irish Marketing Review*, *16*(1), 15-24.

Scupola, A. (2002). Adoption issues of B2B Internet commerce in European SMEs. In *Proceedings of the 35th Hawaii International Conference on Systems Science, 2002*.

Seeman, E. D., & O'Hara, M. (2006). Customer relationship management in higher education: Using information systems to improve the student-school relationship. *Campus - Wide Information Systems*, *23*(1), 24-35. Retrieved January 30, 2006, from ProQuest ABI/INFORM Global database.

Segars, A. H., & Hendrickson, A. R (2000). Value, knowledge, and the human equation: Evolutions of the information technology function in modern organizations. *Journal of Labor Research*, *21*(3), 431-445.

Segerståhl, K., & Jokela, T. (2006, April 22-27). *Usability of interaction patterns.* Paper presented at the CHI 2006 Conference on Human Factors in Computing Systems, Montréal, Canada.

Seidenfaden, L., Kahnwald, N., Kaspar, C., & Ortelbach, B. (2005). *Mediennutzung im digitalen Leben: Active content interfaces, paid content und integrierte geschäftsmodelle*, Goettingen, Germany: Business Village.

Shackelford, R., Cassel, L, Cross, J., Davies, J., Impagliazzo, J., Kamali, R., et al. (2004). *A guide to undergraduate degree programs in computing (Strawman Draft).* Joint Task Force for Computing Curricula 2004. Retrieved June 12, 2004, from http://www.acm.org/education/

Shah, J. R., & Murtaza, M. B. (2005). Effective customer relationship management through Web services. *The Journal of Computer Information Systems, 46*(1), 98-110. Retrieved January 30, 2006, from http://proquest.umi.com.ezproxy.umuc.edu/pqdweb?index=33&did=914802221&SrchMode=1&sid=3&Fmt=4&VInst=PROD&VType=PQD&RQT=309&VName=PQD&TS=1170210789&clientId=8724

Shank, J. K., & Govindarajan, V. (1992, Winter). Strategic cost management and the value chain. *Journal of Cost Management*, 5-21.

Shanks, G. (1999, September 29). *Semiotic approach to understanding representation in information systems.* Paper presented at the Information Systems Foundations Workshop, Sydney, Australia.

Shapiro, C., & Varian, H. R. (1998). *Information rules: A strategic guide to the network economy.* Boston: Harvard Business School Press.

Shapiro, C., & Varian, H. R. (1999). *Information rules.* Boston: Harvard Business School Press.

Sharkey, U., Scott, M., & Acton, T. (2006, June 12-14). *The influence of quality on the success of e-commerce systems.* Paper presented at the Fourteenth European Conference on Information Systems (ECIS 2006), Gothenburg, Sweden.

Sharp, H., Galal, G. H., & Finkelstein, A. (1999, August 30-September 3). *Stakeholder identification in the requirements engineering process.* Paper presented at the Tenth International Conference and Workshop on Database and Expert Systems Applications (DEXA 1999), Florence, Italy.

Sherman, H. D., & Rupert, T. J. (2006). Do bank mergers have hidden or foregone value? Realized and unrealized operating synergies in one bank merger. *European Journal of Operational Resources, 168*(1), 253-268.

Shih, H. P. (2004). An empirical study on predicting user acceptance of e-shopping on the Web. *Information and Management, 41*(3), 351-368.

Shim, S. J., & Min, B. K. (2002). Organizational factors associated with expert systems implementation. *Journal of Computer Information Systems, 42*(4), 71-76.

Silicon Valley Daily. (2006). *HP regains lead in global PC sales.* Retrieved June 11, 2007, from http://www.svdaily.com/gartner1.html

SIM Advocacy Research Team. (2006). *Information technology workforce: Trends and implications 2005-2008.* Society for Information Technology. Retrieved November 5th, 2007, from http://www.simnet.org/Content/NavigationMenu/Programs/Advocacy_and_Communities/Advocacy_Downloads/SIMAdvocacyITWorkforceexecsummary3-10.pdf

Sirkinh, H., & Dean, D. (2000). *Mobile commerce: Winning the on-air consumer.* Retrieved June 11, 2007, from http://www.bcg.com/publications/files/M-Commerce_Nov_2000_Summary.pdf

Slack, N., & Lewis, M. A. (2001). *Operations strategy.* FT Prentice Hall.

Smith, D. M. (2006, May 16). *Web 2.0 offers many opportunities for growth, but few enterprises will immediately adopt all aspects necessary for significant business impact.* Paper presented at the Gartner Symposium/ITxpo, San Francisco.

Smith, H., & Fingar, P. (2003). *Workflow is just a Pi process.* Retrieved June 19, 2008, from www.bpm3.com/picalculus

Smith, R. (2006, June 14-16). *Enterprise mash-ups: An industry case study.* Paper presented at the New York PHP Conference & Expo 2006, The New Yorker Hotel, Manhattan.

Smithson, S., & Hirschheim, R. (1998). Analyzing information systems evaluation: Another look at an old problem. *European Journal of Information System, 7,* 158-174.

Software Engineering Institute. (2002). *Capability maturity model® integration (CMMISM)—version 1.1—staged representation* (Tech. Rep. CMU/SEI-2002-TR-012). Retrieved October 15, 2003, from http://www.sei.cmu.edu/cmmi/

Spiliopoulou, M. (1999). *Tutorial: Data mining for the Web.* Paper presented at the PKDD'99. Prague, Czech Republic.

Spiliopoulou, M., Mobasher, B., Berendt, B., & Nakagawa, M. (2003). A framework for the evaluation of session reconstruction heuristics in Web usage analysis. *INFORMS Journal of Computing - Special Issue on Mining Web-Based Data for E-Business Applications, 15*(2), 171-190.

Spiliopoulou, N., & Faulstich, L. C. (1998). *WUM: A Web utilization miner.* Paper presented at the International Workshop on the Web and Databases, Valencia, Spain (LNCS 1590, pp. 109-115). Springer.

Srinivasan, A. (1985). Alternative measures of system effectiveness: Associations and implications. *MIS Quarterly, 9*(3), 243-253.

Stamper, R. (1992, October 5-8). *Signs, organizations, norms and information systems.* Paper presented at the Third Australian Conference on Information Systems, Wollongong, Australia.

Stansberry, M. (2005). *Yankee group: SOA everywhere by 2006.* Retrieved June 19, 2008, from www.SearchDataCenter.com.

Stathacopoulou, R., Grigoriadou, M., & Magoulas, G. D. (2003). A neurofuzzy approach in student modeling. In *Proceedings of the 9th International Conference on User Modeling, UM2003* (LNAI 2702, pp. 337-342).

Stefani, A., & Xenos, M. (2001, December 7-9). *A model for assessing the quality of e-commerce systems.* Paper presented at the First Panhellenic Conference in Human-Computer Interaction (PC-HCI 2001), Patras, Greece.

Stratopoulos, T., & Dehning, B. (2000). Does successful investment in information technology solve the productivity paradox? *Information & Management, 38*(2), 103-117.

Sullivan, J. R., & Walstrom, K. A. (2001). Consumer perspectives on service quality of electronic commerce Web sites. *Journal of Computer Information Systems, 41*(3), 8-14.

Surowiecki, J. (2005). *The wisdom of crowds* (pp. xviii – xix, Chapters 2-4). Anchor Books.

Suryavanshi, B. S., Shiri, N., & Mudur, S. P. (2005). An efficient technique for mining usage profiles using relational fuzzy subtractive clustering. In *Proceedings of the 2005 International Workshop on Challenges in Web Information Retrieval and Integration (WIRI'05)* (pp. 23-29).

Susman, G., & Evered, R. (1978). An assessment of the scientific merits of action research. *Administrative Science Quarterly, 23*(4), 582-603.

Sweet, P. (2004). *Outsourcing is in PmP research.* Retrieved October 13, 2001, from http://www.conspectus.com/2004/september/ article1.asp

Symbian Limited. (2006). *Fast facts.* Retrieved December 10, 2006, from http://www.symbian.com/about/fastfacts/fastfacts.html

Symonds, M. (1999, June 26-July 2). The Net imperative. *The Economist.* Retrieved December 5, 2007, from http://crab.rutgers.edu/~goertzel/economistnetbusiness.htm

Taghaboni-Dutta, F., & Velthouse, B. (2006, November). RFId technology is revolutionary: Who should be involved in this game of tag. *Academy of Management Perspective*, 65-78.

Tan, P. N., & Kumar, V. (2002). Discovery of Web robot sessions based on their navigational patterns. *Data Mining and Knowledge Discovery, 6*(1), 9-35.

Tang, H-L., Lee, S., & Koh, S. (2001). Educational gaps as perceived by IS educators: A survey of knowledge and skill requirements. *Journal of Computer Information Systems, 41*(2), 76-81.

Tapscott, D. (2001). Rethinking strategy in a networked world. *Strategy + Business, 24*, 8.

Tapscott, D., Lowy, A., & Ticoll, D. (2000). *Plano de Ação para uma economia digital.* Makron Books do Brasil Ltda.

Tavani, H. T. (2004). *Ethics and technology: Ethical issues in an age of information and communication* technology. John Wiley and Sons.

TechMetrx Research. (2001). *Web services adoption & technology choices report.* Retrieved June 19, 2008, from http://www.techmetrix.com/products

TechMetrx Research. (2003, February). *Adoption of Web services & technology choices, version 2 – February 2003 analysis of survey results.*

Teradata. (2004). *Survival of the fastest.* Retrieved June 20, 2008, from http://www.teradata.com/t/page/126753/

Thompson, E, Dunne, M., Radcliffe, J., Herschel, G., Prentice, B., Mertz, S. A., et al. (2006). *Key issues for CRM strategy and implementation* (Gartner Research Issue Brief No. G00137783). Stamford, CT: Gartner, Inc. Retrieved March 8, 2007 from Gartner database.

Thompson, E., & Goldman, M. (2006). *The industries in which CRM consultants and systems integrators deploy Oracle-Siebel and SAP* (Gartner Research Rep. No.G00143194). Retrieved March 27, 2006, from Gartner database.

Tidwell, J. (2006). *Designing interfaces: Patterns for effective interaction design.* O'Reilly Media.

TIMElabs. (2000). *Winning in Mobile eMarkets.* Retrieved February 13, 2007, from http://www.timelabs.de/pdf/TIMElabs_C_mB2B_d.pdf

Timmers, P. (1999). *Electronic commerce: Strategies and models for B2b trading.* New York: Wiley.

Toraskar, K. V., & Lee, P. C. B. (2006). A contextualist approach to the development of e-commerce education: An analysis of key issues and challenges. *Journal of Information Technology Management, XVII*(2), 1-12.

Treber, U., Berg, J. H., & Schwickert, A. C. (2004). *Smart-card-anwendungen am fachbereich wirtschaftswissenschaften der justus-liebig-universitaet giessen* (Arbeitspapiere WI, No. 7/2004). Giessen, Germany: Justus-Liebig-University.

Treese, G. W., & Stewart, L. C. (2003). *Designing systems for Internet commerce* (2nd ed.). Addison-Wesley.

Tsakonas, A., Dounias, G., Vlahavas, I. P., & Spyropoulos, C. D. (2002). *Hybrid computational intelligence schemes in complex domains: An extended review* (LNCS 2308, pp. 494-511).

Turban, E. et al. (2000). *Electronic commerce: A managerial perspective*. NJ: Prentice-Hall Inc.

Turban, E., McLean, E., & Wetherbe, J. (2002). *Information technology for management: Transforming business in the digital economy* (3rd ed.). New York: John Wiley and Sons Inc.

Turban, E., Rainer, R., & Potter, R. (2003). *Introduction to information technology* (3rd ed.). New York: John Wiley & Sons, Inc.

U. S. Department of Labor. (1999). *Skills and tasks for jobs: A SCANS Report for America 2000. Retrieved June 12, 2005, from* http://wdr.doleta.gov/opr/fulltext/document.cfm?docn=6140

U. S. Department of Labor. (2006). *Occupational outlook handbook (OOH), 2006 – 07 edition.* Retrieved July 15, 2007, from http://www.bls.gov/oco/home.htm

Ulfelder, S. (2005). *Apps on tap.* Retrieved May 13, 2005, from http://www.adtmag.com/print.asp?id=10535

Utterback, J., & Abernathy, W. (1975). A dynamic model of product and process innovation. *Omega, 3*(6), 639-656.

Vahidy, T. (2005). *The SOA revolution.* Retrieved June 19, 2008, from www.line56.com/print/default.asp?ArticleID=7099

Vakali, A., Pokorný, J., & Dalamagas, T. (2004). *An overview of Web data clustering practices.* Paper presented at the EDBT Workshops (pp. 597-606).

Van Duyne, D. K., Landay, J., & Hong, J. I. (2003). *The design of sites: Patterns, principles, and processes for crafting a customer-centered Web experience.* Addison-Wesley.

Van Welie, M., & Van Der Veer, C. (2000, April 2-3). *A structure for usability based patterns.* Paper presented at the CHI 2000 Workshop on Pattern Languages for Interaction Design: Building Momentum, The Hague, The Netherlands.

VanLengen, C. (2007). Supply of computer graduates is not binary! In *Proceedings of Information Systems Educators' Conference 2007* (Vol. 24, p. 2714). Chicago: Association of Information Technology Professionals, Foundation for Information Technology Education.

Vegso, J. (2005). Interest in CS as a major drops among incoming freshmen. *Computing Research News, 17*(3). Retrieved December, 5 2007, from http://www.cra.org/CRN/articles/may05/vegso.html

Venkatesh, V., & Davis, F. (2000). A theoretical extension of the technology acceptance model. Four longitudinal field studies. *Management Science, 2*(46), 186-203.

Venners, B. (2003, October 20). Exploring with wiki: A conversation with Ward Cunningham, part I. *Artima developer.* Retrieved January 1, 2006, from http://www.artima.com/intv/wiki.html

Vittori, R. (2004). *Web strategy.* Milano, Italia: Franco Angeli editore.

Von Hellens, L., & Nielsen, S. (2001). Australian women in IT0. *Communications of the Association for Computing Machinery, 41*(7), 46-52.

Voříšek, J. (2000). Nová dimenze systémové integrace—integrace podnikových procesů a znalostí. In *Proceedings of the Systémová integrace'2000, sborník mezinárodní konference* (pp. 195-206). Praha,VŠE. ISBN 80-245-0041-8.

Voříšek, J., & Dunn D. (2001). Management of business informatics—opportunities, threats, solutions. In *Proceedings of Systems Integration 2001 Conference.* Praha: VŠE. ISBN 80-245-0169-4.

Voříšek, J., Pavelka, J., & Vít, M. (2003). *Aplikační služby IS/ICT formou ASP—Proč a jak pronajímat informatické služby.* Praha: Graga Publishing. ISBN 80-247-0620-2.

Wacka, F. (2005). Enterprise blogging in practice. *webpronews.* Retrieved at January 1, 2007, from http://www.webpronews.com/topnews/2005/01/24/enterprise-blogging-in-practice

Wade, M., & Parent, M. (2002). Relationships between job skills and performance: A study of Webmasters. *Journal of Management Information Systems, 18*(3), 71-96.

Wallace, D., Raggett, I., & Aufgang, J. (2002). *Extreme programming for Web projects.* Addison-Wesley.

Walrad, C., & Moss, E. (1993) Measurement: The key to application development quality. *IBM Systems Journal, 32*(3), 445-460.

Webb, G. I., Pazzani, M. J., & Billsus, D. (2001). Machine learning for user modeling. *User Modeling and User-Adapted Interaction, 11,* 19-29. Kluwer.

Webmethods. (2005). *Survey of 480 global 2000 companies.* Retrieved June 19, 2008, from www.Webmethods.com

Webopedia. (2007). *CRM.* Retrieved March 8, 2007 from http://www.webopedia.com/TERM/C/CRM.html

Weil, P., & Vitale, M. R. (2001). *Place to space: Migrating to e-business models.* Harvard Business School Publishing Corporation.

Weill, P., & Ross, J. W. (2004, May). *IT governance: How top performers manage IT decision rights for superior results.* Boston: Harvard Business School Press.

Weiss, M. (2003, September 8-12). *Patterns for Web applications.* Paper presented at the Tenth Conference on Pattern Languages of Programs (PLoP 2003), Urbana.

Wentzlaff, I., & Specker, M. (2006, July 10). *Pattern based development of user friendly Web applications.* Paper presented at the Workshop on Model-Driven Web Engineering (MDWE 2006), Palo Alto.

West, L. A., & Bogumil, W. A. (2001). Immigration and the global IT work force. *Communications of the Association for Computing Machinery, 44*(7), 34-38.

Westbrook, R. (1995). Action research, a new paradigm for research on production and operations management. *International Journal of Operations and Production Management, 15*(12), 6-20.

Wetherbe, J. C. (1979). *Systems analysis for computer-based information systems.* St. Paul: West Publishing CO.

Wicks, A. M., Visich, J. K., & Suhong, L. (2006, Summer). Radio frequency identification applications in hospital environments. *Hospital Topics: Research and Perspective on Healthcare, 84*(3), 3-8.

Wiegers, K. (1996). *Creating a software engineering culture.* Dorset House.

Willcoks, L. (1994). *Information management: Evaluation of information systems investments.* London: Chapman and Hall.

Willcoks, L. P., & Lester, S. (1997). In search of information technology productivity: Assessment issues. *Journal of the Operational Society, 48*, 1082-1094.

Williams, J. (2004, November-December). IT Education: More specialised in 2010. *IT Professional,* 18-20.

Winer, R. S. (2001). A framework for customer relationship management. *California Management Review, 43*(4). Retrieved February 1, 2007 from http://groups.haas.berkeley.edu/fcsuit/PDF-papers/CRM%20paper.pdf

Wohlfahrt, J. (2004). *Akzeptanz und wirkung von mobile-business-anwendungen.* Hamburg, Germany: Dr. Kovač.

Wong, B. (2006). Different views of software quality. In E. Duggan & J. Reichgelt (Eds.), *Measuring information systems delivery quality* (pp. 55-88). Hershey, PA: IGI Global, Inc.

Wong, S. S. C., & Pal, S. (2001). *Mining fuzzy association rules for Web access case adaptation.* Paper presented at the Workshop on Soft Computing in Case-Based Reasoning/International Conference on Case-Based Reasoning (ICCBR_Ol).

WPS. (2007). *Informationen zum WPS.* Retrieved June 1, 2007, from http://www.Web-portal-system.de/

Wymer, S., & Regan, E. (2006). Factors influencing e-commerce adoption and use by small and medium businesses. *Electronic Markets, 15*(4), 438-453.

Xie, X., Miao, G., Song, R., Wen, J.-R., & Ma, W.-Y. (2005, March). Efficient browsing of Web search results on mobile devices based on block importance model. In *Proceedings of the 3rd IEEE International Conference on Pervasive Computing and Communications (PerCom 2005)*, Kauai Island, HI.

Yahoo! Finance. (2007). *Yahoo! finance - business summary.* Retrieved April 14, 2007, from http://finance.yahoo.com/q/pr?s=CI

Yan, Y., & Fang, X. (2005). Technology adoption in e-commerce education: A case study. *Information Systems Education Journal, 3*(40).

Yao, Y. Y. (2005, May 19-21). Web intelligence: New frontiers of exploration. In H. Tarumi, Y. Liand, & T. Yoshida (Eds.), *Proceedings of the 2005 International Conference on Active Media Technology (AMT 2005)*, Takamatsu, Kagawa, Japan (pp. 3-8).

Yaobin, L., & Zhaohua, D., & Bin, W. (2007, June). Tourism and travel electronic commerce in China. *Electronic Markets, 17*(2), 101-112.

Yen, D. C., Chen, K., Lee, S., & Koh, S. (2003). Differences in perceptions of IS knowledge and skills between academia and industry: Findings from Taiwan. *International Journal of Information Management, 23*(16), 507-522.

Yin, R. K. (2003). *Applications of case study research* (2nd ed.). Thousand Oaks, CA: Sage Publication.

Yin, X., & Lee, W. (2004). Using link analysis to improve layout on mobile devices. In *Proceedings of the 13th International conference on World Wide Web*, New York City (pp. 338-344).

Yoder, J., & Barcalow, J. (1997, September 3-5). *Architectural patterns for enabling application security.* Paper presented at the Fourth Conference on Pattern Languages of Programs (PLoP 1997), Monticello.

Zdun, U., & Vogel, O. (2002, July 3-7). *Content conversion and generalisation on the Web: A pattern language.* Paper presented at the Seventh European Conference on Pattern Languages of Programs (EuroPLoP 2002), Irsee, Germany.

Zhao, J. L., & Cheng, H. K. (2005). Web services and process management: A union of convenience or a new area of research? *Decision Support Systems*, *40*, 1-8.

Ziegler, M., & Adam, B. (2001). Mastering mobile media markets. Eschborn, Germany: Diebold Consulting.

Zlilber, S. N., & Vasconcellos, E. (2005). The adoption of e-business from a strategic perspective: The case of the Celta at General Motors Brazil. *International Journal of Automotive Technology and Management (IJATM)*, *5*(4), 411-429.

Zobel, J. (2001). *Mobile business and m-commerce.* Munich: Hanser.

Zschau, O., Traub, D., & Zahradka, R. (2002). *Web content management.* Bonn, Germany: Galileo Press.

Zwass, V. (2003). Electronic commerce and organizational innovation: Aspects and opportunities. *International Journal of Electronic Commerce*, *7*(3), 7-37.

About the Contributors

Latif Al-Hakim lectures in management in the Faculty of Business at the University of Southern Queensland, Australia. His experience spans industry, research and development, and academic institutions. Latif was awarded his undergraduate degree in 1968. His master's (1978) and PhD (1983) were awarded by the University of Wales (UK). Dr. Hakim has published extensively in industrial engineering, information management, and systems modelling. He is the author and editor of seven books, more than 10 chapters in books, and more than 75 papers in various journals and conference proceedings. He is the editor-in-chief of International Journal of Information Quality and Associate Editor International Journal of Networking and Virtual Orgainsations. Latif has also conducted international technology transfer seminars and consulted to a number of major industrial organisations in Australia.

Massimo Memmola is a researcher in the Department of Business Administration, Faculty of Economy, Catholic University of Milan, and a researcher of the Centre for Research and Studies in Healthcare Management (Ce.Ri.S.Ma.S.) and holds the same role at the Association for Managerial and Business Studies (ASAM) in the same University. He teaches courses in planning and control, management accounting, financial analysis, performance measurement, and Internet strategy. He is involved in several research projects in the area of the development of managerial practices in large and small companies and healthcare organizations. He is member of the editorial board of the International Journal of Healthcare Delivery Reform Initiatives (JHDRI). He graduated from Catholic University in 1999, and received a PhD in business administration from Catholic University in 2003.

* * *

Abdu Ali Albur graduated from King Fahad University of Petroleum and Minerals, in Dahran Saudi Arabia with a bachelor's of science degree in physics and from the University of Maryland University College with a master's degree in e-commerce. He has worked as an educational supervisor for 5 years in Saudi Arabia and for 4 years in Bahrain building his expertise in e-trading and computer programming. He is currently the IT manager at the Ministry of Education in the Kingdom of Saudi Arabia, Dhahran, Eastern Province.

Fernando Alonso received a BEng in industrial engineering in 1970 and a BSc in computing in 1972, both from Universidad Politécnica de Madrid (UPM), Spain. Alonso earned his PhD in computing from the UPM in 1985, and his PhD dissertation merited the distinction of a university prize. Alonso's major field of study is software development in the fields of software engineering and knowledge engineering and e-learning. He is a full professor of computing at the UPM's School of Computer Science, where he has worked as a full-time professor and researcher since 1989. He is also the R&D Director of the Centre of Computing and Communications Technology Transfer (CETTICO), and leads the Information and Communications Technology Research Group at the UPM. He has authored or co-authored 15 books on programming and software engineering and over 60 papers published in journals and international conferences.

Giovanna Castellano is an assistant professor at the Computer Science Department, University of Bari, Italy. She received a PhD from the Computer Science Department at the same University in 2001. Her teaching experience includes computer architectures and multimedia publishing. Her research interests embrace artificial neural networks, fuzzy systems, knowledge-based neurocomputing, granular computing systems, metalearning systems, computational intelligence for Web personalization, with special emphasis on neuro-fuzzy modeling. Within these research areas, she is author of more than 100 papers published in scientific journals and international conference proceedings. Her e-mail address is <castellano@di.uniba.it>

Lei Chen received his PhD in computer science and software engineering from the Auburn University in 2007. He is currently an assistant professor in the Department of Computer Science at the Sam Houston State University. His research interests include wireless routing, wireless communications, QoS, and Multimedia.

Kieran Conboy is a lecturer in information systems at NUI Galway, Ireland. His doctoral research focused on agile methods for systems development as well as agility across other disciplines. Kieran's other research interests include systems analysis and management accounting in systems development projects and he is currently leading a €600,000 project in the area. Some of his research has been published in various journals and conferences such as the European Journal of Information Systems, the International Conference in Information Systems (ICIS), the European Conference in Information Systems (ECIS), IFIP 8.6, and the XP200n conference series. Kieran is also the program chair for the European Conference in Information Systems to be held in Galway from June 9-11, 2008, and the program chair for the XP and Agile Development Conference to be held in Limerick from June 10-14, 2008.

Anna Maria Fanelli is a full professor at the Computer Science Department, University of Bari, Italy. Currently, she is involved in teaching computer architecture and computational intelligence metodologies courses for the Computer Science Curriculum. Her scientific activity concerns knowledge-based neurocomputing paradigm, neuro-fuzzy systems, granular computing, and computational Web intelligence. She has published more than 150 papers on the previous topics. She is the director of the Phd School in Informatics. She is Senior Member of the IEEE Computational Intelligence Society and the IEEE System, Man, and Cybernetics Society. She is also member of the INNS (International Neural Network Society), the EUSFLAT (European Society for Fuzzy Logic and Technologies), and the AI*IA (Italian Association for Artificial Intelligence). Her e-mail address is <fanelli@di.uniba.it>

George Feuerlicht is the director of MSc programs at the Faculty of Information Technology, University of Technology, Sydney, Australia. George has published over 60 conference and journal papers on a range of topics in information technology and computer science. George focuses on industry-relevant research problems and is the author of a number of recent publications on the topic of design of service-oriented applications. George has achieved international recognition in the area of design of service-oriented applications through his contributions to leading international conferences and chairing workshops. George holds a PhD from the Imperial College, London University, UK.

Maria Alice Frontini holds a MSc degree in management of technology from Massachusetts Institute of Technology, Boston. She has a graduation degree in production engineering from Polytechnic School of University of São Paulo, Brazil, of which she is a PhD candidate. Her research interests are in business strategy, technology innovation, and information technology. In the past 12 years she has working in strategic consulting in telecommunications, financial services, and retail with a focus on management of technology.

Chiara Frigerio is researcher of organization science and information systems management at Università Cattolica in Milan, where she is responsible of some of research activities of CeTIF (research centre on innovation and management for financial institutions). She got a PhD in information systems at LUISS University in Rome. She wrote some articles on the role of information and communication technology on the organisational mechanisms, in particular on work process and structures. At the moment she is involved also in some European project on the role of collaboration in change management practices. She teaches organization design and information systems at the same university.

Luigi Geppert spent 3 years at the Mathematics Department of Milan Polytechnic, involved in applied research activities, after earning the university degree on Theoretical Physics (1982) at University of Milan. In 1985 he founded an information technology (IT) consulting company realizing software platforms for manufacturing applications. At the end of 80s he joined as a senior partner at Efeso Srl, a management consulting company focused on vertical applications for internal and external logistics. He became CEO of the IT business unit deploying reengineering (BPR) and business process improvement (BPI) projects for ERP and logistics applications. In 1995, he founded his own management consulting company, Minerva Consulting, focused on IT system for BPR and BPI. Since the early 90s he has been using systems thinking and system dynamics approach and tools to accomplish his modeling activities. By using these competencies, he has worked as an acting senior consultant, especially for important enterprises in manufacturing and banking, national wide and international wide. He is also professor of system dynamics for strategy at the Catholic University of Milano at the department of Enterprise Management.

Svenja Hagenhoff has been an assistant professor at the Institute of Information Systems, University of Goettingen since 2002. She finished her doctoral thesis in information systems in 2001. In 2008, she finished her state doctorate (habilitation) and passed the postdoctoral lecture qualification in management and information systems. She was visiting researcher at the University of California Los Angeles, and interim professor at the University of Hildesheim, Germany. At the University of Luebeck, Germany, she was guest lecturer. She holds a master in management from the University of Goettingen, Germany. Svenja Hagenhoff has research experience in the field of Internet economics and information markets, e-learning, knowledge management and innovation management, and service-oriented architectures.

Kevin K.W. Ho is working on his doctorate in information systems at the School of Business and Management of the Hong Kong University of Science and Technology (HKUST). He obtained his BSc (Hons.) and MPhil in chemistry and MSc in information systems management from local universities and an MA in managerial economics from the University of Oklahoma. Kevin is now working in the Hong Kong Police Force as an executive officer. His research interests include electronic auction, information systems strategy, and electronic government. He has published in the Journal of E-Government, Journal of Global Information Management, and the International Journal of Cases on Electronic Commerce.

Wen-Chen Hu received a BE, an ME, an MS, and a PhD all in computer science from the Tamkang University, Taiwan, National Central University, Taiwan, University of Iowa, Iowa City, and University of Florida, Gainesville in 1984, 1986, 1993, and 1998, respectively. He is currently an assistant professor in the Department of Computer Science at the University of North Dakota. Dr. Hu has published over 60 articles on refereed journals, conferences, and books. His current research interests are in handheld computing, electronic and mobile commerce systems, Web technologies, and databases.

Pankaj Kamthan has been teaching in academia and industry for several years. He has also been a technical editor for books, a member of the editorial board of journals, a member of the program committee of conferences, and participated in standards development. His research interests include software quality, markup languages, and knowledge representation.

Christian Kaspar is a consultant at Cap Gemini Consulting since 2007. He worked at his doctoral thesis at the University of Goettingen at the Institute of Information Systems between 2003 and 2006. After finishing his thesis, he was a visiting scholar at the Centre of Digital Enterprise at the University of Auckland, New Zealand. He holds a master in management from University of Munich, Germany. Christian Kaspar has experience in the fields of publishing industry, information markets, and telecommunication industry.

Orla Kirwan is currently completing her PhD at NUI Galway, Ireland. Her research is in the area of the implementation of energy management information systems. She is funded by the Irish Research Council for Science, Engineering and Technology (IRCSET) through the Embark Initiative. She has also completed a B. Comm degree, a higher diploma in research methods and a master's in e-commerce at NUI Galway. She has presented her research at several international conferences and the research presented here is from her master's degree.

Fernando José Barbin Laurindo holds graduation, MSc and PhD degrees in production engineering from Polytechnic School of University of São Paulo, Brazil. He has a graduation degree in law from the University of São Paulo and graduated in an extension course in business administration from Fundação Getúlio Vargas. His research interests are information technology and business strategies, information technology planning, and project management. For 15 years, he had worked in companies of the manufacturing, financial, and service sectors. He is an associate professor of the Production Engineering Department at Polytechnic School - University of São Paulo - EPUSP, where he has worked since 1997.

Susan Leferink is a resident of Montana. She graduated from Montana State University in business management/marketing and has established a broad based career through both private and public sector customer service-oriented work environments. She currently works as a computer support specialist for Montana's Department of Commerce (networking, database administration, web development, and computer training). She is scheduled to graduate with a master's degree in e-commerce from the University of Maryland University College in Spring 2008.

David Lizcano holds a MSc degree with honors in computer science (2006) from the Universidad Politécnica de Madrid. He is a PhD student at the Universidad Politécnica de Madrid's Department of Computer Science, and holds a research grant from the UPM and the European Social Fund under their Research Personnel Training program. He is working as a research associate and PhD candidate at the Computer Networks and Web Technologies Lab (CoNWeT Lab, Universidad Politécnica de Madrid), where he is currently involved in several national and European funded projects relating to Web 2.0 and Enterprise 2.0 technologies and service-oriented architectures (SOA).

Genoveva López holds a MSc and PhD in computer science from the Universidad Politécnica de Madrid. She is now an associate professor of the UPM working with the Department of Computer Languages and Systems and Software Engineering at the UPM's School of Computing. She is a researcher of the Information and Communication Technologies Research Group, focusing on distributed systems development, and Web services and technologies. As a member of this research group, she has participated in 22 publicly funded R&D projects, and has coauthored over 25 publications or scientific documents, 16 papers at congresses, and 10 books on these areas.

Thomas Lyle was born and raised in Philadelphia, Pennsylvania and currently resides in its western suburbs. He received his BS in computer science from The Pennsylvania State University in 1973 and has spent the last 30 plus years working in the private sector. He began working for his current employer, PNC, in the mutual fund services division, as a Systems Architect and Product Development Manager primarily focused on migrating mainframe legacy software packages to client server personal computer software packages initially and then eventually to Internet and e-commerce software packages. Since earning his BS, he has been continually learning new technologies from training provided both internally and externally outside of the university system as university curriculums were playing catch-up with industry. He enrolled in the ECOM/MBA dual degree program mainly because of its online capabilities and because it allows him to spend equal amounts of time on business management tasks as well as technical development tasks.

B. Dawn Medlin is currently serving as an interim chair and associate professor in the Department of Computer Information Systems in the John A. Walker College of Business at Appalachian State University in Boone, North Carolina. She earned her doctorate at Virginia Polytechnic Institute and State University. Her teaching and research activities have been in the area of information technology, specifically in security and privacy issues related to healthcare institutions and retailing organizations. Dr. Medlin's research articles have appeared in journals such as the Information Systems Security: The (ISC)2 Journal and the Journal of Information Privacy and Security, Int. J. Information Technology Management, Journal of Computer Information Systems, Int. Journal of Management, among other refereed journals.

Sílvia Novaes Zilber is a PhD in business administration. She is an electronic engineer with a degree from "Escola Politécnica da USP" (POLI/ USP), with masters and doctors degrees in business administration from FEA/ USP, São Paulo, Brazil. She has published many papers in journals, such as Journal of Technology Management & Innovation and International Journal of Automotive Technology and Management (IJATM). She participated of many international conferences as author, including EUROMA, Academy of Management, IAMOT , ALTEC, and BALAS. She is a researcher and advisor in the areas of e-business, the automotive industry, and strategies for innovation.

Björn Ortelbach works at Henkel as an IT controller. He finished his doctoral thesis in 2007 at the Institute of Information Systems, University of Goettingen. He holds a master in business information systems from the University of Goettingen. Björn Ortelbach has experience in the fields of information markets, publishing industry, and controlling.

Bernhard Ostheimer is postdoctoral research fellow at the University of Giessen, Germany. His current research interests include distributing e-business systems, e-university and e-campus systems, e-learning, and Web content management.

Hsueh-Ieng Pai is a research assistant and a doctoral candidate at Concordia University, Canada. Her research interests include knowledge representation and management, uncertainty reasoning on the Semantic Web, and query optimization.

Giovanna Palumbo received the Sociologist Degree at Universidad Catòlica Andrès Bello, Caracas in 1974, the MBA Degree at Instituto de Estudios Superiores de Administraciòn (IESA), Caracas in 1989, and the master degree on international health care management, economics, and policy at Università Luigi Bocconi, Milano, 2001. Since 2005 she has been project leader of the "Digital Hospital Project" which is being developed at Ospedale Valduce, Centro Riabilitazione Villa Beretta, Lecco. Before arriving to Italy in 2001, she had been general manager of the Center for Managerial Development at IESA, Caracas, and researcher/faculty/consultant at the same institute. From 1989 to 1992 she had been consultant/faculty of the training program on Health Service Management for the Regional Program on Mother-Child Health, at Pan American Health Organization/World Health Organization (PAHO/WHO), Washington/Sao Paulo.

Krassie Petrova is currently a senior research lecturer at the School of Computing and Mathematical Sciences in the faculty of Design and Creative Technologies, Auckland University of Technology (New Zealand). Krassie is serving (since 2004) as Postgraduate Programme Leader at the School, and (since 2003) as editor of the New Zealand-based open access journal Bulletin of Applied Computing and Information Technology. She earned her MSc and postgraduate diplomas at the University of Sofia (MSc) and the University of Auckland. Her teaching, postgraduate supervision, and research activities have been in the discipline areas of information systems and technology, and in curriculum development (e-business, net-centric computing and information security). In the last 10 years, Krassie has published a number of works in research journals, edited books, and refereed conference proceedings (e- and m-Learning, skill and a capability development, mobile business applications).

Stephen Proctor is a director for Computer Sciences Corporation (CSC) located in suburban Washington, DC. He has been in the IT field for many years supporting infrastructure initiatives in commercial and government markets. He currently leads an incredible team providing infrastructure engineering and operations to the US government. He is scheduled to graduate with a master's degree in e-commerce from the University of Maryland University College in summer 2008.

Mahesh S. Raisinghani is an associate professor in the Executive MBA program at the TWU School of Management. He is a certified e-commerce consultant (CEC) and a project management professional (PMP). Dr. Raisinghani was the 2007 G. Ann Uhlir Fellow in Higher Education Administration. He also received: the TWU School of Management's 2005 Best Professor Award for the Most Innovative Teaching Methods; 2002 research award; 2001 King/Haggar Award for excellence in teaching, research and service; and a 1999 UD-GSM Presidential Award. His research has been published in several academic journals such as IEEE Transactions on Engineering Management, Information & Management, Journal of IT Review, Journal of Global IT Management among others, and international/national conferences. Dr. Raisinghani is included in the millennium edition of Who's Who in the World, Who's Who among Professionals, Who's Who among America's Teachers, and Who's Who in Information Technology.

Marcos Reyes holds a MSc degree in computer science from the Universidad Politécnica de Madrid (UPM). Marcos has been with Telefónica I + D since 2000, where he has been involved in several R&D projects related to identity management and business process management systems applied to workforce management and trouble-ticketing systems. In 2005, Marcos was promoted to chief architect and designer of Telefónica I + D's SmartFlow Platform and Enterprise 2.0 projects, applying Web 2.0 concepts to enterprise systems within the Morfeo Open Source (www.morfeo-project.com) community set up by Telefónica I + D. Also, Marcos coordinates several community organizational matters.

Mauro Rossini received the degree in electronics engineering with a specialization in bioengineering from Politecnico di Milano in 1998. He is researcher and consultant for Gait & Movement Analysis Laboratory at Valduce Hospital - Villa Beretta Rehabilitation Center since 1995. Since 2004 he has been responsible for several projects developed inside the HINT@Lecco (Health Innovation Network Technology at Lecco) research network. His main research areas concern gait and movement analysis, robotic rehabilitation, telemedicine, medical informatics in particular for electronic medical record, and introduction of high level technologies in rehabilitation field. The main current projects where he is technical project leader are "Introduction of Electronic Medical Record in a Rehabilitation Hospital" and "Telemedicine to follow up patient at home."

Lutz Seidenfaden is an assistant to a member of the Board of Festo Corporation. He finished his doctoral thesis in 2007 at the Institute of Information Systems, University of Goettingen. He holds a master's in business information systems from the University of Goettingen. Lutz Seidenfaden has experience in the fields of P2P services and architectures, information markets, and publishing industry.

Javier Soriano holds a MSc in computing (1998) and a PhD in computer science (2003) from Universidad Politécnica de Madrid (UPM). He is an associate professor at the UPM's School of Computer Science, Spain, where he teaches both undergraduate and postgraduate courses on computer networks, distributed systems, and emergent Web technologies. He leads the Computer Networks and Web Technologies Lab (CoNWeT Lab), is a member of the UPM's Information and Communications Technology Research Group (ICTG-CETTICO) and collaborates with the Centre of Computing and Communications Technology Transfer (CETTICO). He is coauthor of over 30 journal publications and conference articles, and has coauthored six books (one having won an academic prize) and four book chapters. He is also involved with the Networked European Software & Services Initiative NESSI and has contributed to the NESSI Strategic Research Agenda (SRA).

Axel C. Schwickert is full professor of business administration and information management at the University of Giessen, Germany. His current research interests include engineering, management and operation of Web portals, Web applications and Web site farms, electronic university and e-learning, IT management, as well as IT compliance.

Maria Alessandra Torsello is a PhD student at the Computer Science Department, University of Bari, Italy. She got the degree at the University of Bari in 2004. She presented the degree thesis concerning the recognition of static postures for the human motion analysis. She was interested in the investigation of computational intelligence techniques for the development of self-adaptive learning systems. Her current research activity concerns the design of Web personalization systems by the application of soft computing methodologies, with particular reference to the neuro-fuzzy strategy. Her e-mail address is <torsello@di.uniba.it>

Alessandra Tzannis has a PhD in "marketing strategies for enterprise," XX cycle at the University of Bergamo, with a thesis entitled "SMEs' Internationalization Processes and Cultural Dynamics."

She graduated from the Catholic University of Milan in foreign languages and literature (French-English), addressed "tourism management," and she attended the advanced course "marketing management online" as well. She teaches in modules, tutorials, and masters degree in marketing organized by the same University (locations: Brescia and Milan). She is focused on topics related to the internationalization of small and medium enterprises and the impact of cultural dynamics on the development processes across borders, working on research projects among different universities. Her research areas include international marketing, intercultural marketing, internet marketing, Web strategy, and business-to-business. She has collaborated with Centrimark since 2003. She worked for contributions translations from English into Italian and drafting the index of the Manuale di Management nell'economia digitale by W.G. Scott, M. Stecco, and M. Murtula published by Il Sole 24 Ore, in October 2003.

Jiri Vorisek is a professor and head of the Department of Information Technologies at the University of Economics, Prague and adjunct professor at the Pfeiffer University, Charlotte, North Carolina. He is founder and the president of the Czech Association for Systems Integration, and CEO of consulting company ITG Ltd. He received his PhD in management information systems from the University of Economics, Prague. He specializes in strategic management of information systems, systems integration, outsourcing, and methodologies of IS design and implementation. He is author or coauthor of seven books and dozens of textbooks, articles, and conference papers. He is founder of the systems integration international conferences.

Chyuan-Huei Thomas Yang received an MS in computer science from the New Jersey Institute of Technology in 1991 and a PhD in computer science from the National Tsing Hua University, Taiwan in 2005. He is currently an assistant professor in the Department of Computer Science at the Hsuan Chuang University, Taiwan. His research interests include computer vision, information security, and databases.

Vincent Yen is a professor of management information systems in the Department of Information Systems and Operations Management, Raj Soin College of Business, Wright State University, Dayton, Ohio. He has been teaching a variety of information systems courses since 1988. His research interest in recent years is in methodologies for analysis and design of business processes while keeping abreast with professional and software industry developments. Some of his recent articles reflecting this interest and have appeared in refereed journals and proceedings are titled: "A Node-Centric Analysis of Metagraphs and Its Applications to Workflow Models"; "Web Services and Workflow Modeling"; and "An Integrated Model for Business Process Measurement". He is on the editorial review board of Journal of Organization and End User Computing and Annals of Cases on Information Technology.

Yanjun Zuo received an MS and a PhD in computer science from the University of Arkansas in 2003 and 2005, respectively. He is currently an assistant professor in the Department of Information Systems and Business Education at the University of North Dakota. His research interests include information and computer security, trustworthy computing, and database security.

Index

A

acceptance research 249, 250
action research 236–247
activity-based costing 145, 147, 148, 149, 152, 158, 162, 167, 168, 343, 353
activity-based costing (ABC) 147, 148, 149, 150, 151, 161, 162, 167, 346
advanced health process management (AHPM) 151, 162, 163
automotive industry 298, 299, 311, 313
automotive sector 170–197

B

banking and finance industry 219–235
block importance model 204, 207, 369
blogging 76, 86
business intelligence 89
business process 50, 51, 60, 135, 136, 140, 142, 143, 364
business process management 75, 87, 89, 219, 228
business strategy 170, 174, 176, 179, 186, 190

C

Celta 298–317
central cyber government office (CCGO) 264–276
chief information officer (CIO) 277, 278
Clustering 323
clustering 322, 325, 326, 327, 328, 329, 330, 332
collective intelligence 63, 64, 65–68, 69, 70, 87, 90
communication portal 264–276
connected economy 118, 119, 120, 125, 128
content management 219, 228, 230
content relevance 267, 268, 270, 271, 272, 273
content usability index (CUI) 181, 185, 186
corporate Web site farm 211, 212
CRM vendor selection 106–110
curriculum development 27
customer loyalty 100
customer relationship management (CRM) 94–117
customer service 95, 96, 99, 101, 102, 106, 109, 110, 111, 113
customer service issues 97

D

data fragmentation 96–97
data mining 320, 322, 325, 337, 338, 358, 359
data quality 97
data redundancy 97
Deloitte&Touche 57–58
distribution channel 248–263

E

e-business 118, 298–317
e-commerce 301, 302, 305, 307, 309, 316, 317, 349, 354
e-commerce adoption 236–247
e-commerce applications 2–4
E-commerce capability 98
e-learning 212, 218, 364
e-university 217
electronic government 264–276
Enterprise 2.0 61–93
enterprise collaboration 72, 73, 74, 91, 347
enterprise computing 46, 47, 50
extendable Web content management system (WCMS) 213–216

F

folksonomy 75, 82, 92, 351

G

Global SOA 72
GM Brazil 298–317
gold 127–131

H

hard skills 27, 28, 33
healthcare 145, 146, 147, 151, 152, 155, 157, 158, 166, 167
Hewlett Packard 56, 57–58

I

IBM 56, 57–58
ICT curriculum 27, 31
ICT industry information needs 32
ICT investment 48, 51, 52
ICT jobs 37
ICT products 47, 52, 57
ICT professionals 27, 30, 31, 39, 40
ICT services 51, 52, 53, 54, 57
ICT skills 28, 30–33, 31, 36, 37–41, 44, 361
industry requirements 34, 35
information accuracy 267, 268, 269, 270, 272, 274
information quality 264, 265, 266, 267, 268, 269, 271, 272, 273, 274, 275, 355
information systems (IS) research 240
information timeliness 267, 268, 269, 270, 271, 272, 273, 274
innovation 64, 65, 84, 298, 311, 313, 316, 345
Internet banking 219, 221, 222, 223, 226
Internet marketing 94
Internet presence 177, 178, 179, 180, 181, 183, 185, 187, 188, 189, 191, 192, 194
Internet strategy 186, 187, 190, 195, 197, 348
Intranet 219, 221, 226, 227, 228, 231
IS curriculum 28, 42, 346
IT effectiveness 278–280
IT evaluation 277–297
IT investments 277, 278, 281, 282, 287, 288, 289, 290, 291, 294
IT strategy 280, 281

K

knowledge discovery 322–323
knowledge discovery in data (KDD) 320
knowledge work 65, 68, 71

L

learning management systems (LMS) 208–218
link analysis 204

M

macro-architecture design 10–11
mash-up 74
micro-architecture design 11–15
microbrowsers 199, 201, 204, 205
mobile central processing units 201
mobile commerce 248–263
mobile content services 253–255
mobile Internet 248–263
mobile service formats 251–253
mobile Web, the 198
mobile Web browsing 198, 200

N

navigation pattern discovery 202
neuro-fuzzy 318, 319, 320, 325, 326, 327, 328, 334, 335, 337, 344, 355

O

old economy 121
operational costs 98
organizational structure 298, 300, 301, 302, 304, 305, 307, 310, 312, 314, 315

P

page and form summarization 204
page reformatted 204
paid content 248–263
partner chain 130
pattern language 8, 19, 20, 21, 23, 350, 353, 361, 370
POECAM 2, 5, 6, 15, 16, 17, 18
preprocessing 321–322, 327–328
presentation format 267, 268, 269, 270, 271, 272, 273
process management 145, 147, 151
publishing industry 248–263

R

radio frequency identification (RFId) 145–169
real options 277–297
rehabilitation medicine 146, 150, 166
research framework 26, 28, 35

River Deep Mountain High (RDMH) 239, 242–243, 243
RSS 63, 68, 71, 72, 75, 77, 80, 81, 85, 86, 87, 92, 351

S

SAP 57–58
Semantic Web 1, 16, 17, 19, 20, 350, 353
semiotic quality 6, 16
service oriented architecture 135
service oriented architecture (SOA) 135, 136, 137, 140, 142
silver 127–131
simple object access protocol (SOAP) 135, 142
small to medium size enterprise (SME) 236–247
social capital 68, 71, 83, 84, 86
social network 75, 83, 90
soft computing (SC) 319, 320, 324–326, 334
soft skills 28, 33, 43, 356
software as a service 46, 47
steel 121–126
SUN Microsystems 56
supply chain 119, 120, 121, 122, 123, 124, 125, 126, 127, 128, 129, 130
surveys 138–142

T

technology enabled organisation 216–217
trend 136

U

u-steel.com 121–126
up2gold.com 127–131
usage data gathering 202
usage data preparation 202
user-centered SOA 87
user satisfaction 266, 267, 268, 271, 273, 274

W

Web 2.0 61–93
Web content management system (WCMS) 211
Web content mining 320
Web engineering 1, 3, 5, 19, 20, 21, 22, 350, 353, 358, 363
Web mining 198–207
Web personalization 318–339
Web portal system (WPS) 213, 214, 215, 216, 217, 218, 369
Web recommendation 325, 328, 335, 344
Web services 134–144
Web strategy 118–133, 170–197
Web structure mining 320
Web technologies 152
Web usage mining 198, 200, 201, 202, 203, 204, 205, 206, 320, 321, 322, 324, 325, 334, 335, 336, 337, 338, 340, 352, 353, 359, 362
Wiki 78, 80, 84, 86, 92, 351